Evaluating Derivatives

Evaluating Derivatives

Principles and Techniques of Algorithmic Differentiation

Second Edition

Andreas Griewank

Yachay Tech University
Urcuquí, Imbabura Province,
Ecuador

Andrea Walther

Paderborn University
Paderborn, Germany

Society for Industrial and Applied Mathematics • Philadelphia

10 9 8 7 6 5 4 3 2

Library of Congress Cataloging-in-Publication Data

Griewank, Andreas.
Evaluating derivatives : principles and techniques of algorithmic differentiation. – 2nd ed. / Andreas Griewank, Andrea Walther.
p. cm.
Includes bibliographical references and index.
ISBN 978-0-898716-59-7 (alk. paper)
1. Differential calculus–Data processing. I. Walther, Andrea. II. Title.
QA304.G76 2008
515'.33--dc22

2008021064

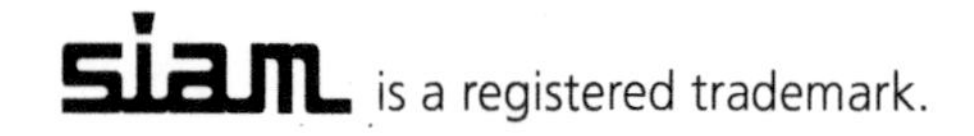

Contents

Rules

Preface

The advent of high-speed computers and sophisticated software tools has made the computation of derivatives for functions defined by evaluation programs both easier and more important. On one hand, the dependence of certain program outputs on certain input parameters can now be determined and quantified more or less automatically, i.e., without the user having to append or rewrite the function evaluation procedure. On the other hand, such qualitative and quantitative dependence analysis is invaluable for the optimization of key output objectives with respect to suitable decision variables or the identification of model parameters with respect to given data. In fact, we may juxtapose the mere *simulation* of a physical or social system by the repeated running of an appropriate computer model for various input data with its *optimization* by a systematic adjustment of certain decision variables and model parameters. The transition from the former computational paradigm to the latter may be viewed as a central characteristic of present-day scientific computing.

Optimization nowadays already forms an important application area for exact derivative calculation using algorithmic differentiation. This includes the provision of gradients and Hessians for the unconstrained case, as, for example, in nonlinear finite element calculations [Wri08], the optimal laser control of chemical reactions [BCL01] or the optimization of low-pass analog filters [Alk98]. In the constrained case, algorithmic differentiation can be used to compute the required Jacobians, Hessians, or Hessian-vector products. Here, the applications areas include chemical engineering [AB04], race car performance [CS$^+$01], and industrial production processes [KW00]. Since 1997 the Network Enabled Optimization Server (NEOS) at Argonne National Laboratory has been using algorithmic differentiation to compute gradients and Jacobians of remotely supplied user code for optimization objectives and constraints [CMM97]. The modelling language AMPL and GAMS have incorporated both modes of algorithmic differentiation to provide first and second derivatives for various optimization solvers. Also multicriteria optimization benefits from exact derivatives, e.g., to optimize medical radiotherapy [JMF06]. For the field of parameter estimation, algorithmic differentiation is used to improve, for example, weather models [GK$^+$06] or the simulation of the climate [KHL06]. The cover shows the independence between various vector quantities in optimal design as discussed in Chapter 15 superimposed on a global sensitivity map. It was provided to us by Patrick

Heimbach of MIT and represents the derivative of the poleward heat transfer with respect to local maritime temperatures.

Moreover, even the pure simulation may be improved by the provision of exact derivative information. In this context, algorithmic differentiation may be used to compute exact Jacobians for the integration of stiff ordinary differential equations (ODEs). Numerical quadrature is probably the first—and numerical integration of ODEs and differential algebraic equations (DAEs) are still the most important—application areas for third and higher-order derivative evaluations. For example, derivatives of arbitrary high-order may be required for the integration of DAEs modeling multibody systems [NP05]. Many computational schemes for nonlinear problems are known to work well if certain higher-order terms, often error estimates, are small. The heuristic procedures employed to estimate these terms are quite unreliable, so the ability to evaluate them up to working accuracy should greatly improve algorithmic performance and robustness. In other words, the availability of selected analytical derivatives can facilitate the intelligent use of adaptive and higher-order methods, which may otherwise be somewhat unreliable or simply wasteful in unfavorable circumstances that could not be properly detected beforehand. Subsequently, the simulation may be subject of a sensitivity analysis, where the required derivatives can be provided exactly by algorithmic differentiation. Here, current research topics cover, e.g., the analysis of financial instruments like options [Gil07]. Algorithmic differentiation is used successfully also in the context data assimilation, e.g., for flood modeling [CD+06].

Of course, our distinction between simulation and optimization is not a sharp dichotomy. Many practical calculations have always incorporated optimization methods, for example, to determine stable equilibria of certain subsystems with phase transitions. However, what may be characteristic for modern optimization calculations is that the number of decision or design variables can grow so large that estimating their effect on the objective function(s) becomes the most expensive task of the overall calculation. This happens in particular when unknown functions like geometrical shapes or distributions of material properties need to be discretized, which leads to a virtually unlimited number of variables or parameters [AC92]. Using a discrete analog of adjoint differential equations, we will be able to calculate the gradient of a scalar-valued function at a temporal complexity not exceeding that of five function evaluations, regardless of the number of independent variables. This is being used extensively in weather forecasting based on four-dimensional data assimilation [Tal08].

This book should be almost invaluable to the designers of algorithms and software for nonlinear computational problems. It should also be especially useful for the users of current numerical software, which usually still places on them the responsibility for providing derivative values, sparsity patterns, and other dependence information. Even without reading the more theoretical sections of the book, the reader should gain the insight necessary to choose and deploy existing algorithmic differentiation software tools to best advantage. Finally, we hope that there will be a significant number of readers who are interested in algorithmic differentiation for its own sake and who will continue or begin to contribute to this growing area of theoretical research and software development.

The earlier books *Automatic Differentiation: Techniques and Applications*, by Louis Rall [Ral81], and *Numerical Derivatives and Nonlinear Analysis*, by Kagiwada et al. [KK+86], cover only the *forward* or *direct* mode of what we call here *algorithmic differentiation.* Most material in this book concerns the "reverse" or "adjoint" mode as well as hybrids between forward and reverse that promise lower operation counts at acceptable storage requirements. Many newer results and application studies were published in the proceedings [GC91], [BB+96], [CF+01], [BC+06], and [BB+08] of the international workshops on algorithmic, or computational, differentiation held in Breckenridge (1991), in Santa Fe (1996), in Nice (2000), in Chicago (2004), and in Bonn (2008). They also contain excellent introductory articles and an extensive bibliography. Apart from some papers of historical interest, we have therefore listed in our bibliography only those sources that are directly pertinent to developments in the text.

Acknowledgments

The plan for writing the original version of this book was hatched with Bruce Christianson. We are indebted to Olaf Vogel, Klaus Röbenack, Mike Giles, and many colleagues, who have commented on the first edition and made suggestions for the second. Section 6.2 on source transformation and section 6.3 on parallelism were largely provided by Christele Faure and Jean Utke, respectively. To help improve the readability for novices in the field, the second author reorganized the book's contents. As with the first edition, Mrs. Sigrid Eckstein composed and revised the LaTeX source with its many tables and diagrams.

Prologue

> *Derivatives of the determinant or those of the smallest singular value cannot be calculated "explicitly" since they are here calculated "numerically", except for very simple cases when they can be calculated "symbolically". Therefore, the first partial derivative is also calculated "numerically".*

This (slightly rearranged) statement has been taken from a monograph on modeling and identification in robotics [Koz98]. For many scientists and engineers this will pass as a common sense, even obvious, statement. Yet, we must take issue with this line of reasoning and hope that after browsing through the present book, the reader will no longer find it entirely convincing either. On the contrary, it will be shown here how the derivatives of much more complicated functions than those mentioned above can in fact be evaluated with working accuracy at a reasonable cost. Since determinants and singular values already exhibit the crucial aspects of our general problem, let us discuss them a little more closely.

At the risk of appearing pedantic and, even worse, of seemingly putting down a fellow author (who probably has other things to worry about) let us examine the meaning of the highlighted adverbs in the quote: *explicitly*, *numerically*, *symbolically*, and again *numerically*. The last, *numerically*, undoubtedly refers to the approximation of derivatives by divided difference, a popular technique that is often called *numerical differentiation* [AB74]. From our point of view this naming convention is unfortunate because it immediately suggests a natural affiliation with other "numerical" processes, specifically, methods for "numerically" evaluating singular values and determinants of a given matrix. In fact, there is no such intrinsic association, and the meaning of the two uses of *numerical* in the quote are almost completely unrelated. The terms *explicitly* and *symbolically*, on the other hand, seem to be used as synonyms for the same property. So how are they contrasted with *numerically*?

The qualifier *numerically* indicates the presence of truncation errors, which lead to a halving of the significant digits under the best of circumstances. This severe loss of precision is unavoidable if all we can find out about the function to be differentiated is its value at given points with a certain accuracy. A black box evaluation procedure of this kind is sometimes called an "oracle" and its values certainly deserve to be labeled "numerical." Fortunately, we usually can find out a lot more about the function, especially when it is evaluated by a computer program in an accessible source code. If treated with the right tools and techniques such evaluation programs represent the function as "explicitly" and "symbolically" as a one-line formula might appear to a high school student.

In fact determinants and singular values can now be evaluated by single statements in MATLAB, Scilab, and various other scientific computing environments. True, these functions are just subroutine calls, where a lot of "numerical" calculations are effected "under the hood." But then the same is true for square root, logarithm, exponential, and other humble intrinsic functions, which have always been considered "explicit" and "symbolic" expressions by everyone. So what makes them special? Probably, only that we are reasonably familiar with their shapes and know their derivatives by heart. How "explicit" are the Bessel functions and other hypergeometric functions?

But suppose we don't have any fancy "symbolic" environment and were never told that the gradient of the determinant is the transpose of the inverse multiplied by the determinant. Then we could just write a Fortran code that generates all possible row permutations of the matrix, computes the products of their diagonal entries, and adds them together with alternating sign. What could be more "symbolic" than that? After all, the summation symbols in mathematical expressions are no more symbolic than loop controls in an evaluation program. The only discernible difference between such a program and what is usually considered an explicit expression is that common subexpressions (here representing subdeterminants) are given a name and treated as program variables. Since these named intermediates can always be eliminated by substitution, they make little difference from an abstract point of view and certainly do not change the mathematical nature of the function.

However, from a practical point of view, the skillful handling of *named intermediates* is crucial for computational efficiency, especially with regards to differentiation. But even without that, "simplifying" a function by substituting all intermediates typically leads to formulas of exponentially growing size whose visual inspection yields no analytical insight whatsoever. For that very reason computer algebra packages represent functions internally as (directed and acyclic) computational graphs rather than expression trees. Apart from named intermediates, we may identify three other aspects of evaluation programs that might dissuade people from considering them as "explicit" or "symbolic" representations of the function in question.

Since the expansion of a determinant involves a huge number of subdeterminants, the program variables containing their values are usually reused for other mathematical variables once the original value is no longer needed. Through the joint allocation of several mathematical variables in the same program variable, a collection of formulas turns into a sequence of assignments that must be executed in a certain order. The overwriting of variables complicates various tasks, notably, symbolic simplification in computer algebra packages or dependence analysis in compilers. However, it still does not take much away from the explicit nature of the evaluation program, since by replicating program variables any sequential program can easily be turned into a single assignment code, where mathematical and program variables need not be distinguished. While joint allocation makes little difference for differentiation in the standard "forward" mode, its proper handling is crucial for the correctness and efficiency of the so-called reverse mode (see section 4.1).

Even if implemented with jointly allocated, named intermediates, the expansion of the determinant as a polynomial involves far too many terms for its efficient evaluation. Hence, one usually begins the evaluation with an LU factorization, which requires pivoting even if calculations could be carried out in exact arithmetic. Pivoting generates *program branches*, so that the determinant function is actually represented by different rational formulas on various parts of its domain, the space of square matrices. In the case of the determinant calculation via an LU, the various pieces fit together at the boundaries so that the resulting function is globally continuous. However, for a general evaluation program with branches, this continuity certainly need not be the case, and is in fact unlikely, unless the programmer has very carefully glued the pieces together. Whether or not one wishes to consider evaluation programs with branches as "explicit" is a question of personal preference. Fortunately, they are "explicit" and "symbolic" enough to allow the calculation of exact directional derivatives and generalized gradients.

Due to their mathematical nature singular values, unlike determinants, cannot be evaluated exactly by finite procedures even in infinite-precision arithmetic. Hence one has to employ *iterative loops*, which may be distinguished from other program branches by the fact that the total number of execution steps is not fixed or even bounded. Now we really do have a problem that better be called "implicit" rather than "explicit." This is not necessarily a serious difficulty as far as differentiation is concerned, because in principle we are in exactly the same situation when it comes to evaluating and differentiating exponentials and the other humble intrinsics mentioned earlier.

There, as in the case of singular values, one knows that the value being approximated iteratively is actually defined as the solution of some mathematical problem like an ODE or a system of algebraic equations. Quite often, the derivative just pops out from these relations once the solution itself has been computed. This is certainly the case for singular values (see Exercise 3.7f). Of course, such implicit differentiation is possible only if we are told about it, i.e., if somebody knows and specifies the mathematical problem being solved by an iterative loop. When this is not the case, we are reduced to differentiating the evaluation program at hand as though the loop were any other branch. Whether this yields reasonable derivative approximations depends on the mathematical nature of the iteration.

In summary, we observe that evaluation programs are distinguished from what is universally accepted as an explicit formula by the following four aspects:

- named intermediates
- joint allocations
- program branches
- iterative loops

The purpose of this book is to show why these program constructs provide challenges for the efficient evaluation of truncation-error-free derivative values and how these difficulties can be satisfactorily resolved in almost all situations. The only exceptions are large iterative or adaptive solvers with unspecified mathematical purpose. Overall, the reader should find the principles and techniques elaborated here very useful for any kind of nonlinear modeling in scientific computing.

Throughout this book we will assume that the input/output relations of interest can be evaluated by a reasonably tight computer program written in a procedural computer language like Fortran or C and their extensions. Here *tight* means that results are usually obtained with several significant digits of accuracy so that it makes sense to view the outputs as mathematical functions of the inputs. These vector-valued functions may be interpreted as composites of the basic arithmetic operations and a few intrinsic functions, including exponentials and trigonometrics. Since these elementary building blocks are in most places differentiable, computing derivatives of the composite function becomes an exercise in applying the chain rule judiciously. By excluding nondeterministic elements like Monte Carlo calculations and assuming some degree of local and directional smoothness, we justify our contention that quantitative sensitivity analysis is best performed by evaluating derivatives.

Due to nonsmooth intrinsics and program branches, functions defined by evaluation procedures may not be everywhere differentiable and may even have some discontinuities. However, we will always be able to compute directional derivatives at least, and it is hoped that such first- and higher-order sensitivity information will be useful for the ensuing nonlinear computation. Directional derivatives can be used, for example, in numerically integrating differential equations with nondifferentiable right-hand sides. Here, higher-order methods can still achieve full accuracy, provided jumps in the derivatives occur only at boundaries of the numerical time steps. This desirable situation may be arranged by implicitly defining the time increments through appropriate break conditions.

A great deal of work in scientific computing has been predicated upon the assumption that derivative information, and particularly second and higher-order derivative values, are expensive to compute and likely to be inaccurate. The techniques and methods of algorithmic differentiation described in this book should therefore have a far wider significance than simply making fast and accurate numerical derivative values available to existing numerical methods. Algorithmic differentiation provides an exciting opportunity to develop algorithms reflecting the true costs of derivatives and to apply them more widely. We believe that the techniques described in this book will be a crucial ingredient for the next generation of scientific software.

Mathematical Symbols

$n[m]$	Number of independent [dependent] variables	15
F	Given vector function	15
$F'(x)$	Jacobian of F	15
$x\,[y]$	Vector of independent [dependent] variables	17
v_i, v	Intermediate variable	18
φ_i, φ, ψ	Elemental function	18
Φ_i	State transformation	18
$\prec$	Data dependence relation	19
$\prec^*$	Transitive closure of $\prec$	19
l	Number of elemental functions	19
$u_i = (v_j)_{j \prec i}$	Elemental argument vector	19
$v_i(x), v(x)$	Intermediate function	20
$E(x; v)$	Extended system	21
c_{ij}	Elemental partial derivatives	21
$E'(x; v)$	Extended Jacobian	21
Ψ	Library of elementals	22
$\mathcal{D}_i$	Elemental domain	23
$\mathcal{F}$	Composite function set	25
e_i	Cartesian basis vector	32
$\dot{x}\,[\dot{y}]$	Tangent vector in domain [range]	32
$\dot{\varphi}_i, \dot{\varphi}, \dot{\psi}$	Elemental tangent function	34
$\dot{v}_i, \dot{v}$	Tangent of intermediate variable	34
$\dot{u} = (\dot{v}_j)_{j \prec i}$	Tangent argument vector	34
$\dot{F} = \dot{F}(x, \dot{x})$	Composite tangent function	35
$\dot{X}\,[\dot{Y}]$	Bundle of domain [range] tangents	36
$\bar{x}\,[\bar{y}]$	Normal vector in domain [range]	38
A_i	Extended Jacobian of elemental	39
$\nabla\varphi_i$	Elemental gradient	39
$\bar{u}_i$	Adjoint of intermediate variable	40
$\bar{u}_i = (\bar{v}_j)_{j \prec i}$	Adjoint argument vector	40
$\bar{X}\,[\bar{Y}]$	Bundle of normals in domain [range]	48
RAM	Random access memory	48
&	Allocation function	61
ꝏ	Reading tape of prevalues	64

○○	Writing tape of prevalues	64
$\bar{\varphi}_i, \bar{\varphi}, \bar{\psi}$	Elemental adjoint function	64
$\bar{F} = \bar{F}(x, \bar{y})$	Composite adjoint function	66
$task(F)$	Computational task on F	73
$eval(F)$	Evaluation task on F	73
$WORK$	Vector complexity	74
$TIME$	Runtime functional	75
$w \in \mathbb{R}^4$	Runtime weight vector	76
$\mathcal{C}_{task}$	Complexity bound	77
ω_{task}	Runtime ratio of task	78
$tang(F)$	Tangent propagation task on F	80
W_{tang}	Tangent propagation cost	81
C_{tang}	Tangent complexity bound	81
$grad(F)$	Adjoint propagation task on F	83
W_{grad}	Adjoint propagation cost	84
C_{grad}	Adjoint complexity bound	84
$\dot{\bar{x}}, \dot{\bar{y}}, \dot{\bar{v}}$	Second-order adjoints	92
$\dot{\bar{F}} = \dot{\bar{F}}(x, \dot{x}, \bar{y}, \dot{\bar{y}})$	Composite second-order adjoint	99
$\dot{\bar{\varphi}}_i, \dot{\bar{\varphi}}, \dot{\bar{\psi}}$	Elemental second-order adjoint	100
$\mathcal{X}_k$	Index domain	147
$\mathcal{Y}_k$	Index range	148
$\hat{n}$	Maximal domain size	149
$\hat{m}$	Maximal domain range	149
$\bar{n}$	Average domain size	149
$\bar{m}$	Average domain range	150
$\dot{\mathcal{X}}_i$	Second index domains	154
$\bar{\mathcal{X}}_i$	Second adjoint domains	156
S	Seed matrix	161
$\mathcal{G}_c$	Column incidence graph	164
W	Weight matrix	171
$\mathcal{G}_r$	Row incidence graph	171
$H(x, \bar{y})$	Hessian of Lagrangian	179
$[j \mapsto i]$	Set of paths from ⓙ to ⓘ	195
$[\mathcal{U} \mapsto \mathcal{W}]$	Set of paths from U to W	195
$\lvert\mathcal{U} \mapsto \mathcal{W}\rvert$	Number of paths in $[U \mapsto W]$	195
$\lVert\mathcal{U} \mapsto \mathcal{W}\rVert$	Total length of paths in $[U \mapsto W]$	196
$c_{\mathcal{P}}$	Path value = product of arc values	196
$c_{[j \mapsto i]}$	Total derivative of v_i w.r.t. v_j	196
$mark(j)$	Markowitz degree of ⓙ	200
$\mathfrak{F}$	Faces	204
$A(C)$	Jacobian set	227
$\mathrm{scarce}(\mathcal{G})$	Scarcity	227
$\dim(\mathcal{A})$	Jacobian dimension	227
$v_{\bar{\imath}} = \bar{v}_i$	Adjoint value in Hessian graph	237
$\square$	Advancing motion/sweep	268

⌦	Recording motion/sweep	268
⌫	Returning motion/sweep	268
⬭	Reversing motion/sweep	269
A_k	Advancing action	280
C_j	Copying action	280
R_j	Reset action	280
D	Decrement action	280
F_k	Taylor coefficient function	300
$\bar{F}_k$	Higher-order adjoint	300
$v(x(t))$	Restriction of $v(x)$ to path $x(t)$	304
$v(t)$	Taylor or Laurent approximant to $v(x(t))$	304
$\mathcal{P}_d$	Polynomials of degree d	309
NaN	Not a number (IEEE)	339
$R\{\ \}$	Root convergence factor	379
$Q\{\ \}$	Quotient convergence factor	379

Chapter 1

Introduction

The basic idea behind the differentiation methodology explored in this book is by no means new. It has been rediscovered and implemented many times, yet its application still has not reached its full potential. It is our aim to communicate the basic ideas and describe the fundamental principles in a way that clarifies the current state of the art and stimulates further research and development. We also believe that algorithmic differentiation (AD) provides many good exercises that require and develop skills in both mathematics and computer science. One of the obstacles in this area, which involves "symbolic" and "numerical" methods, has been a confusion in terminology, as discussed in the prologue. There is not even general agreement on the best name for the field, which is frequently referred to as *automatic* or *computational differentiation* in the literature. For this book the adjective *algorithmic* seemed preferable, because much of the material emphasizes algorithmic structure, sometimes glossing over the details and pitfalls of actual implementations. Readers who are primarily seeking guidance on using AD software should consult the various software sites linked to the Web page `http://www.autodiff.org/`.

Frequently we have a program that calculates numerical values for a function, and we would like to obtain accurate values for derivatives of the function as well. Derivatives play a central role in sensitivity analysis (model validation), inverse problems (data assimilation), and design optimization (simulation parameter choice). At a more elementary level they are used for solving algebraic and differential equations, curve fitting, and many other applications.

Sometimes we seek first-order derivatives, for example, gradient vectors, for a single-target (cost) function, or a Jacobian matrix corresponding to the normals of a set of constraint functions. Sometimes we seek higher-order derivatives, for example, a Hessian times direction vector product or a truncated Taylor series. Sometimes we wish to nest derivatives, for example, getting gradients of the form $\nabla_x \, F(x, f(x), f'(x))$ from programs for f and F.

The techniques described and analyzed in this book are applied to programs that calculate numerical values in order to produce transformed programs that calculate various derivatives of these values with comparable accuracy and efficiency.

Friends and Relations

At this point, it is worth pausing to distinguish AD from some of the things that it is not. A rather crude way of obtaining approximate numerical derivatives of a function f is the divided difference approach that is sketched in Table 1.1.

$$D_{+h} f(x) \equiv \frac{f(x+h) - f(x)}{h} \quad \text{or} \quad D_{\pm h} f(x) \equiv \frac{f(x+h) - f(x-h)}{2h}$$

Table 1.1: How we DO NOT compute derivatives

It is easy to see why these difference quotients do not provide accurate values: If h is small, then cancellation error reduces the number of significant figures in $D_{+h}f$, but if h is not small, then truncation errors (terms such as $h^2 f'''(x)$) become significant. Even if h is optimally chosen, the values of D_{+h} and $D_{\pm h}$ will be accurate to only about $\frac{1}{2}$ or $\frac{2}{3}$ of the significant digits of f, respectively. For second- (and higher-)order derivatives these accuracy problems become acute.

In contrast, AD methods incur no truncation errors at all and usually yield derivatives with working accuracy.

RULE 0

ALGORITHMIC DIFFERENTIATION DOES NOT INCUR TRUNCATION ERRORS.

To illustrate the difference, let us consider the squared Euclidean norm

$$f(x) = \sum_{i=1}^{n} x_i^2 \quad \text{at} \quad x_i = i \quad \text{for} \quad i = 1 \dots n\,.$$

Suppose we try to approximate the first gradient component of $f(x)$ at this point by the difference quotient

$$\frac{1}{h}\Big[f(x + h\,e_1) - f(x)\Big] = \frac{\partial}{\partial x_1} f(x) + h = 2x_1 + h = 2 + h\,,$$

where $e_1 \in \mathbb{R}^n$ is the first Cartesian basis vector and the truncation error term h happens to be exact. No matter how we perform the calculation of $f(x + h\,e_1)$, its value must be represented as a floating point number, which may require a roundoff error of size $f(x + h\,e_1)\varepsilon \approx n^3 \frac{\varepsilon}{3}$, with $\varepsilon = 2^{-54} \backsimeq 10^{-16}$ denoting the machine accuracy (see Exercise 2.2). For $h = \sqrt{\varepsilon}$, as is often recommended, the difference quotient comes with a rounding error of size $\frac{1}{3}n^3\sqrt{\varepsilon} \approx \frac{1}{3}n^3 10^{-8}$. Hence, not even the sign needs to be correct when n is of order 1,000, a moderate dimension in scientific computing. Things are not much better if one selects h somewhat larger to minimize the sum of the roundoff and truncation errors, whose form is normally not known, of course.

In contrast, we will see in Exercise 3.2 that AD yields the ith gradient component of the squared norm as $2\,x_i$ in both its forward and reverse modes.

The last statement is prone to create the impression that somehow derivatives are generated "symbolically." While this is in some sense true, there is in fact a big difference between the techniques advocated here and the ones implemented in current computer algebra packages. In AD tools the algebraic expression $2\,x_i$ is never generated; instead the numerical value of x_i is multiplied by 2 and then returned as the value of the gradient component. This distinction between AD and fully symbolic computing may seem almost intangible on the squared Euclidean norm, but it should become clear in the following example.

Consider the product example due to Speelpenning [Spe80]:

$$f(x) = \prod_{i=1}^{n} x_i = x_1 * x_2 * \cdots * x_n \, .$$

The corresponding gradient has the symbolic form

$$\begin{aligned}
\nabla f(x) &= \left(\frac{\partial f}{\partial x_1}, \frac{\partial f}{\partial x_2}, \ldots, \frac{\partial f}{\partial x_n} \right) = \left(\prod_{j \neq i} x_j \right)_{i=1\ldots n} \\
&= \big(x_2 * x_3 * \cdots \quad \cdots * \ x_i \ * x_{i+1} * \cdots \quad \cdots * x_{n-1} * \ x_n \ , \\
&\qquad x_1 * x_3 * \cdots \quad \cdots * \ x_i \ * x_{i+1} * \cdots \quad \cdots * x_{n-1} * \ x_n \ , \\
&\qquad \cdots\cdots\cdots\cdots\cdots\cdots\cdots\cdots\cdots\cdots\cdots\cdots\cdots\cdots\cdots\cdots \\
&\qquad x_1 * x_2 * \cdots \quad \cdots * x_{i-1} * x_{i+1} * \cdots \quad \cdots * x_{n-1} * \ x_n \ , \\
&\qquad \cdots\cdots\cdots\cdots\cdots\cdots\cdots\cdots\cdots\cdots\cdots\cdots\cdots\cdots\cdots\cdots \\
&\qquad x_1 * x_2 * \cdots \quad \cdots * x_{i-1} * \ x_i \ * \cdots \quad \cdots * x_{n-2} * \ x_n \ , \\
&\qquad x_1 * x_2 * \cdots \quad \cdots * x_{i-1} * \ x_i \ * \cdots \quad \cdots * x_{n-2} * x_{n-1} \big) \, .
\end{aligned}$$

As we have emphasized by the generous allocation of space to this display, symbolic expressions can take up a lot of memory or screen without necessarily being all that useful to look at. Obviously, there are a lot of common subexpressions that can be used repeatedly for evaluating ∇f at any particular point. This is exactly what AD achieves directly, i.e., without first generating a mess of symbolic expressions and then simplifying them again. A key role is played by intermediate values, such as the initial partial products $x_1 * x_2 * \cdots * x_i$, which are computed during the evaluation of $f(x)$ itself (see Exercise 3.6). They get their own identifiers and may be saved for subsequent calculations, especially the evaluation of ∇f and higher derivatives. In short, the result of applying AD to a function is a program for evaluating its derivative, but not a formula.

Like symbolic differentiation, AD operates by systematic application of the chain rule, which should be familiar from elementary differential calculus. However, in the case of AD, the chain rule is applied not to symbolic expressions but to actual numerical values. As demonstrated by Monagan and Neuenschwander one can of course implement AD within a computer algebra system [MN93].

In summary, we may view symbolic differentiation as a fairly close relative with whom we ought to interact more than we usually do. Differencing, on the other hand, is a completely different technique, which is often extremely useful as a consistency check [Wol82] and certainly must be considered a baseline in terms of accuracy and efficiency. Probably infuriated by their undiminished popularity, AD enthusiasts sometimes abuse difference quotients as easy targets and punching bags. To save us from this unprofessional transgression, we note right away:

RULE 1

DIFFERENCE QUOTIENTS MAY SOMETIMES BE USEFUL TOO.

Computer algebra packages have very elaborate facilities for simplifying expressions on the basis of algebraic identities. In contrast, current AD packages assume that the given program calculates the underlying function efficiently, and under this assumption they produce a transformed program that calculates the derivative(s) with the same level of efficiency. Although large parts of this translation process may in practice be automated (or at least assisted by software tools), the best results will be obtained when AD takes advantage of the user's insight into the structure underlying the program, rather than by the blind application of AD to existing code.

Insight gained from AD can also provide valuable performance improvements in handwritten derivative code (particularly in the case of the reverse accumulation AD technique that we shall meet shortly) and in symbolic manipulation packages. In this book, therefore, we shall view AD not primarily as a set of recipes for writing software but rather as a conceptual framework for considering a number of issues in program and algorithm design.

A Baby Example

Let us look at some of the basic ideas of AD. In this part of the introduction we shall give a plain vanilla account that sweeps all the difficult (and interesting) issues under the carpet.

Suppose we have a program that calculates some floating point outputs y from some floating point inputs x. In doing so, the program will calculate a number of intermediate floating point values. The program control flow may jump around as a result of if-statements, do-loops, procedure calls and returns, and maybe the occasional unreconstructed go-to. Some of these floating point values will be stored in various memory locations corresponding to program variables, possibly overwriting a previous value of the same program variable or reusing a storage location previously occupied by a different program variable that is now out of scope. Other values will be held for a few operations in a temporary register before being overwritten or discarded. Although we will be concerned with many aspects of these computer-related issues later, for the moment we wish to remain more abstract.

Our first abstraction is that of an *evaluation trace.* An evaluation trace is basically a record of a particular run of a given program, with particular specified values for the input variables, showing the sequence of floating point values calculated by a (slightly idealized) processor and the operations that computed them. Here is our baby example: Suppose $y = f(x_1, x_2)$ has the formula

$$y = [\sin(x_1/x_2) + x_1/x_2 - \exp(x_2)] * [x_1/x_2 - \exp(x_2)]$$

and we wish to compute the value of y corresponding to $x_1 = 1.5, x_2 = 0.5$. Then a compiled computer program may execute the sequence of operations listed in Table 1.1.

Table 1.2: An Evaluation Trace of Our Baby Example

v_{-1}	$=$	x_1	$=$	1.5000		
v_0	$=$	x_2	$=$	0.5000		
v_1	$=$	v_{-1}/v_0	$=$	1.5000/0.5000	$=$	3.0000
v_2	$=$	$\sin(v_1)$	$=$	sin(3.0000)	$=$	0.1411
v_3	$=$	$\exp(v_0)$	$=$	exp(0.5000)	$=$	1.6487
v_4	$=$	$v_1 - v_3$	$=$	$3.0000 - 1.6487$	$=$	1.3513
v_5	$=$	$v_2 + v_4$	$=$	$0.1411 + 1.3513$	$=$	1.4924
v_6	$=$	$v_5 * v_4$	$=$	$1.4924 * 1.3513$	$=$	2.0167
y	$=$	v_6	$=$	2.0167		

The evaluation trace, shown in Table 1.2, contains a sequence of mathematical variable definitions. Unlike program variables, mathematical variables can normally not be assigned a value more than once, so the list actually defines a sequence of mathematical functions as well as a sequence of numerical values. We stress that this is just a baby example to illustrate what we mean by applying the chain rule to numerical values. While we are doing this, we wish you to imagine a much bigger example with hundreds of variables and loops repeated thousands of times. The first variables in the list are the input variables; then the intermediate variables (denoted by v_i with $i > 0$) follow, and finally, so does the output variable (just one in this example). Each variable is calculated from previously defined variables by applying one of a number of simple operations: plus, minus, times, sin, exp, and so forth. We can imagine obtaining a low-level evaluation trace by putting a tiny spy camera into the computer's CPU and recording everything that goes past. The evaluation trace does not correspond directly to the expression for f as it was originally written, but to a more efficient evaluation with repeated subexpressions evaluated once and then reused.

It is common for programmers to rearrange mathematical equations so as to improve performance or numerical stability, and this is something that we do not wish to abstract away. The evaluation trace therefore deliberately captures the precise sequence of operations and arguments actually used by a particular implementation of an algorithm or function. Reused subexpressions will be algorithmically exploited by the derivative program, as we shall see.

However, numerical algorithms are increasingly being programmed for parallel execution, and the evaluation trace is (apparently) a serial model. Although a scheduler would spot that any left-hand side can be evaluated as soon as all the corresponding right-hand sides are available, and the evaluation trace could be annotated with an allocation of operations to processors, it is usually more convenient to use a slightly different representation of the evaluation trace called the "computational graph." Figure 1.1 displays the computational graph for our example.

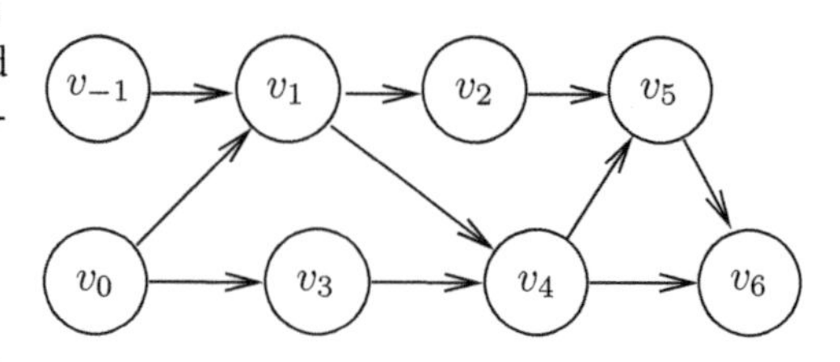

Figure 1.1: Computational Graph of Table 1.2

We shall formalize the concepts of evaluation procedures and computational graphs in Chapter 2. The evaluation trace is a linearization of the computational graph. Our baby example lacks many of the features of a real program. Any real modeling function will also contain nested loops, lots of procedure calls to solve systems of equations, and so on. Consequently, the evaluation trace recording the execution of a self-respecting program may be billions of lines long, even if the text of the program itself is quite short. Then there is no prospect of the entire evaluation trace actually being written down (or even fitting into a computer memory) at all.

Forward Mode by Example

Suppose we want to differentiate the output variable y with respect to x_1. (We probably want the derivative with respect to x_2 as well, but let us keep things simple for the moment.) In effect, we are regarding x_1 as the only independent variable and y as a dependent variable.

One way of getting what we want is to work out the numerical value of the derivative of every one of the variables on the list with respect to the independent variable x_1. So, we associate with each variable v_i another variable, $\dot{v}_i = \partial v_i/\partial x_1$. Applying the chain rule mechanically to each line in the evaluation trace, in order, tells us how to assign the correct numerical value to each $\dot{v}_i$. Clearly, $\dot{v}_{-1} = 1.0, \dot{v}_0 = 0.0$ and (for example) since $v_1 = v_{-1}/v_0$, we must have

$$\begin{aligned}\dot{v}_1 &= (\partial v_1/\partial v_{-1})\dot{v}_{-1} + (\partial v_1/\partial v_0)\dot{v}_0 = (\dot{v}_{-1} - v_1 * \dot{v}_0)/v_0\\ &= (1.0000 - 3.0000 * 0.0000)/0.5000 = 2.0000\,.\end{aligned}$$

Similarly, $v_2 = \sin(v_1)$, so

$$\dot{v}_2 = (\partial v_2/\partial v_1)\dot{v}_1 = \cos(v_1) * \dot{v}_1 = -0.9900 * 2.0000 = -1.9800\,.$$

Augmenting the evaluation trace for our example gives the derived trace in Table 1.3. Imagining a longer calculation with lots of outputs, at the end we would still have that $\dot{y}_i = \partial y_i/\partial x_1$ for each y_i. The total floating point operation

Table 1.3: Forward-Derived Evaluation Trace of Baby Example

$v_{-1} = x_1$	$= 1.5000$		
$\dot{v}_{-1} = \dot{x}_1$	$= 1.0000$		
$v_0 \quad = x_2$	$= 0.5000$		
$\dot{v}_0 \quad = \dot{x}_2$	$= 0.0000$		
$v_1 \quad = v_{-1}/v_0$	$= 1.5000/0.5000$	$=$	3.0000
$\dot{v}_1 \quad = (\dot{v}_{-1} - v_1 * \dot{v}_0)/v_0$	$= 1.0000/0.5000$	$=$	2.0000
$v_2 \quad = \sin(v_1)$	$= \sin(3.0000)$	$=$	0.1411
$\dot{v}_2 \quad = \cos(v_1) * \dot{v}_1$	$= -0.9900 * 2.0000$	$=$	-1.9800
$v_3 \quad = \exp(v_0)$	$= \exp(0.5000)$	$=$	1.6487
$\dot{v}_3 \quad = v_3 * \dot{v}_0$	$= 1.6487 * 0.0000$	$=$	0.0000
$v_4 \quad = v_1 - v_3$	$= 3.0000 - 1.6487$	$=$	1.3513
$\dot{v}_4 \quad = \dot{v}_1 - \dot{v}_3$	$= 2.0000 - 0.0000$	$=$	2.0000
$v_5 \quad = v_2 + v_4$	$= 0.1411 + 1.3513$	$=$	1.4924
$\dot{v}_5 \quad = \dot{v}_2 + \dot{v}_4$	$= -1.9800 + 2.0000$	$=$	0.0200
$v_6 \quad = v_5 * v_4$	$= 1.4924 * 1.3513$	$=$	2.0167
$\dot{v}_6 \quad = \dot{v}_5 * v_4 + v_5 * \dot{v}_4$	$= 0.0200 * 1.3513 + 1.4924 * 2.0000$	$=$	3.0118
$y \quad = v_6$	$= 2.0100$		
$\dot{y} \quad = \dot{v}_6$	$= 3.0110$		

count of the added lines to evaluate $\partial y/\partial x_1$ is a small multiple of that for the underlying code to evaluate y.

This is the basic forward mode of AD, which we shall examine in sections 3.1 and 4.5. It is called "forward" because the derivatives values $\dot{v}_i$ are carried along simultaneously with the values v_i themselves. The problem that remains to be addressed is how to transform a program with a particular evaluation trace into an augmented program whose evaluation trace also contains exactly the same extra variables and additional lines as in the derived evaluation trace. This transformation can be done by a preprocessor or by systematic use of operator overloading. Chapter 6 examines this task in some detail.

We can use exactly the same code to evaluate $\partial y/\partial x_2$ as well: the only change is to set $\dot{x}_1 = 0.0, \dot{x}_2 = 1.0$ at the beginning, instead of the other way round, obtaining $\dot{y} = -13.7239$. However, it may be more efficient to redefine the $\dot{v}_i$ to be vectors (rather than scalars) and to evaluate several partial derivatives at once. This approach is particularly efficient when, because of the structure of the problem, the vectors of partial derivatives are sparse. We shall examine this vector-mode approach in Chapters 7 and 8.

Reverse Mode by Example

An alternative to the forward approach is the *reverse* or *adjoint* mode, which we shall study in detail in sections 3.2 and 4.6 as well as in Chapter 5. Rather than choosing an input variable and calculating the sensitivity of every intermediate variable with respect to that input, we choose instead an output variable and calculate the sensitivity of that output with respect to each of the intermediate variables. As it turns out adjoint sensitivities must naturally be computed backward, i.e., starting from the output variables. We use the term "reverse mode" for this technique because the label "backward differentiation" is well established for certain methods to integrate stiff ODEs [HNW96].

For our example there is only one output variable, namely, y. Hence we associate with each variable v_i another variable $\bar{v}_i = \partial y/\partial v_i$, called the *adjoint variable.* (Strictly speaking, this is an abuse of derivative notation. What we mean is that $\bar{v}_i = \partial y/\partial \delta_i$, where δ_i is a new independent variable added to the right-hand side of the equation defining v_i.) Adding a small numerical value δ_i to v_i will change the calculated numerical value of y by $\bar{v}_i\,\delta_i$ to first order in δ_i.

Once again, the evaluation procedures can be mechanically transformed to provide a procedure for the computation of the adjoint derivatives $\bar{v}_i = \partial y/\partial v_i$. This time, in order to apply the chain rule, we must work through the original evaluation trace backwards. By definition $\bar{y} = 1.0$, and (for example) since the only ways in which v_1 can affect y are via the definitions $v_4 = v_1 - v_3$ and $v_2 = \sin(v_1)$ in which v_1 appears on the right-hand side, we must have

$$\bar{v}_1 = \bar{v}_4 + \bar{v}_2 * \cos(v_1) = 2.8437 + 1.3513 * (-0.9900) = 1.5059\ .$$

If you are troubled by this handwaving derivation, simply accept the results for the time being or consult section 3.2 directly. Although we could keep all the contributions in a single assignment to $\bar{v}_1$, it is convenient to split the assignment into two parts, one involving $\bar{v}_4$ and one involving $\bar{v}_2$:

$$\begin{aligned} \bar{v}_1 &= \bar{v}_4 && = 2.8437\ , \\ \bar{v}_1 &= \bar{v}_1 + \bar{v}_2 * \cos(v_1) && = 2.8437 + 1.3513 * (-0.9900) = 1.5059\ . \end{aligned}$$

This splitting enables us to group each part of the expression for $\bar{v}_1$ with the corresponding line involving v_1 in the original evaluation trace: $v_4 = v_1 - v_3$ and $v_2 = \sin(v_1)$, respectively. The split does mean (in contrast to what has been the case so far) that some of the adjoint variables are incremented as well as being initially assigned.

The complete set of additions to the evaluation trace (corresponding to all the lines of the original) is given in Table 1.4. Once again, the augmented evaluation trace is generated by a program for calculating derivatives. This time the augmented trace is called the reverse trace. Note that the adjoint statements are lined up vertically underneath the original statements that spawned them.

As with the forward propagation method, the total floating point operation count of the added lines is a small multiple of that for the underlying code to evaluate y. Note that the value 3.0118 obtained for $\bar{x}_1 = dy/dx_1$ agrees to four digits with the value obtained for $\partial y/\partial x_1 = \dot{y}$ with $\dot{x}_1 = 1$ in the forward mode. However, this time we have obtained accurate values (to within roundoff error) for both $\bar{x}_1 = \partial y/\partial x_1$ and $\bar{x}_2 = \partial y/\partial x_2$ together. This situation remains true when we imagine a larger example with many more independent variables:

Table 1.4: Reverse-Derived Trace of Baby Example

$$
\begin{aligned}
v_{-1} &= x_1 = 1.5000 \\
v_0 &= x_2 = 0.5000 \\
v_1 &= v_{-1}/v_0 = 1.5000/0.5000 = 3.0000 \\
v_2 &= \sin(v_1) = \sin(3.0000) = 0.1411 \\
v_3 &= \exp(v_0) = \exp(0.5000) = 1.6487 \\
v_4 &= v_1 - v_3 = 3.0000 - 1.6487 = 1.3513 \\
v_5 &= v_2 + v_4 = 0.1411 + 1.3513 = 1.4924 \\
v_6 &= v_5 * v_4 = 1.4924 * 1.3513 = 2.0167 \\
y &= v_6 = 2.0167 \\
\bar{v}_6 &= \bar{y} = 1.0000 \\
\bar{v}_5 &= \bar{v}_6 * v_4 = 1.0000 * 1.3513 = 1.3513 \\
\bar{v}_4 &= \bar{v}_6 * v_5 = 1.0000 * 1.4924 = 1.4924 \\
\bar{v}_4 &= \bar{v}_4 + \bar{v}_5 = 1.4924 + 1.3513 = 2.8437 \\
\bar{v}_2 &= \bar{v}_5 = 1.3513 \\
\bar{v}_3 &= -\bar{v}_4 = -2.8437 \\
\bar{v}_1 &= \bar{v}_4 = 2.8437 \\
\bar{v}_0 &= \bar{v}_3 * v_3 = -2.8437 * 1.6487 = -4.6884 \\
\bar{v}_1 &= \bar{v}_1 + \bar{v}_2 * \cos(v_1) = 2.8437 + 1.3513 * (-0.9900) = 1.5059 \\
\bar{v}_0 &= \bar{v}_0 - \bar{v}_1 * v_1/v_0 = -4.6884 - 1.5059 * 3.000/0.5000 = -13.7239 \\
\bar{v}_{-1} &= \bar{v}_1/v_0 = 1.5059/0.5000 = 3.0118 \\
\bar{x}_2 &= \bar{v}_0 = -13.7239 \\
\bar{x}_1 &= \bar{v}_{-1} = 3.0118
\end{aligned}
$$

Even if our output y depended upon a million inputs x_i, we could still use this reverse or adjoint method of AD to obtain all 1 million components of $\nabla_x y$ with an additional floating point operation count of between one and four times that required for a single evaluation of y. For vector-valued functions one obtains the product of its Jacobian transpose with a vector $\bar{y} = (\bar{y}_i)$, provided the adjoints of the y_i are initialized to $\bar{y}_i$.

Attractions of the Reverse Mode

Although this "something-for-nothing" result seems at first too good to be true, a moment's reflection reassures us that it is a simple consequence of the fact that there is only one dependent variable. If there were several outputs, we would need to rerun the adjoint code (or redefine the $\bar{v}_i$ to be vectors) in order to obtain the complete Jacobian. However, in many applications the number of output variables of interest is several orders of magnitude smaller than the number of inputs. Indeed, problems that contain a single objective function depending upon thousands or millions of inputs are quite common, and in such cases the reverse method does indeed provide a complete, accurate set of first-order derivatives (sensitivities) for a cost of between one and four function evaluations. So much for gradients being computationally expensive!

Of course, we still have to find a way to transform the original program (which produced the initial evaluation trace) into a program that will produce the required additional lines and allocate space for the adjoint variables. Since we need to process the original evaluation trace "backward," we need (in some sense) to be able to transform the original program so that it runs backward. This program reversal is in fact the main challenge for efficient gradient calculation.

We conclude this subsection with a quick example of the way in which AD may produce useful by-products. The adjoint variables $\bar{v}_i$ allow us to make an estimate of the effect of rounding errors on the calculated value of evaluation-trace-dependent variable y. Since the rounding error δ_i introduced by the operation that computes the value of v_i is generally bounded by $\varepsilon|v_i|$ for some machine-dependent constant ε, we have (to first order) that the calculated value of y differs from the true value by at most

$$\varepsilon \sum_i |\bar{v}_i \, v_i| \, .$$

Techniques of this kind can be combined with techniques from interval analysis and generalized to produce self-validating algorithms [Chr92].

Further Developments and Book Organization

The book is organized into this introduction and three separate parts comprising Chapters 2–6, Chapters 7–11, and Chapters 12–15, respectively. Compared to the first edition, which appeared in 2000 we have made an effort to bring the material up-to-date and make especially Part I more readable to the novice. Chapter 2 describes our model of computer-evaluated functions. In the fundamental Chapter 3 we develop the forward and reverse modes for computing dense derivative vectors of first order. A detailed analysis of memory issues and the computational complexity of the forward and reverse mode differentiation is contained in Chapter 4. Apart from some refinements and extensions of the reverse mode, Chapter 5 describes how a combination of reverse and forward differentiation yields second-order adjoint vectors. In the context of optimization calculations these vectors might be thought of as products of the Hessian of the Lagrangian function with feasible directions $\dot{x}$. The weight vector $\bar{y}$ plays the role of the Lagrangian multipliers. The resulting second-order derivative vectors can again be obtained at a cost that is a small multiple of the underlying vector function. It might be used, for example, in truncated Newton methods for (unconstrained) optimization. The sixth chapter concludes Part I, which could be used as a stand-alone introduction to AD. Here we discuss how AD software can be implemented and used, giving pointers to a few existing packages and also discussing the handling of parallelism.

Part II contains some newer material, especially concerning the NP completeness issue and has also been substantially reorganized and extended. The point of departure for Part II is the observation that, in general, Jacobians and Hessians cannot always be obtained significantly more cheaply than by computing a family of matrix-vector or vector-matrix products, with the vector ranging over a complete basis of the domain or range space. Consequently, we will find that (in contrast to the situation for individual adjoint vectors) the cost ratio for square Jacobians or Hessians relative to that for evaluating the underlying function grows in general like the number of variables. Besides demonstrating this worst-case bound in its final Chapter 11, Part II examines the more typical situation, where Jacobians or Hessians can be obtained more cheaply, because they are sparse or otherwise structured. Apart from dynamically sparse approaches (Chapter 7) we discuss matrix compression in considerable detail (Chapter 8). Chapter 9 opens the door to generalizations of the forward and reverse modes, whose optimal application is provably NP hard by reduction to ensemble computation. The actual accumulation of Jacobians and Hessians is discussed in Chapter 10. Chapter 11 considers reformulations of a given problem based on partial separability and provides some general advice to users about problem preparation.

Part III, called "Advances and Reversals," contains more advanced material that will be of interest mostly to AD researchers and users with special needs (very large problems, third and higher derivatives, generalized gradients), in addition to the mathematically curious. Note that in AD, "reversals" are not a bad thing—quite the opposite. While the basic reverse mode for adjoint calculations has an amazingly low operations count, its memory consumption can be excessive. In Chapter 12 we consider various ways of trading space and time in running the reverse mode. Chapter 13 details methods for evaluating higher derivatives, and Chapter 14 concerns the differentiation of codes with nondifferentiabilities (i.e., strictly speaking, most of them). The differentiation of implicit functions and iterative processes is discussed in Chapter 15.

Most chapters contain a final section with some exercises. None of these are very difficult once the curious terminology for this mixture of mathematics and computing has been understood.

There are some slight changes in the notation of this second edition compared to the first. The most important one is that the adjoint vectors overlined with a bar are no longer considered as row vectors but as column vectors and thus frequently appear transposed by the appropriate superscript. The other change, which might cause confusion is that in the final Chapter 15 on implicit and iterative differentiation the implicit adjoint vector is now defined with a positive sign in equation (15.8).

Historical Notes

AD has a long history, with many checkpoints. Arguably, the subject began with Newton and Leibniz, whose approach to differential equations appears to envisage applying the derivative calculations to numbers rather than symbolic formulas. Since then the subject has been rediscovered and reinvented several times, and any account of its history is necessarily partisan.

Something very like the reverse method of AD was being used to design electrical circuits in the 1950s [DR69, HBG71], and Wilkinson's backward error analysis dates to the beginning of the 1960s. Shortly thereafter the availability of compilers led to pressure for an algorithmic transformation of programs to compute derivatives. Early steps in this direction, using forward mode, were taken by Beda et al. [BK+59], Wengert [Wen64], Wanner [Wan69], and Warner [War75] with developments by Kedem [Ked80] and Rall [Ral84]. For more application-oriented presentations see Sargent and Sullivan [SS77], Pfeiffer [Pfe80], and Ponton [Pon82]. Subsequently, the potential of the reverse mode for automation was realized independently by a number of researchers, among them Ostrovskii, Volin, and Borisov [OVB71], Linnainmaa [Lin76], Speelpenning [Spe80], Cacuci et al. [CW+80], Werbos [Wer82], Baur and Strassen [BS83], Kim et al. [KN+84], Iri, Tsuchiya, and Hoshi [ITH88], and Griewank [Gri89]. There is also a close connection to back propagation in neural networks [Wer88]. Checkpointing and preaccumulation were discussed by Volin and Ostrovskii as early as 1985 [VO85]. Computer technology and software have continued to develop, and the ability to build algorithmic tools to perform differentiation is now within the reach of everyone in the field of scientific computing.

Part I

Tangents and Gradients

Chapter 2

A Framework for Evaluating Functions

In this chapter we address the question, What can be differentiated by our computational techniques? The formal answer is that we wish to differentiate a more or less arbitrary vector-valued

$$\text{function} \quad F : \mathcal{D} \subset \mathbb{R}^n \longmapsto \mathbb{R}^m \,.$$

Naturally, if all we are told about a function F is that its domain $\mathcal{D}$ is an open subset of $\mathbb{R}^n$ and its range $F(\mathcal{D})$ a subset of $\mathbb{R}^m$, then differentiation cannot proceed beyond the formal act of appending the symbol F with one or more primes. This notation is nothing more than an assertion of existence, stating specifically that the

$$\text{Jacobian} \quad F' : \mathcal{D} \subset \mathbb{R}^n \longmapsto \mathbb{R}^{m \times n}$$

is a well-defined matrix-valued function on $\mathcal{D}$.

Since we are studying practical, constructive ways of evaluating derivatives, the underlying function F must somehow be given *explicitly*. In other words, it must be described to us in terms of constituent components that we already know how to evaluate and differentiate. It is often stated or tacitly assumed that this means having the function specified by an *algebraic formula* or *closed-form expression*. In contrast, supplying a computer program for evaluating a vector function is often viewed as a somewhat grungy, definitely nonanalytical way of specifying a function. As we have argued in the prologue, this supposed dichotomy between symbolic and numerical function specifications simply does not exist. Moreover, defining functions by suitable evaluation procedures or the corresponding computational graphs is just as analytical and typically more economical than the provision of lengthy formulas.

The key concept we introduce in this chapter is that of a straight-line *evaluation procedure*. We define evaluation procedures from a largely mathematical point of view. The *instantiation* of an evaluation program for a certain set of input parameters with the nomination of independent and dependent variables maps it into an evaluation procedure, since then the control flow is fixed. For example, we may think of a general partial differential equation (PDE) solver program being turned into an evaluation procedure by fixing the geometry and its discretization and selecting certain boundary values as independents and some interior values (or integrals thereof) as dependents. Hence, one might view instantiation as conditional compilation with loop unrolling.

Given specific values of the independents, the execution of an evaluation procedure produces a trace, which can be visualized as an evaluation graph. As a matter of fact, the nodes in the evaluation graph are often still viewed as variables rather than constants, so one may prefer to label the transition from procedures to graphs as *substitution* rather than *expansion*. In this process the code for each subprocedure (starting from the bottom of the calling hierarchy) is substituted or in-lined into the next higher level until one obtains a one-level evaluation procedure called the trace. Hence, we have the following relation between evaluation program, evaluation procedure, and the corresponding evaluation trace.

$$\text{Program} \xrightarrow{\text{instantiation}} \text{Evaluation Procedure} \xrightarrow{\text{expansion}} \text{Trace}$$

Evaluation procedures may overwrite variable values and call subprocedures, but they are restricted to be *straight-line* in that neither loops nor branches nor recursion nor input/output other than parameter passing is allowed. Programs that contain these more general constructs will be called in this book *evaluation programs.* In *evaluation procedures,* subprocedures may be called repeatedly but not recursively, so that the calling hierarchy is acyclic but not necessarily a tree. In other words, one might view an evaluation procedure as an unrolled Fortran 77 program without any branches or input/output. This analogy also captures the global static allocation of all variables that will be assumed for evaluation procedures.

2.1 The Lighthouse Example

Let us consider the function $F : \mathbb{R}^4 \longmapsto \mathbb{R}^2$ that describes the position $y = F(x)$ of the spot at which the beam from a lighthouse hits a quay wall of a certain shape. We consider at first a straight wall with slope γ and a horizontal distance ν from the lighthouse (see Fig. 2.1). These two parameters and the angular velocity ω, as well as the time t, form the vector of independent variables

$$x = (x_1, x_2, x_3, x_4) = (\nu, \gamma, \omega, t) .$$

Using some planar geometry, we find that the coordinates $y = (y_1, y_2)$ of the light spot are given by the algebraic formulas

$$y_1 = \frac{\nu * \tan(\omega * t)}{\gamma - \tan(\omega * t)} \quad \text{and} \quad y_2 = \frac{\gamma * \nu * \tan(\omega * t)}{\gamma - \tan(\omega * t)} . \tag{2.1}$$

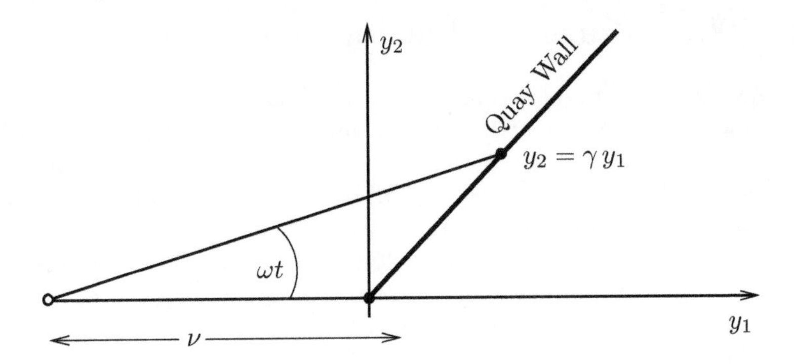

Figure 2.1: Lighthouse Geometry

Since the two dependents differ only in the ratio $\gamma = y_2/y_1$, it would obviously be uneconomical to store and manipulate the expressions for y_1 and y_2 separately. Moreover, there is the common subexpression $\tan(\omega * t)$, whose evaluation is likely to dominate the cost of calculating y_1 and y_2. The repeated occurrence of $\tan(\omega * t)$ could be avoided by expanding the fractions with the reciprocal $\cot(\omega * t)$, yielding the seemingly simpler expression $y_1 = \nu/[\gamma * \cot(\omega * t) - 1]$. However, this is definitely not a good idea because the cotangent is undefined at the (presumably initial) time $t = 0$. On the basis of this small example we make the following sweeping generalizations.

- Explicit algebraic expressions have common subexpressions.

- Symbolic simplifications can lead to numerical instabilities.

If there are many common subexpressions and we cannot expect to prevent their repeated occurrence safely, how can we avoid manipulating them many times during the evaluation of a function and its derivatives? The answer is simply that we give them a name (or a pointer to an object in memory) and treat them as intermediate variables or even functions in their own right.

In other words, we no longer look for an explicit expression for the dependent variables $(y_i)_{1 \le i \le m}$ directly in terms of the independent variables $(x_j)_{1 \le j \le n}$, but prefer to deal with evaluation procedures like the following:

$$v = \tan(\omega * t); \quad y_1 = \nu * v/(\gamma - v); \quad y_2 = \gamma * y_1 \,.$$

Mathematicians are likely to object that the pair of two-level formulas given earlier yields much more insight into the functional properties of y_1 and y_2 than the three statements involving the intermediate variable v. However, if the functional dependence between the x and the y is of any significant complexity, the resulting algebraic formula is likely to run over several lines or pages and thus to provide little analytical insight.

2.2 Three-Part Evaluation Procedures

Suppose we go further and decompose the lighthouse function into even smaller atomic operations. In practice this decomposition happens anyway when a function is evaluated on a computer. In our example a compiler might generate the sequence of statements listed in Table 2.1. The actual calculations are carried out in the midsection of the box.

Table 2.1: Lighthouse Procedure

v_{-3}	$=$	$x_1 = \nu$
v_{-2}	$=$	$x_2 = \gamma$
v_{-1}	$=$	$x_3 = \omega$
v_0	$=$	$x_4 = t$
v_1	$=$	$v_{-1} * v_0$
v_2	$=$	$\tan(v_1)$
v_3	$=$	$v_{-2} - v_2$
v_4	$=$	$v_{-3} * v_2$
v_5	$=$	v_4/v_3
v_6	$=$	v_5
v_7	$=$	$v_5 * v_{-2}$
y_1	$=$	v_6
y_2	$=$	v_7

The first section copies the current values of the independent variables to internal variables $(v_{-3},\, v_{-2},\, v_{-1},\, v_0)$, and the third section copies the resulting values $(v_6,\, v_7)$ into the dependent variables $(y_1,\, y_2)$. The seemingly unproductive assignment $v_6 = v_5$ was introduced to make the dependent variables $y_1 = v_6$ and $y_2 = v_7$ mutually independent. In talking about operations and assignments, we have already tacitly interpreted the contents of Table 2.1 as a procedural code. Later we shall emphasize this point of view, but so far one may also view the midsection of the box as an unordered collection of mathematical identities. That will no longer be possible when several values are successively placed in the same memory location, as could be done here with v_1, v_3, and v_6. Suitable addressing functions will be discussed in section 4.1.

In general, we shall assume that all quantities v_i calculated during the evaluation of a function at a particular argument are numbered such that

$$[\underbrace{v_{1-n}, \ldots\ldots\ldots, v_0}_{x}, v_1, v_2, \ldots\ldots\ldots\ldots, v_{l-m-1}, v_{l-m}, \underbrace{v_{l-m+1}, \ldots\ldots\ldots\ldots\ldots, v_l}_{y}] \,.$$

Each value v_i with $i > 0$ is obtained by applying an *elemental function* φ_i (of a nature discussed later) to some set of arguments v_j with $j < i$, so that we may write

$$v_i = \varphi_i(v_j)_{j \prec i} \,, \tag{2.2}$$

where the dependence relation $\prec$ is defined in the following subsection. For the application of the chain rule in its standard form it is sometimes useful to associate with each elemental function φ_i the *state transformation*

$$\mathbf{v}_i = \Phi_i(\mathbf{v}_{i-1}) \quad \text{with} \quad \Phi_i \,:\, \mathbb{R}^{n+l} \to \mathbb{R}^{n+l} \,, \tag{2.3}$$

where

$$\mathbf{v}_i \equiv \left(v_{1-n},\, \ldots,\, v_i\,,\, 0,\, \ldots,\, 0\right)^{\top} \in \mathbb{R}^{n+l} \,.$$

In other words Φ_i sets v_i to $\varphi_i(v_j)_{j\prec i}$ and keeps all other components v_j for $j \neq i$ unchanged. For brevity, we may concatenate the v_j on which φ_i depends to the argument vector

$$u_i \equiv (v_j)_{j\prec i} \in \mathbb{R}^{n_i} \,. \tag{2.4}$$

The Dependence Relation $\prec$ and the Computational Graph

In (2.2) the precedence relation $j \prec i$ means that v_i depends directly on v_j, as is typically the case for one or two indices $j < i$. In section 5.1 we shall allow for the possibility $i \prec i$, which means that the statement (2.2) becomes *iterative*. On the lighthouse problem we have, for example, $-3 \prec 4$ and $2 \prec 4$ but $0 \not\prec 4$. Note, however, that $0 \prec 1 \prec 2 \prec 4$ implies $0 \prec^* 4$, where $\prec^*$ is the *transitive closure* [SW85] of $\prec$ and thus a proper partial ordering of all indices $i = 1-n \ldots l$. These data dependence relations can be visualized as an acyclic graph, which we will call the corresponding *computational graph* [Bau74]. The set of vertices are simply the variables v_i for $i = 1-n \ldots l$, and an arc runs from v_j to v_i exactly when $j \prec i$. The roots of the graph represent the independent variables $x_j = v_{j-n}$ for $j = 1 \ldots n$, and the leaves represent the dependent variables $y_{m-i} = v_{l-i}$ for $i = 0 \ldots m-1$. We draw the graph from the roots on the left to the leaves on the right. The computational graph for the lighthouse problem is shown in Fig. 2.2. Here and later, the graph description is used only as an added illustration; all methods and results can be understood and proven without reference to graph terminology.

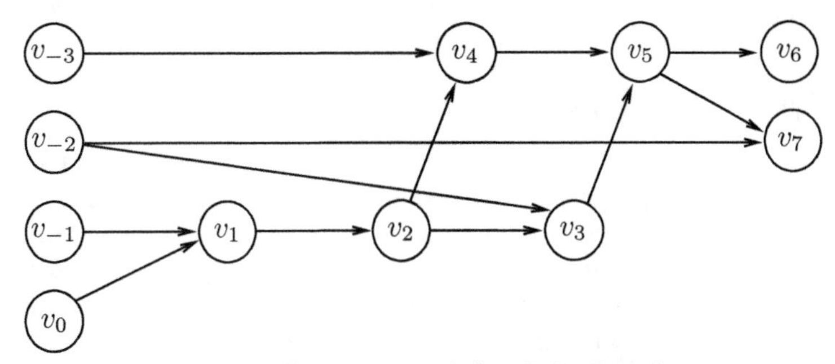

Figure 2.2: Computational Graph for Lighthouse

Throughout this book we shall be concerned with functions whose evaluation at a particular argument can be represented by an evaluation procedure of the general form given in Table 2.2.

Table 2.2: General Evaluation Procedure

v_{i-n}	$=$	x_i	i	$=$	$1 \ldots n$
v_i	$=$	$\varphi_i(v_j)_{j\prec i}$	i	$=$	$1 \ldots l$
y_{m-i}	$=$	v_{l-i}	i	$=$	$m-1 \ldots 0$

We will sometimes represent the execution of a given evaluation procedure by a subroutine call

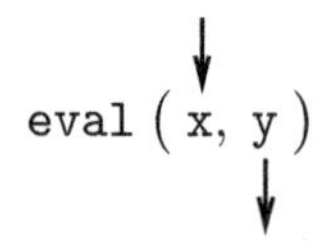

where the vertical arrows indicate that the arrays $\texttt{x} = x \in \mathbb{R}^n$ and $\texttt{y} = y \in \mathbb{R}^m$ are input and output parameters, respectively. Of course this notation does not preclude that some variables are input-output parameters, nor that the evaluation routine `eval` calls other subroutines.

Evaluation procedures are convenient for our purposes because they furnish a mathematical representation of F as a composite function of its elemental constituents φ_i to which the rules of calculus can be unambiguously applied. Various differential operators may then be applied, yielding derived and adjoint evaluation procedures of a closely related structure.

Obviously, any evaluation procedure uniquely defines a vector function $y = F(x)$ whose properties are determined by the precedence relation between the v_i and the properties of the elemental constituents φ_i. Thus each value $v_i = v_i(x)$ can also be interpreted as an *intermediate function* $v_i(x)$ of the independent variable vector $x \in \mathbb{R}^n$. As will be seen below, most *elemental* functions tend to be rather *elementary*, but our framework allows for the inclusion of computationally more intensive elementals as discussed at the end of this subsection.

Code Quality Independence

To the extent that one is used to thinking of functions as abstract mappings, it might appear inappropriate that AD hinges so strongly on their particular representation as a composite of certain elemental functions. Since this decomposition is by no means unique, we are faced with a certain arbitrariness: two evaluation procedures representing the same mathematical function may have widely varying stability and efficiency properties with respect to differentiation. However, this effect is not specific to differentiation; it applies to the basic task of evaluating the function at any given argument. Moreover, in our experience, the issues of code quality and AD can be largely separated from each other. Hence, we formulate the following working hypothesis.

RULE 2

WHAT'S GOOD FOR FUNCTION VALUES IS GOOD FOR THEIR DERIVATIVES.

To be little more technical, we claim that derived evaluation procedures and programs are as good (or bad) as the underlying original evaluation procedures and programs. In the next section we formulate our minimal requirements on the library of elemental functions from which the constituent components φ_i are drawn.

The Extended System $E(x; v) = 0$

The evaluation procedure listed in Table 2.2 can also be encapsulated in the nonlinear system of equations

$$0 = E(x; v) \equiv (\varphi_i(u_i) - v_i)_{i=1-n\ldots l} \, , \tag{2.5}$$

where the first n components of E are defined as the initialization functions

$$\varphi_i(u_i) \equiv x_{i+n} \quad \text{for} \quad i = 1 - n \ldots 0 \, . \tag{2.6}$$

Moreover, we may assume without loss of generality that the dependent variables $y_{m-i} = v_{l-i}$ for $0 \leq i < m$ are mutually independent of each other, so we have that

$$i < 1 \quad \text{or} \quad j > l - m \quad \text{implies} \quad c_{ij} \equiv 0 \, , \tag{2.7}$$

where from now on

$$c_{ij} \equiv c_{ij}(u_i) \equiv \frac{\partial \varphi_i}{\partial v_j} \quad \text{for} \quad 1 - n \leq i, j \leq l \, . \tag{2.8}$$

These derivatives will be called *elemental partials* throughout. The Jacobian of E with respect to the $n + l$ variables v_j for $j = 1 - n \ldots l$ is then given by the unitary lower triangular matrix

$$E'(x; v) = (c_{ij} - \delta_{ij})_{j=1-n\ldots l}^{i=1-n\ldots l} = C - I$$

$$= \left[\begin{array}{cccccccccc}
-1 & 0 & \cdots & 0 & 0 & \cdots & 0 & 0 & \cdots & 0 \\
0 & \ddots & \ddots & \vdots & \vdots & & \vdots & \vdots & & \vdots \\
\vdots & \ddots & \ddots & 0 & \vdots & & \vdots & \vdots & & \vdots \\
0 & \cdots & 0 & -1 & 0 & \cdots & 0 & 0 & \cdots & 0 \\
\times & \cdots & \times & -1 & 0 & \cdots & 0 & 0 & \cdots & 0 \\
\vdots & & \vdots & \times & \ddots & \ddots & \vdots & \vdots & & \vdots \\
\vdots & & \vdots & \vdots & \ddots & \ddots & 0 & \vdots & & \vdots \\
\times & \cdots & \times & \times & \cdots & \times & -1 & 0 & \cdots & 0 \\
\times & \cdots & \times & \times & \cdots & \times & -1 & 0 & \cdots & 0 \\
\vdots & & \vdots & \vdots & & \vdots & 0 & \ddots & \ddots & \vdots \\
\vdots & & \vdots & \vdots & & \vdots & \vdots & \ddots & \ddots & 0 \\
\times & \cdots & \times & \times & \cdots & \times & 0 & \cdots & 0 & -1
\end{array}\right] \tag{2.9}$$

$$= \left[\begin{array}{ccc} -I & 0 & 0 \\ B & L - I & 0 \\ R & T & -I \end{array}\right] \begin{array}{l} \\ \big\} \, l \\ \end{array} \, , \quad \text{with the first block column of width } \underbrace{}_{n} \text{ and the last of width } \underbrace{}_{m}$$

Here δ_{ij} denotes the Kronecker Delta and $c_{ij} \equiv 0$ for the indices $i < 1$ by definition of the initialization functions. Since the unitary lower triangular matrix

$-E'(x;v) = I - C$ can never be singular, the implicit function theorem guarantees that all v_i are uniquely defined by $E(x;v) = 0$ as implicit functions of the variables x. This applies particularly to the last m components, which we denote by $y = F(x)$ as before. More mathematically oriented readers may prefer to think of the basic evaluation procedure in terms of the triangular nonlinear system $E(x;v) = 0$ and its Jacobian $E' = C - I$. Many relations and algorithms can be derived directly from $E(x;v) = 0$, though more computational considerations are not as easy to discuss in this context.

The strictly lower *triangular matrix* C of order $n + l$ and the resulting *extended Jacobian* $C - I$ occur quite frequently in the literature on the reverse mode (see, e.g., [FF99] and [Hor92]). In Chapter 9 we will use the extended Jacobian $E' = C - I$ to obtain more general "cross-country elimination methods" for computing the "compact" Jacobian $F'(x)$ as a Schur complement.

Rounding Up the Usual Suspects

The arithmetic operations and some of the intrinsic functions available in most high-level programming languages are listed in Table 2.3. While the basic arithmetic operations $+$, $-$, and $*$ are indispensable, the choice of the other elemental functions making up our elemental library is quite flexible. As we shall indicate in section 13.4, the rules of differential calculus can be extended from the field of real or complex numbers to any commutative ring of scalars that contains a neutral element with respect to multiplication. In other words, the set Ψ of elemental functions must at least contain the binary operations $+$ and $*$, the unary sign switch $-$, and the initialization to some constant scalar c. One may call this minimal set the *polynomial core* because it allows the evaluation of any polynomial function by a suitable evaluation trace.

Table 2.3: Elemental Operators and Functions

	Essential	Optional	Vector
Smooth	$u+v, u*v$	$u-v, u/v$	$\sum_k u_k * v_k$
	$-u, c$	$c*u, c \pm u$	$\sum_k c_k * u_k$
	$\mathrm{rec}(u) \equiv 1/u$	u^k	
	$\exp(u), \log(u)$	u^c, $\arcsin(u)$	
	$\sin(u), \cos(u)$	$\tan(u)$	
	$\|u\|^c, c > 1$	...	
	...	...	
Lipschitz	$\mathrm{abs}(u) \equiv \|u\|$	$\max(u,v)$	$\max\{\|u_k\|\}_k$, $\sum_k \|u_k\|$
	$\\|u,v\\| \equiv \sqrt{u^2+v^2}$	$\min(u,v)$	$\sqrt{\sum_k u_k^2}$
General	$\mathrm{heav}(u) \equiv \begin{cases} 1 \text{ if } u \geq 0 \\ 0 \text{ else} \end{cases}$	$\mathrm{sign}(u)$ $(u > 0)?v,\ w$	
Convention: k integer; u, v, w, u_k, v_k variables; c, c_k constants			

Since all univariate intrinsics such as $\sin(u)$ and even the reciprocal function $\mathrm{rec}(u) = 1/u$ are actually evaluated by a sequence of polynomial operations, possibly following a table look-up, one might be tempted to conclude that for practical purposes there is no need to introduce elemental functions outside the polynomial core. However, this is a false conclusion because table look-ups and subsequent iterations involve program branches that are so far not allowed in our evaluation procedures. They could be introduced in the form of the Heaviside function, the sign function, or a ternary conditional assignment, which are all listed in the last row of Table 2.3.

Unfortunately, these elemental functions are patently nondifferentiable, thus apparently destroying any prospect of evaluating derivatives of composite functions by applying the chain rule to the derivatives of their elemental constituents. Even though we shall see in Chapter 14 that nonrecursive programs with branches are still in some sense directionally differentiable, initially we wish to consider only smooth elemental functions. For the time being, this requirement also eliminates the absolute value and norm functions listed in the Lipschitz category of Table 2.3.

2.3 Elemental Differentiability

Let us impose the following condition.

Assumption (ED): ELEMENTAL DIFFERENTIABILITY

All elemental functions φ_i are d times continuously differentiable on their open domains $\mathcal{D}_i$, i.e., $\varphi_i \in C^d(\mathcal{D}_i)\,,\; 0 \le d \le \infty$.

The assumption is obviously satisfied for the polynomial operations, which are analytic everywhere on their domains $\mathbb{R}$ or $\mathbb{R}^2$. For the division $\varphi_i(v,u) = v/u$, the axis $u = 0$ must be excluded from the plane, so that $\mathcal{D}_i = \mathbb{R} \times \{\mathbb{R} \setminus \{0\}\}$. Sometimes it is convenient to conceptually decompose the division into a reciprocal $\mathrm{rec}(u) = 1/u$ and a subsequent multiplication by v. Then all binary operations are commutative and globally analytic, while $\mathrm{rec}(u)$ can be treated like any other univariate function that is not defined everywhere. As an immediate, though not altogether trivial, consequence, we obtain the following theoretical result.

Proposition 2.1 (CHAIN RULE)
Under Assumption (ED) *the set $\mathcal{D}$ of points $x \in \mathcal{D}$ for which the function $y = F(x)$ is well defined by the evaluation procedure Table* 2.2 *forms an open subset of $\mathbb{R}^n$ and $F \in C^d(\mathcal{D})\,, 0 \le d \le \infty$.*

Proof. By assumption we must have $u_i = u_i(x) \in \mathcal{D}_i$ for $i = 1 \ldots l$. Because of the assumed openness of the $\mathcal{D}_i$ and the differentiability of the φ_i, it follows from the chain rule that all v_i, including the dependents $y_i = F_i(x)$, are in fact C^d functions on some neighborhood of x. ∎

Loosely speaking, the proposition asserts that wherever an evaluation procedure can be executed, it is $d \geq 0$ times differentiable. Naturally, the maximal open domain $\mathcal{D}$ may be empty because, at any point $x \in \mathbb{R}^n$, one of the elemental functions φ_i gets an argument vector $u_i = (v_j)_{j \prec i}$ that lies outside its domain $\mathcal{D}_i$. Then one may return a NaN as discussed in section 14.2.

Since F inherits continuous differentiability from its constituent elements, we may use the commutativity of differentiation with respect to several variables or directions. Obviously, stronger properties like Hölder or even Lipschitz continuity of the dth derivative are also inherited by F from its elements. We have not stated this fact formally because mere continuity of derivatives is good enough for our main purpose, namely, the application of the chain rule.

When $d = \infty$, which holds for most test problems in the numerical optimization literature, the proposition asserts the inheritance of real analyticity (C^∞). Many larger engineering codes, however, involve the look-up of interpolated data and we can expect the following situation. If the interpolation functions are splines or are otherwise stitched together from local functions, C^d continuity across the boundaries is rarely achieved with $d = 2$, quite possibly with $d = 1$, and probably with $d = 0$. Even if the first or second derivatives do not match exactly across boundaries, one may expect the discrepancies to be small enough for local Taylor expansions to provide reasonable approximations for optimization and other numerical purposes. Another possible source of a finite degree of differentiability d are powers of the absolute value function $\varphi(u) = |u|$.

Fractional Powers and Singularities

We have listed u^c for fractional exponents c as an optional element because it equals $\exp[c * \log(u)]$ provided $u > 0$. That begs the question of why we have thus excluded the origin from the domain of u^c and, in particular, for the square root $u^{0.5} = \sqrt{u}$. Formally, including 0 would violate our condition that the elemental domains $\mathcal{D}_i$ have to be open. This requirement makes some numerical sense because arbitrary small perturbations of the argument can lead to a situation where the square root—and thus the whole evaluation procedure—cannot be executed, at least in real arithmetic. Hence, we might claim that the underlying evaluation process is poorly determined whenever a square root of zero is taken, so that we, the "differentiators," cannot reasonably be expected to provide meaningful derivatives at such marginal points. Unfortunately, this argument is not always convincing for examples when the square root occurs in the context of a function such as

$$f(x) = \sqrt{x^6} = |x|^3 \quad \text{with} \quad f'(x) = 3\,|x|\,x\,.$$

This function is globally defined and can be safely evaluated via $u = x^6$ and $y = \sqrt{u}$ or via $u = \mathrm{abs}(x)$ and $y = u^3$ even in low-precision arithmetic. Our problem is that the derivative value $f'(0) = 0$ cannot be obtained by applying the chain rule to either of these two small evaluation procedures because neither the square root nor the *modulus* has a finite derivative at the origin. As we will see in Chapter 14, one can still save the day by performing directional differentiation in this and related cases.

For the time being we must introduce $\varphi(x) = |x|^c$ with $c > 1$ as an elemental function and provide extra "microcode" for its derivative

$$\varphi'(x) = c\,x\,|x|^{c-2} \in C^{\lfloor c-1 \rfloor}(\mathbb{R}) .$$

Clearly, the code for φ and φ' must contain tests on the sign and possibly the size of x. This situation is comparable to the reason for introducing the exponential and other intrinsic transcendentals as elemental functionals. They could be decomposed further into evaluation algorithms, but these would contain branches and thus not be amenable to differentiation by the chain rule. Hence, we may formulate the following corollary of Proposition 2.1.

Corollary 2.1 (Composite Differentiation)
If derivatives up to order $d > 0$ of a vector function F are needed on a domain $\mathcal{D}$, the elemental functions φ_i and their open domains $\mathcal{D}_i$ must be chosen such that $\varphi_i \in C^d(\mathcal{D}_i)$ and the resulting evaluation procedure is well defined at all $x \in \mathcal{D}$.

The assertion of the corollary can be trivially satisfied by defining F itself as an elemental function. Obviously, this is not very helpful, since the techniques of AD are defined at the level of evaluation procedures. A nested approach is certainly possible, but in general we want to keep the elemental functions small so that their differentiation "by hand" is not too laborious.

In the remainder of the book we shall normally assume that the elemental functions φ_i are drawn from some fixed library

$$\Psi \equiv \{c, \pm, *, \exp, \sin, \dots\} , \tag{2.10}$$

whose members $\varphi \in \Psi$ satisfy Assumption (ED) for some common $d \geq 1$. Hence the last two categories of Table 2.3 are generally excluded so as to ensure that all composite functions F are at least once continuously differentiable. The nondifferentiable and even discontinuous situation arising from appending norms and the Heaviside function to Ψ will be analyzed in Chapter 14. Given the library Ψ, we denote by

$$\mathcal{F} \equiv \mathrm{Span}[\Psi] \subset C^d \tag{2.11}$$

the set of all functions that can be defined as evaluation procedures with elemental components in Ψ.

2.4 Generalizations and Connections

Generalization to Vector-Valued Elementals

In general we may allow the v_i with $i > 0$ to be vectors of some dimension

$$m_i \equiv \dim(v_i) \geq 1 \quad \text{for} \quad i = 1 \dots l - m .$$

However, there is no harm in considering them as scalars, which makes the basic methods of AD easier to understand. As m_i is by definition the range dimension of the elemental function φ_i, allowing $m_i > 1$ means that we may treat basic linear algebra subroutines or other subprograms as single elemental functions. This interpretation can have conceptual and practical advantages. In particular, it becomes possible to interpret a whole evaluation procedure as a (super)elemental function that can be part of an evaluation procedure at a higher level. In this way evaluation procedures become hierarchical structures whose various levels can be subjected to different differentiation techniques. For the most part we shall consider only the two-level situation, but some techniques and results turn out to be fully recursive over arbitrarily many levels. Independent and dependent variables are always assumed to be scalars, so that

$$i < 1 \quad \text{or} \quad i > l - m \quad \text{implies} \quad m_i = 1 \,.$$

To avoid notational complications, we may treat each v_i as one data item. Thus the domain dimension of the elemental function φ_i is given by

$$n_i \equiv \sum_{j \prec i} m_j \quad \text{for} \quad i = 1 \ldots l \,. \tag{2.12}$$

By breaking vector-valued elemental functions into their scalar components, one can always achieve $m_i = \dim(v_i) = 1$ for the sake of conceptional and notational simplicity. The only drawback of this simplifying assumption is that efficiency may be lost in complexity theory and computational practice when several of the component functions involve common intermediates. For example, in the lighthouse problem, it would not be a good idea to treat the two component functions $y_1(x)$ and $y_2(x)$ as two elemental functions, since the common intermediate $\tan(\omega * t)$ would be duplicated. To avoid this, one may either break the two down further, as we have done above, or combine them into one elemental with $n_i = 4$ independents and $m_i = 2$ dependents in the context of a larger function evaluation. In general, allowing vector-valued elemental functions makes it possible to extend a given library of elemental functions by any of its composites defined in the form of an evaluation procedure.

Overwriting, Recursion, and Round-Off

For any program of reasonable size, the values of variables are overwritten frequently when they are no longer needed to reduce the overall memory requirement of the program. Therefore, AD has to cope with overwriting of intermediate variables. However, the handling of overwrites complicates the derivation of the forward and reverse modes considerably. For this reason, we introduce basic versions of the forward and reverse modes without any overwriting in sections 3.1 and 3.2, respectively. Appropriate strategies to incorporate the possibility of overwrites are derived in sections 4.1–4.3. Iterative processes are an important ingredient for a wide range of computer programs. A detailed discussion of the corresponding effects for differentiation is contained in Chapters 12, 14, and 15.

As shown there, directional derivatives can be evaluated for any evaluation procedure not containing a recursion. For recursive processes this statement is not longer valid, as illustrated by the following example.

Consider the piecewise linear continuous odd function $f : [-1, 1] \mapsto [-0.5, 0.5]$ that has the values

$$f(2^{-j}) = 2^{-j} \quad \text{if} \quad j \text{ odd} \quad \text{and} \quad f(2^{-j}) = 0.1 * 2^{-j} \quad \text{if} \quad j \text{ even} \tag{2.13}$$

for all integers $j > 0$. In between these values at the binary powers we interpolate by straight lines whose slope alternates between -0.8 and 1.9. To obtain an odd function we have to set $f(0) = 0$ and $f(x) = -f(-x)$ for $-1 \leq x < 0$. The resulting piecewise linear continuous function is displayed in Fig. 2.3.

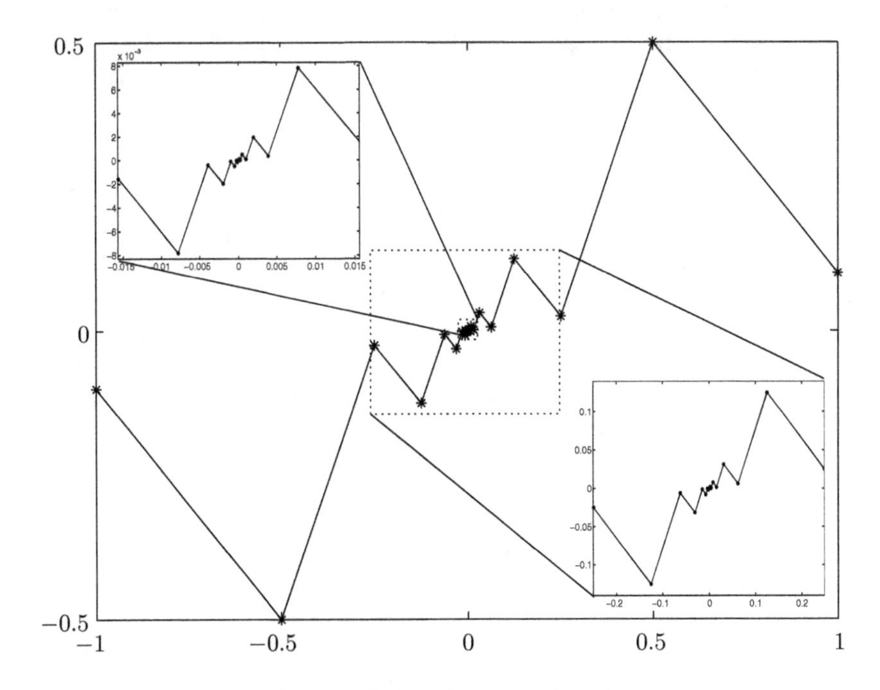

Figure 2.3: Development of $f(x)$

The function $f(x)$ has the Lipschitz constant 1.9, which implies by Rademacher's theorem that it is almost everywhere differentiable. Here, that is anyway obvious with all binary powers $x = \pm 2^{-j}$ for $j > 0$ and $x = 0$ being the exceptional points where f is not differentiable. However, whereas f is still piecewise differentiable according to Definition (PD) on page 342 on all neighborhoods excluding $x = 0$, this is no longer true at the origin itself. As a consequence our analysis in sections 14.2 and 14.4 concerning the existence of essential selection functions, directional derivatives, and even generalized Jacobians does not apply. Of course, the function could be modified such that the set of such essential nondifferentiabilities is much larger and even dense in its domain.

As easy as the function looks its evaluation by a suitable evaluation procedure is not completely obvious. The most elegant way would be to call eval(x, 1.0) with eval(x, s) defined recursively, for example, as the C function

```
eval(x,s)
  { if(x==0.0) return 0.0;
    else if (x <  0) return -eval(-x,0);
    else if (x >  s/2) return 0.9*s-0.8*x;
    else if (x >= s/4) return -9*s/20+1.9*x;
    else return eval(x, s/4);
  }
```

Let us assume that s has been declared as a floating point number of a certain format, whose smallest representable positive number is 2^{-m}. Then we find that the recursion depth of the program above will be maximally m, i.e., something like 62 in current double precision arithmetic. However, in exact arithmetic an unbounded recursion is possible by choosing x arbitrary close to 0.

To avoid recursion of arbitrary depth one may evaluate the relevant binary power directly in C as $s =$ pow(4, floor($\log_4(x)$)) for $x > 0$ and then set the value to $0.9s - 0.8x$ or $-9s/20 + 1.9x$ depending on whether $x > s/2$ or not. Here the function floor(y) is supposed to yield the largest lower integer bound of any real number y. In this way we would obtain an evaluation procedure that has the structure prescribed in this chapter for the whole book. However, not only does it employ conditional assignments, but more importantly the elementary function $\log_4(x) = \ln(x)/\ln(4)$ is evaluated arbitrarily close to the boundary of its domain where it does not attain a well-defined finite value. In conclusion our general result on piecewise and directional differentiability developed in Chapter 14 does not apply because we have either an infinite recursion or an elementary function at the boundary of its domain.

From another point of view the undesirable lack of local definition arises not from the infinite slope of $\log_4$ but from the seemingly innocuous function floor( ). Due to its infinite number of jumps it is not semi-algebraic in that its epigraph $\{(x, y) \in \mathbb{R}^2, y \geq \texttt{floor}(x)\}$ is not a semi-algebraic set. By contrast semi-algebraic sets are defined by a finite set of polynomial inequalities. The family of such sets is closed under finite unions and intersections as well as projections and other linear mappings. The theory of such families has recently attracted considerable interest and is, for example, surveyed in the book *Algorithms in Real Algebraic Geometry* [BPS06]. Functions are called semi-algebraic if their epigraph is semi-algebraic and have the remarkable property that in the neighborhood of any argument they stratify into a finite number of polynomial functions on algebraic domains with nonempty relative interior. This stratification allows for the existence and computation of Puiseux expansions along regular arcs in the domain. Our proposal of generalized Laurent expansion developed in section 14.3 pursues a similar approach for a more general class of subanalytic functions including the exponential, the sinusoidal functions, and even the logarithm. For a survey of such o-minimal function classes see the paper by Coste [Cos00]. Functions that are subanalytic and locally Lipschitz are called tame. It has recently been shown by Bolte, Daniilidis, and Lewis [BDL08] that all tame functions F are semismooth in that

$$\lim_{\|s\|\to 0} \big\|F(x+s) - F(x) - F'(x+s)s\big\|/\|s\| = 0$$

where $F'(x+s)$ may represent any generalized Jacobian of F at $x+s$. While this property looks like a minor variation on Fréchet differentiability, it is in fact much weaker and thus more prevalent. Moreover, it still ensures superlinear convergence of Newton's method, provided all generalized Jacobians are nonsingular. In other words, when excluding such special constructs our evaluation procedures always produce semismooth functional relations, which makes them amenable to local linearization. This recent observation should have a strong impact on the way we develop the theory and tools of AD in the near future. In particular we should always provide at least one element of the generalized derivative in the sense of Clark using the techniques foreshadowed in section 14.3. The connection between the theory of semi-algebraic and subanalytic functions and the properties of our evaluation procedures have to be analyzed in more detail in the future.

As customary even in wide areas of applied mathematics, we will mostly ignore roundoff effects in this book and work under the tacit assumption that functions and derivatives can be evaluated with infinite precision. That is just a little ironic since, if this hypothesis were indeed valid, derivatives could be evaluated with arbitrary precision by divided differences, and all the techniques advocated in this book concerning the forward mode would be rather useless. However, the reverse mode and various cross-country techniques would still be useful, since they do also reduce the total count of real number operations. Section 3.4 contains a brief backward roundoff analysis for both modes of algorithmic differentiation. Furthermore we always work under the assumption that the cost of arithmetic operations between real scalars and their transfer to and from memory is homogeneous, i.e., does not depend on the actual value of the real number. Of course this assumption is not really valid for multi- or variable precision-number formats, which are provided in computer algebra environments and are also the basis of some advanced complexity studies.

2.5 Examples and Exercises

Exercise 2.1 (*Coordinate Transformation in* $\mathbb{R}^3$)
Consider the mapping from the Cartesian coordinates x_1, x_2, x_3 to the corresponding spherical polar coordinates y_1, y_2, y_3. Write a suitable evaluation trace and draw the computational graph.

Exercise 2.2 (*Difference Quotients on Norm*)
Consider the example $f(x) = \sum_{i=1}^{n} x_i^2$ with $x_i = i$ for $i = 1 \ldots n$ and implement it in single or double precision.
a. Examine the absolute errors $\big[f(x + h\,e_1) - f(x)\big]/h - 2$ for $n = 10^j$ and $h = 10^{-k}$, where $2 = 2\,x_1 = \frac{\partial f}{\partial x_1}$ is the first gradient component. Observe for

which j and k the difference underflows to zero and determine the best possible approximation for $j = 4$. Check whether the order of summation or prescaling of the components by γ, so that $f(x)$ is calculated as $\gamma^2 f(\frac{x}{\gamma})$, makes any difference.

b. Check that at the given point $f(x)$ will be obtained exactly in IEEE double-precision arithmetic for n ranging up to $\approx 300{,}000$. Show that the total error between the computed difference quotient and the derived first gradient component may be as large as $\sqrt{4n^3/3} \cdot 10^{-8}$. Compare the theoretical relations with the numerical results obtained in part **a**.

Exercise 2.3 (*Recursive Norm*)

Express the Euclidean norm $\|r\| \equiv \left(\sum_k w_k^2\right)^{\frac{1}{2}}$ recursive in terms of the bivariate function $\|u, w\|$. Compare the gradients obtained by differentiating $\|r\|$ directly with those resulting from the reverse mode applied to the recursive representation in terms of $\|u, w\|$.

Chapter 3

Fundamentals of Forward and Reverse

In this chapter we calculate first derivative vectors, called *tangents* and *gradients* (or normals), in the forward and reverse modes of AD, respectively. It is assumed that a given vector function F is the composition of a sequence of once continuously differentiable elemental functions $(\varphi_i)_{i=1\ldots l}$, as discussed in Chapter 2. In other words we assume that F is given as an evaluation procedure.

As is typical for AD, we obtain truncation-error-free numerical values of derivative vectors rather than the corresponding symbolic expressions. Furthermore, in the general case these values depend on certain *seed directions* ($\dot{x}$, $\dot{X}$) and *weight functionals* ($\bar{y}$, $\bar{Y}$) rather than representing derivatives of particular dependent variables y_i with respect to particular independent variables x_j which we call *Cartesian* derivatives. In other words we obtain directional derivatives of scalar functions that are defined as weighted averages of the vector function F as shown in Fig. 3.1.

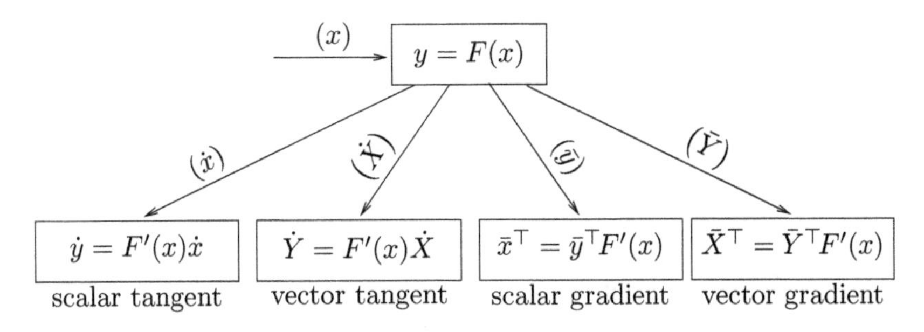

Figure 3.1: Basic Calculations of Tangents and Gradients

If complete Jacobians, Hessians, or higher derivative tensors are desired, they can be constructed from the derivatives obtained for seed directions and weight vectors that are Cartesian basis vectors or identity matrices in their respective spaces. Moreover, as described in Chapters 7 and 8, sparsity can be exploited efficiently in this process. We believe that complete multidimensional arrays of higher derivatives are rarely needed in practical algorithms. Hence their immediate contraction to lower dimensional objects makes sense in terms of computational economy.

To illustrate the basic AD approach, let us consider again the lighthouse example introduced at the beginning of Chapter 2. Suppose we want to calculate the speed at which the light spot moves along the quay as a function of time t with the parameters ν, γ, ω fixed. This means we vary the independent variable vector x along the straight line

$$x = x(t) = \big(\nu,\, \gamma,\, \omega,\, 0\big)^\top + t\, e_4$$

where $e_4 = (0,0,0,1)^\top$ is the fourth Cartesian basis vector and hence $\dot{x} = e_4$ for all t.

3.1 Forward Propagation of Tangents

To calculate the first derivative of the resulting curve $y(t) \equiv F(x(t))$, we merely have to differentiate each single statement in the evaluation procedure Table 2.1 and obtain for the lighthouse example the tangent procedure displayed in Table 3.1.

Table 3.1: Tangent Procedure

$$\begin{array}{lll}
v_{-3} = & x_1 = \nu\,; & \dot{v}_{-3} = \dot{x}_1 = 0 \\
v_{-2} = & x_2 = \gamma\,; & \dot{v}_{-2} = \dot{x}_2 = 0 \\
v_{-1} = & x_3 = \omega\,; & \dot{v}_{-1} = \dot{x}_3 = 0 \\
v_0 = & x_4 = t\,; & \dot{v}_0 = \dot{x}_4 = 1 \\
\hline
v_1 & = & v_{-1} * v_0 \\
\dot{v}_1 & = & \dot{v}_{-1} * v_0 + v_{-1} * \dot{v}_0 \\
v_2 & = & \tan(v_1) \\
\dot{v}_2 & = & \dot{v}_1 / \cos^2(v_1) \\
v_3 & = & v_{-2} - v_2 \\
\dot{v}_3 & = & \dot{v}_{-2} - \dot{v}_2 \\
v_4 & = & v_{-3} * v_2 \\
\dot{v}_4 & = & \dot{v}_{-3} * v_2 + v_{-3} * \dot{v}_2 \\
v_5 & = & v_4 / v_3 \\
\dot{v}_5 & = & (\dot{v}_4 - \dot{v}_3 * v_5)/v_3 \\
v_6 & = & v_5 \\
\dot{v}_6 & = & \dot{v}_5 \\
v_7 & = & v_5 * v_{-2} \\
\dot{v}_7 & = & \dot{v}_5 * v_{-2} + v_5 * \dot{v}_{-2} \\
\hline
y_1 & = & v_6\,; \qquad \dot{y}_1 = \dot{v}_6 \\
y_2 & = & v_7\,; \qquad \dot{y}_2 = \dot{v}_7
\end{array}$$

Those values and operations that depend only on x might have already been calculated before the direction $\dot{x}$ was selected. In contrast, values that are dependent on $\dot{x}$ must be recalculated for each newly selected input tangent $\dot{x}$. For example, one might observe that because of the surf, the quay slope γ is not exactly constant but some periodic function of time, say,

$$\gamma(t) = \gamma_0(1 + \sin(\omega_0 t)/100)\,.$$

Then the speed $\dot{y}$ of the light spot at one time t is obtained by differentiating $y(t) = F(x(t))$ along the curve

$$x = x(t) = \big(\nu,\, \gamma_0\,(1 + \sin(\omega_0\, t)/100),\, \omega,\, t\big)^\top$$

with the domain tangent

$$\dot{x} = \dot{x}(t) = \big(0,\, \gamma_0\, \omega_0\, \cos(\omega_0\, t)/100,\, 0,\, 1\big)^\top\,.$$

One can easily see that the formula obtained for $y(t)$ by eliminating $x(t)$ symbolically is already quite complicated and that its differentiation with respect to t yields a rather unwieldy formula. Since this algebraic expression provides little insight anyway, one might as well be content with an algorithm for computing $\dot{y}$ accurately at any given value of x and $\dot{x}$.

That purpose is achieved by the code listed in Table 3.1, where we merely have to change the initialization of $\dot{x}_2$ to $\dot{x}_2 = \dot{\gamma} = \gamma_0\, \omega_0\, \cos(\omega_0\, t)/100$ for the given value of t. Obviously the transition from the original evaluation procedure Table 2.1 to the derived procedure is completely mechanical. Every original statement is succeeded by a corresponding, derived statement so that the length of the derived evaluation procedure is essentially twice that of the original. Geometrically, we may visualize the computation of first derivatives along a smooth curve $x(t)$, as shown in Fig. 3.2.

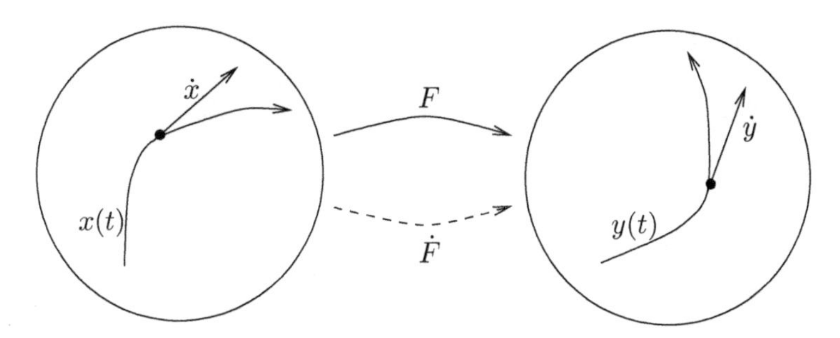

Figure 3.2: Mapping of Tangent $\dot{x}$ into Tangent $\dot{y}$ by Function $\dot{F}$

According to the chain rule we have

$$\dot{y}(t) = \frac{\partial}{\partial t} F(x(t)) = F'(x(t))\, \dot{x}(t) \;, \tag{3.1}$$

where $F'(x) \in \mathbb{R}^{m\times n}$ for the argument $x \in \mathcal{D}$ denotes the Jacobian matrix of F as in Chapter 2. Looking at the matrix-vector product representation $\dot{y} = F'(x)\, \dot{x}$, one might be tempted to think that the calculation $\dot{y}$ is performed most naturally by first evaluating the full Jacobian $F'(x)$ and then multiplying it with the direction $\dot{x}$.

However, this approach is quite uneconomical, except when computing $\dot{y}$ for very many tangents $\dot{x}$ at a fixed point x. This situation may arise when the Jacobian-vector products are used to iteratively approximate the Newton step Δx as a solution of the linear system $F'(x)\Delta x = -F(x)$. On the other hand, if only a few tangents (i.e., Jacobian-vector products) are needed, computing the Jacobian is not at all appropriate. Instead we employ the tangent evaluation procedure Table 3.2 which is obtained by simply differentiating Table 2.2.

Tangent Recursion and Procedure

In Table 3.2 we abbreviate $u_i = (v_j)_{j\prec i} \in \mathbb{R}^{n_i}$ as before, in (2.4). The fact that the elemental partials c_{ij} introduced in (2.8) can be evaluated as real numbers at the current point before they enter into the multiplication prescribed by the

Table 3.2: Tangent Recursion for General Evaluation Procedure

v_{i-n}	$\equiv$	x_i	$i = 1 \ldots n$
$\dot{v}_{i-n}$	$\equiv$	$\dot{x}_i$	
v_i	$\equiv$	$\varphi_i(v_j)_{j \prec i}$	$i = 1 \ldots l$
$\dot{v}_i$	$\equiv$	$\sum_{j \prec i} \frac{\partial}{\partial v_j} \varphi_i(u_i)\, \dot{v}_j$	
y_{m-i}	$\equiv$	v_{l-i}	$i = m - 1 \ldots 0$
$\dot{y}_{m-i}$	$\equiv$	$\dot{v}_{l-i}$	

chain rule is a key distinction from fully symbolic differentiation. There, the elementary partials $\partial\varphi_i(u_i)/\partial v_j$ would be represented as algebraic expressions and their combination could yield partial derivative expressions of rapidly growing complexity.

In contrast, applying the chain rule to floating point numbers causes some roundoff but makes memory and runtime a priori bounded, as analyzed in the following chapter. Each elemental assignment $v_i = \varphi_i(u_i)$ in the original evaluation procedure of Table 2.2 spawns the corresponding tangent operation

$$\dot{v}_i = \sum_{j \prec i} \frac{\partial}{\partial v_j} \varphi_i(u_i) * \dot{v}_j = \sum_{j \prec i} c_{ij} * \dot{v}_j \; . \tag{3.2}$$

Abbreviating $\dot{u}_i = (\dot{v}_j)_{j \prec i}$ in agreement with (2.4), we may write the tangent operation as

$$\dot{v}_i = \dot{\varphi}_i(u_i, \dot{u}_i) \equiv \varphi_i'(u_i)\dot{u}_i \; ,$$

where

$$\dot{\varphi}_i \, : \, \mathbb{R}^{2n_i} \mapsto \mathbb{R}$$

will be called the *tangent function* associated with the elemental φ_i.

So far we have always placed the tangent statement yielding $\dot{v}_i$ after the original statement for the underlying value v_i. This order of calculation seems natural and certainly yields correct results as long as there is no overwriting. Then the order of the $2\,l$ statements in the middle part of Table 3.2 does not matter at all. However, when v_i shares a location with one of its arguments v_j, the derivative value $\dot{v}_i = \dot{\varphi}_i(u_i, \dot{u}_i)$ will be incorrect after $v_i = \varphi_i(u_i)$ has been updated, a fact that will be studied in more detail in section 4.1. Therefore, source preprocessors, for example, ADIFOR [BC$^+$92] and Tapenade [HP04], always put the derivative statement ahead of the original assignment. On the other hand, for most univariate functions $v = \varphi(u)$ it is much better to obtain the undifferentiated value first and then use it in the tangent function $\dot{\varphi}$.

Hence, from now on we will list φ and $\dot{\varphi}$ side by side with a common bracket to indicate that they should be evaluated simultaneously, sharing intermediate results. The most important cases are listed in Table 3.3. It should be noted that the additional tangent operations $\dot{\varphi}$ for all the nonlinear univariate cases, except $\varphi(u) = \sin(u)$, are quite cheap compared to φ itself.

In analogy to the definition of $\dot{\varphi}$ we set for the overall tangent function

$$\dot{y} = \dot{F}(x, \dot{x}) \equiv F'(x)\,\dot{x}$$

where $\dot{F} : \mathbb{R}^{n+n} \longmapsto \mathbb{R}^m$, and we write the combination

$$[y, \dot{y}] = \left[F(x), \dot{F}(x, \dot{x})\right]$$

whenever possible.

Here again we assume implicitly that optimal use is made of common subexpressions in evaluating F and $\dot{F}$. In fact, except when F is a linear function it rarely makes sense to evaluate $\dot{F}$ without evaluating F at the same time. The notion of a "tangent code" or "tangent model," which is fairly popular in meteorology, usually also includes the evaluation of the underlying function F itself. To evaluate $\dot{F}$ together with F itself, we may execute the tangent procedure of Table 3.4, which is just a more concise version of the recursion of Table 3.2.

Table 3.3: Tangent Operations

φ	$[\varphi, \dot{\varphi}]$
$v = c$	$v = c$ $\dot{v} = 0$
$v = u \pm w$	$v = u \pm w$ $\dot{v} = \dot{u} \pm \dot{w}$
$v = u * w$	$\dot{v} = \dot{u} * w + u * \dot{w}$ $v = u * w$
$v = 1/u$	$v = 1/u$ $\dot{v} = -v * (v * \dot{u})$
$v = \sqrt{u}$	$v = \sqrt{u}$ $\dot{v} = 0.5 * \dot{u}/v$
$v = u^c$	$\dot{v} = \dot{u}/u; v = u^c$ $\dot{v} = c * (v * \dot{v})$
$v = \exp(u)$	$v = \exp(u)$ $\dot{v} = v * \dot{u}$
$v = \log(u)$	$\dot{v} = \dot{u}/u$ $v = \log(u)$
$v = \sin(u)$	$\dot{v} = \cos(u) * \dot{u}$ $v = \sin(u)$

Table 3.4: General Tangent Procedure

$[v_{i-n}, \dot{v}_{i-n}]$	$=$	$[x_i, \dot{x}_i]$	$i = 1 \ldots n$
$[v_i, \dot{v}_i]$	$=$	$[\varphi_i(u_i), \dot{\varphi}_i(u_i, \dot{u}_i)]$	$i = 1 \ldots l$
$[y_{m-i}, \dot{y}_{m-i}]$	$=$	$[v_{l-i}, \dot{v}_{l-i}]$	$i = m-1 \ldots 0$

Basic Complexity Results for Tangent Propagations

As seems obvious from the lighthouse example and Table 3.3 the effort for evaluating $[F, \dot{F}]$ is always a small multiple of that for evaluations F itself. More precisely, it will be shown in section 4.5 for some temporal complexity measure $TIME$ that

$$TIME\{F(x),\, F'(x)\dot{x}\} \le \omega_{tang}\, TIME\{F(x)\}$$

with a constant $\omega_{tang} \in [2, 5/2]$. Furthermore, it is possible to derive for the memory requirement of the forward mode the estimate

$$MEM\{F(x),\, F'(x)\dot{x}\} \;=\; 2\, MEM\{F(x)\}$$

where MEM may count the number of floating point locations needed. The same holds for other sensible spatial complexity measures, as discussed in section 4.1.

The procedure of Table 3.4 requires the evaluation of each elemental pair $[\varphi_i, \dot{\varphi}_i]$ once. It should be noted in particular that when φ_i is multivariate ($n_i > 1$), evaluating the directional derivative $\dot{\varphi}_i(u_i, \dot{u}_i)$ may be considerably cheaper than first forming the gradient $\varphi_i' \in \mathbb{R}^{n_i}$ and subsequently multiplying it by $\dot{u}_i$, as suggested by the original tangent procedure Table 3.2. This effect occurs almost certainly when φ_i is a more complex elemental function such as a BLAS routine [BD$^+$02]. Tangent and gradient propagation can be performed fully recursively in that each φ_i may again be defined by an evaluation procedure of exactly the same structure as in Table 2.2. In that scenario we have to think of the v_i as local variables that are not visible from outside the procedure. The input vector x and the output vector y can be thought of as calling parameters with x remaining unchanged, which corresponds to a call by value.

Amortizing Overhead in Vector Mode

Rather than keeping the elemental partials c_{ij} for subsequent tangent propagations, we may apply them immediately to propagate a collection of $p \geq 1$ tangents. In other words, we can calculate the matrix product

$$\dot{Y} \;=\; F'(x)\dot{X} \;\in\; \mathbb{R}^{m\times p} \quad \text{for} \quad \dot{X} \in \mathbb{R}^{n\times p}\,.$$

For this purpose the intermediate derivatives $\dot{v}_j \in \mathbb{R}$ must be replaced by vectors $\dot{V}_j \in \mathbb{R}^p$, but otherwise the tangent procedure can be applied unchanged. The simultaneous propagation of multiple tangent vectors has numerous applications, especially the cheap evaluation of sparse Jacobians and partially separable gradients, which will be discussed in Chapter 8. In this vector mode we simply write $\dot{F}(x, \dot{X}) \equiv F'(x)\dot{X}$, so that the meaning of the function symbol $\dot{F}$ adjusts to the format of its second argument.

As will also be proven in section 4.5, one can derive for the vector forward mode and some temporal complexity measure $TIME$ the estimate

$$TIME\{F(x),\, F'(x)\dot{X}\} \;\leq\; \omega_{tangp}\, TIME\{F(x)\}$$

for $\omega_{tangp} \in [1 + p,\, 1 + 1.5p]$ and $\dot{X} \in \mathbb{R}^{n\times p}$. Hence, the work for evaluating an Jacobian-matrix product grows linearly in the number of columns of $\dot{X}$. Furthermore, the temporal complexity to evaluate the product $F'(x)\dot{X}$ can be considerably reduced in comparison to the scalar version of the forward mode, where p Jacobian-vector products have to be evaluated separately. Using the vector version the range of the coefficient in the upper bound can be reduced from $[2p,\ (5/2)p]$ for p Jacobian-vector products to $[1+p,\ 1+1.5p]$ for the vector version evaluating one Jacobian-matrix product. For the memory requirement one obtains

$$MEM\{F(x),\, F'(x)\dot{X}\} \;=\; (1+p)\; MEM\{F(x)\}$$

as shown in section 4.5.

Calling Sequences for Tangent Propagation

Due to the structure of the forward propagation of tangents, the calling sequence of corresponding software routines is quite standard. As in section 2.2 we assume that the evaluation $y = F(x)$ can be affected by the call `eval(x, y)` with `x` $= x \in \mathbb{R}^n$ as input and `y` $= y \in \mathbb{R}^m$ as output. For the first-order scalar forward mode to compute $\dot{y} = F'(x)\dot{x}$ one has to add `dx` $= \dot{x} \in \mathbb{R}^n$ as input array and `dy` $= \dot{y} \in \mathbb{R}^m$ as output array, yielding the calling structure

```
            ↓  ↓
eval_tang ( x, dx, y, dy )
                   ↓  ↓
```

Throughout we assume that `eval_tang` is *primaly consistent* in that it affects x and y in exactly the same way as `eval`. This convention is not adhered to by all AD tools, especially in the reverse mode to be discussed later.

For the vector version of the first-order forward mode to compute $\dot{Y} = F'(x)\,\dot{X}$ with `x` $= x \in \mathbb{R}^m$ and `dX` $= \dot{X} \in \mathbb{R}^{n\times p}$ as input arrays and `y` $= y \in \mathbb{R}^m$ and `dY` $= \dot{Y} \in \mathbb{R}^{m\times p}$ as output arrays one obtains the calling structure

```
             ↓  ↓
eval_tangp ( x, dX, y, dY )
                    ↓  ↓
```

Note that some AD tools still allow only the scalar version with $p = 1$. As discussed in Chapter 6 there are various software technologies for producing `eval_tang` or `eval_tangp` from a given routine `eval`, and our naming conventions are of course quite arbitrary. However, the structure of the information flow between the calling parameters is obvious. This is definitely not the case for the reverse mode as described in following section.

3.2 Reverse Propagation of Gradients

The cost of evaluating derivatives by propagating them forward through the evaluation procedure will increase linearly with p, the number of domain directions $\dot{x}$ along which we want to differentiate. While this may at first seem natural or even inevitable, AD can benefit from a small number of dependent variables just as much as from a small number of independent variables. In particular, the gradient of a single dependent variable can be obtained for a fixed multiple of the cost of evaluating the underlying scalar-valued function. The same is true for the gradient of a weighted sum of the components of a vector-valued function F, namely,

$$\bar{x}^\top \;\equiv\; \nabla[\bar{y}^\top F(x)] \;=\; \bar{y}^\top F'(x) \in \mathbb{R}^n \,. \tag{3.3}$$

Here, $\bar{y}$ is a fixed vector that plays a dual role to the domain direction $\dot{x}$ considered in section 3.1. There we computed $\dot{y} = F'(x)\dot{x} = \dot{F}(x, \dot{x})$ as a function of x and $\dot{x}$. Here we evaluate

$$\bar{x}^\top = \bar{y}^\top F'(x) \equiv \bar{F}(x, \bar{y})$$

as a function of x and $\bar{y}$. Again we usually assume that F and $\bar{F}$ are evaluated together.

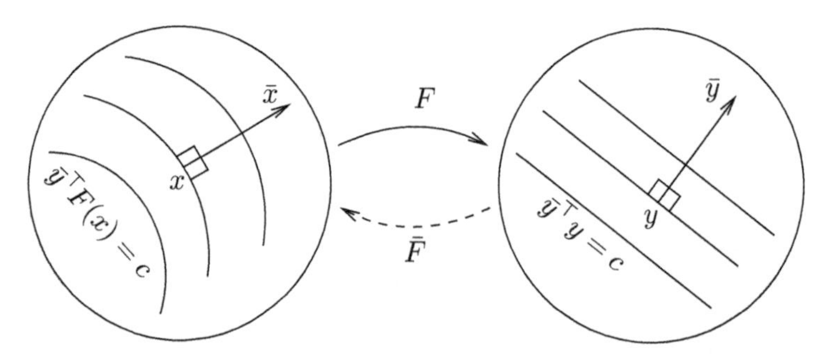

Figure 3.3: Mapping of Normal $\bar{y}$ into Normal $\bar{x}$ by Function $\bar{F}$

Geometrically, $\bar{y}$ and $\bar{x}$ are *normals* or *cotangents*, as depicted in Fig. 3.3. The hyperplane $\bar{y}^\top y = c$ in the range of F has the inverse image $\{x : \bar{y}^\top F(x) = c\}$. By the implicit function theorem this set is a smooth hypersurface with the normal $\bar{x}^\top = \bar{y}^\top F'(x)$ at x, provided $\bar{x}$ does not vanish. Whereas the dotted quantities were propagated forward through the evaluation procedure, the barred quantities will be propagated in the opposite, or reverse, direction. When $m = 1$, then $F = f$ is scalar-valued and we obtain for $\bar{y} = 1 \in \mathbb{R}$ the familiar gradient $\nabla f(x) = \bar{y}^\top F'(x)$.

Derivation by Matrix-Product Reversal

As its name suggests, the reverse mode of AD is in some sense a backward application of the chain rule. The term *backward differentiation* should be avoided in this context because it is well established as a method for discretizing ODEs for their numerical integration [GK98]. We therefore use the term *reverse differentiation*. There are at least three ways in which the reverse mode can be introduced and its mathematical correctness established. To some casual observers, the reverse mode always remains a little mysterious [DS96]. The mystery rests in part upon the fact that the chain rule is usually taught as a rather mechanical procedure for propagating derivatives forward. Also, since the standard chain rule assumes three separate layers of variables connected by two vector functions, its application to evaluation procedures requires some notational transformations. None of this rewriting is terribly deep, but it can certainly be confusing enough to hinder any intuitive understanding.

Without explicitly referring to the chain rule, we will first derive the procedure for computing $\bar{x}$ as defined by (3.3) for given x and $\bar{y}$ from the tangent evaluation procedure of Table 3.2 and the identity

$$\boxed{\bar{y}^\top \dot{y} \equiv \bar{x}^\top \dot{x}} \tag{3.4}$$

which must hold for all pairs $[\dot{x}, \dot{y}]$ with $\dot{y} = F'(x)\dot{x}$.

Using the state transformations Φ_i as introduced in (2.3) we may rewrite the mapping from x to $y = F(x)$ as the composition

$$y = Q_m \Phi_l \circ \Phi_{l-1} \circ \cdots \circ \Phi_2 \circ \Phi_1 \big(P_n^\top x\big) \tag{3.5}$$

where

$$P_n \equiv [I, 0, \ldots, 0] \in \mathbb{R}^{n \times (n+l)} \quad \text{and} \quad Q_m \equiv [0, \ldots, I] \in \mathbb{R}^{m \times (n+l)} \tag{3.6}$$

are the matrices that project an arbitrary $(n+l)$-vector onto its first n and last m components, respectively. Labeling the elemental partials c_{ij} as in (2.8), we then obtain for $i = 1, 2, \ldots, l-1, l$ the state Jacobians

$$A_i \equiv \Phi_i' = \begin{bmatrix} 1 & 0 & \cdot & 0 & \cdot & \cdot & \cdot & 0 \\ 0 & 1 & \cdot & 0 & \cdot & \cdot & \cdot & 0 \\ \vdots & \vdots & \vdots & \vdots & \vdots & \vdots & \vdots & \vdots \\ 0 & 0 & \cdot & 1 & \cdot & \cdot & \cdot & 0 \\ c_{i\,1-n} & c_{i\,2-n} & \cdot & c_{i\,i-1} & 0 & \cdot & \cdot & 0 \\ 0 & 0 & \cdot & \cdot & \cdot & 1 & \cdot & 0 \\ \vdots & \vdots & \vdots & \vdots & \vdots & \vdots & \vdots & \vdots \\ 0 & 0 & \cdot & \cdot & \cdot & \cdot & \cdot & 1 \end{bmatrix} \in \mathbb{R}^{(n+l)\times(n+l)}$$

where the c_{ij} occur in the $(n+i)$th row of A_i. The square matrices A_i are lower triangular and may also be written as rank-one perturbations of the identity, namely,

$$A_i = I + e_{n+i}\big[\nabla\varphi_i(u_i) - e_{n+i}\big]^\top \tag{3.7}$$

where e_j denotes the jth Cartesian basis vector in $\mathbb{R}^{n+l}$. Differentiating (3.5) we obtain by the chain rule the transformation from $\dot{x}$ to $\dot{y}$ defined by (3.1) as the matrix-vector product

$$\dot{y} = Q_m A_l A_{l-1} \ldots A_2 A_1 P_n^\top \dot{x}\,. \tag{3.8}$$

In other words, the multiplication by $P_n^\top \in \mathbb{R}^{(n+l)\times n}$ embeds $\dot{x}$ into $\mathbb{R}^{(n+l)}$ corresponding to the first part of the tangent recursion of Table 3.2. The subsequent multiplications by the A_i generate one component $\dot{v}_i$ at a time, according to the middle part, and finally, Q_m extracts the last m components as $\dot{y}$ corresponding to the third part of Table 3.2.

By comparison with (3.1) we have in fact a product representation of the full Jacobian

$$F'(x) = Q_m A_l A_{l-1} \dots A_2 A_1 P_n^\top \in \mathbb{R}^{m\times n} \tag{3.9}$$

By transposing the product we obtain the adjoint relation

$$\bar{x} = P_n A_1^\top A_2^\top \dots A_{l-1}^\top A_l^\top Q_m^\top \bar{y} \tag{3.10}$$

Since

$$A_i^\top = I + [\nabla\varphi_i(u_i) - e_{n+i}]e_{n+i}^\top \tag{3.11}$$

we see that the transformation of a given vector $(\bar{v}_j)_{1-n\le j\le l}$ by multiplication with $A_i^\top$ represents an incremental operation. In detail, one obtains for $i = l \dots 1$ the operations

- for all j with $i \neq j \not\prec i$: $\bar{v}_j$ is left unchanged,
- for all j with $i \neq j \prec i$: $\bar{v}_j$ is incremented by $\bar{v}_i\, c_{ij}$,
- subsequently $\bar{v}_i$ is set to zero.

Hence, at the very end of the product calculation (3.10) only the first n adjoint values $\bar{v}_i$ with $1 - n \le i \le 0$ are nonzero and then assigned to the components $\bar{x}_i$ of $\bar{x}$ by the projection P_n.

Incremental and Nonincremental Recursion

Using the C-style abbreviation $a \mathrel{+}\equiv b$ for $a \equiv a + b$, we may rewrite the matrix-vector product as the adjoint evaluation procedure listed in Table 3.5. It is assumed as a precondition that the adjoint quantities $\bar{v}_i$ for $1 \le i \le l$ have been initialized to zero, as indicated by the box on top. The zeroing out of the $\bar{v}_i$ after their appearance on the right-hand side has been omitted for the sake of simplicity. It will be enforced later when overwriting is allowed. As indicated by the range specification $i = l \dots 1$, we think of the incremental assignments as being executed in reverse order, i.e., for $i = l, l-1, l-2, \dots, 1$. Only then is it guaranteed that each $\bar{v}_i$ will reach its full value before it occurs on the right-hand side. We can combine the incremental operations affected by the adjoint of φ_i to

$$\bar{u}_i \mathrel{+}= \bar{v}_i \cdot \nabla\varphi_i(u_i) \quad \text{where} \quad \bar{u}_i \equiv \left(\bar{u}_j\right)_{j\prec i} \in \mathbb{R}^{n_i} \ . \tag{3.12}$$

As φ_i can be a whole evaluation procedure itself adjoint procedures are often viewed and applied as incremental.

Alternatively, one can directly compute the value of the adjoint quantity $\bar{v}_j$ by collecting all contributions to it as a sum ranging over all successors $i \succ j$. This approach yields the nonincremental adjoint evaluation procedure listed in Table 3.6. An obvious disadvantage of this nonincremental form is that information is needed that is not directly available from the collection of elemental functions φ_i and their arguments $u_i = (v_j)_{j \prec i}$. Instead, one has to compile for each variable v_j a list of all elemental functions φ_i that depend on it. Such forward pointers from an index j to its successors $i \succ j$ are usually hard to come by, whereas the backward pointers from an index i to its predecessors $j \prec i$ are explicitly given by the elemental decomposition of F. In other words, the nonincremental version requires global information, whereas the incremental one can proceed locally by looking at one φ_i at a time.

Table 3.5: Incremental Adjoint Recursion, No Overwrites!

$\bar{v}_i$	$\equiv$	0	$i = 1-n \ldots l$
v_{i-n}	$\equiv$	x_i	$i = 1 \ldots n$
v_i	$\equiv$	$\varphi_i(v_j)_{j \prec i}$	$i = 1 \ldots l$
y_{m-i}	$\equiv$	v_{l-i}	$i = m-1 \ldots 0$
$\bar{v}_{l-i}$	$\equiv$	$\bar{y}_{m-i}$	$i = 0 \ldots m-1$
$\bar{v}_j$	$+\equiv$	$\bar{v}_i \frac{\partial}{\partial v_j} \varphi_i(u_i)$ for $j \prec i$	$i = l \ldots 1$
$\bar{x}_i$	$\equiv$	$\bar{v}_{i-n}$	$i = n \ldots 1$

Table 3.6: Nonincremental Adjoint Recursion, No Overwrites!

v_{i-n}	$\equiv$	x_i	$i = 1 \ldots n$
v_i	$\equiv$	$\varphi_i(v_j)_{j \prec i}$	$i = 1 \ldots l$
y_{m-i}	$\equiv$	v_{l-i}	$i = m-1 \ldots 0$
$\bar{v}_{l-i}$	$\equiv$	$\bar{y}_{m-i}$	$i = m-1 \ldots 0$
$\bar{v}_i$	$\equiv$	$\sum_{j \succ i} \bar{v}_j \frac{\partial}{\partial v_i} \varphi_j(u_j)$	$i = l-m \ldots 1-n$
$\bar{x}_i$	$\equiv$	$\bar{v}_{i-n}$	$i = n \ldots 1$

As is apparent in Tables 3.5 and 3.6, the calculation of the adjoint vector $\bar{x}^\top = \bar{y}^\top F'(x)$ proceeds in two stages: a *forward sweep* up to the double line and a *return sweep* below it. Throughout the book we will refer to the combination of a *forward sweep* and a subsequent *return sweep* as a *reverse sweep*. In the literature the term *reverse sweep* is sometimes used to describe what we call here the *return sweep* only. So far, the forward sweep is identical to Table 2.2 and effects only the function evaluation. However, we will see later that additional preparations may need to be taken during the forward sweep for the subsequent return sweep. In particular, such preparatory forward sweeps may save certain intermediate results for the return sweep, which might otherwise be lost because of overwriting. In any case, these *recording sweeps*, as we will call them, and the subsequent return sweeps must come in corresponding pairs that properly communicate with each other. These issues will be discussed in Chapters 4, 5, and 12.

Adjoint Procedures for the Lighthouse Example

In the lighthouse example the evaluation procedure Table 2.1 yields the incremental adjoint procedure listed in Table 3.7. After execution of the original evaluation procedure displayed on the left, the adjoint quantities $\bar{v}_7$ and $\bar{v}_6$ are initialized to the user-specified values $\bar{y}_2$ and $\bar{y}_1$, respectively. Subsequently, all original assignments are taken in reverse order and, depending on whether they have one or two arguments, are replaced by one or two corresponding incremental assignments. At the very end the accumulated adjoint values of $\bar{v}_{i-4}$ for $i = 1 \ldots 4$ are assigned to the adjoint values $\bar{x}_i$ of the independent variables x_i. They represent the gradient of $\bar{y}_1\, y_1 + \bar{y}_2\, y_2$ with respect to ν, γ, ω, and t. This incremental return sweep is listed in the center of Table 3.7; the corresponding nonincremental version is listed on the right. The latter turns out to be slightly more concise.

Table 3.7: Adjoint Procedure for Lighthouse Example, No Overwrites!

Forward Sweep
$v_{-3} = x_1 = \nu$
$v_{-2} = x_2 = \gamma$
$v_{-1} = x_3 = \omega$
$v_0 = x_4 = t$
$v_1 = v_{-1} * v_0$
$v_2 = \tan(v_1)$
$v_3 = v_{-2} - v_2$
$v_4 = v_{-3} * v_2$
$v_5 = v_4 / v_3$
$v_6 = v_5$
$v_7 = v_5 * v_{-2}$
$y_1 = v_6$
$y_2 = v_7$

Incremental Return
$\bar{v}_7 = \bar{y}_2$
$\bar{v}_6 = \bar{y}_1$
$\bar{v}_5 \mathrel{+}= \bar{v}_7 * v_{-2}$
$\bar{v}_{-2} \mathrel{+}= \bar{v}_7 * v_5$
$\bar{v}_5 \mathrel{+}= \bar{v}_6$
$\bar{v}_4 \mathrel{+}= \bar{v}_5 / v_3$
$\bar{v}_3 \mathrel{-}= \bar{v}_5 * v_5 / v_3$
$\bar{v}_{-3} \mathrel{+}= \bar{v}_4 * v_2$
$\bar{v}_2 \mathrel{+}= \bar{v}_4 * v_{-3}$
$\bar{v}_{-2} \mathrel{+}= \bar{v}_3$
$\bar{v}_2 \mathrel{-}= \bar{v}_3$
$\bar{v}_1 \mathrel{+}= \bar{v}_2 / \cos^2(v_1)$
$\bar{v}_{-1} \mathrel{+}= \bar{v}_1 * v_0$
$\bar{v}_0 \mathrel{+}= \bar{v}_1 * v_{-1}$
$\bar{x}_4 = \bar{v}_0$
$\bar{x}_3 = \bar{v}_{-1}$
$\bar{x}_2 = \bar{v}_{-2}$
$\bar{x}_1 = \bar{v}_{-3}$

Nonincremental Return
$\bar{v}_7 = \bar{y}_2$
$\bar{v}_6 = \bar{y}_1$
$\bar{v}_5 = \bar{v}_7 * v_{-2} + \bar{v}_6$
$\bar{v}_4 = \bar{v}_5 / v_3$
$\bar{v}_3 = -\bar{v}_5 * v_5 / v_3$
$\bar{v}_2 = \bar{v}_4 * v_{-3} - \bar{v}_3$
$\bar{v}_1 = \bar{v}_2 / \cos^2(v_1)$
$\bar{v}_0 = \bar{v}_1 * v_{-1}$
$\bar{v}_{-1} = \bar{v}_1 * v_0$
$\bar{v}_{-2} = \bar{v}_7 * v_5 + \bar{v}_3$
$\bar{v}_{-3} = \bar{v}_4 * v_2$
$\bar{x}_4 = \bar{v}_0$
$\bar{x}_3 = \bar{v}_{-1}$
$\bar{x}_2 = \bar{v}_{-2}$
$\bar{x}_1 = \bar{v}_{-3}$

To check consistency between the tangent and adjoint procedures, let us calculate the derivative of y_1 with respect to $x_4 = t$ at the initial time $t = 0$.

In other words, for a particular value of x, we confirm that the first element of the last column $F'(x)e_4$ of the Jacobian F' is the same as the last element of its first row $e_1^\top F'(x)$. Starting the tangent procedure Table 3.1 with $\dot{x} = e_4$, we obtain the values

$$\begin{array}{llllll} v_1 = 0, & v_2 = 0, & v_3 = \gamma, & v_4 = 0, & v_5 = 0, & v_6 = 0\,, \\ \dot{v}_1 = \omega, & \dot{v}_2 = \omega, & \dot{v}_3 = -\omega, & \dot{v}_4 = \omega\,\nu, & \dot{v}_5 = \omega\,\nu/\gamma, & \dot{v}_6 = \omega\,\nu/\gamma\,. \end{array}$$

Since $y_1 = v_6$, we have obtained the value $\dot{y}_1 = \omega\nu/\gamma$ for the desired derivative. Alternatively, we may apply the adjoint procedure of Table 3.7 starting with $\bar{y} = e_1$, i.e., $\bar{v}_6 = \bar{y}_1 = 1$ and $\bar{v}_7 = \bar{y}_2 = 0$, to successively obtain the adjoint values

$$\bar{v}_6 = 1,\; \bar{v}_5 = 1,\; \bar{v}_4 = 1/\gamma,\; \bar{v}_3 = 0,\; \bar{v}_2 = \nu/\gamma,\; \bar{v}_1 = \nu/\gamma,\; \bar{v}_0 = \omega\,\nu/\gamma\,.$$

For brevity, we have omitted the intermediate adjoints that do not affect the target value $\bar{v}_0 = \bar{x}_4$, which does turn out to be equal to the value of $\dot{v}_6 = \dot{y}_1$ calculated above.

Adjoint Values as Lagrange Multipliers

So far, we have treated the $\bar{v}_j$ merely as auxiliary quantities during the reverse matrix multiplication. One may also view or derive the $\bar{v}_i$ as Lagrange multipliers of the "equality constraint" $v_i - \varphi_i(u_i) = 0$ with respect to the objective function $\bar{y}^\top y$. Let us consider all v_j for $j = 1-n \ldots l$ as independent optimization variables subject to the side constraints

$$-E(x;\,v) \;=\; (v_i - \varphi_i(u_i))_{i=1-n\ldots l} = 0$$

with $\varphi_i(u_i) \equiv x_{i+n}$ for $i = 1-n \ldots 0$. This extended system was originally introduced in (2.5) of section 2.2. Now we may view

$$\gamma \;\equiv\; \bar{y}^\top y \;=\; \sum_{i=0}^{m-1} \bar{y}_{m-i}\, v_{l-i} \tag{3.13}$$

as an objective function. For fixed x and $\bar{y}$ the Karush–Kuhn–Tucker conditions [KT51, NS96] require that the transpose of the negative Jacobian $E' = -\partial E/\partial v$ multiplied by some vector of Lagrange multipliers $\bar{v} = (\bar{v}_i)_{i=1-n\ldots l}$ must equal the gradient of γ.

In particular, we obtain for the jth component of this vector identity by inspection of (3.13)

$$\bar{v}_j - \sum_{i \succ j} \bar{v}_i\, c_{ij} \;=\; \begin{cases} 0 & \text{if } \; j \le l-m \\ \bar{y}_{m-l+j} & \text{otherwise} \end{cases}$$

By comparison with Table 3.6, we see that these are exactly the relations we derived as the nonincremental form of the adjoint recursion, using the chain rule via matrix-product reversal. The $\bar{v}_j$ with $j \le 0$ represent the sensitivities of γ

with respect to the initialization constraints $v_j - x_{j+n} = 0$. They are therefore the components of the reduced gradient of γ with respect to the x_j with $j \leq n$, if these are considered variable after all. The interpretation of adjoint values as Lagrange multipliers has been used extensively by Thacker in four-dimensional data assimilation [Tha91]. We will use the $\bar{v}_j$ to estimate roundoff errors in section 3.4.

Treating the defining relations $v_i = \varphi_i(u_i)$ as side constraints or equations may have some advantages. For example the sparsity pattern of the extended Jacobian may allow the direct computation of Newton steps at a reduced cost. In optimal control many intermediates v_i occur as discretized state values along a solution trajectory. Solution methods are called *direct* if they treat the v_i as proper optimization variables subject to the additional constraints $v_i - \varphi_i(u_i) = 0$. Following this direct approach one may, for example, attack the discretized optimal control problem with a sparse sequential quadratic programming solver. Methods that treat only the state variables at endpoints of the trajectory (and possibly some switching points) as proper variables or functions in the optimization process are called *indirect*. Moreover, there are many in-between formulations, for example those based on multiple shooting. In many cases adjoints $\bar{v}_i$ of intermediates v_i will be implicitly evaluated or approximated in one way or another, even though the programmer may be unaware of this fact.

3.3 Cheap Gradient Principle with Examples

Looking at Tables 3.5 and 3.6 representing the reverse mode in its incremental and nonincremental form and their instantiation in Table 3.7 for the light house example, we note that the growth in the number of elementary instructions is quite comparable to that for the forward mode. Taking into account memory moves, we will find in section 4.6 that the growth in computational complexity relative to the original evaluation procedure is in fact a bit larger, so that this time

$$TIME\{F(x), \bar{y}^\top F'(x)\} \leq \omega_{grad}\, TIME\{F(x)\} \tag{3.14}$$

for a constant $\omega_{grad} \in [3, 4]$. Hence, if $m = 1$, i.e., $f = F$ is scalar-valued then the cost to evaluate the gradient ∇f is bounded above by a small constant times the cost to evaluate the function itself. This result is known as the *cheap gradient principle.*

Adjugate as Gradient of the Log-Determinant

To get a feeling for this somewhat counterintuitive result and to see that the ratio $\omega_{grad} \in [3, 4]$ is fairly optimal, we look at a classical example from statistics, namely, the function $f(X) \equiv \ln|\det(X)|$ for $X \in \mathbb{R}^{\hat{n}\times\hat{n}}$ which occurs in maximum likelihood estimation. When X is restricted to being symmetric, $f(X)$ is used also as a self-concordant barrier function for enforcing positive definiteness of X in interior point methods [OW97]. Hence we have either $n = \hat{n}^2$ or

$n = \hat{n}(\hat{n}+1)/2$ variables. It has been known for a long time that the gradient of $f(X)$ is given by

$$\nabla(f(X)) = \text{adj}(X) = \det(X)\big[X^{-1}\big]^{\top} \in \mathbb{R}^{\hat{n}\times\hat{n}}$$

which is the matrix of signed cofactors of X [DM48, Dwy67]. The last equation holds only for nonsingular X but uniquely determines $\nabla f(X)$ as a multilinear vector polynomial in X. Since transposition and division by $\det(X)$ incurs a negligible effort, the cheap gradient principle guarantees that any finite algorithm for evaluating the determinant of a general square matrix can be transformed into an algorithm for computing the inverse with essentially the same temporal complexity.

Any numerically inclined individual will evaluate the determinant via an LU factorization of X, which costs almost exactly $\hat{n}^3/3$ multiplications and a similar number of additions or subtractions. The $\hat{n}-1$ divisions needed to form the reciprocals of the pivot elements can be neglected. In the symmetric case this effort is essentially halved. To compute the inverse on the basis of the factorization $X = LU$ we have to solve $\hat{n}$ linear systems $L\,(Ua_i) = e_i$ with the right-hand side being a Cartesian basis vector $e_i \in \mathbb{R}^{\hat{n}}$ for $i = 1 \ldots \hat{n}$. In solving $L\,z_i = e_i$ one can use the fact that the first $i-1$ components of the intermediate vector $z_i \in \mathbb{R}^{\hat{n}}$ must vanish so that the effort for the forward substitution to compute z_i is only $(\hat{n}-i+1)^2/2$ multiplications and the same number of additions or subtractions. Summing over $i = 1 \ldots \hat{n}$ we get a total of $\hat{n}^3/6$ such operations. In the general case the subsequent backward substitution to solve $U\,a_i = z_i$ allows no extra savings and we have a total effort of $\hat{n} * \hat{n}^2/2 = \hat{n}^3/2$. Adding the effort for the LU factorization, the forward substitution, and the backward substitution, we get a total effort of $\hat{n}^3 + \mathcal{O}(\hat{n}^2)$ multiplications followed by an addition or subtraction. This effort for computing the inverse of X and hence the gradient of its log-determinant is exactly three times that for computing the latter by itself, which nicely conforms with our bound (3.14).

Similarly, when X is restricted to being symmetric we need to calculate only the last $\hat{n} - i + 1$ components of the solution a_i to $U\,a_i = z_i$, which means that the backward substitution, like the forward substitution and the Cholesky factorization, costs only $\hat{n}^3/6$. Hence we have in the symmetric case a total of $\hat{n}^3/2$ arithmetic operations for X^{-1}, which is again exactly three times the effort for computing the log-determinant by itself. In either case, all three processes, i.e., factorization, forward substitution, and backward substitution, can be coded such that the innermost loop consists only of SAXPYs $b[i] -= a[i] * c$ for $i = 1, \ldots$. That means for each arithmetic operation one value $a[i]$ must be fetched, the factor c can stay in memory, and $b[i]$ must be fetched and stored, unless special hardware allows incrementation in a significant number of storage locations. In any case the number and structure of memory accesses is exactly proportional to the operations count, which means that the cost ratio 3 is valid even if data transfers are accounted for. This is not true in the following example, which can also be derived from the determinant.

Speelpenning's Example

Even though that would be rather inefficient from a computational point of view one might consider evaluating $\det(X)$ by its explicit representation, namely, computing for all $n!$ permutations π of n elements the product of the elements $X_{i,\pi(i)}$ for $i = 1 \ldots n$ and then summing them with the appropriate signs. This requires exactly an effort of $n! * n$ multiplications and additions or subtractions assuming no use is made of any common subexpressions. Differentiating each one of these additive terms separately we face the problem of calculating the gradient of a product of the form

$$\tilde{f}(x_1, \ldots, x_n) = x_1 * x_2 * \cdots * x_n$$

where we have left off the permutation for simplicity. The task of differentiating $\tilde{f}$ is known in the AD literature as Speelpenning's example [Spe80]. As already mentioned in the introduction, the n algebraic expressions

$$\frac{\partial \tilde{f}}{\partial x_i} = \prod_{j \neq i} x_j \quad \text{for} \quad i = 1 \ldots n$$

suggest almost the same effort for calculating any one of the n gradient components as for the evaluation of $\tilde{f}$ itself. That would of course be $n-1$ rather than 3 or 4 times as costly as evaluating $\tilde{f}$ itself.

Now let us apply the reverse mode. Starting from the natural, forward evaluation loop

$$v_0 = 1;\ v_i = v_{i-1} * x_i \quad \text{for} \quad i = 1 \ldots n\,; \quad y = v_n$$

we obtain the incremental form of the reverse mode in Table 3.8.

Table 3.8: Incremental Reverse Mode on Speelpenning

v_0	$=$	1	
v_i	$=$	$v_{i-1} * x_i$	$i = 1 \ldots n$
y	$=$	v_n	
$\bar{v}_n$	$=$	1	
$\bar{x}_i$	$+=$	$\bar{v}_i * v_{i-1}$	$i = n \ldots 1$
$\bar{v}_{i-1}$	$+=$	$\bar{v}_i * x_i$	

Here we have assumed that $\bar{v}_i = 0 = \bar{x}_i$ for $i = 1 \ldots n$ initially and the reader can easily verify that the final values of the $\bar{x}_i$ are indeed the gradient components of $\tilde{f}$. Now let us examine the computational complexity. The original function evaluation in the forward sweep requires n multiplications without any additions or subtractions. The return sweep requires $2n$ multiplications and the same number of additions. Hence the complexity ratio is exactly 3 if we count only multiplications but is actually 5 if we consider additions as equally costly. The latter reckoning is much more appropriate on modern processor architectures than the purely multiplicative complexity measures of classical complexity analysis. However, because the Speelpenning example is in fact a so-called SUE (single usage expression), where each independent and intermediate occurs only once as an argument, one can eliminate the additions by employing the nonincremental reverse in Table 3.9. Now the ratio in the operations count is down to 3, but we cannot just ignore that for each i we need to fetch x_i, $\bar{v}_{i-1}$, and $\bar{v}_i$ from memory and store both $\bar{v}_{i-1}$ and $\bar{x}_i$. That would mean 5 memory accesses on the return sweep in addition to 3 on the forward sweep and thus a total of 8. Since $8 < 3 * 3$ this suggests that the number of memory accesses for gradient and function would also conform to the complexity ratio 3.

Table 3.9: Nonincremental Return Sweep on Speelpenning Example

$\bar{v}_n$	$=$	1	
$\bar{x}_i$	$=$	$\bar{v}_i * v_{i-1}$	$i = n \ldots 1$
$\bar{v}_{i-1}$	$=$	$\bar{v}_i * x_i$	

Actually this interpretation is rather optimistic because the original function evaluation can put all v_i in the same memory location. While we can do that also for $\bar{v}_i$ the total number of memory accesses for the nonincremental reverse mode will still be $3 + 1$ and thus 4 times as much as for the function itself. Since on modern machines fetching or storing a floating point number can take much more time than multiplying or adding them, memory management is really crucial for algorithmic differentiation. Therefore we will discuss various trade-offs between storage and computation in Chapters 4 and 12.

The Speelpenning example and the calculation of determinants based on LU factorizations or using expansion by rows or columns are very good test cases for AD tools. On the first problem even source transformation tools have a very hard time realizing the theoretically predicted runtime ratio of 3 to 5. The differentiation of log-determinant calculations is a good test case for the memory management of AD tools. Curiously enough ADOL-C [GJU96] obtains a runtime ratio of less than 2 if the determinant expansion by rows or columns is coded up as a recursion with just about as many C-function calls as arithmetic operations. For this kind of highly dynamic code little can be done at compile-time in the way of optimization so that the added cost of the reverse mode is comparatively small. The log-determinant calculation based on an LU factorization creates $n^3/3$ intermediate values, which are all stored by naive AD implementations. As we will see in the subsequent chapter and especially in Exercise 4.6 most of these intermediates do not occur as arguments to nonlinear operations and the storage requirement can actually be reduce to $\mathcal{O}(n^2)$. In general, one can derive the following basic bound for the total memory requirement.

For the computation of the partial derivatives c_{ij} one usually needs the intermediate values v_j which may have been overwritten in the forward sweep. Therefore, one may have to save the values of all intermediate variables v_j on a sequentially accessed data structure. This strategy will be explained in more detail in section 4.2. For the time being we just state the resulting spatial complexity estimate, namely,

$$SAM\{F(x),\, \bar{y}^\top F'(x)\} \;\sim\; OPS\{F(x)\}\,,$$

where SAM stands for sequentially accessed memory. This memory requirement is proportional to the number ℓ of elemental functions evaluated to compute $F(x)$ and thus may cause problems if ℓ is large. However, there is a variety of strategies to reduce this memory requirement of the reverse mode in its basic form. These approaches will be described in detail in section 12. The amount of randomly accessed memory needed by the reverse mode is given by

$$RAM\{F(x),\, \bar{y}^\top F'(x)\} \;=\; 2\, RAM\{F(x)\}\,,$$

which corresponds nicely to the memory increase in the forward mode.

Amortizing Overhead in Vector mode

In correspondence to the scalar reverse mode we have in vector reverse mode for collections of $q \geq 1$ weight vectors $\bar{Y} \in \mathbb{R}^{m\times q}$,

$$\bar{X}^\top = \bar{F}(x,\bar{Y}) \quad \text{with} \quad \bar{F}(x,\bar{Y}) \equiv \bar{Y}^\top F'(x)\,. \tag{3.15}$$

As in the tangent case, the vector mode promises a better amortization of memory usage and other overhead costs. For its realization the intermediate adjoints $\bar{v}_j \in \mathbb{R}$ must be replaced by vectors $\bar{V}_j \in \mathbb{R}^q$, but otherwise the adjoint procedure of Tables 3.5 and 3.6 can be applied unchanged. The propagation of adjoint vectors has numerous applications, especially the cheap evaluation of sparse Jacobians by compression, which will be discussed in Chapter 8.

For the vector reverse mode, one obtains the temporal complexity estimate

$$TIME\{F(x),\, \bar{Y}^\top F'(x)\} \;\leq\; \omega_{gradq}\, TIME\{F(x)\}$$

for a constant $\omega_{gradq} \in \big[1+2q,\, 1.5+2.5q\big]$. Hence, the temporal complexity grows again linearly in the number of weight vectors. The savings that are possible due to the vector version reduces the range of the upper bound for ω_{gradq} from $\big[3q,\, 4q\big]$ for q separate vector-Jacobian products to $\big[1+2q,\, 1.5+2.5q\big]$ for a matrix-Jacobian product. Once more, one obtains for the sequential accessed memory required by the vector reverse mode that the space for saving and restoring intermediate values is exactly the same as in the scalar mode, so that

$$SAM\{F(x),\, \bar{Y}^\top F'(x)\} \;\sim\; OPS\{F(x)\}\,.$$

However, the randomly accessed memory grows in that

$$RAM\{F(x),\, \bar{Y}^\top F'(x)\} \;=\; (1+q)\; RAM\{F(x)\}\,.$$

Calling Sequences for Reverse Propagation

As described earlier in this subsection, the reverse propagation of gradients consists of one forward sweep and one return sweep to compute the derivative values. Naturally, the calling sequence for a routine `eval_dual` combining the two sweeps provided by the corresponding AD tools reflects this structure. Hence, the computation $\bar{x} \mathrel{+}= \bar{F}(x, \bar{y})$ is usually performed for the input arrays $\texttt{x} = x \in \mathbb{R}^n$ and $\texttt{by} = \bar{y} \in \mathbb{R}^m$ yielding the output array $\texttt{y} = y \in \mathbb{R}^m$ and the input-output array $\texttt{bx} = \bar{x} \in \mathbb{R}^n$ by a call of the form

```
eval_dual ( bx, x, by, y )
```

Here, the dotted vertical arrow indicates that `bx` enters only additively into the calculation of the output value of `bx`. In other words `bx` is updated incrementally. Whether or not this is actually the case for any particular implementation has to be checked. Of course it is a rather trivial matter to convert a nonincremental procedure `eval_dual` that computes `bx` from scratch, i.e., implicitly assumes `bx` $= 0$ on entry, to one that is incremental. All one has to do is to save `bx` initially and increment that value to `bx` upon exit. This weak dependence barely affects the principal observation that the information flow for the adjoint quantities `bx` and `by` is opposite that of the corresponding primal quantities `x` and `y`. The following distinction is more critical.

Whereas there is no question that the values of `x` on entry determine the point at which the adjoining is happening, there is no general consensus on what `x` and correspondingly `y` should represent on exit to the procedure `eval_dual(...)`. One clear and for the most part attractive concept is that `eval_dual(...)` should be *primaly consistent* in that its effects on the "primal" arguments `x` and `y` are exactly the same as those of the underlying `eval(...)`, including the possibility of overlap and aliasing between `x` and `y`. This concept is adhered to in the adjoint utilities of ADOL-C, for example.

Especially if reverse mode AD is implemented as a multilevel process, one may prefer another kind of dual procedure that we will call *primaly constant*. Here the inversion of all elementary transformations is considered an integral part of the dual evaluation so that `x` and `y` have on exit from `eval_dual(...)` exactly the same value as on entry. So in effect `y` drops completely out of the calling sequence and we have merely

```
eval_dual ( bx, x, by)
```

The dotted vertical line under `by` indicates that `by` should be set to zero on exit, which is consistent with the adjoining of individual assignments as will be described in Table 4.3. Generally, the user should be aware that there is no agreement between various AD tools as to the status of `y` and `by` on exit from an adjoint code. For notational simplicity we have assumed throughout the book that `by` and `bY` stay unchanged even though that is slightly inconsistent with an incremental interpretation of adjoints.

Converting a primaly constant dual into a primaly consistent one requires an additional call to `eval(...)` after the call to `eval_dual(...)`, which may or may not be considered a serious impediment to efficiency. Hence a good AD tool should provide both versions but nothing whose input-output characteristics are undefined and obscure to the users.

For the vector version of the first-order reverse mode to compute the matrix $\bar{X}^\top = \bar{Y}^\top F'(x)$ for input arrays $\mathtt{x} = x \in \mathbb{R}^m$, $\mathtt{bY} = \bar{Y} \in \mathbb{R}^{m\times q}$, $\mathtt{y} = y \in \mathbb{R}^m$ as pure output array, and $\mathtt{bX} = \bar{X} \in \mathbb{R}^{n\times q}$ as input-output array, one obtains the calling structure

```
          ⋮   ↓   ↓
eval_dualq ( bX, x, bY, y )
             ↓          ↓
```

Here again the dotted line indicates that the matrix `bX` is incremented so that its values on entry enters additively into its value on exit.

3.4 Approximate Error Analysis

Proponents of AD commonly claim that this technique of applying the chain rule yields derivative values with "working accuracy." Indeed, once an AD tool has been applied successfully without transformation or compilation errors we have never heard of complaints concerning the precision of the derivative values obtained. In this section we will justify the general optimism by estimating the error caused by floating point arithmetic.

Nowadays, numerical calculations are usually carried out in double precision where $\varepsilon \approx 10^{-14}$ and the decimal exponent ranges from $-e$ to e for $e \approx 308$. Hence we may assume without too much loss of generality that no underflows or overflows occur so that the fundamental relation

$$\tilde{v} = u \circ w\big(1 + \tilde{\varepsilon}\big) \quad \text{with} \quad |\tilde{\varepsilon}| \le \varepsilon$$

is in fact valid throughout for $\circ \in \big\{+\,,-\,,*\,,/\big\}$. While IEEE arithmetic also enforces $\tilde{v} = \varphi(u)(1+\tilde{\varepsilon})$ for $\varphi(u) = \sqrt{u}$ and $\varphi(u) = \ln u$, this desirable relation cannot be guaranteed for other elementals, in particular the exponential $v = \exp(u)$, the trigonometrics $v = \sin(u)$, $v = \cos(u)$, and other periodic functions. There, a severe loss of accuracy is unavoidable for arguments u that are quite large but whose size $|u|$ is still well within the range $[0, 10^e]$. Fortunately, this effect occurs for the elementals $\varphi(u)$ at least as strongly as for their derivatives $\varphi'(u)$, as one can see very clearly for the exponential $\varphi(u) = \exp(u) = \varphi'(u)$ and the power $\varphi(u) = u^c$ where $\varphi'(u) = c * \varphi(u)/u$.

For simplicity we assume from now on optimal rounding such that the floating point values of the elemental functions and derivatives satisfy

$$\tilde{v}_i = (1+\varepsilon_i)\,\varphi(\tilde{u}_i) \quad \text{and} \quad \tilde{c}_{ij} = (1+\varepsilon_{ij})\frac{\partial \varphi_i}{\partial v_j}(\tilde{u}_i)\,,$$

where $|\varepsilon_i| \le \varepsilon \ge |\varepsilon_{ij}|$ for all i, j and $\tilde{u}_i = (\tilde{v}_j)_{j \prec i}$. Now let us consider what happens when we apply the forward differentiation algorithm listed in Table 3.2. Then we obtain for the floating point values $\tilde{\dot{v}}_i$ of the directional derivatives $\dot{v}_i$

$$\tilde{\dot{v}}_i = \sum_{j \prec i} (1+\tilde{\varepsilon}_{ij})^{n_i} (1+\varepsilon_{ij})\, c_{ij}\,(\tilde{u}_i)\,\tilde{\dot{v}}_j \equiv \sum_{j \prec i} (1+\dot{\varepsilon}_{ij}) c_{ij}\,(\tilde{u}_i)\,\tilde{\dot{v}}_j\,. \tag{3.16}$$

Here, n_i is the number of arguments of φ_i and the $\dot{\varepsilon}_{ij}$ satisfy

$$(1-\varepsilon)^{1+n_i} \le 1+\dot{\varepsilon}_{ij} \le (1+\varepsilon)^{1+n_i} \quad \Longrightarrow \quad \dot{\varepsilon}_{ij} = \mathcal{O}((1+n_i)\,\varepsilon)\,.$$

In the spirit of backward error analysis we slightly modify the $\varphi_i(u_i)$ to $\tilde{\varphi}_i(u_i)$ such that the resulting values of $\tilde{F}(x)$ and $\tilde{F}'(x)\dot{x}$ are exactly equal to the vectors $\tilde{y}$ and $\tilde{\dot{y}}$ obtained in finite precision. To this end we set simply

$$\tilde{\varphi}_i(v_j)_{j \prec i} \equiv (1+\varepsilon_i)\,\varphi_i\Big(\big[(1+\dot{\varepsilon}_{ij})\,v_j + (\varepsilon_i - \dot{\varepsilon}_{ij})\,\tilde{v}_j\big]/(1+\varepsilon_i)\Big)_{j \prec i}\,. \tag{3.17}$$

In other words we slightly shift the arguments and rescale them and the function φ_i itself by a factor of size $1+\mathcal{O}(\varepsilon)$. Then one can easily check that for all i exactly

$$\tilde{v}_i \equiv \tilde{\varphi}_i(\tilde{u}_i) \quad \text{and} \quad \tilde{\dot{v}}_i = \tilde{\varphi}_i'(\tilde{u}_i)\,\tilde{\dot{u}}_i \quad \text{with} \quad \tilde{\dot{u}}_i \equiv (\tilde{\dot{v}}_j)_{j \prec i}\,.$$

Similarly in the reverse mode we have instead of (3.16) for the approximate adjoints $\tilde{\bar{v}}_j$ obtained in finite precision the nonincremental representation

$$\tilde{\bar{v}}_j \equiv \sum_{i \succ j} \tilde{\bar{v}}_i \cdot (1+\bar{\varepsilon}_{ij})\, c_{ij}\,(\tilde{u}_i) \quad \text{with} \quad (1-\varepsilon)^{1+\tilde{n}_j} \le 1+\bar{\varepsilon}_{ij} \le (1+\varepsilon)^{1+\tilde{n}_j}\,.$$

Here $\tilde{n}_j$ counts the number of times v_j occurs as an argument, i.e., the number of successors $i \succ j$. Replacing in (3.17) the $\dot{\varepsilon}_{ij}$ by these $\bar{\varepsilon}_{ij}$ we obtain the perturbed elementals $\tilde{\varphi}_i$ for which the mathematical results are exactly those obtained in floating point arithmetic for the original elementals $\varphi_i(u_i)$. Hence we conclude boldly that

RULE 3

THE JACOBIAN-VECTOR AND JACOBIAN TRANSPOSED VECTOR PRODUCTS CALCULATED IN THE FORWARD AND REVERSE MODE, RESPECTIVELY, CORRESPOND TO THE EXACT VALUES FOR AN EVALUATION PROCEDURE WHOSE ELEMENTALS ARE PERTURBED AT THE LEVEL OF THE MACHINE PRECISION.

Assuming that all φ_i are locally at least once Lipschitz continuously differentiable one can show by induction for given $x \in \mathbb{R}^n$ and $\bar{y} \in \mathbb{R}^m$ fixed and ε theoretically variable that

$$\tilde{v}_i = v_i + \mathcal{O}(\varepsilon)\,, \;\dot{\tilde{v}}_i = \dot{v}_i + \mathcal{O}(\varepsilon)\,, \quad \text{and} \quad \bar{\tilde{v}}_i = \bar{v}_i + \mathcal{O}(\varepsilon)\,.$$

The estimation of the leading constants in the $\mathcal{O}(\varepsilon)$ terms in a forward error analysis is quite complicated. However, it follows from the role of the $\bar{v}_i$ as Lagrange multipliers that $\tilde{y} \equiv \left(\tilde{v}_{l-m+i}\right)_{i=1\ldots m}$ satisfies

$$\begin{aligned}\left|\bar{y}^\top(\tilde{y}-y)\right| &= \left|\sum_{i=1}^{l} \bar{v}_i[\tilde{v}_i - \varphi_i(\tilde{u}_i)]\right| + \mathcal{O}(\varepsilon^2)\\ &= \left|\sum_{i=1}^{l} \bar{v}_i \varepsilon_i \tilde{v}_i\right| + \mathcal{O}(\varepsilon^2) \le \varepsilon \sum_{i=1}^{l} |\bar{\tilde{v}}_i \tilde{v}_i| + \mathcal{O}(\varepsilon^2)\,.\end{aligned}$$

In other words for the inaccuracy in the weighted result $\bar{y}^\top y = \bar{y}^\top F(x)$ caused by roundoff we obtain the error constant $\sum_i |\bar{\tilde{v}}_i \tilde{v}_i|$ with respect to the relative machine precision ε. It has been used by Stummel [Stu80], [ITH88], and others as a condition number for the problem of computing $y = F(x)$ using the given evaluation procedure.

3.5 Summary and Discussion

Blinded by the counterintuitive nature of the cheap gradient result, we and other researchers in AD tacitly assumed for a long time that it was optimal. Here "optimality" was roughly supposed to mean that $\bar{x}^\top = \bar{y}^\top F'(x)$ could not be computed any "cheaper" than by propagating adjoint values backward. Fortunately, we do not have to formulate this assertion more precisely because it is patently wrong, as shown by example in Exercise 3.5**d**. There we find that the reverse mode can be undercut by a factor of up to 2 in terms of arithmetic operations. Since this reduction is not dramatic and the natural structure of the code is lost by the "faster" alternative, we may still view the reverse mode as the method of choice for calculating gradients (more generally, adjoints). The same applies for the forward mode if one wishes to calculate tangents. To conclude, let us highlight the difference between the forward and reverse modes on another class of examples.

The Pyramid Example

With φ_{ij} representing any family of nonlinear binary functions, consider the scalar-valued function $y = f(x)$ defined by the double loop in Table 3.10.

Here, there are n layers of variable values $\{v_{ij}\}_{j \le n-i+1}$ starting with the independents $\{v_{1j} = x_j\}_{j=1\ldots n}$ and culminating with the scalar-dependent $y = v_{n1}$. Each value $v_{i+1\,j}$ is obtained as a nonlinear function of its left and right predecessors $v_{i\,j}$ and $v_{i\,j+1}$. The computational graph of this function is an isosceles

Table 3.10: Pyramid Example Procedure

for $j = 1 \ldots n$
$\quad v_{1\,j} = x_j$
for $i = 1 \ldots n-1$
$\quad$ for $j = 1 \ldots n-i$
$\quad\quad v_{i+1\,j} = \varphi_{i\,j}(v_{i\,j},\, v_{i\,j+1})$
$y = v_{n\,1}$

Table 3.11: Pyramid Dependence

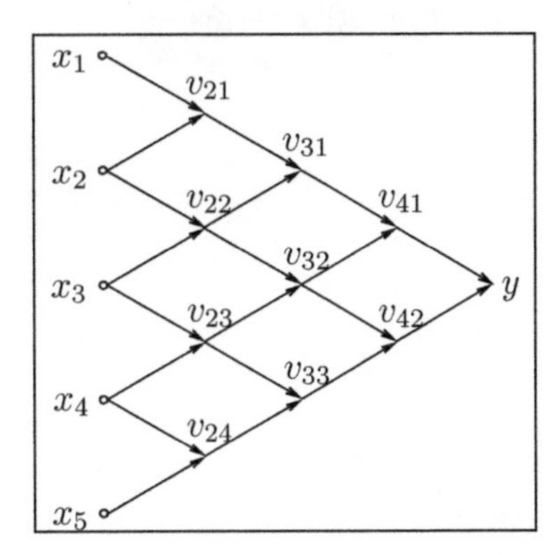

triangle tiled into $(n-1)(n-2)/2$ diamonds and $n-1$ small triangles. The case $n = 5$ is depicted in Table 3.11. Note that we have kept the double subscripts to maintain the structure as much as possible. Similarly, we denote the partial of $v_{i+1\,j}$ with respect to its left predecessor v_{ij} by c_{ij} and that with respect to its right predecessor $v_{i\,j+1}$ by $d_{i\,j}$. To compute the gradient $\nabla f(x) \in \mathbb{R}^n$, we may employ the vector forward mode displayed in Table 3.12 or the scalar reverse mode displayed in Table 3.13.

Table 3.12: Vector Forward on Pyramid

for $j = 1 \ldots n$
$\quad \dot{V}_{1j} = e_j$
for $i = 1 \ldots n-1$
$\quad$ for $j = 1 \ldots n-i$
$\quad\quad \dot{V}_{i+1\,j} = c_{ij} * \dot{V}_{ij} + d_{i\,j} * \dot{V}_{i\,j+1}$
$\dot{y} = \dot{V}_{n\,1}$

Table 3.13: Reverse on Pyramid

$\bar{v}_{n1} = 1$
for $i = n-1 \ldots 1$
$\quad$ for $j = n-i \ldots 1$
$\quad\quad \bar{v}_{ij} \mathrel{+}= \bar{v}_{i+1\,j} * c_{ij}$
$\quad\quad \bar{v}_{i\,j+1} \mathrel{+}= \bar{v}_{i+1\,j} * d_{ij}$
for $j = n \ldots 1$
$\quad \bar{x}_j = \bar{v}_{1\,j}$

In both Tables 3.12 and 3.13 we have listed only the derivative calculation and ignored the calculations that yield the values v_{ij} and the elemental partials c_{ij} and d_{ij}. The key difference is that the $\dot{V}_{i\,j}$ are n-vectors, whereas the $\bar{v}_{i\,j}$ are merely scalars. Even exploiting the sparsity of the $\dot{V}_{i\,j}$, one cannot get the operations count down to the level of the reverse mode. As will be shown in Exercise 3.1, the operations count for the vector forward mode is $n^3/3 + O(n^2)$ multiplications and half this many additions. In contrast, the reverse mode gets by with just $n^2 + O(n)$ multiplications and the same amount of additions. Also, even when $v_{i+1\,j}$ and $\dot{V}_{i+1\,j}$ overwrite $v_{i\,j}$ and $\dot{V}_{i\,j}$, then the memory requirement for the forward mode is still $n^2/4 + O(n)$ compared with $n^2 + O(n)$ for the reverse mode.

Suppose we apply the two modes symbolically in that the elemental partials

c_{ij} and d_{ij} are manipulated as variables rather than numbers. Then each $\dot{V}_{\tilde{\imath},\tilde{\jmath}}$ is a polynomial of degree $\tilde{\imath}-1$ in the c_{ij} and d_{ij} with $i < \tilde{\imath}$ and $j < \tilde{\jmath}$. Clearly, the number of additive terms doubles with each application of the chain rule. The same is true for the reverse mode. However, if the two procedures were executed in a computer algebra system like Maple, *Mathematica*, or Reduce, the resulting 2^n additive terms would not actually be formed until the user asked for them to be printed on the screen. Numerical tests were made using Maple version 10 and $\varphi_{ij}(u,v) = \sin(u) * \cos(v)$, so that no algebraic simplification of the original function $y = f(x_1, \ldots, x_n)$ was possible.

Figure 3.4 shows the runtime and the memory required to generate a symbolic representation of the gradient for the pyramid example and $n = 6, \ldots, 20$. The lines of generated code are also given. To get a better impression of the results, we included the time in seconds and the memory required by the AD tool ADOL-C to generate a corresponding internal representation for the gradient calculation. Note that Figure 3.4 has a log-log scale. For smaller problems there were some significant improvements, but the memory and runtime requirements of the expression simplification process grew exponentially with n. While Maple cannot shift automatically from the forward to the reverse mode, Michael Monagan has provided forward and reverse routines in the shareware library of Maple (see [MR96]).

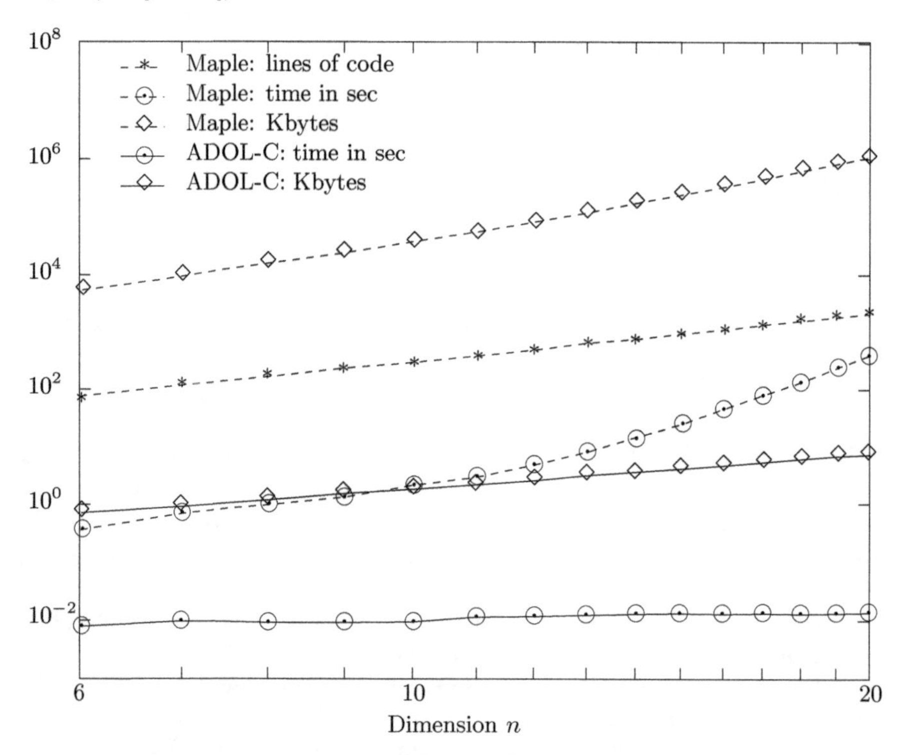

Figure 3.4: Temporal and Spatial Complexities on Pyramid

Implications for Iterative Equation Solving

Most methods for the iterative solution of linear or nonlinear equations are based on the evaluation of a sequence of Jacobian-vector products and possibly also vector-Jacobian products (see, e.g., [Saa03]). The latter are avoided by "transpose-free" methods on the assumption that they are hard to come by, in contrast to Jacobian-vector products. Even in the linear case, matrices are often stored such that Jacobian-vector products are easier to calculate than matrix-vector products involving the transpose, especially on a parallel machine.

In the nonlinear case Jacobian-vector products can be approximated by difference quotients

$$[F(x+\varepsilon\dot{x})-F(x)]/\varepsilon \;\approx\; F'(x)\dot{x}$$

at the cost of one extra function evaluation. In contrast, it is still a widely held belief that "there is no matrix-free approximation to the transpose-vector product." Now we know better, namely, that the reverse mode yields vector-Jacobian products $\bar{y}^\top F'(x)$ without even coming close to forming a matrix and of course without any approximation error. Moreover, the operations count is proportional to that of Jacobian-vector products $F'(x)\dot{x}$ and thus also that of the residual function $F(x)$ itself. This realization should change the paradigms of iterative equation solving even though we do have to keep the memory requirement of the basic reverse mode in mind.

Moreover, as of now, the AD-generated vectors $F'(x)\dot{x}$, and even more so $\bar{y}^\top F'(x)$, are considerably more expensive than a single evaluation of the residual $F(x{+}\varepsilon\dot{x})$ at the neighboring point. There is some hope that this situation can be changed when certain comparatively costly intermediate results are saved during the initial evaluation of $F(x)$ and then reused during subsequent evaluations of $F'(x)\dot{x}$ and/or $\bar{y}^\top F'(x)$ for a sequence of directions $\dot{x}$ and weights $\bar{y}$ at the same point x. At exactly which cost ratios the evaluation of vector-Jacobian products pay off or when the accuracy of Jacobian-vector products justifies their higher cost compared to difference quotients can of course not be stated in general. Everything hinges on the resulting variations in the speed of convergence, which are very hard to predict for most realistic problems.

It is well known that iterative equation solvers are crucially dependent on good scaling, which typically requires preconditioning. Unfortunately, neither AD nor any other technique can provide the required scaling information at a cost significantly below that of evaluating the full Jacobian or Hessian. This conclusion follows as a corollary of the cheap gradient result by the following observation of Morgenstern [Mor85] and Miller, Ramachandran, and Kaltofen [MRK93].

Consider the vector function

$$F(x) = Y^\top AB\,x \;:\; \mathbb{R}^n \longmapsto \mathbb{R}^n$$

with Y, A, and B general $n \times n$ matrices. Obviously, we have the Jacobian

$$F'(x) = Y^\top AB \in \mathbb{R}^{n\times n}$$

and the trace

$$Tr(F'(x)) = Tr(Y^\top AB),$$

which we may consider as absolutely minimal scaling information about F' and thus F itself.

Applying the cheap gradient result to differentiation of the trace with respect to Y, we find that computing

$$\nabla_Y Tr(Y^\top AB) = AB$$

costs no more than $\omega_{grad} \leq 4$ times as much as evaluating $Tr(Y^\top AB)$. Disregarding storage costs, we derive that

$$\frac{\text{TIME}\{Tr[F'(x)]\}}{\text{TIME}\{F'(x)\}} \geq \frac{\text{TIME}\{AB\}/\omega_{grad}}{\text{TIME}\{Y^\top AB\}} \geq \frac{1}{2\omega_{grad}} \geq \frac{1}{8}.$$

The penultimate inequality follows from the observation that, whatever scheme for computing dense matrix products is being used, the ternary product $Y^\top AB$ can always be obtained at a cost not exceeding that of two binary products.

Since the same argument is valid for the gradient $(Y^\top AB + B^\top A^\top Y)x$ of $f(x) = x^\top Y^\top ABx$, we can conclude that traces of Jacobians and Hessians are, in general, not much cheaper to evaluate than the full derivative matrices themselves. Clearly, any useful conditioning information is likely to cost even more. It appears that this "*traces are no cheaper than matrices*" principle extends to sparse or otherwise structured Jacobians and Hessians, but we are currently not aware of a generalized result in this direction. Nevertheless, we conclude tentatively that preconditioning information can only be gained for significantly lower cost than the whole Jacobian or Hessian if structural information, in addition to the evaluation program, is available. Combining the good news about transpose-vector products with the bad news about preconditioning, we formulate the following rule.

RULE 4

PRODUCTS OF TRANSPOSE-JACOBIANS WITH VECTORS ARE CHEAP, BUT THEIR TRACE IS AS EXPENSIVE AS THE WHOLE JACOBIAN.

3.6 Examples and Exercises

Exercise 3.1 (*Pyramid Example*)

Consider the pyramid example given in section 3.5.

a. Show by induction that the gradient $\dot{V}_{ij}$ has exactly i nonzero components and that its computation from $\dot{V}_{i-1,j}$ and $\dot{V}_{i-1,j+1}$ involves exactly $2(i-1)$ multiplications. Conclude that the total operations count for Table 3.12 is $n^3/3 + O(n^2)$ multiplications and $n^3/6 + O(n^2)$ additions.

b. Show that the reverse mode according to Table 3.13 requires only $n^2 + O(n)$ multiplications and the same amount of additions.

Exercise 3.2 (*Differentiation of Norm*)
Write an evaluation procedure for the sum of square function $f(x)$ already considered in Exercise 2.2. Show that the corresponding tangent and adjoint procedures yield $\partial f/\partial x_i = 2\,x_i$ with working accuracy, as asserted in the introduction.

Exercise 3.3 (*Forward Differentiation of Division*)
Consider the division operation $v = u/w$, which may be split into the pair of statements $t = 1/w$ and $v = u * t$.
a. Write down the succession of tangent operations as given in Table 3.3 and count the number of multiplications and additions or subtractions.
b. Show that by eliminating $\dot{t}$ and extracting common factors the number of multiplications incurred in part **a** can be reduced by 2, yielding the procedure $t = 1/w, v = u * t, \dot{v} = (\dot{u} - v * \dot{w}) * t$.
c. Observe that the low operations cost according to part **b** cannot be achieved without the temporary t.

Exercise 3.4 (*Differentiation of Linear Solvers*)
Generalize the previous exercise to the case $V = \varphi(U, W) \equiv W^{-1}U$ with W as a nonsingular square matrix of size n and V, U as rectangular matrices with m columns of length n. Now the temporary $T = W^{-1}$ may be thought of as the explicit inverse or a suitable factorization of W, so that its calculation in either case involves $O(n^3)$ operations.
a. Show by implicit differentiation of the identity $W\,T = I$ that $\dot{T} = -T\,\dot{W}\,T$, and substitute this expression into the product rule $\dot{V} = T\,\dot{U} + \dot{T}U$ to obtain $\dot{V} = \dot{\varphi}(U,\, W,\, \dot{U},\, \dot{W}) = T(\dot{U} - \dot{W}\,V)$. Note that the calculation of $\dot{V}$ essentially requires $2n^2m$ operations once T is available, as discussed above. Conclude that when $m \ll n$ the evaluation of $\dot{\varphi}$ comes essentially for free on top of φ.
b. On the linear space of square matrices of order n the trace

$$\langle A,\, B\rangle \;\equiv\; Tr(A^{\top}B)$$

represents an inner product with the corresponding norm

$$\langle A,\, A\rangle \;=\; Tr(A^{\top}A) \;=\; \|A\|_F^2$$

being the Frobenius norm. For the function $T = \varphi(I, W) \equiv W^{-1}$, derive from $\dot{T} = -T\,\dot{W}\,T$ and the required identity $Tr(\bar{T}\,\dot{T}) = Tr(\bar{W}\,\dot{W})$ for all $\dot{W}$, the value of $\bar{W}$ for given W and $\bar{T}$.
c. Use the fact that $\bar{T} = U\bar{V}$ together with the result from part **b** to show that the adjoint procedure $\bar{\varphi}(U, W, \bar{V})$ corresponding to $V = W^{-1}U = TU$ is given by the procedure $S = \bar{V}T$; $\bar{U}$+= S; $\bar{W}$−= VS. Draw the corresponding conclusion to part **a** for $\bar{\varphi}$ in place of $\dot{\varphi}$.

Exercise 3.5 (*Premature Conclusion*)
Consider an evaluation procedure that contains no additions and subtractions, so that none of the elemental partials $c_{ij} = \partial\varphi_i/\partial v_j$ with $j \prec i$ has one of the

special values in $\{-1, 0, +1\}$. Moreover, assume that there are no dead ends in that all intermediates v_k with $1 \le k \le l-m$ depend on at least one independent variable x_j and impact at least one dependent variable y_i. Finally, assume that no dependent is directly calculated from an independent, so that none of the c_{ij} is itself an element of the Jacobian. Then each value c_{ij} affects the value of the Jacobian element $\partial y_i / \partial x_j$ and must therefore enter somehow into any procedure for evaluating $F'(x)$ at least once. Let the gradient $\nabla f(x) = \bar{F}(x, 1)$ of a scalar-valued function $f = F$ with $m = 1$ be calculated by the adjoint procedure of Table 3.5 with $\bar{y}_1 = 1$. Then each elemental partial c_{ij} enters exactly once, namely, as a factor in a multiplication. In the incremental form each one of these multiplications is followed by an addition.

a. Conclude that the number of arithmetic operations used to accumulate ∇f from the elemental partials in the reverse mode cannot exceed the minimal number needed to calculate ∇f by a factor of more than 4.

For some $d \gg 1$ and positive constants w_i, z_i, b_i for $i = 1 \ldots d$, consider the function

$$f(x) \equiv \ln \left| \sum_{i=1}^{d} w_i (\exp(z_i * \sin(x)) - b_i)^2 \right| .$$

We might think of $\sin(x)$ as a parameter restricted to $[-1, 1]$ that should be chosen to minimize the discrepancies between the values $\exp(z_i * \sin(x))$ and the b_i as measured by the logarithm of a suitably weighted sum. To optimize such an exponential fit, one obviously needs the scalar-valued derivative $f'(x)$. To satisfy the special assumptions made above, let us consider the $\psi_i(u) \equiv (u - b_i)^2$ and the weighted sum $\psi_0(v) = \sum_{i=1}^{d} w_i v_i$ as elemental functions.

b. Write a procedure for evaluating f using the elementaries ψ_i for $i = 0, 1 \ldots d$, and draw the computational graph.

c. Derive the corresponding adjoint procedure for accumulating ∇f from the elemental partials. Count the number of multiplications and additions used in the incremental form. Verify that the number of multiplications corresponds exactly to the number of edges, which is $4d + O(1)$.

d. Rewrite the formula for $f(x)$ using the ψ_i, and differentiate it by hand. Show that by keeping the common factors $\cos(x)$ and $1/\psi_0$ out of the internal sum, the number of multiplications needed for the calculation can be kept to $3d + \mathcal{O}(1)$.

e. Show that the tangent procedure for calculating $f'(x) = \dot{f}(x, \dot{x})$ with $\dot{x} = 1$ involves exactly as many multiplications as the reverse mode and is therefore also less economical than the procedure derived by hand in part **d**.

Exercise 3.6 (*Speelpenning's Product*)

a. Write an explicit expression for evaluating the ith gradient component

$$g_i(x) \equiv \partial f(x) / \partial x_i .$$

Estimate the number of multiplications needed to evaluate the n gradient components one by one without using shared intermediates.

b. Write a procedure for evaluating $f(x)$ using $n-1$ multiplications. Write the corresponding tangent and adjoint procedures. Verify that the number of multiplications in the adjoint code is a small multiple of n.
c. It will be established in Lemma 7.1 that when all elementals in Ψ are unary or binary, the intermediate v_i (in our notation) can depend nontrivially on at most $i+1$ independent variables. Apply this result to any sequence of arithmetic operations for evaluating the value f and the n gradient components g_i of Speelpenning's product. Show that it must involve at least $l \geq 2n-2$ operations. Then show that this lower bound can be achieved when all x_i may be assumed to be nonzero by constructing a suitable procedure of minimal length $l = 2n-2$.

Exercise 3.7 (***Determinants, Inverses, and Linear Systems***)
Consider the function $f(X) \equiv \ln|\det(X)|$, where $X \in \mathbb{R}^{\hat{n}\times\hat{n}}$ is a square matrix whose $n \equiv \hat{n}\times\hat{n}$ entries X_{ij} are considered as independent variables. This function $f(X)$ has been used as a penalty term for maintaining positive definiteness of X when it is restricted to being symmetric [OW97].
a. Show that, provided $\det(X) \neq 0$ so that $f(X)$ can be evaluated, the matrix $G \in \mathbb{R}^{\hat{n}\times\hat{n}}$ of corresponding gradient components $G_{ij} = G_{ij}(X) \equiv \partial f(X)/\partial X_{ij}$ is the transpose of the inverse X^{-1}. Hence the cheap gradient principle guarantees that any finite algorithm for evaluating the determinant of a general square matrix can be transformed into an algorithm for computing the inverse with essentially the same temporal complexity.
b. Examine whether the last assertion is still true when X is restricted to having a certain sparsity pattern (for example, tridiagonal). Explain your conclusion.
c. For a constant matrix $A \in \mathbb{R}^{n\times n}$ and a vector $b \in \mathbb{R}^n$, determine the gradient of

$$f(x) \equiv \ln|\det(A + b\,x^\top)| \quad \text{at} \quad x = 0\,.$$

d. Write an evaluation procedure that evaluates $f(x)$ for $n=2$ by first computing the entries of $A + b\,x^\top$, then applying the explicit determinant formula for 2×2 matrices, and finally taking the logarithm of the determinants modulus. Show that the resulting adjoint procedure for evaluating $\nabla f(0)$ is not optimal.
e. Suppose that $A + b\,x^\top$ has for all x near the origin n simple nonzero real eigenvalues $\lambda_j(x)$ so that $f(x) = \sum_j \ln|\lambda_j(x)|$. Determine $\nabla_x \lambda_j$ at $x=0$ by implicit differentiation of the identities $(A + b\,x^\top)v_j = \lambda_j\,v_j$ and $(A + b\,x^\top)^\top w_j = \lambda_j\,w_j$. Here $v_j, w_j \in \mathbb{R}^n$ denote the corresponding left and right eigenvectors normalized such that $v_j^\top w_j = 1$ for $j = 1\ldots n$. Write an evaluation procedure for $f(x)$ treating the λ_j as elemental functions. Derive and interpret the corresponding adjoint procedure for evaluating $\nabla f(0)$.

Chapter 4

Memory Issues and Complexity Bounds

4.1 Memory Allocation and Overwrites

So far we have interpreted the original evaluation procedure of Table 2.2 exclusively from a mathematical point of view. In particular, we have assumed that each variable v_i occurs exactly once on the left-hand side and that its assignment does not affect any other variable value. Such single-assignment code could be evaluated in any order that ensures that all v_j are defined before they occur on the right-hand side as components of some argument vector $u_i \equiv (v_j)_{j \prec i}$.

In virtually all practical evaluation procedures, most values are kept in local or temporary variables that are reused many times so that the total number of memory locations used is much smaller than the number of elementary operations. In the literature on AD, the distinction between mathematical variables and program variables has often been blurred to avoid extensive formalism. We adhere to this tradition by considering the v_i as mathematical variables but allowing them to share memory locations. Consequently, an assignment to some v_i may invalidate (overwrite) another variable v_j with $j < i$.

The Allocation Function &

Without giving up the unique variable names v_i, we shall assume that a preprocessor or compiler has generated an *addressing scheme* "&" that maps the variables v_i into subsets of an integer range

$$\&i \equiv \&v_i \subset \mathcal{R} \equiv \{1, 2 \ldots r\} \quad \text{for} \quad i = 1 \ldots l \, . \tag{4.1}$$

We refer to the elements of $\mathcal{R}$ as *locations*. It is essential that none of these locations be reused before its current value v_i has been referenced as an argument for the last time. In the lighthouse example we might use the allocation given in Table 4.1. Renaming the variables accordingly, one might use instead of Table 2.1 the sequence of assignments listed in Table 4.2.

Hence we see that the programmer may already allocate variables by naming.

Table 4.1: Allocation for Lighthouse

& \ i	1	2	3	4	5	6	7
1	v_1	v_1	v_3	v_3	v_3	v_6	v_6
2	–	v_2	v_2	v_2	v_5	v_5	v_5
3	–	–	–	v_4	v_4	v_4	v_7

Table 4.2: Renamed Lighthouse

$v_1 = x_3 * x_4;$	$v_2 = \tan(v_1);$
$v_1 = x_2 - v_2;$	$v_3 = x_1 * v_2;$
$v_2 = v_3/v_1;$	$v_3 = v_2 * x_2;$
$y_1 = v_2;$	$y_2 = v_3$

To make sure that no location is reused before its current value has been referenced as an argument, we require that

$$\boxed{j \prec i, \quad j < k \leq i \quad \Rightarrow \quad \&v_k \neq \&v_j \ .} \tag{4.2}$$

In other words, if v_j is an argument of v_i, it may not be overwritten by any v_k with $k \leq i$. We refer to this relation between the data dependence $\prec$ and the allocation function & as *forward compatibility*. In the context of the nonincremental reverse mode, we use the dual concept of *reverse compatibility*, namely, the condition that

$$\boxed{j \prec i, \quad j \leq k < i \quad \Rightarrow \quad \&v_i \neq \&v_k \ .} \tag{4.3}$$

If both conditions are met, we call the allocation scheme & *two-way compatible* with the dependence relation $\prec$. In Exercise 4.1 we study a little example problem with forward- and reverse-compatible allocation schemes.

We say that v_i overwrites v_j if it is the first variable occupying a location previously used by v_j. Note that v_i may not overwrite one of its arguments v_j with $j \prec i$. In section 4.3 the condition (4.2) or (4.3) is relaxed to allow for $\&v_i = \&v_j$ even when $j \prec i$. Then we will call & weakly forward or reverse compatible, respectively. Our standard assumption on & is (strong) forward compatibility.

Assuming that all locations in $\mathcal{R}$ are used at least once, we find that the maximal number of live variables (i.e., overlapping lifespans) is simply r. In each line of the evaluation procedure given in Table 2.2, the arguments v_j may occur without any particular pattern. Hence, we must generally assume that the accesses to the corresponding memory locations $\&v_j$ are more or less random. Therefore, we write

$$RAM\{F(x)\} \ \equiv \ r \ \equiv \ \max\{k \mid k \ = \ \&v_i, 1 \leq i \leq l\} \ . \tag{4.4}$$

Here RAM stands for randomly accessed memory, in contrast to sequentially accessed memory, SAM, which we separately account for in the context of adjoint calculations. The key question about SAM is whether or not it can be accommodated in internal memory or spills over onto disk. In the latter case the speed of adjoint calculations can be slowed down by orders of magnitudes.

Unless otherwise specified, it is implicitly assumed that the v_i and certain associated derivative objects are allocated according to an addressing scheme &. Hence they may overwrite other values that are no longer needed.

Aliasing and Derivative Consistency

Even if the forward compatibility condition (4.2) is not met, the loop listed in Table 2.2 still uniquely defines y as some function $F(x)$, albeit not the same as in the compatible case. As an extreme possibility one might even consider the case in which & maps all variables v_i into a single location. While this possibility does not seem to make much practical sense, there are certainly situations where (within some subroutine) certain array indices or pointer values may or may not be the same, perhaps depending on the prior control flow. Even if such aliasing arises by mistake, for example, because array boundaries are exceeded by runtime indices, we would still like the differentiation process to be *consistent.* More specifically, we want the derivative values to be those of the functional relation $y = F(x)$ actually computed, whether or not that is the one intended by the programmer. To this end we will strive to make all our differentiation rules *alias-safe* even though that may occasionally require a little more conceptual or computational effort.

For example, in Table 1.3 the first two statements in the central part, namely,

$$\begin{aligned} v_1 &= v_{-1}/v_0 \ , \\ \dot{v}_1 &= (\dot{v}_{-1} - v_1 * \dot{v}_0)/v_0 \ , \end{aligned}$$

are not alias-safe. If $\&v_0 = \&v_1$ the denominator in the derivative statement will no longer be correct once v_1 and hence v_0 have been assigned their new value. Here, as always in the forward mode, one can easily fix the situation by simply switching the order of the two statements, thus computing derivatives before updating the corresponding values. In reverse mode it is generally a bit more difficult but still possible to achieve alias-safety (see Proposition 4.2). In either mode alias-safety means that differentiation can proceed locally, i.e., line by line, without any need for a nonlocal analysis of variable interdependence.

Allocation of Directional Derivatives

We assume that the directional derivatives $\dot{v}_i$ are allocated just like the v_i, for example, by the addressing scheme $\&\dot{v}_i = r + \&v_i$. Hence, we obtain the result

$$RAM\{F(x), F'(x)\dot{x}\} \ = \ 2RAM\{F(x)\} \ .$$

In other words, we have to double the randomly accessed memory and can allocate $\dot{v}_i$ parallel to v_i. Hence, $\dot{v}_i$ overwrites $\dot{v}_j$ exactly when v_i overwrites v_j and the lifespans of the $\dot{v}_i$ are exactly the same as those of the corresponding v_i. As a consequence, the values $\dot{y}$ obtained are indeed the correct tangent for the original $F(x)$. When the forward mode is implemented by overloading as described in section 6.1 the derivatives $\dot{v}_i$ are usually allocated right next to v_i in a common data structure. In contrast, source transformation tools usually allocate the $\dot{v}_i$ in separate arrays as sketched in section 6.2. Under the forward compatibility condition, both y and $\dot{y}$ are identical to the values obtained without any overwriting, i.e., with an injective allocation function &.

In contrast, when the allocation & is arbitrary, the functional relation $y = F(x)$ is likely to be altered, but $\dot{y} = \dot{F}(x, \dot{x})$ will be adjusted accordingly. Formally, we obtain the following result.

Proposition 4.1 (General Consistency of Tangents)
Suppose all φ_i are once continuously differentiable on some neighborhood of their current argument. Then the tangent procedure of Table 3.4 yields consistent results in that

$$\dot{y} = \frac{\partial}{\partial t} y(x + t\,\dot{x})\Big|_{t=0}$$

whenever the allocation of the v_i is consistent in that $\&v_i = \&v_j$ if and only if $\&\dot{v}_i = \&\dot{v}_j$.

Proposition 4.1 holds under even more general assumptions where the precedence relation $j \prec i$ does not imply $j < i$, so that some arguments v_j may not be properly initialized but contain some arbitrary default values. Provided these values are considered as constants the directional derivatives $\dot{y}$ still come out right. As we will see, the same is true by Proposition 4.2 for the adjoint vectors $\bar{x}$ calculated in the section 3.2

4.2 Recording Prevalues on the Tape QO

Just as in the case of tangent propagation, we can generate the adjoint recursion of Table 3.5 more or less line by line from the underlying evaluation procedure of Table 2.2. However, this time in contrast with the tangent forward case, we have to take overwriting of variables explicitly into account. The reason is that the v_i are needed on the way back as part of the arguments u_i for the derivatives $c_{ij}(u_i)$. Alternatively, or in addition, one may store the elemental partials c_{ij}.

In either case the extra memory can be organized strictly sequentially, since the data are needed on the way back in exactly the same order they were generated on the way forward. We can therefore store or, as we prefer to say, "record" these data on a single tape denoted by QO. In our context a tape is simply a last in first out (LIFO) data structure that can contain a large amount of data. We use the notation $\rightarrowtail$ QO to indicate that something is being pushed onto the global tape, and correspondingly $\leftarrowtail$ QO to indicate that something is being popped off it. Since the left side of any elemental assignment $v_i = \varphi_i(u_i)$ may share memory with other variables, we record its previous value on the tape before the elemental operation. On the way back, we may then recover this *prevalue* before the corresponding adjoint calculation. Thus, we obtain the pair of operations listed in Table 4.3.

After v_i has been reset from the tape as part of the ith adjoint operation, the values of all variables v_j must be exactly the same as they were just before the ith elemental function φ_i was called on the way forward. Using the n_i-vector $\bar{u}_i$ adjoint to $u_i = (v_j)_{j \prec i}$ as defined in (2.4) and (3.12), we may write the middle line of Table 4.3 as

$$\bar{u}_i \mathrel{+}= \bar{\varphi}_i(u_i, \bar{v}_i) \equiv \bar{v}_i * \varphi'(u_i)^\top \quad \text{with} \quad \bar{\varphi}_i : \mathbb{R}^{n_i+1} \mapsto \mathbb{R}^{n_i} \; .$$

Table 4.3: Adjoining a Single Assignment (Noniterative)

Forward :

v_i	$\rightarrowtail$	$\circ\!\circ$
v_i	$=$	$\varphi_i(v_j)_{j \prec i}$

Return :

v_i	$\leftarrowtail$	$\circ\!\circ$	
$\bar{v}_j$	$+=$	$\bar{v}_i * c_{ij}$	for $j \prec i$
$\bar{v}_i$	$=$	0	

The notation $\bar{\varphi}_i$ corresponds closely to the one we introduced for the tangent operation $\dot{\varphi}_i$ in section 3.1. Just as we observed for the tangent operations listed in Table 3.3, we find that the new value $v_i = \varphi_i(u_i)$ is often useful in carrying out the corresponding adjoint operations. There we varied the order in which function and tangent calculations were performed so as to combine efficiency with the correct handling of overwrites. Unfortunately the latter goal cannot be so easily achieved in the reverse mode.

Like the corresponding tangent operations, the adjoint operations for the elementary cases are mostly cheaper than two of the underlying elementary functions. The tape operations ensure that the argument u_i of $\bar{\varphi}_i$ shall have the same value it had as an argument of φ_i because any possible overwrites will have been undone on the way back. At first it might seem more natural to save the argument vector u_i after the call to φ_i and to recover it before the call to $\bar{\varphi}_i$. This is a valid approach and has, in fact, been implemented in some AD packages. However, we prefer the "copy-on-write" method embodied in Table 4.3, for the following reasons.

In the case of the elemental functions listed in Table 2.3, we always have one single result variable, but often two (and sometimes more) argument variables. For general elemental functions we can still expect fewer result variables than argument variables. Hence, storing the prevalues of the results rather than the values of the arguments generally requires much less tape space. When φ_i is a subroutine, there may be common blocks or global structures that it depends on, therefore storing all of them would require extensive memory. In contrast, it is usually easier to determine which variables are being changed by a subroutine φ_i. When storing the prevalues, one obtains the complexity result

$$SAM\{F(x), \bar{y}^\top F'(x)\} \sim l \,,$$

where SAM denotes the sequentially accessed memory and l the number of elemental functions evaluated to calculate $F(x)$. There is a large variety of possibilities to reduce this memory requirement of the reverse mode in its basic form; see section 12.

As with the dotted quantities $\dot{v}_i$ in the tangent propagation procedure, we can allocate the adjoints $\bar{v}_i$ so that $\&\bar{v}_i = r + \&v_i$. This yields the similar complexity estimate

$$RAM\{F(x), \bar{y}^\top F'(x)\} = 2RAM\{F(x)\} \,,$$

for randomly accessed memory.

Using the $\bar{\varphi}_i$ notation we can rewrite the adjoint procedure of Table 3.5 in the more concise form of Table 4.4. The box on the top of Table 4.4 represents a precondition that ought to hold initially, and the one on the bottom represents a postcondition that will be enforced by the procedure. In going from Table 3.5 to Table 4.4 we have allowed the $\bar{v}_{i-n}$ to be set to some nonzero initial values $\bar{x}_i$. Consequently, the overall effect of executing the procedure listed in Table 4.4 will be

$$\bar{x} \mathrel{+}= \bar{F}(x,\bar{y}) \quad \text{with} \quad \bar{F}(x,\bar{y}) \equiv F'(x)^\top \bar{y} : \mathbb{R}^{n+m} \mapsto \mathbb{R}^n \,. \tag{4.5}$$

This slight generalization costs no extra arithmetic operations and makes the task at hand, namely, evaluating F and $\mathrel{+}= \bar{F}(x,\bar{y})$, largely homogeneous in the following sense. Except for the additional tape operations, exactly the same task evaluating φ_i and $\mathrel{+}= \bar{\varphi}_i$ must be performed for each elemental constituent of F, as being performed for F itself.

Table 4.4: Adjoint Evaluation Procedure with Tape and Overwrites

$\bar{v}_i$	$\equiv$	0	$i = 1 \ldots l$
$[v_{i-n}, \bar{v}_{i-n}]$	$=$	$[x_i, \bar{x}_i]$	$i = 1 \ldots n$
v_i	$\rightarrowtail$	$\bullet\circ$	$i = 1 \ldots l$
v_i	$=$	$\varphi_i(u_i)$	
y_{m-i}	$=$	v_{l-i}	$i = m-1 \ldots 0$
$\bar{v}_{l-i}$	$=$	$\bar{y}_{m-i}$	$i = 0 \ldots m-1$
v_i	$\leftarrowtail$	$\circ\bullet$	
$\bar{u}_i$	$\mathrel{+}=$	$\bar{\varphi}_i(u_i, \bar{v}_i)$	$i = l \ldots 1$
$\bar{v}_i$	$=$	0	
$\bar{x}_i$	$=$	$\bar{v}_{i-n}$	$i = n \ldots 1$

Lives and Times of Adjoint Pairs

The addressing scheme $\&\bar{v}_i = r + \&v_i$ implies, by (4.2), the adjoint compatibility condition

$$\boxed{j \prec i, j < k \leq i \quad \Rightarrow \quad \&\bar{v}_k \neq \&\bar{v}_j \,.}$$

Loosely speaking, we may say that $\bar{v}_i$ has the same lifespan $[i,k]$ as v_i, with v_k the first variable overwriting v_i. However, it lives through that bracket of statement counters backwards.

More specifically, we have juxtaposed in Table 4.5 the corresponding events that happen in the "lives" of any particular pair $v_i, \bar{v}_i$. Strictly speaking, v_i also has a life on the way back, but alas a boring one. Having been saved onto the tape at the kth forward step, it gets recovered at the same counter on the way back and then is available for evaluating partials $c_{j\tilde{\imath}}$ with $\tilde{\imath} \prec j \succ i$; that is, both v_i and $v_{\tilde{\imath}}$ belong to the same u_j. After the ith backward step, the value v_i is no longer needed and may be overwritten. An alternative to this *value taping* approach is to compute and tape the partials c_{ij} on the way forward and to recover them on the way back.

Table 4.5: Corresponding Events for Intermediate v_i and Its Adjoint $\bar{v}_i$

Counter	i	All $j \succ i$	First $\&k = \&i$
Fate of v_i	Birth by evaluation $v_i = \varphi_i(u_i)$	Appearance as argument $v_j = \varphi_j(\ldots, v_i, \ldots)$	Death by overwrite $v_k = \varphi_k(u_k)$
Fate of $\bar{v}_i$	Death after dispatch $\bar{u}_i$+= $\bar{v}_i * \varphi_i'(u_i)^\top$	Value incrementation $\bar{v}_i$+= $\bar{v}_j * c_{ji}$	Birth as zero $\bar{v}_i = 0$

Allowing Arbitrary Aliasing

If one does not know whether the result v_i occupies the same location as some of its arguments v_j with $j \prec i$, we may utilize a temporary variable $\bar{t}_i \in \mathbb{R}^{n_i}$ and use the modification of Table 4.3 given in Table 4.6

Table 4.6: Adjoining a General Assignment

Forward :	Return :
$v_i \rightarrowtail$ ꝏ $v_i = \varphi_i(v_j)_{j \prec i}$	$v_i \leftarrowtail$ ꝏ $\bar{t}_i = \bar{\varphi}_i(u_i, \bar{v}_i)$ $\bar{v}_i = 0;\ \bar{u}_i \mathrel{+}= \bar{t}_i$

One can easily check that Table 4.6 is consistent with Table 4.3, 5.1, or 5.3 when the allocation & makes the assignment $v_i = \varphi_i(u_i)$ noniterative, general iterative, or incremental, respectively. Substituting the right-hand side of Table 4.6 into Table 4.4, we obtain the more general procedure listed in Table 4.7. Table 4.8 instantiates Table 4.7 for the most important elemental functions. Adjoints of general iterative operations will be considered in section 5.1.

So far, we have skirted the question of how v_i may receive any value before it occurs on the left-hand side of the ith statement. Because of our single assignment notation, this can happen only in the guise of another variable v_k with $k < i$ that shares the same location $\&v_k = \&v_i$. Otherwise, we assume that all values are initialized to some real value, which may be leftovers from previous program executions.

Although, thus far we have still assumed that the precedence relation $\prec$ and the addressing function & satisfies the forward compatibility condition (4.2), the resulting functional relations $y = F(x)$ and $\bar{x} = F'(x)^\top \bar{y}$ are the same regardless of the specific memory allocation scheme used.

Table 4.7: Adjoint Evaluation Procedure with Tape and Iterative Elementals

$[v_{i-n}, \bar{v}_{i-n}]$	$=$	$[x_i, \bar{x}_i]$	i	$=$	$1 \ldots n$
v_i	$\rightarrowtail$	$\circ\!\circ$	i	$=$	$1 \ldots l$
v_i	$=$	$\varphi_i(u_i)$			
y_{m-i}	$=$	v_{l-i}	i	$=$	$m-1 \ldots 0$
$\bar{v}_{l-i}$	$=$	$\bar{y}_{m-i}$	i	$=$	$0 \ldots m-1$
v_i	$\leftarrowtail$	$\circ\!\circ$	i	$=$	$l \ldots 1$
$\bar{t}_i$	$=$	$\bar{\varphi}_i(u_i, \bar{v}_i)$			
$\bar{v}_i$	$=$	0			
$\bar{u}_i$	$+=$	$\bar{t}_i$			
$\bar{x}_i$	$=$	$\bar{v}_{i-n}$	i	$=$	$n \ldots 1$

Table 4.8: Reverse Operations of Elementals with Restores

Forward	Return	
$v \rightarrowtail \circ\!\circ\,;\quad v = c$	$v \leftarrowtail \circ\!\circ\,;$	$\bar{v} = 0$
$v \rightarrowtail \circ\!\circ\,;\quad v = u + w$	$\bar{t} = \bar{v}\,;$	$\bar{v} = 0$
	$\bar{u} \mathrel{+}= \bar{t}\,;$	$\bar{w} \mathrel{+}= \bar{t}$
	$v \leftarrowtail \circ\!\circ$	
$v \rightarrowtail \circ\!\circ\,;\quad v = u * w$	$v \leftarrowtail \circ\!\circ\,;$	$\bar{t} = \bar{v}$
	$\bar{v} = 0\,;$	$\bar{u} \mathrel{+}= \bar{t} * w$
	$\bar{w} \mathrel{+}= \bar{t} * u$	
$v \rightarrowtail \circ\!\circ\,;\quad v = 1/u$	$\bar{t} = (\bar{v} * v) * v\,;$	$\bar{v} = 0$
	$\bar{u} \mathrel{-}= \bar{t}\,;$	$v \leftarrowtail \circ\!\circ$
$v \rightarrowtail \circ\!\circ\,;\quad v = \sqrt{u}$	$\bar{t} = 0.5 * \bar{v}/v\,;$	$\bar{v} = 0$
	$\bar{u} \mathrel{+}= \bar{t}\,;$	$v \leftarrowtail \circ\!\circ$
$v \rightarrowtail \circ\!\circ\,;\quad v = u^c$	$\bar{t} = (\bar{v} * c) * v\,;$	$v \leftarrowtail \circ\!\circ$
	$\bar{v} = 0\,;$	$\bar{u} \mathrel{+}= \bar{t}/u$
$v \rightarrowtail \circ\!\circ\,;\quad v = \exp(u)$	$\bar{t} = \bar{v} * v\,;$	$\bar{v} = 0$
	$\bar{u} \mathrel{+}= \bar{t}\,;$	$v \leftarrowtail \circ\!\circ$
$v \rightarrowtail \circ\!\circ\,;\quad v = \log(u)$	$v \leftarrowtail \circ\!\circ\,;$	$\bar{t} = \bar{v}/u$
	$\bar{v} = 0\,;$	$\bar{u} \mathrel{+}= \bar{t}$
$v \rightarrowtail \circ\!\circ\,;\quad v = \sin(u)$	$v \leftarrowtail \circ\!\circ\,;$	$\bar{t} = \bar{v} * \cos(u)$
	$\bar{v} = 0\,;$	$\bar{u} \mathrel{+}= \bar{t}$

Now suppose the precedence relation $\prec$ is completely arbitrary in that v_i may even depend on v_j with $j > i$. Because we assume that all v_j have initially some default value, the upper half of Table 4.7 can still be executed unambiguously no matter how the v_i are allocated. However, while allowing any addressing function & for the v_i, we assume that the n independents x_i and the m dependents y_i are separately allocated scalar variables. Even under these rather weak assumptions the adjoint procedure remains consistent in the following sense.

Proposition 4.2 (General Consistency of Adjoints)
Suppose all φ_i are once continuously differentiable on some neighborhood of their current argument. Then the combined forward and return sweep specified by Table 4.7 is consistent in that

$$\bar{x}^\top += \nabla_x \left[\bar{y}^\top y(x)\right]$$

whenever the allocation of the v_i and their adjoints $\bar{v}_i$ is consistent in that

$$\&v_i = \&v_j \iff \&\bar{v}_i = \&\bar{v}_j \ .$$

Proof. We prove the assertion by reinterpreting the computation according to Table 4.7 such that it satisfies our standard forward compatibility assumption (4.2). Let us first replace all components v_j of some argument u_i by the v_k with maximal $k < i$ for which $\&j = \&k$. Since v_j and v_k share the same location they must also have the same value at all stages of the calculation. If no such k exists the initial value of v_j enters into the calculation by default and we may define one extra "artificial" independent variables x_i with $i < 1$ and the corresponding v_{i-n} with the common location $\&(i-n) = \&j$ for each such j. Hence we formalize their initialization as $v_{i-n} = x_i$ with $i < 1$ and must also add the corresponding adjoint assignment $\bar{x}_i += \bar{v}_{i-n}$. These values are of no interest regarding the assertion of the theorem and have no impact on the adjoints of the original independents. By the renumbering and the introduction of artificial variables we have ensured forward compatibility and also that $j \prec i$ implies $j < i$. However, some or even all assignments may still be iterative in that $j \prec i$ and $\&j = \&i$. Thus we may introduce separately allocated temporary variables t_i of the same dimension as v_i and replace $v_i = \varphi(u_i)$ by the pair of assignments $t_i = \varphi(u_i); v_i = t_i$. Adjoining these yields exactly the four statements in the lower part of Table 4.7, assuming again that t_i play no role before or afterward so that the prevalue of t_i need not be saved and the adjoint $\bar{t}_i$ need not be incremented. ■

Proposition 4.2 means that adjoints can always be generated line by line without any need to consider dependencies between program variables in a larger context in order to ensure correctness. Hence, we obtain

Rule 5

Adjoining can be performed line by line irrespective of dependencies and aliasing.

Naturally, nonlocal analysis may be quite useful to improve efficiency.

4.3 Selective Saves and Restores

According to Table 4.4, one *prevalue* must be stored and then later retrieved for each elemental function, including simple additions and multiplications. Since these arithmetic operations are executed extremely fast on modern computer platforms, the transfer of data from and to the tape may slow down both sweeps of the adjoint procedure significantly. As we observe at the end of section 4.6 the whole computation may become memory bound even though we have ensured that the access pattern is strictly sequential in both directions. Whether this problem occurs depends strongly on the system architecture and in particular on the way in which the tape is implemented. Here, we will discuss another obvious remedy, namely, reducing the amount of data that needs to be stored and retrieved.

Consider the following function:

$$f(x) = \exp(x_1 * x_n) + \sum_{i=2}^{n} x_i * x_{n-i+1} \, .$$

This convolution example can be evaluated by the procedure listed in Table 4.9. Here we have adhered to our notational convention of giving each left-hand side a new variable name, but the intermediates v_i for $i = 1 \ldots n$ are all assumed to be mapped into the same location. Otherwise, the computation is not even well defined, due to the incremental nature of the loop body.

Table 4.9: Adjoint of Convolution with $\&v_i = \&v_1$, $\&\bar{v}_i = \&\bar{v}_1$ for $i > 0$

Forward :	Return :
for $i = 1 \ldots n$	$\bar{v}_{n+1} = \bar{y}$
$\quad v_{i-n} = x_i$	for $i = n \ldots 2$
$v_1 = v_0 * v_{1-n}$	$\quad \bar{v}_{1-i} \mathrel{+}= \bar{v}_{i+1} * v_{i-n}$
$v_2 = \exp(v_1)$	$\quad \bar{v}_{i-n} \mathrel{+}= \bar{v}_{i+1} * v_{1-i}$
for $i = 2 \ldots n$	$\quad v_{i+1} \mathrel{-}= v_{1-i} * v_{i-n}$
$\quad v_{i+1} \mathrel{+}= v_{i-n} * v_{1-i}$	$\bar{v}_1 \mathrel{+}= \bar{v}_2 * v_2$
$y \; = v_{n+1}$	$\bar{v}_0 \mathrel{+}= \bar{v}_1 * v_{1-n}$
	$\bar{v}_{1-n} \mathrel{+}= \bar{v}_1 * v_0$
	for $i = n \ldots 1$
	$\quad \bar{x}_i = \bar{v}_{i-n}$

Now we claim that the right-hand side of Table 4.9 represents an appropriate return sweep even though no intermediate variables have been saved, and the scalar function $y = f(x)$ is certainly nonlinear. Normally, we would have to put the (pre)values $v_1 \ldots v_n$ on the tape before they are overwritten by the variables $v_2 \ldots v_{n+1}$, respectively, since all of them reside in the same location. Fortunately, v_1 is not needed and v_2, like the other v_i with $i \geq 2$, can be recomputed on the way back. Let us examine these assertions a bit more closely.

As we have already noted in Table 4.8, the adjoint operation associated with the exponential and some other elementary functions involves only the value (here v_2) but not the argument (here v_1). On the other hand, v_1 need not be available as the value of the previous statement $v_1 = v_0 * v_{1-n}$ either, because the adjoint of the multiplication operation is defined in terms of the arguments (here the independent variables v_0 and v_{1-n}). For the same reasons we need not save the product values $v_{1-i} * v_{i-n}$ for $i = 2 \ldots n$, even if they were placed into some named local variable. Here we have effectively treated $v \mathrel{+}= \varphi(u, w)$ with $\varphi(u, w) = u * w$ as a single incremental operation. As for ordinary multiplications, the adjoint operation needs the arguments u and w (here also independent variables). Neither the old value of v, which may be considered a hidden argument, nor the new value of v is needed for adjoining the statement $v \mathrel{+}= \varphi(u, w)$. They need not be stored because they occur only linearly. In general we only have to worry about nonlinear arguments of a function and obtain the following rule for selective storage.

RULE 6

SAVE ONLY THE VALUES OF POWERS OR EXPONENTIALS
AND THE ARGUMENTS OF OTHER NONLINEAR ELEMENTALS.

Specifically, an intermediate value need only be available on the way back if it occurs as the result of a reciprocal, square root, power, or exponential, or if it occurs as a nonlinear argument in some other elemental function. The decision of whether and when a variable needs to be available because either of the two conditions given above is satisfied can only rarely be made locally. Of course, variables could be saved whenever and wherever either of the two conditions is satisfied. However, then variables might be stored repeatedly, which defeats the purpose of our whole space-saving exercise.

A simple runtime solution is to append a status flag to each memory location. This flag is initialized according to the first "value" criterion when a new value is computed and updated according to the second "argument" criterion when the value appears as an argument. Whenever a location is overwritten the status flag determines whether the prevalue must be saved on the tape. Otherwise, a special token can be used to indicate that the variable need not be restored on the way back. For efficiency one may emulate this decision process at compile-time using a technique known as abstract interpretation. Naturally, one could theoretically combine both approaches by keeping status flags only for those variables whose status could not be satisfactorily resolved at compile-time.

In a related approach of Hascoët et al. [HNP05], the number of stored prevalues is reduced by a "to-be-recorded" attribute that is determined for each intermediate variable v_i.

Inverse Computation versus Recomputation

We have endowed the convolution example above with a little nonlinear twist in the form of the exponential. We need its value v_2 for the adjoint operation

$\bar{v}_1 += \bar{v}_2 * v_2$, yet did not bother to save it. In the reverse sweep on the right-hand side of Table 4.9, the variable v_2 is recomputed by recalculating the variables v_i backward from the final v_{n+1} left over from the forward sweep. This purpose is achieved by the inverse statements $v -= \varphi(u, w)$, which, due to their decremental nature, should generally be numerically quite safe and cheap. Unfortunately, there is a potential for catastrophic cancellation in special circumstances, as we shall see in Exercise 4.5.

A list of iterative operations and their adjoints with inversion is given in Table 4.10. These adjoint operations can be used extensively for the gradient calculation of the well known function $f(X) = \ln|\det(X)|$ when the computation of determinant is based on an LU factorization. This will be examined in more detail in Exercise 4.6.

Table 4.10: Forward and Reverse of Incrementals with Inverses

Forward			Reverse								
v	$+=$	u	$\bar{u}$	$+=$	$\bar{v}$;	v	$-=$	u			
v	$-=$	u	$\bar{u}$	$-=$	$\bar{v}$;	v	$+=$	u			
v	$*=$	c	$\bar{v}$	$*=$	c ;	v	$/=$	c			
v	$=$	$1/v$	$\bar{v}$	$*=$	$-v * v$;	v	$=$	$1/v$			
v	$+=$	$u * w$	$\bar{u}$	$+=$	$\bar{v} * w$;	$\bar{w}$	$+=$	$\bar{v} * u$;	v	$-=$	$u * w$
v	$*=$	u	v	$/=$	u ;	$\bar{u}$	$+=$	$\bar{v} * v$;	$\bar{v}$	$*=$	u
v	$/=$	u	$\bar{v}$	$/=$	u ;	$\bar{u}$	$-=$	$\bar{v} * v$;	v	$*=$	u
v	$=$	$\sqrt{v}$	$\bar{v}$	$*=$	$0.5/v$;	v	$=$	$v * v$			

The idea of reducing tape storage by actually inverting certain floating point calculations was credited by Shiriaev [Shi93] to Matiyasevich [Mat86]. Apparently it has not been used extensively, probably because of the uncertain gains and for fear of numerical inaccuracies. These contingencies cannot arise when intermediate quantities are recomputed forward by using exactly the same sequence of elemental operations as in the original calculation. In the return sweep on the convolution example we can delete the inverse computation $v_{i+1} -= v_{1-i} * v_{i-n}$ within the loop and instead insert—right after loop completion—the two statements $v_1 = v_0 * v_{1-n}$ and $v_2 = \exp(v_1)$. This modification would eliminate any potential for numerical inconsistencies and also reduce the runtime when n is sufficiently large.

Already, even in this puny example, it becomes clear that there is a wide range of possibility for treating individual variables. Preferably at compile-time, but possibly only during the (first) forward sweep, each variable must be placed into one of the following categories of variables:

- not needed backwards
- saved on the tape
- recomputed forward
- recomputed inverse

Obviously, the "best" classification of individual variables cannot, in general, be decided independently of their neighbors in the computational graph. In fact, we have implicitly added a third condition for being needed on the way back, namely, being required for restoring other needed variables either through forward or inverse recomputations. Many current AD tools employ rather simple strategies to avoid getting bogged down in too many choices. Most of them save on the tape anything that looks like it might possibly be needed on the way back. Forward recomputation is used extensively in the reversal of evolutionary simulations by checkpointing, as discussed in Chapter 12. It is also employed for vertical checkpointing, where subroutines are executed repeatedly in forward recomputation mode and once in adjoint mode. Then their status is effectively no longer static but changes depending on the call mode of the subroutines. For adjoining the propagation of seismic waves within a certain domain it is sufficient to tape the evolution of values on the boundary as the strictly elastic wave equation is reversible in the interior of the domain [Coe01].

4.4 A Temporal Complexity Model

One key advantage of AD techniques is that they allow a priori bounds on the cost of evaluating certain derivative objects in terms of the cost of evaluating the underlying function itself. Such bounds can be derived quite easily if computational cost is measured simply by counting additions, multiplications, and other arithmetic operations. Unfortunately, this classical complexity measure is increasingly inappropriate on modern computer systems, where accesses to the memory hierarchy play a crucial role in determining the runtime at various stages of a calculation. Moreover, vectorization and parallelism in superscalar CPUs or multiprocessor systems make any realistic complexity modeling quite difficult and, in any case, highly hardware- and software-dependent.

In this section we develop a flexible but slightly naive concept of vector-valued complexity measures. They are assumed to be additive with regard to the execution of evaluation procedures and subadditive with respect to other, related tasks. Our intention is to provide a framework for complexity analysis which can be extended and adapted to more realistic measures in specific contexts. By appropriately weighting the components of our vector-valued complexity we can derive the scalar measure $TIME$ already used in sections 3.1 and 3.2. Users who are mainly interested in the actual algorithms for computing derivatives may prefer to skip the remainder of this chapter on a first reading.

Additive Tasks and Complexities

Let us examine the complexity of certain computational tasks in connection with the composite function F, which we will denote as $task(F)$. Throughout, the basic task $eval(F)$ of evaluating F at a given argument will serve as a reference; that is, we shall bound the cost of the derived computational tasks by multiples of the cost for evaluating the composite function.

The sequential execution of an evaluation procedure is an additive method in that each elemental function needs to be evaluated exactly once at a corresponding argument. The fundamental methods of AD can also be performed in an "element-by-element" fashion, and we will therefore also refer to them as *additive*. The advantage of additive tasks is that they require little knowledge about the interaction between the various elements and yield rather simple complexity bounds. Certain overall results, like the computation of the complete Jacobian F', can be formulated alternatively as additive or nonadditive tasks. The latter versions are potentially much more efficient in terms of operation counts, but tend to involve more overhead and may suffer from less regular memory access patterns (see Chapter 9).

For each additive task $task(F)$, we will denote the related task that needs to be performed for each elemental function by $task'(\varphi)$. Typically $task'$ is not exactly the same as $task$ because some additional merging of elemental results is required. When $task'$ is exactly the same as $task$ we refer to the computational task as homogeneous. To get reasonably sharp and simple bounds, we shall assume that additive tasks have (sub)additive complexity, with evaluation being strictly additive.

Assumption (TA): TEMPORAL ADDITIVITY OF TASK

$$\begin{aligned} WORK\{task(F)\} &\leq \sum_{i=1}^{l} WORK\{task'(\varphi_i)\}\ , \\ WORK\{eval(F)\} &= \sum_{i=1}^{l} WORK\{eval(\varphi_i)\}\ . \end{aligned}$$

Here $WORK\{task(F)\} \in \mathbb{R}^{\dim(WORK)}$ may denote any vector (or scalar) measure of temporal complexity for performing some computational task on the vector function $F \in \mathcal{F}$ defined in (2.11). For example, the $\dim(WORK)$ components may separately count the number of multiplications, additions, and accesses to certain levels of the memory hierarchy. Our only requirement is that the components be nonnegative and that the additivity Assumption (TA) be satisfied.

A Four-Component Complexity Measure

Even though this four-component measure represents a rather crude point of view, we will use it for illustrative purposes throughout the book

$$WORK\{task\} \equiv \begin{pmatrix} MOVES \\ ADDS \\ MULTS \\ NLOPS \end{pmatrix} \equiv \begin{pmatrix} \text{\#of fetches and stores} \\ \text{\#of additions and subtractions} \\ \text{\#of multiplications} \\ \text{\#of nonlinear operations} \end{pmatrix} . \quad (4.6)$$

Since the division usually takes a lot longer than the other arithmetic operation, we treat it as a reciprocal followed by a multiplication. The reciprocal is counted among the nonlinear operations, which consist otherwise of intrinsic function evaluations such as exponentials and trigonometrics. Probably much more important for most calculations is the number of accesses to various parts of the memory hierarchy, which is here assumed to be completely flat. In other words, our four-component complexity measure makes no distinction between reading and writing data from and to registers, caches, and remote storage devices. Hence, we use a simplified model of a random access machine [Tar83].

It is important to distinguish between an addition or multiplication as a computational cost and a polynomial operator as an elemental function. For example, the multiplication operator $w \equiv \varphi(u,v) = u * v$ is viewed as an elemental function involving two fetches (of u, v) and one store (of w), so that in this case

$$WORK\{eval(\varphi)\} = (3,0,1,0)^\top .$$

Similarly, we attribute one fetch and one store to the univariate functions so that, for example, $WORK\{eval(\psi)\} = (2,0,0,1)^\top$ for $\psi(u) = \sin(u)$. Again, the concept of an elemental operator, or function, $v = \varphi(u)$ encompasses more than just its properties as a mathematical mapping. With c representing initialization to a constant, we obtain the complexities listed in Table 4.11 for the basic computational task of evaluation.

The fact that the table entries could be arranged to form the unitary upper triangular matrix W_{eval} is incidental but (we hope) makes it a little easier to interpret and remember. For any function F whose evaluation procedure is made up of l_1 initializations, l_2 additions, l_3 multiplications, and l_4 intrinsics, we obtain by Assumption (TA) the linear relation

Table 4.11: Elementary Complexities

eval	c	$\pm$	$*$	ψ
MOVES	1	3	3	2
ADDS	0	1	0	0
MULTS	0	0	1	0
NLOPS	0	0	0	1

$$WORK\{eval(F)\} \equiv W_{eval}\,|F| \equiv \begin{bmatrix} 1 & 3 & 3 & 2 \\ 0 & 1 & 0 & 0 \\ 0 & 0 & 1 & 0 \\ 0 & 0 & 0 & 1 \end{bmatrix} \begin{bmatrix} l_1 \\ l_2 \\ l_3 \\ l_4 \end{bmatrix} . \tag{4.7}$$

Here $|F| \equiv (l_1, l_2, l_3, l_4)^\top$ might be called the elemental frequency vector of F. Later we shall consider corresponding matrices W_{task} for the evaluation of tangents, gradients, and other additive computational tasks. By comparing them to W_{eval}, one obtains a relative measure of computational complexity.

The Runtime Functional

In general, we would like to be able to make the vector complexity measure *WORK* largely independent of any particular computing platform and its current load. The actual runtime may then be estimated as a functional:

$$TIME\{task(F)\} \;=\; w^\top WORK\{task(F)\} . \tag{4.8}$$

Here w is a vector of dim($WORK$) positive weights, which are typically quite system-dependent. For example, the components of w may represent the number of clock cycles needed for a multiplication, for an addition, or for fetching and storing a data item at a certain level of the memory hierarchy. From Assumption (TA) we immediately derive

$$TIME\{eval(F)\} = \sum_{i=1}^{l} TIME\{eval(\varphi_i)\} \tag{4.9}$$

and

$$TIME\{task(F)\} \leq \sum_{i=1}^{l} TIME\{task'(\varphi_i)\} \tag{4.10}$$

which means that the runtime of evaluation is strictly additive and the runtime of other additive tasks is subadditive. In the case of the four-component measure we use additions as units and assume that

$$w^\top = (\mu, 1, \pi, \nu) \quad \text{with} \quad \mu \geq \pi \geq 1\,, \quad \text{and} \quad \nu \geq 2\pi\,. \tag{4.11}$$

In other words, we make the reasonable assumption that a nonlinear operation is more expensive than two multiplications and that a memory access is at least as slow as a multiplication which is in turn at least as slow as an addition. Because of the assumed nonnegativity of all complexity components, we may also view $TIME$ as a weighted L_1 norm of $WORK$. This interpretation allows us to use the associated induced matrix norms as scalar measures for the complexity of certain tasks relative to plain evaluation. First-time readers and those who are not concerned about the tightness and realism of complexity estimates are advised to assume dim($WORK$) $= 1$, so that $TIME = w\,WORK$ is simply some scalar complexity measure. For example, we might simply count all multiplicative operations, the classical unit in computational complexity theory.

Dividing (4.10) by (4.9), we obtain the important bound

$$\boxed{\begin{aligned} \frac{TIME\{task(F)\}}{TIME\{eval(F)\}} &\leq \frac{\sum_{i=1}^{l} TIME\{task'(\varphi_i)\}}{\sum_{i=1}^{l} TIME\{eval(\varphi_i)\}} \\ &\leq \sup_{\varphi \in \Psi} \frac{TIME\{task'(\varphi)\}}{TIME\{eval(\varphi)\}}\,. \end{aligned}} \tag{4.12}$$

Obviously the right-hand side must be bounded if Ψ consists of finitely many elemental functions. Then we have a uniform bound on the runtime of $task(F)$ relative to that of $eval(F)$ under Assumption (TA) on the task. In the following subsection we develop conditions for such boundedness for the vector complexity $WORK$ without assuming finiteness of Ψ.

Bounded Complexity

We say that a given *task* has *bounded complexity* on $\mathcal{F}$ as defined in (2.11) if there exists a square matrix C_{task} such that

$$WORK\{task(F)\} \leq C_{task}\, WORK\{eval(F)\} \quad \text{for all} \quad F \in \mathcal{F}\,, \tag{4.13}$$

where $\leq$ denotes componentwise ordering in the vector case. We call any such a matrix C_{task} a complexity bound of the task. Obviously evaluation itself has by definition bounded complexity with $C_{eval} = I$, the identity matrix of order $\dim(WORK)$.

For any given complexity measure $WORK$ and any particular task, the relation (4.13) is true for all elements in a convex set of square matrices. In any case, we (mostly) have to be content with upper bounds that are derived from a sequence of inequalities and, hence, far from minimal.

The property of bounded complexity is quite useful but by no means indispensable for a computational task. Suppose some computational task is not bounded on a given set $\mathcal{F} = \text{Span}[\Psi]$. Then the ratio between some component of $WORK\{task(F)\}$ and the same component of $WORK\{eval(F)\}$ can be arbitrarily large. In other words, performing the particular task on F may be infinitely more expensive than just evaluating F. This is not necessarily disastrous, however, since there might be a quadratic dependence rather than a linear bound, which we assume for tasks with bounded complexity.

For example, evaluating a full Jacobian matrix is not a task with bounded complexity, even if Ψ is restricted to the polynomial core. However, it will be seen that the task of evaluating a certain number of p columns or q rows of the Jacobian has bounded complexity. Here p or q must be viewed as a fixed parameter of the task. These results yield estimates for the complexity of the full Jacobian that are proportional to n, the number of independent variables, or m, the number of dependent variables. Under the reasonable assumption that the components of $WORK\{eval(F)\}$ are bounded below by some multiple of n or m over all $F \in \mathcal{F}$, we conclude that $WORK\{eval(F')\}$ is bounded by some quadratic form in $WORK\{eval(F)\}$. To mark the distinction to such polynomial complexity growth, we could have labeled our concept of *bounded complexity* more specifically as *linearly bounded complexity.*

Boundedness and Composites

Bounded complexity can be easily derived for all additive tasks. Suppose a particular task has bounded complexity on some library Ψ. This assumption is trivially satisfied if Ψ is finite, or at least the set of value pairs

$$[WORK\{eval(\varphi)\}, WORK\{task'(\varphi)\}]_{\varphi\in\Psi} \tag{4.14}$$

is finite. This slight generalization is necessary because, strictly speaking, even our polynomial core library is infinite, since it contains initialization to arbitrarily many real numbers. Naturally all these initializations are assumed to have identical complexity, so the set (4.14) is definitely finite in that case. As an immediate consequence we obtain the following property.

Assumption (EB): ELEMENTAL TASK BOUNDEDNESS

There exists a matrix $C_{task}(\Psi)$ such that

$$WORK\{task'(\varphi)\} \leq C_{task}(\Psi)\, WORK\{eval(\varphi)\} \quad \text{for all} \quad \varphi \in \Psi$$

In contrast to the finiteness of (4.14), this weaker assumption may still be satisfied when Ψ contains parameterized families of elemental functions, for example, inner products of variable dimension. Therefore, we will use it as our main assumption regarding the performance of a particular task on the elemental functions. Now we can easily derive bounded complexity.

Proposition 4.3 (BOUNDEDNESS OF ADDITIVE TASKS)

Assumptions (TA) *and* (EB) *imply bounded complexity on* $\mathcal{F}$ *in the sense of* (4.13) *with* $C_{task} = C_{task}(\Psi)$. *Moreover, we have for any runtime functional* (4.8) *and* $F \in \mathcal{F}$

$$TIME\{task(F)\} \leq \omega_{task}\, TIME\{eval(F)\}\ ,$$

where

$$\omega_{task} \equiv \max_{\varphi \in \Psi} \frac{w^\top WORK\{task'(\varphi)\}}{w^\top WORK\{eval(\varphi)\}} \leq \| D\, C_{task}\, D^{-1} \|_1\ ,$$

with $D \equiv diag(w)$ *and* $\| \ \|_1$ *denoting the* 1*-norm of matrices.*

Proof. By elementary inequalities and the nonnegativity of all vector components we find

$$\begin{aligned} WORK\{task(F)\} &\leq \sum_{i=1}^{l} WORK\{task'(\varphi_i)\} \\ &\leq \sum_{i=1}^{l} C_{task}(\Psi)\, WORK\{eval(\varphi_i)\} \\ &= C_{task}\, WORK\{eval(F)\}\ . \end{aligned}$$

As a result of the assumed additivity, we have, furthermore,

$$\begin{aligned} \frac{TIME\{task(F)\}}{TIME\{eval(F)\}} &\leq \frac{\sum_{i=1}^{l} TIME\{task'(\varphi_i)\}}{\sum_{i=1}^{l} TIME\{eval(\varphi_i)\}} \\ \leq \max_{1 \leq i \leq l} \frac{TIME\{task'(\varphi_i)\}}{TIME\{eval(\varphi_i)\}} &\leq \max_{\varphi \in \Psi} \frac{w^\top WORK\{task'(\varphi)\}}{w^\top WORK\{eval(\varphi)\}} \\ \leq \max_{\varphi \in \Psi} \frac{w^\top C_{task} WORK\{eval(\varphi)\}}{w^\top WORK\{eval(\varphi)\}} &\leq \max_{0 \neq z \in \mathbb{R}^{\dim(w)}} \frac{\| D\, C_{task} z \|_1}{\| D z \|_1} \end{aligned}$$

which concludes the proof by definition of $\| \ \|_1$. ∎

The proposition essentially says that the boundedness of an additive task on the library Ψ carries over to its span $\mathcal{F}$. This is stated differently in Rule 7.

RULE 7

> ADDITIVE TASKS WITH BOUNDED COMPLEXITY ON ELEMENTALS ARE ALSO BOUNDED ON COMPOSITE FUNCTIONS.

For any additive vector complexity and any corresponding runtime functional, we find that the penalty factor for performing a certain derived task compared to evaluation is uniformly bounded. In the case of the four-component complexity, we obtain from Proposition 4.3 for ω_{task} the convenient representation

$$\omega_{task} \;\equiv\; \max_{1\leq i\leq 4} \frac{w^\top W_{task}\, e_i}{w^\top W_{eval}\, e_i} \;, \tag{4.15}$$

where $e_i \in \mathbb{R}^4$ is the ith Cartesian basis vector.

Interval Evaluation as an Additive Task

To illustrate the additivity assumption and our complexity measure, we consider the task of evaluating F on an interval. All variables v are replaced by closed intervals $[\inf v, \sup v]$, and unary operators $v = \psi(u)$ are extended so that

$$\psi(u) \in [\inf v, \sup v] \quad \text{for all} \quad u \in [\inf u, \sup u]\;.$$

This inclusion may be an overestimate in that the lower bound $\inf v$ and the upper bound $\sup v$ are not as tight as possible. Similarly, one finds for the additive operators

$$\begin{aligned}
[\inf u, \sup u] + [\inf v, \sup v] &= [\inf u + \inf v, \sup u + \sup v]\\
[\inf u, \sup u] - [\inf v, \sup v] &= [\inf u - \sup v, \sup u - \inf v]
\end{aligned}$$

and for the multiplication operator

$$\begin{aligned}
&[\inf u, \sup u] * [\inf v, \sup v]\\
=\; &[\min(\inf u * \inf v, \inf u * \sup v, \sup u * \inf v, \sup u * \sup v),\\
&\max(\inf u * \inf v, \inf u * \sup v, \sup u * \inf v, \sup u * \sup v)]\;.
\end{aligned}$$

Again, we interpret division as a multiplication following the univariate operator $\text{rec}(u) = 1/u$. For our purposes we need not be concerned with the difficulties that arise when an interval containing the origin occurs in the denominator.

For a detailed discussion of interval arithmetic, see [Moo79] and [Moo88]. Although this may lead to rather sloppy bounds, interval evaluation is usually performed as an additive task, which we will denote with the subscript *intev*. As discussed in [KM81] and [AH83], we have approximately the complexities listed in Table 4.12, where the ψ in the last line represents any univariate function from Ψ.

The matrix C_{intev} on the left is the relative complexity obtained by multiplying the matrix W_{intev} from the right by W_{eval}^{-1}, as defined in (4.7):

$$W_{intev} = \begin{bmatrix} 2 & 6 & 6 & 4 \\ 0 & 2 & 4 & 0 \\ 0 & 0 & 4 & 0 \\ 0 & 0 & 0 & 2 \end{bmatrix},$$

$$C_{intev} = \begin{bmatrix} 2 & 0 & 0 & 0 \\ 0 & 2 & 4 & 0 \\ 0 & 0 & 4 & 0 \\ 0 & 0 & 0 & 2 \end{bmatrix}.$$

Table 4.12: Interval Complexity

intev	c	$\pm$	$*$	ψ
MOVES	2	6	6	4
ADDS	0	2	4	0
MULTS	0	0	4	0
NLOPS	0	0	0	2

Finally, we obtain for any runtime vector $w = (\mu, 1, \pi, \nu)$ satisfying (4.11) with $D = diag(w)$

$$\|DC_{intev}\, D^{-1}\|_1 = \left\| \begin{matrix} 2 & 0 & 0 & 0 \\ 0 & 2 & 4/\pi & 0 \\ 0 & 0 & 4 & 0 \\ 0 & 0 & 0 & 2 \end{matrix} \right\|_1 = \max\{2, 4(1+1/\pi)\} \leq 8\,.$$

The actual value of ω_{intev} as given by (4.15) is

$$\omega_{intev} = \max\left\{\frac{2\mu}{\mu}, \frac{6\mu+2}{3\mu+1}, \frac{6\mu+4+4\pi}{3\mu+\pi}, \frac{4\mu+2\nu}{2\mu+\nu}\right\} = \frac{6\mu+4+4\pi}{3\mu+\pi} \leq 4\,.$$

The gap between ω_{intev} and its bound $\|D\,C_{intev}D^{-1}\|_1$ is here quite significant. It is caused by the fact that all arithmetic operations occur jointly with memory moves, which somewhat reduces the penalty for replacing single floating point values with intervals. A similar but less pronounced effect occurs for the derivative calculations.

4.5 Complexity of Tangent Propagation

A simple but powerful observation about computing $\dot{y} = F'(x)\dot{x}$ via Table 3.4 by evaluating $[F(x), \dot{F}(x,\dot{x})]$ is that exactly the same task, namely, evaluating $[\varphi_i(u_i), \dot{\varphi}_i(u_i, \dot{u}_i)]$, must be performed for each elemental function once. In our terminology this means that (forward) tangent propagation is an additive and even homogeneous computational task, which we will denote by the label *tang*. Hence we may apply the reasoning discussed in section 4.4 about computational complexity to conclude that tangent propagation has bounded complexity in that for some square matrix $C_{tang}(\Psi)$ of order $\dim(WORK)$ and all $F \in \mathcal{F}$,

$$WORK\{tang(F)\} \leq C_{tang}(\Psi)\, WORK\{eval(F)\}\,.$$

To obtain a specific matrix $C_{tang} = C_{tang}(\Psi)$, we consider again the four-component complexity measure $WORK$ introduced in (4.6), which accounts separately for the numbers of moves (fetches or stores), additions, multiplications, and nonlinear operations/intrinsics.

As in the case of interval evaluation we now derive separate complexities for the four elemental operator classes $c, \pm, *$, and ψ, where ψ may be any nonlinear univariate function. Considering evaluation of the elemental functions as part of the tangent task, we obtain Table 4.13. Constant initialization and the corresponding tangent operation $\dot{v} = 0$ both require a single store. All binary arithmetic operators involve one floating point operation and three memory accesses to fetch the two arguments and store the one result. For the addition and subtraction operators the corresponding tangent operator $\dot{v} = \dot{u} \pm \dot{w}$ has exactly the same complexity. In the case of the multiplication operator $v = u * w$, the tangent operator $\dot{v} = w * \dot{u} + u * \dot{w}$ involves three floating point operations and five memory accesses, including the fetching of u and v, which is already done for the calculation of w itself. Finally, in the case of nonlinear operators like $v = \sin(u)$, the tangent operator $\dot{v} = \cos(u) * \dot{u}$ involves one nonlinear operation, one multiplication, one extra fetch, and one extra store.

Table 4.13: Elemental Tangent Complexity

$tang$	c	$\pm$	$*$	ψ
$MOVES$	$1+1$	$3+3$	$3+3$	$2+2$
$ADDS$	0	$1+1$	$0+1$	$0+0$
$MULTS$	0	0	$1+2$	$0+1$
$NLOPS$	0	0	0	$1+1$

Table 4.14: Original and Tangent Operations with Associated Work ($MOVES$, $ADDS$, $MULTS$, $NLOPS$)

Original	Work	Tangent	Work
$v = c$	$(1,0,0,0)$	$\dot{v} = 0$	$(2,0,0,0)$
$v = u \pm w$	$(3,1,0,0)$	$\dot{v} = \dot{u} \pm \dot{w}$	$(6,2,0,0)$
$v = u * w$	$(3,0,1,0)$	$\dot{v} = w * \dot{u} + u * \dot{w}$	$(6,1,3,0)$
$v = \psi(u)$	$(2,0,0,1)$	$\dot{v} = \psi'(u) * \dot{u}$	$(4,0,1,2)$

Table 4.14 illustrates the associated work for the tangent propagation in a slightly different way. Here we have split all the entries into the count for plain evaluation and the extra effort for the first derivative calculation. Now C_{tang} can be determined as $W_{tang}W_{eval}^{-1}$, with W_{eval} as defined in (4.7) and W_{tang} given by Table 4.13:

$$W_{tang} \equiv \begin{bmatrix} 2 & 6 & 6 & 4 \\ 0 & 2 & 1 & 0 \\ 0 & 0 & 3 & 1 \\ 0 & 0 & 0 & 2 \end{bmatrix}, \qquad C_{tang} \equiv \begin{bmatrix} 2 & 0 & 0 & 0 \\ 0 & 2 & 1 & 0 \\ 0 & 0 & 3 & 1 \\ 0 & 0 & 0 & 2 \end{bmatrix}. \tag{4.16}$$

To derive a scalar complexity bound, we assume that the runtime is determined by a vector of weights $w^\top = (\mu, 1, \pi, \nu)$ satisfying the inequalities (4.11). Then we find, according to (4.15),

$$\omega_{tang} = \max\left\{\frac{2\mu}{\mu}, \frac{6\mu+2}{3\mu+1}, \frac{6\mu+1+3\pi}{3\mu+\pi}, \frac{4\mu+\pi+2\nu}{2\mu+\nu}\right\} \in [2, 5/2]\,. \tag{4.17}$$

An effort of a bit more than two function evaluations for computing one directional derivative in addition to the function itself seems quite reasonable because the same cost is incurred if one uses a divided-difference approximation

$$\dot{y} \approx [F(x+\varepsilon\dot{x}) - F(x)]/\varepsilon + \mathcal{O}(\varepsilon)\,.$$

However, the cost per divided difference is almost halved if one "shoots" divided differences in several (say $p \geq 1$) directions $\dot{x}$ from the same base point x. Correspondingly, one may propagate p tangents simultaneously using the vector forward mode, a task we denote by the label *tangp*. Now we obtain Table 4.15 and the complexity matrices

Table 4.15: Tangent Vector Complexity

tangp	c	$\pm$	$*$	ψ
MOVES	$1+p$	$3+3p$	$3+3p$	$2+2p$
ADDS	0	$1+p$	$0+p$	$0+0$
MULTS	0	0	$1+2p$	$0+p$
NLOPS	0	0	0	2

$$W_{tangp} \equiv \begin{bmatrix} 1+p & 3+3p & 3+3p & 2+2p \\ 0 & 1+p & p & 0 \\ 0 & 0 & 1+2p & p \\ 0 & 0 & 0 & 2 \end{bmatrix},$$

$$C_{tangp} \equiv \begin{bmatrix} 1+p & 0 & 0 & 0 \\ 0 & 1+p & p & 0 \\ 0 & 0 & 1+2p & p \\ 0 & 0 & 0 & 2 \end{bmatrix}.$$

The runtime ratio is, according to (4.15),

$$\omega_{tangp} = \max\left\{\frac{(1+p)\mu}{\mu}, \frac{(1+p)(3\mu+1)}{3\mu+1}, \frac{3(1+p)\mu+p+(1+2p)\pi}{3\mu+\pi}, \right.$$
$$\left.\frac{2(1+p)\mu+p\pi+2\nu}{2\mu+\nu}\right\} \in [1+p, 1+1.5p]\,. \tag{4.18}$$

This suggests that even in vector mode, tangent propagation may be about 50% more expensive than differencing. However, this is definitely the worst-case scenario. If memory accesses dominate the runtime in that $\mu \gg \pi$ or there are not many multiplications compared with additions and nonlinear intrinsics, then the ratio is close to p. Also, on-chip parallelism may hide some extra arithmetic operations from the runtime.

Experimental timings for propagating vectors of $p = 1\ldots 27$ derivative components through the code of a nuclear reactor model are shown in Fig. 4.1.

The original Fortran code was preprocessed by the differentiation tool ADIFOR [BC+96], which generates portable Fortran codes to calculate derivatives. Evaluation of the Jacobian product by AD as a vector of 27 directional derivatives takes only half of the runtime of the divided-difference method. Furthermore, there is only a surprisingly small increase of the memory requirement compared with the undifferentiated code, when $p = 0$.

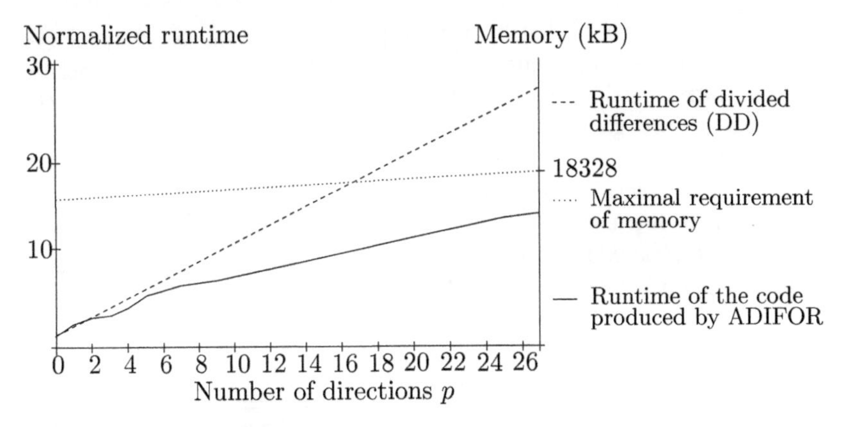

Figure 4.1: Runtime Comparison of the Divided Difference Method and AD on Reactor Model ATHLET [Ges95]

In general the random access requirement of vector-forward is

$$RAM\{tangp(F)\} = (1 + p)RAM\{eval(F)\} ,$$

where RAM is defined in section 4.1. Of course this proportionality applies only to the code that actually evaluates F.

4.6 Complexity of Gradient Propagation

In this section we consider the complexity of the simultaneous evaluation of $F(x)$ and $\bar{F}(x, \bar{y})$, a task we denote by $grad(F)$. By inspection of Table 4.4 we observe that $grad(F)$ requires performing the following subtasks for each element φ_i. On the way forward, v_i must be stored on the tape and $\varphi_i(u_i)$ evaluated. On the way back $\bar{\varphi}_i(u_i, \bar{v}_i)$ needs to be evaluated and its result incremented to $\bar{u}_i$ and, finally, v_i restored from the tape. We denote the combined effort for the two tape operations by $WORK\{v_i \leftrightarrow \text{QO}\}$. This cost may be set to zero if there is no overwriting. We denote by $WORK\{\bar{\varphi}_i(u_i, \bar{v}_i)\}$ the additional effort required to calculate $\bar{\varphi}_i(u_i, \bar{v}_i)$ and increment this to $\bar{u}_i$, on the assumption that access is available to an evaluation of $\varphi_i(u_i)$, including all intermediate variable values. In any case, we denote the combination of the three subtasks discussed above as $grad'(\varphi_i)$ and set

$$\begin{aligned} & WORK\{grad'(\varphi_i)\} \\ = \;& WORK\{\varphi_i(u_i)\} + WORK\{\bar{\varphi}_i(u_i, \bar{v}_i)\} + WORK\{v_i \leftrightarrow \text{QO}\} . \end{aligned} \tag{4.19}$$

Since the task *grad* is additive as defined in Assumption (TA), Proposition 4.3 ensures that for some bounding matrix $C_{grad}(\Psi)$ of order $\dim(WORK)$ and all $F \in \mathcal{F}$,

$$WORK\{grad(F)\} \leq C_{grad}(\Psi)\, WORK\{eval(F)\}\,.$$

To obtain a specific matrix $C_{grad}(\Psi)$, we consider again the four-component complexity measure introduced in (4.6). As in the case of tangent propagation we examine four basic operators: constant initialization, addition/subtraction, multiplication, and nonlinear intrinsic with their computational cost as listed in Table 4.16. For the time being we neglect the last term in (4.19), which will be considered separately at the end of this section.

Table 4.16: Original and Adjoint Operations with Associated Work (*MOVES*, *ADDS*, *MULTS*, *NLOPS*)

Operation	Work
$v = c$	$(1,0,0,0)$
$\bar{v} = 0$	$(2,0,0,0)$
$v = u \pm w$	$(3,1,0,0)$
$\bar{u} \mathrel{+}= \bar{v}$; $\bar{w} \mathrel{\pm}= \bar{v}$; $\bar{v} = 0$	$(9,3,0,0)$
$v = u * w$	$(3,0,1,0)$
$\bar{u} \mathrel{+}= \bar{v} * w$; $\bar{w} \mathrel{+}= \bar{v} * u$; $\bar{v} = 0$	$(11,2,3,0)$
$v = \psi(u)$	$(2,0,0,1)$
$\bar{u} \mathrel{+}= \bar{v} * \psi'(u)$; $\bar{v} = 0$	$(7,1,1,2)$

Constant initialization and the corresponding adjoint operation $\bar{v} = 0$ both require a single store. All binary arithmetic operators involve one floating point operation and three memory accesses to fetch the two arguments and store the one result. For the addition or subtraction operator $v = u \pm w$, the corresponding adjoint operator

$$\bar{u} \mathrel{+}= \bar{v}\,;\quad \bar{w} \mathrel{\pm}= \bar{v}\,;\quad \bar{v} = 0$$

involves three fetches, three stores, and two additions/subtractions. In the case of the multiplication operator $v = u * w$, the associated adjoint operator

$$\bar{u} \mathrel{+}= \bar{v} * w\,;\quad \bar{w} \mathrel{+}= \bar{v} * u\,;\quad \bar{v} = 0$$

involves two more fetches and two more multiplications than for an addition. Finally, in the case of nonlinear operators, such as $v = \sin(u)$, the adjoint operation pair $\bar{u} \mathrel{+}= \bar{v} * \cos(u); \bar{v} = 0$ involves one nonlinear operation, one multiplication, one addition, three fetches, and two stores.

Including evaluation of the φ_i on the way forward as part of the gradient task, we obtain Table 4.17. Now $C_{grad}(\Psi)$ can be determined as $W_{grad} W_{eval}^{-1}$, with W_{eval} as defined in (4.7):

$$W_{grad} \equiv \begin{bmatrix} 2 & 9 & 11 & 7 \\ 0 & 3 & 2 & 1 \\ 0 & 0 & 3 & 1 \\ 0 & 0 & 0 & 2 \end{bmatrix}, \qquad C_{grad} \equiv \begin{bmatrix} 2 & 3 & 5 & 3 \\ 0 & 3 & 2 & 1 \\ 0 & 0 & 3 & 1 \\ 0 & 0 & 0 & 2 \end{bmatrix}. \tag{4.20}$$

Again using w as defined in (4.11), we obtain, according to (4.15),

$$\omega_{grad} = \max\left\{\frac{2\mu}{\mu}, \frac{9\mu+3}{3\mu+1}, \frac{11\mu+2+3\pi}{3\mu+\pi}, \frac{7\mu+1+\pi+2\nu}{2\mu+\nu}\right\} \in [3,4]\,. \quad (4.21)$$

Under the assumption $\pi \geq 1$, but without any restriction on the memory access time μ, the bound grows to 5, a more conservative value that was used in the survey paper [GR89]. For the classical multiplicative measure that neglects additions and memory accesses altogether, the bound drops to the lower value 3, which has been used by many authors. That bound is exactly the same in the tangent mode.

A slight reduction can also be achieved if one implements the nonincremental adjoint recursion listed in Table 3.6. Then one saves the initialization of $\bar{v}_i$ to zero for all $i = 1-n, \ldots, l-m$, and the first incrementation becomes an assignment. As we will see in Exercise 4.4 this ensures the validity of the bound 4 without any restriction on μ, provided $\nu \geq \pi \geq 1$.

Table 4.17: Gradient Complexity

grad'	c	$\pm$	$*$	ψ
MOVES	$1+1$	$3+6$	$3+8$	$2+5$
ADDS	0	$1+2$	$0+2$	$0+1$
MULTS	0	0	$1+2$	$0+1$
NLOPS	0	0	0	$1+1$

Table 4.18: Complexity of Gradient Vector

gradq	c	$\pm$	$*$	ψ
MOVES	$1+q$	$3+6q$	$5+6q$	$3+4q$
ADDS	0	$1+2q$	$0+2q$	$0+q$
MULTS	0	0	$1+2q$	$0+q$
NLOPS	0	0	0	2

To amortize the effort for the forward sweep, let us now consider the task of propagating q gradients using the vector reverse mode, denoted by *gradq*. Here the amortization effects are larger than in the case of tangents because all the taping operations need to be performed only once, irrespective of q. We obtain the complexities listed in Table 4.18 and the following complexity matrices:

$$\begin{aligned} W_{gradq} &\equiv \begin{bmatrix} 1+q & 3+6q & 5+6q & 3+4q \\ 0 & 1+2q & 2q & q \\ 0 & 0 & 1+2q & q \\ 0 & 0 & 0 & 2 \end{bmatrix}, \\ C_{gradq} &\equiv \begin{bmatrix} 1+q & 3q & 2+3q & 1+2q \\ 0 & 1+2q & 2q & q \\ 0 & 0 & 1+2q & q \\ 0 & 0 & 0 & 2 \end{bmatrix}. \end{aligned} \quad (4.22)$$

Again using the assumptions (4.11), we obtain the scalar bound

$$\begin{aligned} \omega_{gradq} = \max\bigg\{&(1+q), (1+2q), \frac{(5+6q)\mu+2q+(1+2q)\pi}{3\mu+\pi}, \\ &\frac{(3+4q)\mu+q(1+\pi)+2\nu}{2\mu+\nu}\bigg\} \in [1+2q,\, 1.5+2.5q]\,. \end{aligned} \quad (4.23)$$

For additions and subtractions the ratio is slightly smaller than for multiplications. Specifically it is given by $[1+2q+(3+6q)\mu]/(3\mu+1) \leq 3+2(q-1)$. This suggests that unless the computation is dominated by multiplications, gradients calculated in a vector cost about twice the effort of evaluating the underlying function F by itself. That is certainly a satisfactory situation. When q adjoint components are propagated simultaneously we have the random access memory requirement $RAM\{gradq(F)\} = (1+q)RAM\{eval(F)\}$. In contrast the sequential access requirement $SAM\{gradq(F)\} \sim OPS\{eval(F)\}$ does not depend on q at all.

The Cost of Accessing QO

So far we have completely neglected the last term in (4.19), namely, $WORK\{v_i \leftrightarrow \text{QO}\}$. The nice thing about this recording effort is that it is completely predictable with exactly two such sequential memory moves required per scalar elemental, one on the way forward and one on the return. If the code is known to be single assignment the taping is not necessary at all. Otherwise, as long as the QO can be accommodated in main memory, a super scalar chip with an efficient cache may well perform the sequential memory moves so fast "on the side" that they do not affect the runtime at all. To enhance this hiding effect a good AD tool may flag some prevalues as not needed for the return sweep, as discussed in section 4.3.

However, such selective saves and restores are not likely to help much once the tape becomes so large that it must be stored on disk, at least in parts. Suppose $\tau \geq 1$ represents the ratio between $w^\top WORK\{v_i \leftrightarrow \text{QO}\}/2$ and μ, i.e., the time to move a real value from main memory onto tape or back compared to moving it between main memory and the CPU. Here the time in the denominator may be calculated at the burst rate, i.e., the maximal bandwidth of the disk without the latency of asking for individual data entries. Nevertheless, this disk access ratio τ is at least 10 and is more likely to be in the hundreds on modern computing platforms.

In Fig. 4.2 we have plotted the runtime ratio between $(f, \nabla f)$ and f as a function of the dimension n for Speelpenning's product as specified in Exercise 3.6. Function and gradient were hand-coded in C, compiled with gcc, and executed on a Fujitsu Siemens Esprimo P5615 with Athlon 64 X2 3800+ processor. The tape QO was implemented as a dynamically allocated array in C. The good news is that up to a dimension of about 100 million independent variables the ratio of the elapsed times is virtually constant equal to 3.2. This shows that our upper bound of 4 for ω_{grad} is indeed rather conservative. However, the bad news is that once the save array no longer fits into main memory in its entirety, the number of swaps grows steadily and the runtime ratio grows rapidly into the hundreds.

There are two main lessons to be drawn and remedies to be investigated. First, one should avoid going out to disk if at all possible, for example, by employing checkpointing and other modifications of the reverse mode that are discussed in Chapter 12. Second, one should strive to make better (re)use of the

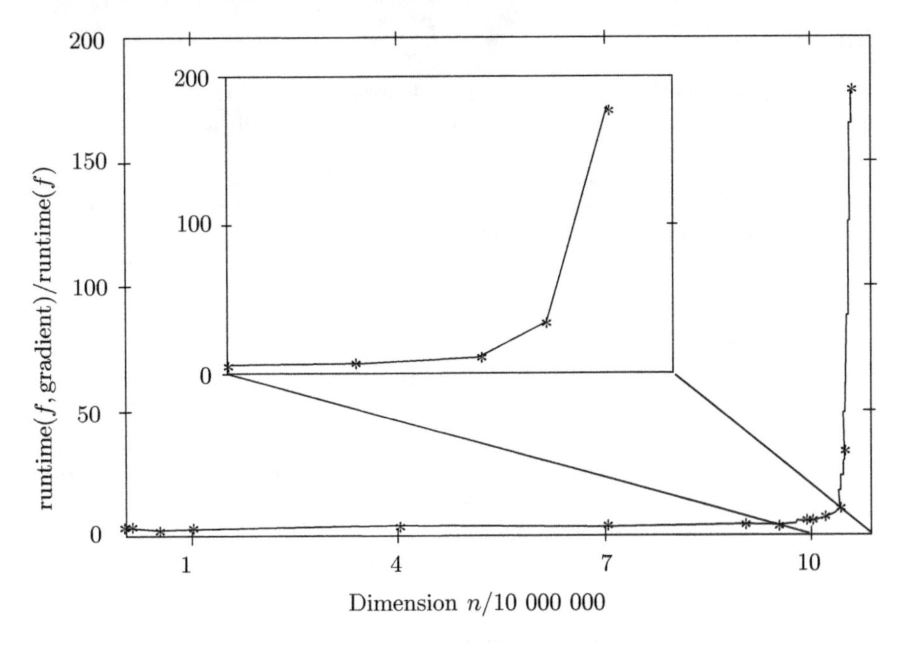

Figure 4.2: Elapsed-time Ratio for Scalar Adjoints on Speelpenning

data streaming to and from disk, for example, by employing the vector mode to propagate q adjoint components at a time. To get an idea of how large this q must be to balance computation with communication we may use the following heuristic estimates.

Suppose $\mu = 1 = \pi$, which means that multiplications and main memory accesses take about as much time as an addition. Then we may add up all columns in Tables 4.17 and 4.18 and adding $2\tau\mu = 2\tau$ for the tape operations to the latter we obtain for the addition and multiplication operator the ratios

$$\frac{8q+4+2\tau}{12} \approx \frac{2}{3}\left(\frac{1}{2}+q+\frac{\tau}{4}\right) \quad \text{and} \quad \frac{10q+6+2\tau}{16} \approx \frac{5}{8}\left(\frac{3}{5}+q+\frac{\tau}{5}\right).$$

Hence we see that the number of adjoint components q should be at least as large as about a quarter of the access ratio τ, if disks come into play at all. Of course, this is only an extremely rough estimate, which is why we hesitate to formulate it as a rule. On one hand more selective save strategies may reduce the number of save and restores on disk, but on the other hand clever register allocation between individual elemental operations may reduce the number of fetches and stores from and to main memory. Also, it must be noted that increasing q merely improves the overall balance between computation and communication, but it does not reduce the danger of memory boundedness on the recording sweep.

Conclusion

Lest the reader overlook the fundamental importance of the inequality (4.21) for the scalar-valued case $F = f$ with $m = 1$, we restate it as follows:

RULE 8

> OPERATION COUNTS AND RANDOM ACCESS MEMORY REQUIREMENTS FOR THEIR GRADIENTS ARE BOUNDED MULTIPLES OF THOSE FOR THE FUNCTIONS.

We refer to this very satisfactory result as the *cheap gradient principle.* The temporal complexity of memory accesses are accounted for, but the availability of temporary storage proportional to the function evaluation time is assumed. Since the memory requirement is a serious concern, we will describe in Chapter 12 trade-offs between temporal and spatial complexity that include the possibility of limiting the growth in both costs to a logarithm of the runtime for evaluating f. For the basic reverse mode described here we note that

RULE 9

> THE SEQUENTIAL ACCESS MEMORY REQUIREMENT OF THE BASIC REVERSE MODE IS PROPORTIONAL TO THE TEMPORAL COMPLEXITY OF THE FUNCTION.

By sequential accessed memory we mean the tape ⌷⌷, which is written once forward and read once backward. Hence there should be no cache misses at all.

4.7 Examples and Exercises

Exercise 4.1 (*Allocation Compatibility*)
Consider the evaluation procedure on the right.

a. Draw the computational graph and find a forward-compatible allocation function

$$f : [v_1, v_2, v_3, v_4, v_5] \longmapsto [1, 2, 3] \, .$$

Show that one more location is necessary for an allocation that is also reverse compatible and hence two-way compatible.

$$\begin{array}{lcl} v_1 &=& c_{10} * v_0 \\ v_2 &=& c_{21} * v_1 \\ v_3 &=& c_{30} * v_0 \\ v_4 &=& c_{43} * v_3 \\ v_5 &=& c_{52} * v_2 + c_{54} * v_4 \end{array}$$

b. Write the corresponding tangent procedure starting with $\dot{v}_0 = 1$. Compute an expression for $\dot{v}_5 = \partial v_5 / \partial v_0$ in terms of the coefficients $c_{10}, c_{21}, c_{30}, c_{43}, c_{52}$, and c_{54}.

c. Write the corresponding nonincremental adjoint, and compute an expression for $\partial v_5 / \partial v_0$, this time starting with $\bar{v}_5 = 1$. Check that the resulting $\bar{v}_0$ is identical to $\dot{v}_5$ as computed in part **b**.

d. Rewrite all three procedures by replacing variables v_j by $w_{f(j)}$, $\dot{v}_j$ by $\dot{w}_{f(j)}$, and $\bar{v}_j$ by $\bar{w}_{f(j)}$, where f is the forward-compatible allocation function developed in part **a**. Recompute the derivatives forward and backward, and check which results are the same as before.

e. Write down the incremental adjoint to the evaluation procedure with the overwrites written in step **d**. Show that it still yields the correct derivative $\partial v_5/\partial v_0$.

Exercise 4.2 (***Forward Differentiation of Division***)
Reconsider again the division operation $v = u/w$, which may be split into the pair of statements $t = 1/w$ and $v = u * t$ as in Exercise 3.3 on page 57.
a. Check that the tangent procedure obtained in part **b** in Exercise 3.3 is still alias-safe, that is, correct when $\&v \equiv \&w$, $\&v \equiv \&u$, $\&u \equiv \&w$, or even $\&v \equiv \&u \equiv \&w$.
b. Observe that neither the low operations cost according to part **b** in Exercise 3.3 nor the alias-safety according to part **a** in Exercise 3.3 can be achieved without the temporary t.

Exercise 4.3 (***Speelpenning's Product***)
Consider again the function $f(x) \equiv \prod_{i=1}^{n} x_i$, as in Exercise 3.6 on page 58. Compute the complexity of the adjoint procedure written in part **b** in Exercise 3.6 and the minimal scheme constructed in Exercise 3.6**c** with regard to the four-component measure (4.6). Compare the resulting runtimes under the assumption (4.11).

Exercise 4.4 (***Complexity of Nonincremental Reverse***)
In the nonincremental form Table 3.6, one saves—compared to the incremental version of Table 3.5—for each $i = l \ldots 1$ two memory moves and one addition to set $\bar{v}_i = 0$ and then increment $\bar{v}_i$ for the first time. To account for this saving, one may subtract a 2 from all entries in the first row of the W_{grad} given in (4.20) and a 1 from each entry in its second row. Calculate the resulting $C_{grad} = W_{grad} W_{eval}^{-1}$ and show that the runtime ratio ω_{grad} for the nonincremental form of reverse is bounded by (4.21) under assumption (4.11) and by 4 without any restriction on the memory access time μ.

Exercise 4.5 (***Inverse Recomputation on Riemann***)
In your favorite computer language write a program that sums up the fractional power series

$$y = f(x) = \sum_{k=0}^{n-1} x^{h_k} \quad \text{with} \quad h_k \equiv h_0 + \sum_{j=1}^{k} 1.0/j$$

after reading in the independent variable x and the parameters n, h_0.
a. Using the elementary relation $\sum_{j=1}^{k} 1.0/j \geq \ln(1+k)$, examine the relation to the Riemannian zeta function and establish convergence of the series as $n \to \infty$ for x positive and below some bound x_*.
b. Verify the convergence numerically by evaluating f for $x = 0.3$, $h_0 = 0.001$, and $n = 10^i$ with $i = 6 \ldots 10$ on your favorite computing platform with 64-bit double precision numbers. Repeat the calculation using only 32-bit single precision and compare the results.

c. Using Table 3.3, extend your evaluation program with tangent operations to calculate $\dot{y}$ for given $\dot{x} = 1$. Observe the convergence of $\dot{y}$ in double and single precision for the same values of x and n as in part **b**.
d. Write two adjoint programs, one that saves and restores the exponents h_k for $k = 0 \ldots n-1$ into a local array h_s of size n and one that recomputes the h_k using the inverse relation $h_{k-1} = h_k - 1.0/k$. For n and x as above, compare the resulting $\bar{x}$ for $\bar{y} = 1$ with each other and with the values $\dot{y}$ obtained by the tangent procedure. Notice that the applicability of the save version is limited to values of n for which the array h_s can be allocated on your platform.
e. Since the terms in the series are positive and their size decreases monotonically, summing them from the beginning is actually disadvantageous from the numerical point of view. To avoid this problem rewrite your program in analogy to the Horner scheme for the evaluation of polynomials. This time the partial products for y satisfy a recurrence that can be inverted by hand. Redo parts **b**, **c**, and **d** to verify that much better accuracy is obtained, especially in the single-precision versions.

Exercise 4.6 (*Determinant via LU Factorization*)

Consider the LU factorization of a square matrix $X = (x_{ij})$ of size n that is assumed to require no pivoting. Performing all calculations in place, we have the procedure on the right for calculating the determinant. In our terminology, for each index pair (i, j), the $\min(i, j+1)$ different values attained by x_{ij} during the factorization process represent distinct variables that are allocated in the same storage location.

```
y = 1
for k = 1...n
  y *= x[k][k]
  for i = k+1...n
  x[i][k] /= x[k][k]
    for j = k+1...n
      x[i][j] -= x[i][k] * x[k][j]
```

a. Applying Rule 6, argue that the intermediate values computed in the innermost loop and assigned repeatedly to x_{ij} for $j > k+1$ are not needed on the way back. Conclude that the tape size can be kept to $O(n^2)$, which is an order of magnitude smaller than the operations count.
b. Write an adjoint procedure with this desirable property. Apply it to a particular matrix such as $X = I + b\,a^\top$, with $a, b \in \mathbb{R}^n$ componentwise positive to ensure that no pivoting is necessary. Verify that in agreement with Exercise 3.7**a** for $\bar{y} = 1$, the resulting matrix $\bar{X} = (\bar{x}_{ij})$ is the transpose of the inverse X^{-1} multiplied by $\det(X)$, which equals $1 + b^\top a$ for the particular choice above.
c. Observe that all other intermediate variables that are needed on the way back can be recomputed using the inverse computations listed in Table 4.10, and write a corresponding adjoint procedure without any stack operations. Compare its results and runtimes to those from **b**.

Chapter 5

Repeating and Extending Reverse

Since the adjoining of (scalar) function evaluation procedures yields their gradients so cheaply, one might hope that adjoining the gradient procedure once more would yield the Hessian matrix at a similarly low complexity. Unfortunately, this expectation cannot be realized, because the individual gradient components are not separate functions. Instead, their evaluation is thoroughly intertwined, especially when they are calculated in the reverse mode. On the positive side, we observe that adjoints of adjoints can be realized as tangents of adjoints, so that code sizes and runtimes need not grow exponentially through nesting, contrary to what one might have expected.

To show how adjoints of adjoints can be obtained as tangents of adjoints, we consider a twice differentiable vector function $F : \mathbb{R}^n \longmapsto \mathbb{R}^m$ at some point x in the interior of its domain. Given a weight vector $\bar{y} \in \mathbb{R}^m$, we may compute the first-order adjoint in the top box of Fig. 5.1.

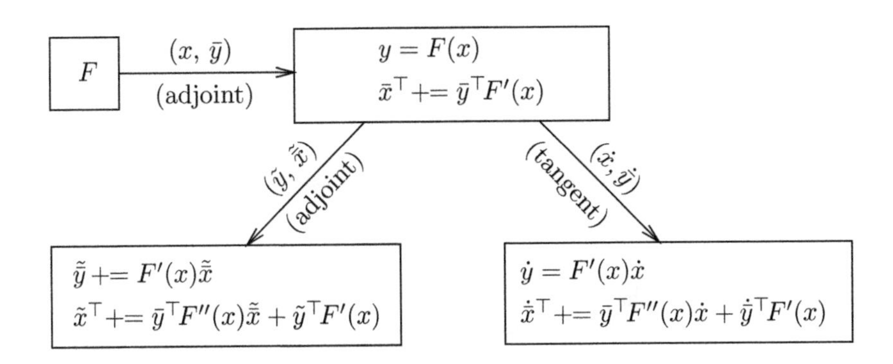

Figure 5.1: Adjoint of Adjoint and Tangent of Adjoint

Given weight vectors $\tilde{y} \in \mathbb{R}^m$ and $\tilde{\bar{x}} \in \mathbb{R}^n$ for the outputs y and $\bar{x}$ of the top box, we may differentiate with respect to the inputs $(x, \bar{y})$ and obtain the second-order adjoint in the bottom left box

$$\bar{y}^\top F''(x)\,\tilde{\bar{x}} \;\equiv\; \frac{\partial}{\partial\alpha}\,\bar{y}^\top F'(x+\alpha\,\tilde{\bar{x}})\Big|_{\alpha=0} \;\in \mathbb{R}^n\,.$$

Alternatively, we may simply differentiate the top box in some direction $(\dot{x}, \dot{\bar{y}})$ with $\dot{x} \in \mathbb{R}^n$ and $\dot{\bar{y}} \in \mathbb{R}^m$. The resulting tangent function is given in the bottom right box with $\dot{y} \in \mathbb{R}^m$ and $\dot{\bar{x}} \in \mathbb{R}^n$. We see immediately that

$$\boxed{\dot{x} \,\equiv\, \tilde{\bar{x}},\; \dot{\bar{y}} \,\equiv\, \tilde{y} \;\Longrightarrow\; \tilde{\bar{y}} \,=\, \dot{y},\; \tilde{x} \,=\, \dot{\bar{x}}\,.}$$

Here we have assumed that the result vectors $\tilde{\bar{y}}, \tilde{x}, \dot{y}$, and $\dot{\bar{x}}$ had all been initialized to zero, so that their incrementation is equivalent to an assignment.

Rather than simply matching the terms in the expression for $\tilde{x}$ and $\dot{\bar{x}}$ one may use the following explanation for the equivalence of tangent and adjoint differentiation the second time around. Replacing the incremental assignments with equalities, we find that the vector function

$$G(\bar{y},\, x) \equiv \left[F(x)^\top,\, \bar{y}^\top F'(x)\right]^\top \equiv \nabla_{\bar{y},x}\left[\bar{y}^\top F(x)\right]$$

maps $\mathbb{R}^{m\times n}$ into itself and is in fact the gradient of the Lagrangian $\bar{y}^\top F(x)$. Consequently, the Jacobian of G is the symmetric Hessian

$$G'(\bar{y},\, x) = \nabla^2_{\bar{y},x}\left[\bar{y}^\top F(x)\right] = \begin{bmatrix} 0 & F'(x) \\ F'(x)^\top & \bar{y}^\top F''(x) \end{bmatrix},$$

where

$$\bar{y}^\top F''(x) \equiv \sum_{i=1}^{m} \bar{y}_i\, F_i''(x) \in \mathbb{R}^{n\times n}\,.$$

Due to the symmetry of the matrix $G'(\bar{y}, x)$, multiplying it from the left by $(\tilde{y}, \tilde{\bar{x}})$ gives the same result as multiplying it from the right by the same vector even if it is disguised as $(\dot{\bar{y}}, \dot{x})$. Hence we can recommend the following rule.

RULE 10

NEVER GO BACK MORE THAN ONCE
(AFTER ALL WEIGHT VECTORS ARE KNOWN).

In other words, nested adjoints can be evaluated as directional derivatives of a single adjoint, once all weight vectors are known.

The qualification that the weight vectors must be known applies specifically to $\dot{x} \equiv \tilde{\bar{x}}$. If these are not known a priori, the first-order adjoint may have to be performed twice, once undifferentiated yielding $\bar{x}$, and then once with derivatives in the direction $\dot{x} = \tilde{\bar{x}}$, which may have been chosen as a function of $\bar{x}$. Consider, for example, the computation of the nested gradient

$$\nabla_x\left(\frac{1}{2}\,\|\nabla_x f(x)\|^2\right) = \nabla^2 f(x)\nabla f(x) = \bar{y}^\top f''(x)\,\dot{x}$$

for $\bar{y} = 1$ and $\dot{x} = \nabla f(x)$. Here we evaluate the adjoint $\bar{f}(x, 1)$ to obtain the direction $\dot{x} = \nabla f(x)$, which can then be ploughed back into a second adjoint computation, this time carrying along first derivatives. Before considering such second-order adjoints we extend our framework to evaluating procedures with incremental statements, so that adjoints themselves become differentiable by our methods.

In Chapter 3 we derived the reverse (or adjoint) mode primarily as a dual procedure to the forward (or tangent) mode. We already noted that under realistic assumptions considerably more complications arise. Here we examine some of those complications partly motivated by their occurrence in first-order adjoints. However, these generalizations are valid and may be useful, irrespective of whether one actually wants to obtain second-order adjoints.

5.1 Adjoining Iterative Assignments

For the sake of simplicity we assumed in section 2.2 that the variables v_i on the left-hand sides of the elemental assignments $v_i = \varphi_i(u_i)$ never occurred before, either in previous statements or on the current right-hand side. Actually, this single assignment property was reduced to a notational fiction when, in section 4.4, we allowed several v_i to be mapped into the same memory location by the addressing function &. However, the compatibility requirement (4.2) specifically excluded the possibility $\&v_i = \&v_j$ when $j \prec i$. We now remove this latter restriction even on the level of the variable names themselves. In particular, we allow incremental statements such as

$$v_i \mathrel{+}= v_k \quad \text{meaning} \quad v_i = v_i + v_k \ ,$$

which occur frequently in the adjoint procedure given in Table 4.4. Consequently, adjoint procedures themselves become eligible for further differentiation in either mode.

Let us consider more general assignments of the form

$$v_i = \varphi_i(v_i, \tilde{u}_i) = \varphi_i(u_i)$$

with $\varphi_i : \mathbb{R}^{n_i} \mapsto \mathbb{R}$ satisfying the differentiability Assumption (ED) and its argument partitioned according to

$$\tilde{u}_i = (v_j)_{i \neq j \prec i} \ , \quad u_i = (v_i, \tilde{u}_i) \ .$$

We call such assignments *iterative* because the argument v_i occurs also on the left-hand side and is therefore altered. By allowing $i \prec i$ but still requiring

$$j \prec i \ \Rightarrow \ j \leq i \ ,$$

we can keep the dependence relation $\prec$ transitive. Consequently, its transitive closure $\prec^*$ is now a reflexive partial ordering, but the one-to-one correspondence of procedure statements and intermediate variables is lost. Note that the forward compatibility condition (4.2) is still valid, since there we excluded the possibility of $i = j$ explicitly.

Now let us derive the adjoints of iterative assignments. We may disguise the iterative nature of the assignment by rewriting it as the statement pair

$$t_i = v_i \,,$$
$$v_i = \varphi_i(t_i, \tilde{u}_i) \,.$$

Here t_i is an auxiliary variable that is assumed never to occur outside these two statements. Then the usual adjoining recipe of Table 3.5 yields the following statements:

$$\begin{aligned} \bar{t}_i &\mathrel{+}= \bar{v}_i * c_{ii} \,, \\ \bar{v}_j &\mathrel{+}= \bar{v}_i * c_{ij} \quad \text{for} \quad i \neq j \prec i \,, \\ \bar{v}_i &= 0 \,, \\ \bar{v}_i &\mathrel{+}= \bar{t}_i \,. \end{aligned}$$

Since $\bar{t}_i = 0$ initially, we can eliminate t_i and $\bar{t}_i$ again to obtain Table 5.1 as a generalization of Table 4.3.

Table 5.1: Adjoining an Iterative Assignment

Forward :	Return :
$v_i \rightarrowtail$ ◘◘	$v_i \leftarrowtail$ ◘◘
$v_i = \varphi_i(u_i)$	$\bar{v}_j \mathrel{+}= \bar{v}_i * c_{ij}$ for $i \neq j \prec i$
	$\bar{v}_i \mathrel{*}= c_{ii}$

Exactly when the diagonal partial $c_{ii} = \partial\varphi_i/\partial v_i$ vanishes, the assignment is noniterative, and Table 5.1 reduces to Table 4.3. In practical terms the new adjoining rule requires that the adjoint $\bar{v}_i$ of the left-hand side be updated last. In the noniterative case this means simply setting it to zero.

Adjoining an Evolution

An important class of problems exists in which the c_{ii} are nonzero matrices. Suppose we have an explicit evolution of the form

$$v = \phi_i(v, x) \quad \text{for} \quad i = 1, 2 \ldots l \,.$$

Here the vector $v \in \mathbb{R}^m$ may be thought of as the "state" at various "times" i, which is updated by the nonlinear transformations $\phi_i : \mathbb{R}^{m+n} \longmapsto \mathbb{R}^m$. The extra argument x represents parameters, which may also determine the initial state, so that

$$v = \phi_0(x) \quad \text{at} \quad i = 0 \quad \text{and} \quad y = v \quad \text{at} \quad i = l \,,$$

where we consider the final state as a vector of dependent variables. Then, in agreement with Table 5.1, we obtain the pair of programs shown in Table 5.2. As one can see, the row vector $\bar{v}^\top$ is multiplied from the right by the square matrix $\partial\phi_i/\partial v$, which is the Jacobian of the new state with respect to the old state.

Table 5.2: Adjoining an Explicit Evolution

Forward :	Return :
$v = \phi_0\,(x)$	$\bar{v} = \bar{y}\,;\, \bar{x} = 0$
for $i = 1 \ldots l$	for $i = l \ldots 1$
$\quad v \rightarrowtail$ QO	$\quad v \leftarrowtail$ QO
$\quad v = \phi_i(v, x)$	$\quad \bar{x}^\top \mathrel{+}= \bar{v}^\top \dfrac{\partial}{\partial x}\phi_i(v, x)$
	$\quad \bar{v}^\top = \bar{v}^\top \dfrac{\partial}{\partial v}\phi_i(v, x)$
$y = v$	$\bar{x}^\top \mathrel{+}= \bar{v}^\top \dfrac{\partial}{\partial x}\phi_0(x)$

Adjoining Incremental Assignments

The most important class of iterative assignments are the *incremental* ones, where

$$v_i \mathrel{+}= \psi_i(u_i)\,, \quad \text{meaning} \quad v_i = v_i + \psi_i\,(v_j)_{i \neq j \prec i}\,.$$

Then the full right-hand side $\varphi_i \equiv v_i + \psi_i(u_i)$ has the partial $c_{ii} = 1$ with respect to v_i, and Table 5.1 simplifies to Table 5.3, where $\bar{\psi}_i(u_i, \bar{v}_i) = (\bar{v}_i\, c_{ij})_{j \prec i}$ as before.

Table 5.3: Adjoining an Incremental Assignment

Forward :	Return :
$v_i \rightarrowtail$ QO	$v_i \leftarrowtail$ QO
$v_i \mathrel{+}= \psi_i(u_i)$	$\bar{u}_i^\top \mathrel{+}= \bar{\psi}_i(u_i, \bar{v}_i)$

Whenever we write $v_i \mathrel{+}= \psi_i(u_i)$ it will be implicitly assumed that u_i does not include v_i.

5.2 Adjoints of Adjoints

In analogy with linear functional analysis one might expect that adjoining is, under certain conditions, a reflexive process in that applying it twice leads back to the original object. Moreover, one would expect certain objects to be self-adjoint and thus invariant with respect to the process of adjoining. To discuss this issue, we need to clarify what we consider to be the object being adjoined once or several times.

At the level $F'(x)^\top \bar{y}$ of mathematical mapping we go from $y = F(x)$ to $\bar{x}^\top \mathrel{+}= \bar{F}(x, \bar{y})$. Here $\bar{F}(x, \bar{y}) = \bar{y}^\top F'(x)$ is a vector function with $n = \dim(x)$ components and $n + m = n + \dim(y)$ input variables. If we consider all the latter as independents, adjoining once more leads to the relations in the left-hand box of Fig. 5.1. Clearly, each level of full adjoining introduces the next higher derivative tensor of F, so that the mappings become more and more involved, with the dimension of their domains growing rapidly. This is exactly what happens if one repeatedly processes a given evaluation program by an adjoint generator such as Tapenade or TAF. Even when $F(x)$ is polynomial, this growth process does not stop because the total polynomial degree remains constant. This can be seen easily on the scalar example $y = F(x) = x^2$ with $n = 1 = m$.

While second derivative matrices are crucial in optimization and other numerical applications, complete third and higher derivative tensors are rarely of practical interest (in our experience). Moreover, as we observed at the beginning of this chapter, the repeated forward and backward sweeping caused by adjoining an evaluation program more than once (with fixed weight vectors) can always be avoided. Loosely speaking, a sequence of combined forward and return sweeps can be replaced by one such pair, with higher derivatives being propagated in both directions. As a consequence, the computational complexity of higher derivatives grows quadratically rather than exponentially with its degree. Moreover the corresponding memory requirement grows only linearly with the degree. Also, we may uniquely specify all derivative quantities using only one overbar and arbitrarily many dots. These claims will be substantiated in Chapter 13 on Taylors and tensors.

The Linear Case

Finally, let us consider a vector function F all of whose constituent elementals are linear, so that

$$v_i = \sum_{j \prec i} c_{ij} * v_j \quad \text{for} \quad i = 1 \ldots l\,. \tag{5.1}$$

Such linear transformations arise in particular as tangent mappings $\dot{G}(z, \dot{z}) = G'(z)\dot{z}$, with z considered constant and the tangent $\dot{z}$ viewed as variable. If we set $x = \dot{z}$ and $y = \dot{G}(z, x)$, the tangent procedure Table 3.4 takes the form (5.1) with $\dot{v}_i$ replaced by v_i.

Therefore, an automatic way of doing so may be quite handy and efficient. For example, if we have some method for solving linear systems $Ay = x$ for a class of matrices $A \in \mathcal{A}$, then adjoining turns this method into one for solving linear systems $A^\top \bar{x} = \bar{y}$. In terms of the matrix inverse we have

$$y = F(x) \equiv A^{-1}x \quad \Rightarrow \quad \bar{x} \mathrel{+}= \bar{F}(\bar{y}) \quad \text{with} \quad \bar{F}(\bar{y}) = \left(A^{-1}\right)^\top \bar{y}\,.$$

Naturally, taking the adjoint of the adjoint will give us back the original linear map, albeit in the incremental form $y \mathrel{+}= A^{-1}x$. Hence we may consider the two procedures as dual to each other. Algorithmically, the adjoining happens at the level of the evaluation (here, solution) procedure without the inverse A^{-1} or its transpose ever being formed. In Exercise 5.1 we interpret algorithms for solving Vandermonde systems and their transposes as examples of such dual linear procedures.

Since for tangent mappings the $c_{ij} = c_{ij}(z)$ do not depend on the intermediates v_i, none of the latter needs to be saved. In other words, we consider the constant multipliers c_{ij} to be part of the procedural code. Consequently, the return sweep to (5.1) takes the nonincremental form

$$\bar{v}_j = \sum_{i \succ j} \bar{v}_i * c_{ij} \quad \text{for} \quad j = l - m \ldots 1 - n \,. \tag{5.2}$$

Clearly, adjoining (5.2) again gives us back (5.1) with $\bar{\bar{v}}_i$ taking the place of v_i. Thus, the transformation from a linear procedure to its nonincremental adjoint is a reflexive process. In writing (5.2) we have tacitly assumed that there is no overwriting (& is an injection). This assumption can be relaxed by requiring two-way compatibility between $\prec$ and & in that

$$i \neq j \prec i \implies \&v_i \neq \&v_j \quad \text{and} \quad \&v_j \neq \&v_k \neq \&v_i \quad \text{for} \quad j < k < i \,. \tag{5.3}$$

This condition is equivalent to combining forward and reverse compatibility, as originally introduced in section 4.1. Hence we can formulate the following result.

Proposition 5.1 (Reflexive Pairs of Nonincremental Procedures)
Under the compatibility condition (5.3), *the loops* (5.1) *and* (5.2) *yield the same (consistent) results as without any overwriting.*

In Exercise 4.1 it is shown on a small example that without reverse compatibility consistency may be lost in that wrong values for $\bar{x}$ can be obtained. As we have noted before, the nonincremental adjoint usually cannot be determined statement by statement. On the way back, adjoint values $\bar{v}_i$ must be kept until one reaches the smallest statement number j with $j \prec i$. In the incremental adjoint version all contributions $\bar{v}_i * c_{ij}$ are dispatched immediately.

As shown in Proposition 4.2 the incremental approach allows much greater freedom in the addressing scheme and even allows for iterative statements (i.e., $i \prec i$). The incremental adjoint of (5.1) is given by the nested loop listed on the right side of Table 5.4, which is just a specialization of the right side of Table 5.1 without the tape operations.

Table 5.4: Adjoint Pair of Linear Incremental Procedures

Forward :	Return :
for $i = 1, 2 \ldots l$	for $i = l, l-1 \ldots 1$
$\quad v_i \mathrel{*}= c_{ii}$	$\quad \bar{v}_j \mathrel{+}= \bar{v}_i * c_{ij}$ for $i \neq j \prec i$
$\quad v_i \mathrel{+}= c_{ij} * v_j$ for $i \neq j \prec i$	$\quad \bar{v}_i \mathrel{*}= c_{ii}$

Adjoining once more, we obtain the loop listed on the left side of Table 5.4, where we have replaced $\bar{\bar{v}}_i$ by v_i. By comparison with (5.1) we see that this loop is just an incremental rewrite: the two loops are functionally equivalent in that they represent the same incremental mapping $y \mathrel{+}= F(x)$.

The two procedures in Table 5.4 contain just two kinds of statements:

$$v \mathrel{*}= c \quad \text{and} \quad v \mathrel{+}= u * c \,. \tag{5.4}$$

The corresponding adjoints are

$$\bar{v} *= c \quad \text{and} \quad \bar{u} += \bar{v} * c \, .$$

Hence we may call the first statement, $v *= c$, *self-adjoint* and the second, $v += u * c$, *reflexive*, as $\bar{\bar{v}} += \bar{\bar{u}} * c$ has the same form as the original statement. Consequently, any evaluation procedure made up exclusively of these kinds of statements must be reflexive or even self-adjoint. Hence, we obtain the following rule.

RULE 11

INCREMENTAL ADJOINTS OF LINEAR PROCEDURES ARE REFLEXIVE IRRESPECTIVE OF ALLOCATION.

Rather than developing an elaborate formalism, let us consider an example of an incremental linear procedure and its adjoint. The procedure listed in Table 5.5 is an incrementally written version of Algorithm 4.6.1 of Golub and van Loan [GV96]. It computes the coefficients of a polynomial of degree $n-1$ given its values at the n abscissas λ_i for $i = 1 \ldots n$. These are considered as constants, whereas the f_i represent the polynomial values on entry and the polynomial coefficients on exit. In other words, the procedure performs the mapping

Table 5.5: Polynomial Interpolation

$$\begin{array}{l} \text{for} \quad j = 1 \ldots n \\ \qquad \text{for} \quad i = n, n-1 \ldots j+1 \\ \qquad\qquad f_i -= f_{i-1} \\ \qquad\qquad f_i *= \dfrac{1}{(\lambda_i - \lambda_{i-j})} \\ \text{for} \quad j = n-1, n-2 \ldots 1 \\ \qquad \text{for} \quad i = j, j+1 \ldots n-1 \\ \qquad\qquad f_i -= \lambda_j \, f_{i+1} \end{array}$$

$$f \longmapsto A^{-1} f \quad \text{with} \quad A = \left(\lambda_i^{j-1}\right)_{i=1\ldots n}^{j=1\ldots n} \, .$$

Since $v -= c * \dot{u}$ is equivalent to $v += (-c) * \dot{u}$, all statements in Table 5.5 are of the special form (5.4) and are thus reflexive or even self-adjoint. In Exercise 5.1 it will be verified that adjoining Table 5.5 yields an incremental linear procedure that performs the mapping

$$b \longmapsto \left(A^{-1}\right)^{\top} b \, . \tag{5.5}$$

The actual computation is equivalent to Algorithm 4.6.2 in Golub and van Loan [GV96]. Rather than deriving it by linear algebra arguments, we obtain it here automatically using a line-by-line transformation.

5.3 Tangents of Adjoints

In optimization calculations one is often interested not only in first, but also in second derivatives. While we have seen that gradients have essentially the same complexity as the underlying scalar-valued functions for $m = 1$, the corresponding Hessians are, in the worst case, about n times as expensive. This fact will be demonstrated by example in Chapter 11. For now, we simply calculate second derivative vectors of the form

$$\dot{\bar{x}} \mathrel{+}= \dot{\bar{F}}(x, \dot{x}, \bar{y}) \quad \text{with} \quad \dot{\bar{F}}(x, \dot{x}, \bar{y}) \equiv \bar{y}^\top F''(x)\dot{x} \in \mathbb{R}^n \,. \tag{5.6}$$

In other words, the so-called second-order adjoint vector $\dot{\bar{x}}$ in the domain space of F is obtained by contracting the second derivative tensor $F''(x) \in \mathbb{R}^{m\times n\times n}$ once along a domain tangent $\dot{x} \in \mathbb{R}^n$ and once by a weight vector $\bar{y}$ on the range of F. As always in AD, this contraction occurs implicitly, so that the full derivative tensor is never formed and the computational complexity is orders of magnitude smaller. In the optimization context one should think of $\dot{x}$ as a feasible direction and $\bar{y}$ as a vector of Lagrange multipliers so that $\bar{y}^\top F''(x) \in \mathbb{R}^{n\times n}$ is the Hessian of the Lagrange function.

To compute $\dot{\bar{x}}$, we consider it as the directional directive of $\bar{x}^\top = \bar{y}^\top F'(x)$. Hence more generally we have

$$\dot{\bar{x}}^\top = \bar{y}^\top F''(x)\dot{x} + \dot{\bar{y}}^\top F'(x) \equiv \dot{\bar{F}}(x, \dot{x}, \bar{y}, \dot{\bar{y}})^\top \,. \tag{5.7}$$

The extra term $\dot{\bar{y}}^\top F'(x)$ vanishes if $\dot{\bar{y}} \in \mathbb{R}^m$ is selected as zero. Allowing $\dot{\bar{y}}$ to be nonzero has the benefit of making the evaluation of $\dot{\bar{F}}$ a homogeneous task in that it essentially requires the evaluation of the corresponding elemental functions $\dot{\bar{\varphi}}_i(u_i, \dot{u}_i, \bar{v}_i, \dot{\bar{v}}_i)$, as defined below.

To obtain a procedure for evaluating $\dot{\bar{F}}$, we apply the tangent propagation approach described in section 3.1 to the nonincremental version Table 3.6 for evaluating $\bar{F}$. In contrast to the incremental versions, this form satisfies the assumptions we placed on the general evaluation procedure of Table 2.2, provided the φ_i are twice differentiable. Then the expression

$$\bar{v}_j \equiv \sum_{i\succ j} \bar{v}_i \frac{\partial}{\partial v_j}\varphi_i(u_i) = \sum_{i\succ j} \bar{v}_i\, c_{ij} \tag{5.8}$$

is for each $j \in \{1-n, \ldots, l-m\}$ a once differentiable elementary function. By directional differentiation we obtain

$$\dot{\bar{v}}_j \equiv \sum_{i\succ j}\left(\dot{\bar{v}}_i \frac{\partial\varphi_i}{\partial v_j} + \bar{v}_i \frac{\partial\dot{\varphi}_i}{\partial v_j}\right), \tag{5.9}$$

where

$$\frac{\partial\varphi_i}{\partial v_j} \equiv \frac{\partial}{\partial v_j}\varphi_i(u_i) = c_{ij}(u_i) \tag{5.10}$$

and

$$\frac{\partial\dot{\varphi}_i}{\partial v_j} \equiv \sum_{k\prec i} c_{ijk}\,\dot{v}_k \quad \text{with} \quad c_{ijk} = \frac{\partial^2}{\partial v_j\,\partial v_k}\varphi_i(u_i)\,. \tag{5.11}$$

Assuming that all $\dot{\bar{v}}_i$ with $1 \le i \le l-m$ are initialized to zero, we can compute their final value incrementally, as was done for the $\bar{v}_i$ in Table 4.4. For brevity we combine all incremental contributions to the components of $\dot{\bar{u}}_i \equiv (\dot{\bar{v}}_j)_{j \prec i}$ in the vector function

$$\dot{\bar{\varphi}}_i(u_i,\, \dot{u}_i, \bar{v}_i,\, \dot{\bar{v}}_i) \equiv \bar{\varphi}_i(u_i,\, \dot{\bar{v}}_i) + \bar{v}_i \sum_{k \prec i} c_{ijk}\, \dot{v}_k\ . \tag{5.12}$$

Finally, we can allow overwrites, as we did in Table 4.4, thus obtaining the complete tangent adjoint procedure listed in Table 5.6. Alternatively, we could have derived Table 5.6 directly from Table 4.4 by generalizing the tangent propagation techniques of section 3.1 to evaluation procedures that involve incremental additions and tape operations.

Table 5.6: Second-Order Adjoint Procedure

$[v_{i-n}, \dot{v}_{i-n}]$	$=$	$[x_i, \dot{x}_i]$	
$[\bar{v}_{i-n}, \dot{\bar{v}}_{i-n}]$	$=$	$[\bar{x}_i, \dot{\bar{x}}_i]$	$i = 1 \ldots n$
$[v_i, \dot{v}_i]$	$\rightarrowtail$	●○	
$[v_i, \dot{v}_i]$	$=$	$[\varphi_i(u_i), \dot{\varphi}_i(u_i, \dot{u}_i)]$	$i = 1 \ldots l$
$[y_{m-i}, \dot{y}_{m-i}]$	$=$	$[v_{l-i}, \dot{v}_{l-i}]$	$i = m-1 \ldots 0$
$[\bar{v}_{l-i}, \dot{\bar{v}}_{l-i}]$	$=$	$[\bar{y}_{m-i}, \dot{\bar{y}}_{m-i}]$	$i = 0 \ldots m-1$
$[v_i, \dot{v}_i]$	$\leftarrowtail$	○●	
$[\bar{u}_i, \dot{\bar{u}}_i]$	$\mathrel{+}=$	$[\bar{\varphi}_i(u_i, \bar{v}_i), \dot{\bar{\varphi}}_i(u_i, \dot{u}_i, \bar{v}_i, \dot{\bar{v}}_i)]$	$i = l \ldots 1$
$[\bar{v}_i, \dot{\bar{v}}_i]$	$=$	$[0, 0]$	
$[\bar{x}_i, \dot{\bar{x}}_i]$	$=$	$[\bar{v}_{i-n}, \dot{\bar{v}}_{i-n}]$	$i = n \ldots 1$

The two tables are certainly closely related. One should also note that the first half of Table 5.6 is identical to the tangent procedure of Table 3.4 except for the tape operations. That suggests that tangent adjoints can also be obtained as adjoint tangents. However, considering $\dot{y} = \dot{F}(x, \dot{x}) = F'(x)\dot{x}$ as a mapping from $\mathbb{R}^{n \times n}$ to $\mathbb{R}^m$ and adjoining with a weight vector $\dot{\bar{y}}$ yields the relations $\bar{x}^\top = \dot{\bar{y}}^\top F''(x)\dot{x}$ and $\dot{\bar{x}}^\top = \dot{\bar{y}}^\top F'(x)$, which are not consistent with our definitions above. While the general formulas for second-order adjoints may seem complicated, their actual instantiation for our four basic operation classes is quite simple.

In Table 5.7 we compile the tangent, adjoint, and second-order adjoint operations associated with constant initialization, addition/subtraction, multiplication, and univariate intrinsics, respectively. The second and fourth columns contain the cumulative counts according to our four-component complexity measure. Here "cumulative" means that the lower degree operations to the left and above are included, thus yielding the total count for the tangent, adjoint, and second-order adjoint modes. The memory access counts are based on the assumption that the operations in each block row are executed together on the same sweep. For example, in the first and second-order adjoint operations for multiplication, the value w occurs twice but needs to be read from memory only once.

Table 5.7: Original, Tangent, Adjoint, and Second-Order Adjoint Operations with Associated Work (*MOVES*, *ADDS*, *MULTS*, *NLOPS*)

$v = c$	$(1,0,0,0)$	$\dot{v} = 0$	$(2,0,0,0)$
$\bar{v} = 0$	$(2,0,0,0)$	$\dot{\bar{v}} = 0$	$(4,0,0,0)$
$v = u \pm w$	$(3,1,0,0)$	$\dot{v} = \dot{u} \pm \dot{w}$	$(6,2,0,0)$
$\bar{u} \mathrel{+}= \bar{v}$ $\bar{w} \mathrel{\pm}= \bar{v}$ $\bar{v} = 0$	$(9,3,0,0)$	$\dot{\bar{u}} \mathrel{+}= \dot{\bar{v}}$ $\dot{\bar{w}} \mathrel{\pm}= \dot{\bar{v}}$ $\dot{\bar{v}} = 0$	$(18,6,0,0)$
$v = u * w$	$(3,0,1,0)$	$\dot{v} = w * \dot{u} + u * \dot{w}$	$(6,1,3,0)$
$\bar{u} \mathrel{+}= \bar{v} * w$ $\bar{w} \mathrel{+}= \bar{v} * u$ $\bar{v} = 0$	$(11,2,3,0)$	$\dot{\bar{u}} \mathrel{+}= \dot{\bar{v}} * w + \bar{v} * \dot{w}$ $\dot{\bar{w}} \mathrel{+}= \dot{\bar{v}} * u + \bar{v} * \dot{u}$ $\dot{\bar{v}} = 0$	$(22,7,9,0)$
$v = \psi(u)$	$(2,0,0,1)$	$\dot{v} = \psi'(u) * \dot{u}$	$(4,0,1,2)$
$\bar{u} \mathrel{+}= \bar{v} * \psi'(u)$ $\bar{v} = 0$	$(7,1,1,2)$	$\dot{\bar{u}} \mathrel{+}= \dot{\bar{v}} * \psi'(u) + \bar{v} * \psi''(u) * \dot{u}$ $\dot{\bar{v}} = 0$	$(14,3,5,4)$

Again, one can amortize some of the overhead by computing the matrix

$$\dot{\bar{X}}^\top \mathrel{+}= \bar{y}^\top F''(x)\dot{X} \in \mathbb{R}^{n\times p}$$

for some seed matrix $\dot{X} \in \mathbb{R}^{n\times p}$ and $\dot{\bar{y}} = 0 \in \mathbb{R}^m$. This might be considered a one-sided projection of the Hessian of the Lagrange function $\bar{y}^\top F$ onto a subspace of feasible directions spanned by the columns of $\dot{X}$. For this purpose the scalar values $\dot{v}_i \in \mathbb{R}$ and $\dot{\bar{v}}_i \in \mathbb{R}$ in Table 5.6 have to be replaced by vectors $\dot{V} \in \mathbb{R}^p$ and $\dot{\bar{V}}_i \in \mathbb{R}^p$, respectively. Alternatively, one may replace the adjoint seed $\bar{y}$ by some matrix $\bar{Y} \in \mathbb{R}^{m\times q}$ and compute the matrix

$$\dot{\bar{X}}^\top \mathrel{+}= \bar{Y}^\top F''(x)\dot{x} \in \mathbb{R}^{q\times n} .$$

Here the scalars $\bar{u}_i$ and $\dot{\bar{u}}_i$ in Table 5.6 have to be replaced by vectors $\bar{U}_i \in \mathbb{R}^q$ and $\dot{\bar{U}}_i \in \mathbb{R}^q$, respectively. Otherwise the second-order adjoint procedure remains in either case unchanged.

Calling Sequences for Second-Order Adjoints

For the computation of second-order adjoints given by

$$\dot{\bar{x}}^\top = \bar{y}^\top F''(x)\dot{x} + \dot{\bar{y}}^\top F'(x) ,$$

the arrays $\mathtt{x} = x \in \mathbb{R}^n$, $\mathtt{dx} = \dot{x} \in \mathbb{R}^n$, $\mathtt{by} = \bar{y} \in \mathbb{R}^m$, and $\mathtt{dby} = \dot{\bar{y}} \in \mathbb{R}^m$ serve as inputs. The output arrays are given by $\mathtt{y} = y \in \mathbb{R}^m$ and $\mathtt{dy} = \dot{y} \in \mathbb{R}^m$. There are two input-output arrays, bx $= \bar{x} \in \mathbb{R}^n$ and dbx $= \dot{\bar{x}} \in \mathbb{R}^n$, that enter into the calculation only linearly. Hence, one obtains the following prototype of a calling sequence for a routine eval_soad to compute second-order adjoints:

```
eval_soad ( bx, dbx, x, dx, by, dby, y, dy ) .
```

As on page 49 in Section 3.3 one must distinguish a primaly consistent version where y and dy have on exit exactly the same values as the ones produced by a call to eval_tang introduced on page 37 and a primaly constant one where they keep their input values and may thus be left of as arguments altogether. The same applies to the vector versions of the second-order adjoints, where we will leave off y and its direct derivative dy or dY completely. Changing some argument dimensions and assuming dby $= 0$ or dbY $= 0$ we obtain the drivers

```
eval_soadp ( bx, dbX, x, dX, by )        eval_soadq ( bX, dbX, x, dx, bY )
```

with bx, x, by, and y as before, dX $\in \mathbb{R}^{n\times p}$, bY $\in \mathbb{R}^{m\times q}$, bX $\in \mathbb{R}^{n\times q}$ and dbX $\in \mathbb{R}^{n\times p}$ or dbX $\in \mathbb{R}^{n\times q}$.

5.4 Complexity of Second-Order Adjoints

Following our standard argument, we note that the task of evaluating the vector quadruple

$$[y, \dot{y}, \bar{x}, \dot{\bar{x}}] = [F(x),\ \dot{F}(x,\dot{x}),\ \bar{F}(x,\bar{y}),\ \dot{\bar{F}}(x,\dot{x},\bar{y},\dot{\bar{y}})]$$

is again an additive task, which will be denoted by *soad*(F). By inspection of Table 5.6 we observe that *soad*(F) requires for each elemental function φ the evaluation of the pair $[\varphi(u_i), \dot{\varphi}_i(u_i, \dot{u}_i)]$ on the way forward and the pair $[\bar{\varphi}(u_i, \bar{v}_i), \dot{\bar{\varphi}}_i(u_i, \dot{u}_i, \bar{v}_i, \dot{\bar{v}}_i)]$ on the way back, the transfer of $[v_i, \dot{v}_i]$ to and from the tape QO, and the resetting of $[\bar{v}_i,\ \dot{\bar{v}}_i]$ to zero. We denote the combination of these four subtasks by *soad'*(φ_i) and set

$$\begin{aligned} WORK\{soad'(\varphi_i)\} = &\ WORK\{\varphi_i\} + WORK\{\dot{\varphi}_i\} + WORK\{\bar{\varphi}_i\} \\ &+ WORK\{\dot{\bar{\varphi}}_i\} + WORK\{[v_i, \dot{v}_i] \leftrightarrow \text{QO}\}\,. \end{aligned}$$

Proposition 4.3 again guarantees the existence of a matrix bound $C_{soad}(\Psi)$ so that

$$WORK\{soad(F)\} \leq C_{soad}(\Psi)\, WORK\{eval(F)\}\,.$$

To obtain a specific matrix $C_{soad}(\Psi)$ for the four-component complexity measure, we again examine the four basic operators. By inspecting Table 5.7 we obtain the following complexity matrices:

$$W_{soad} \equiv \begin{bmatrix} 4 & 18 & 22 & 14 \\ 0 & 6 & 7 & 3 \\ 0 & 0 & 9 & 5 \\ 0 & 0 & 0 & 4 \end{bmatrix}, \qquad C_{soad} \equiv \begin{bmatrix} 4 & 6 & 10 & 6 \\ 0 & 6 & 7 & 3 \\ 0 & 0 & 9 & 5 \\ 0 & 0 & 0 & 4 \end{bmatrix}. \tag{5.13}$$

Once more, using the assumptions (4.11) and the representation (4.15), we obtain the scalar bound

$$\omega_{soad} = \max\left\{\frac{4\mu}{\mu}, \frac{18\mu+6}{3\mu+1}, \frac{22\mu+7+9\pi}{3\mu+\pi}, \frac{14\mu+3+5\pi+4\nu}{2\mu+\nu}\right\} \in [7,10]. \tag{5.14}$$

When propagating p tangents and p second-order adjoints simultaneously using the vector version of the second-order adjoint mode, one obtains the following complexity results:

soadp	const	add/sub	mult	ψ
MOVES	$2+2p$	$12+6p$	$11+11p$	$7+7p$
ADDS	0	$3+3p$	$2+5p$	$1+2p$
MULTS	0	0	$3+6p$	$1+4p$
NLOPS	0	0	0	4

The resulting runtime ratio is, according to (4.15),

$$\begin{aligned} \omega_{soadp} = \max\Big\{ & 2+2p, \frac{(12+6p)\mu+3+3p}{3\mu+1}, \\ & \frac{(11+11p)\mu+2+5p+(3+6p)\pi}{3\mu+\pi}, \\ & \frac{(7+7p)\mu+1+2p+(1+4p)\pi+4\nu}{2\mu+\nu}\Big\} \in [4+3p, 4+6p]. \end{aligned} \tag{5.15}$$

The reduction in the runtime ratio from an upper bound in $[7p, 10p]$ to an upper bound in $[4+3p, 4+6p]$ is caused by the fact that values that are independent of the directions contained in $\dot{X}$ are reused instead of recomputed. Hence, similar to the runtime reductions that can be achieved by using the vector forward mode of AD instead of the scalar forward mode for computing first derivatives, a decrease of the computing time needed for directional second derivatives can be achieved by using a vector version. It should be noted that for the p-version with p primal directions making up $\dot{X}$ the *SAM* requirement grows by the factors $(p+1)/2$ compared to the computation of a scalar second-order adjoint according to Table 5.6. In contrast the q-version involving q dual directions has the same SAM requirement as the scalar method but uses slightly more operations as we will see in Exercise 5.5.

For comparing the complexity bounds for scalar second-order adjoints given in (5.14) and for the p-vector version given in (5.15) we computed the stated runtime ratios using the well-established driver hess_vec of ADOL-C to compute p Hessian-vector products and the recent driver hess_mat of ADOL-C to compute one Hessian-matrix product. The equality constrained optimization problems dtoc2 and eigena2 out of the CUTEr test set collection [GOT03] serve as test cases. For dtoc2 the matrix $\dot{X}$ has 6 columns independent of the size of the problem. For the numerical tests we set the number of variables to $n = 1485$ and the number of constraints to $m = 990$. To get an impression also for a higher number of columns, we set for eigena2 the number of independent variables to $n = 2550$ and the number of constraints $m = 1275$, resulting in a matrix $\dot{X}$ with 52 columns.

The achieved runtime ratios are illustrated by Figure 5.2. First of all, the expected linear behavior in dependence on the number of vectors and columns, respectively, is clearly visible. Furthermore, the line with the larger slope belongs to the scalar second-order adjoint computation evaluating p Hessian-vector products. Hence, so far the theory is verified by the numerical examples, since the vector version of the second-order adjoint requires significantly less runtime.

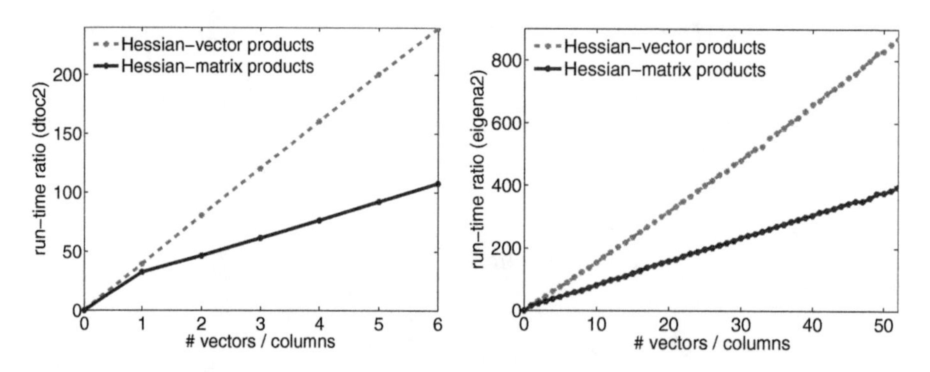

Figure 5.2: Hessian-vector and Hessian-matrix Products

Additionally, one can examine the slopes of the lines in more detail. For that purpose the slopes are stated in Table 5.8. As can be seen, the scalar mode is almost three times slower than the theory predicted for the dtoc2 example, whereas the vector mode is only about two times slower than the theory. For the eigena2 example, the function evaluation is a little bit more complicated and the situation changes considerably in favor of ADOL-C. Here, the runtime needed by the scalar mode

Table 5.8: Slopes Obtained from Runtime Results

	dtoc2	eigena2
scalar *soad*	28.6	16.5
vector *soad*	13.2	7.5

is only about a factor 3/2 larger than expected. Hence, the operator-overloading tool ADOL-C comes almost close to the theory. The same is true for the vector version of the second-order adjoint, where the slope is close to the theoretical bound 6.

5.5 Examples and Exercises

Exercise 5.1 (*Vandermonde Systems*)
Consider the incremental linear procedure listed in Table 5.5. Form its adjoint by reversing the order of all blocks and loops and replacing each statement by its adjoint according to Table 5.4. In doing so treat the λ_i as constants. Check that the resulting procedure obtains the final adjoints $\bar{f}_i = b_i$ for $i = 1 \dots n$ as in (5.5) from their initial values by performing essentially the same calculation as Algorithm 4.6.2 from Golub and Van Loan [GV96].

Exercise 5.2 (*Limited-Memory BFGS*)
In the previous exercise we saw that the tasks of matrix inversion and linear system solving can be recast as gradient calculations for determinants as functions of suitably chosen independent variables. This interpretation might be useful to simplify coding and check correctness. A similar effect occurs in computing steps for limited-memory Broyden–Fletcher–Goldfarb–Shanno (BFGS) methods, as suggested by Gilbert and Nocedal [GN93]. Given any approximation H to the inverse of the Hessian matrix $\nabla^2 f$, one may calculate the corresponding quasi-Newton step as a gradient of the quadratic form $z = g^\top H\, g/2$. Here it is very useful to interpret some first-level basic linear algebraic subroutine (BLAS) operations as elemental. Also, one may apply Rule 6 to determine whether or not an intermediate quantity needs to be saved on the tape.

a. Consider the inner product φ and the incremental saxpy ψ defined by

$$v = \varphi(u) \equiv \sum_{i=1}^{n} c_i\, u_i = c^\top u \quad \text{and} \quad w \mathrel{+}= \psi(u, \alpha) \quad \text{with} \quad \psi(u, \alpha) = \alpha\, u\,.$$

Here the c_i are assumed to be constant, but the scalar α and of course the vector u are variable. Write down the corresponding derived operations

$$\dot{\varphi}(u, \dot{u}),\ \dot{\psi}(u, \alpha, \dot{u}, \dot{\alpha}),\ \bar{\varphi}(u, \bar{v}),\ \bar{\psi}(u, \alpha, \bar{w}), \quad \text{and} \quad \dot{\bar{\psi}}(u, \alpha, \dot{u}, \dot{\alpha}, \bar{w}, \dot{\bar{w}})\,.$$

b. Given vectors $y_i, s_i \in \mathbb{R}^n$ with $y_i^\top s_i = 1$ for $i = 0 \dots l-1$, consider the evaluation procedure on the left-hand side of the following table:

Left	Right
$v = g$	$v = g$
$w = 0$	for $i = l-1 \dots 0$
for $i = l-1 \dots 0$	$\quad u_i = s_i^\top v$
$\quad u = s_i^\top v$	$\quad v \mathrel{-}= u * y_i$
$\quad w \mathrel{+}= u^2$	for $i = 0 \dots l-1$
$\quad v \mathrel{-}= u * y_i$	$\quad u = u_i - y_i^\top v$
$w \mathrel{+}= v^\top v$	$\quad v \mathrel{+}= u * s_i$
$z = w/2$	

Examine whether the intermediate values of the vectors v or the scalars u and w need to be saved or recomputed. For those that are needed, decide whether to store or recompute them. Assuming that the final value z is the only dependent variable and that $\bar{z} = 1$, write down a forward and reverse sweep with the appropriate tape operations and recomputations. Simplify the combined code in view of the fact that only the quasi-Newton step $\bar{g}$ is of interest and compare the result with the formulas from [GN93] listed on the right-hand side above. Here we have assumed for simplicity that $H_0 = I$ and $y_i^\top s_i = 1$ as before, which can always be achieved by a suitable rescaling of these data.

Exercise 5.3 (*Nested Adjoints*)
Consider the first-order adjoint

$$\begin{array}{lll} v_{i-n} = x_i & i = 1 \ldots n \\ v_i = \varphi_i(u_i) & i = 1 \ldots l \\ y_{m-i} = v_{l-i} & i = m-1 \ldots 0 \end{array} \qquad \begin{array}{lll} \bar{v}_{l-i} & = \bar{y}_{m-i} & i = 0 \ldots m-1 \\ \bar{u}_i & \mathrel{+}= \bar{v}_i \, \varphi_i'(u_i) & i = l \ldots 1 \\ \bar{x}_i & = \bar{v}_{i-n} & i = 1 \ldots n \end{array}$$

Assume that $\bar{v}_i = 0$ initially, and disallow overwriting so that & is an injection. View $(x, \bar{y})$ as independent and $(y, \bar{x})$ as dependent variables of the procedure above. Associate with $(y, \bar{x})$ weights $\tilde{y}$ and $\tilde{\bar{x}}$, and write down the adjoint with respect to x and $\bar{y}$, yielding

$$\tilde{\bar{y}} = \frac{\partial}{\partial \bar{y}} \left(\tilde{y}^\top y + \tilde{\bar{x}}^\top \bar{x} \right) \quad \text{and} \quad \tilde{x} = \frac{\partial}{\partial x} \left(\tilde{y}^\top y + \tilde{\bar{x}}^\top \bar{x} \right) .$$

For each intermediate pair $(v_i, \bar{v}_i)$ there will be dual quantities $(\tilde{v}_i, \tilde{\bar{v}}_i)$, which are assumed to be zero initially and then incremented on the reverse sweep through the program listed above. Examine those statements with tilde quantities on the left-hand side, and show how they can all be pulled forward into the original program, so that only one double sweep forward and backward is left.

Exercise 5.4 (*Computing Second Derivatives*)
Recall the coordinate transformation in Exercise 2.1. Using the procedure specified in Table 5.6 with the individual second-order adjoint operations listed in Table 5.7, evaluate

$$\left(\frac{\partial^2 y_2}{\partial x_1 \partial x_j} \right)_{j=1\ldots 3} \quad \text{at} \quad x = (1,\, 0,\, -1) \, .$$

Note that, except for $\bar{y}_2 = 1$, all other adjoint quantities, including $\dot{\bar{y}}_2$, must be initialized to zero.

Exercise 5.5 (*Complexity of q-Version*)
Modify the complexity analysis performed in Section 5.4 for the q-vector adjoint mode sketched as an alternative after Table 5.7. To that end you have to compile the appropriate complexity table and upper bound corresponding to (5.15) in terms of q, the number of adjoint directions.

Chapter 6

Implementation and Software

The techniques explained in Chapters 3 and 5 reveal a sequence of operations that the transformed program must perform in order to calculate the correct derivative values. The insight that the AD framework gives us is valuable even if we propose to carry out the program transformation by modifying the source code manually.

However, manual transformation is time consuming and prone to error. It is also hard to manage; if the underlying program is subject to continual change, then keeping track of these changes, in order to maintain the integrity of the transformed version, is also time consuming. Therefore, it is usually desirable to automate at least partially the process of transformation. Regardless of how automatically, i.e., with what level of automation, program transformation is done there is a trade-off between the sophistication of the transformation process and the efficiency with respect to time and space bounds of the transformed program. As a rule, a general-purpose AD tool will not produce transformed code as efficient as that produced by a special-purpose translator designed to work only with underlying code of a particular structure, since the latter can make assumptions, often with far-reaching consequences, whereas the former can only guess.

In many cases, an unsophisticated approach suffices to produce AD code that is within a constant factor of the optimal performance bounds. In this situation, subsequent effort is devoted just to reducing the value of the constant. In other cases, a careful analysis of the structure of the code will reveal that several orders of magnitude can be gained by the use of more sophisticated techniques, such as preaccumulation (see section 10.2), or by the careful exploitation of by-products of the underlying computation, such as LU decompositions, in the derivative process (see Exercise 4.6). Such transformation processes may involve a substantial amount of human intervention, including some modification of the code, or a much greater degree of sophistication in the design of the automated tools.

For example, our primary motivation might be the wish to explore new models or algorithms to see whether they merit further investigation. In this case the principal benefit of AD is the ability to provide accurate derivative information for newly coded numerical functions while avoiding the labor traditionally associated with developing accurate derivative evaluation code. Hence the emphasis will be on the ease of the transformation process, rather than on the efficiency of the transformed program.

Later, when our preferred model is stable, we may be willing to spend more effort designing our code to produce derivative code that runs in the shortest possible time. Later still, if our approach is a success and we wish to apply it to larger and larger problems, we may be willing to spend more effort in developing a special-purpose AD tool designed to work only with this model and in encapsulating our insights about the fundamental structure of the corresponding programs.

The two basic computer science concepts employed in one guise or another by AD implementers are operator overloading and source transformation performed by compiler generators, sometimes also called compiler-compilers. Although operator overloading and source transformation may look very different to the user, the end product is in both cases an object code, which we have called `eval_der.obj` in Fig. 6.1.

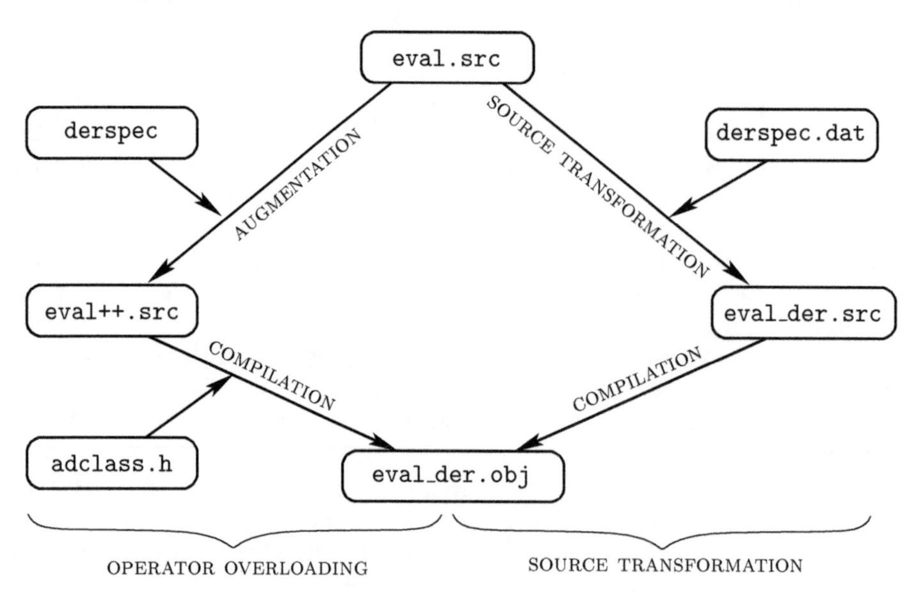

Figure 6.1: From Function Sources to Derived Object Files

There we start with a source code `eval.src` that is either transformed into another source `eval_der.src` by a preprocessor or augmented to a program `eval++.src` by the user before being compiled. Either modification requires the selection of independent and dependent variables as well as the kind of derivative objects and differentiation modes the programmer wishes to specify. We have represented this information by the boxes `derspec` and `derspec.dat`, but it can of course be imported in many different ways.

Subsequently either the source `eval_der.src` or the source `eval++.src` must be compiled, together with header files defining the new type of variable and their operations included in the second case. In either scenario the resulting object file `eval_der.obj` may then be linked with a library of AD routines to yield (possibly only a part of) an executable that evaluates derivative values `dy` $= \dot{y}$, `bx` $= \bar{x}$, and `dbx` $= \dot{\bar{x}}$ for given points x, directions `dx` $= \dot{x}$, adjoints `by` $= \bar{y}$, etc., as depicted in Fig. 6.2. The execution may generate a considerable amount of temporary data, which may be kept in internal arrays or flow out to scratch files.

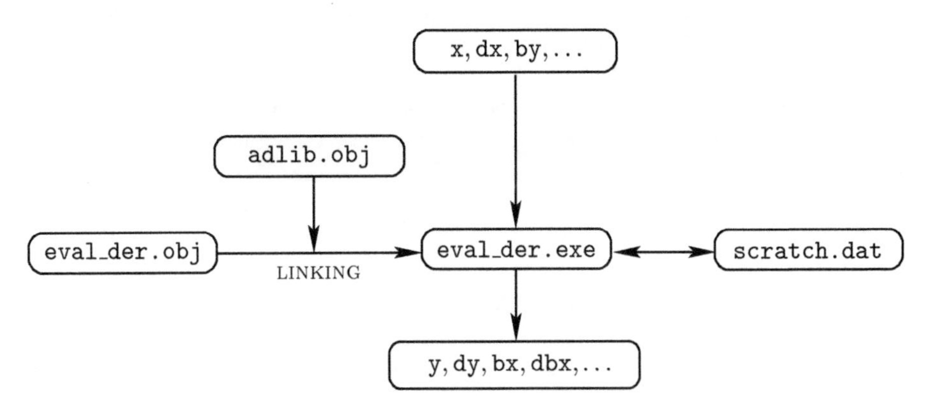

Figure 6.2: Linking and Executing Derived Object Files

There is no better way to understand AD than to implement one's own "baby" AD tool. But it is also worth pointing out that for many small- to medium-size problems, particularly those where derivative calculation does not account for a high proportion of the runtime requirements, a simple AD tool is often perfectly adequate and has the added merit that it contains no unnecessary features or bugs that the programmer did not personally put there.

Active and To-Be-Recorded Variables

The most basic task for the differentiation of a given evaluation code is the identification of all variables that need to be *active* in the following sense. In the computational graph discussed in section 2.2 all vertices lying on a path between the minimal (independent) nodes and the maximal (dependent) nodes are *active* because their values enter into the functional relation between at least one such pair of values. Moreover, unless an active vertex enters only as argument into linear operations, its value will also impact some partial derivative, namely, entry of the overall Jacobian. As discussed in section 4.3, it will then have to be saved during the forward sweep or recomputed during the return sweep. Strictly speaking, the attribute *active* or *to-be-recorded* should be assigned to each value computed during the evaluation of a code. In practice this decision is usually taken for a computer variable, namely, a symbol within a certain scope, or at least for each one of its occurrences as a left-hand side of an assignment. This simplification usually means that some values are treated as active, even though that may not be necessary. Erring on the side of caution in this way forgoes some potential gain of efficiency but entails no loss of correctness and only rarely affects numerical accuracy. Naturally, variables or values that are not selected as *active* are called *passive* throughout this book.

In the overloading approaches described in section 6.1 activity must be decided for each program variable, which is then retyped accordingly. A simple way to avoid this tedious task is to redefine in the evaluation code all `doubles` to the active type `adoubles`. Here we assume tacitly that all active calculations are performed in double precision. This brute-force modification is used, for example, in the Network Enabled Optimization Server (NEOS), which employs ADOL-C for the differentiation of user codes written in C. The downside is that many passive variables may be appended with derivative fields whose value remains zero but that are nevertheless involved in many (trivial) calculations. This effect can be partly overcome by using an activity analysis at runtime as used, for example, by the newer versions of the C/C++ tools ADOL-C and CppAD.

With a language translation approach, a great deal more can be done to automate the dependence analysis required to determine which variables have derivatives associated with them. Of course, where it is impossible to determine at compile-time whether two variable values depend on each other, for example, because array indices or pointers are manipulated in a complex way at runtime, the translator must make a conservative assumption or rely on user-inserted directives.

6.1 Operator Overloading

Overloading is supported by computer languages such as C++, Ada, and Fortran 90. Certain considerations regarding the use of operator overloading influence the choice of which language is appropriate. The main issues regarding language capabilities for the purpose of AD by overloading are

- whether the assignment operator can be overloaded;
- what level of control the user is permitted to exercise over the management of dynamically allocated memory;
- whether user-written constructor functions can be automatically called to initialize user-defined data types; and
- whether user-written destructor functions can be automatically called to do housekeeping when variables go out of scope and are deallocated.

By now, a large number of AD tools exist based on overloading in ADA (see, e.g., [BBC94]), Fortran 90 (see, e.g., [DPS89] and [Rho97]), C++ (see, e.g., [Mic91], [GJU96], [BS96]), [Bel07]), and MATLAB (see, e.g., [RH92] and [CV96]). There are also some C++ implementations (see, e.g., [ACP01]) using expression templates [Vel95] to generate derivative code that is optimized at least at the statement level. In terms of runtime efficiency the selection of a suitable compiler and the appropriate setting of flags may be as important as the choice of language itself. Naturally, we cannot make any recommendation of lasting validity in this regard as software environments and computing platforms are subject to continual change.

The examples given below are coded in a subset of C++ but are designed to be easy to reimplement in other languages that support operator overloading. Reading this chapter does not require a detailed knowledge of C++, since we deliberately do not exploit the full features of C++. In particular, the sample code that we give is intended to illustrate the corresponding algorithm, rather than to be efficient.

Simple Forward Implementation

In this section we describe a simple approach to implementing the basic forward mode described in section 3.1. The implementation presented here is based on the tapeless forward-mode of the AD tool ADOL-C. It propagates a single directional derivative for any number of independent or dependent variables.

Implementing a Forward-Mode Tool

To build a simple forward-mode package, we need to carry out the following tasks:

- Define the new data type or class that contains the numerical values of v_i and $\dot{v}_i$. Here, we will use the new class

```
class  adouble
{  double  val;
   double  dot;      }
```

- Define arithmetic operations on the `adouble` class corresponding to the usual floating point operations on scalars. These overloaded operations must manipulate the second part of the `adouble` corresponding to $\dot{v}_i$ correctly, according to the chain rule; see, for instance, Table 3.3. Some languages allow the programmer to define several different procedures with the same name, but operating on or returning arguments of different types. For example, we may define a function `sin` that operates on an `adouble` and returns another `adouble`.

```
adouble sin (adouble a)
{  adouble b;
   b.val = sin(a.val);
   b.dot = cos(a.val) * a.dot;
   return b;        }
```

For certain programming languages operator overloading allows the programmer the same freedom with symbols corresponding to the built-in operations. For example, we might define the following multiplication of `adoubles`.

```
adouble operator* (adouble a, adouble b)
{  adouble c;
   c.val = a.val * b.val;
   c.dot = b.val * a.dot + a.val * b.dot;
   return c;        }
```

In the case of binary operations, we also need "mixed-mode" versions for combining `adoubles` with constants or variables of type `double`.

To reduce the likelihood of errors, one may define the derivative field `dot` as "private" so that the user can access it only through special member functions.

- Define some mechanism for initializing `adouble` components to the correct values at the beginning, for example, by providing the member functions `setvalue()` and `setdotvalue()`. These functions should also be used to initialized the derivative values of the independent variables. For extracting variable and derivative values of the dependent variables, member functions like `getvalue()` and `getdotvalue()` should be available. Uninitialized derivative fields may default to zero.

In some applications one may wish to deliberately assign a `double` to an `adouble`. Then one needs corresponding facilities.

- Conversions from `doubles` to `adoubles` may occur implicitly with the `dot` value of the `adouble` set to zero. For example, a local variable `a` may be initialized to zero by an assignment `a=0` and later used to accumulate an inner product involving active variables. In C++ [Str86] the assignment operator may invoke a suitable constructor, that is, a user-supplied subroutine that constructs an `adouble` from given data. Here we may define in particular the following constructor.

```
adouble::adouble (double value)
{  val = value;
   dot = 0;      }
```

This constructor sets the derivative field of an `adouble` to zero whenever it is assigned the value of a passive variable. Another approach is to define assignments of the form `x = b` by overloading the assignment operator.

```
adouble& adouble::operator= (double b)
{  val = b;
   dot = 0;
   return *this;    }
```

Note that this is the only point at which we have proposed to overload the assignment operator. Some languages do not allow the assignment operator to be overloaded. Then such implicit conversions cannot be performed, and one has to ensure full type consistency in assignments.

Using the Forward-Mode Tool

To apply our forward-mode AD implementation to a particular numerical program, we need to make a number of modifications to the code that evaluates the function. Minimizing these modifications is one of the design goals of AD implementation, since rewriting large pieces of legacy code is not only tedious but also error prone. The changes that cannot be avoided are the following.

Changing Type of Active Variables

Any floating point program variable whose derivatives are needed must be active and thus redeclared to be of type `adouble` rather than of type `double`.

Initializing Independents and Their Derivatives

We also need a way of initializing independent variables. In the C++ package ADOL-C [GJU96] the binary shift operator was overloaded such that the statement `x<<=b` has the effect of specifying `x` an independent variable and initializing its value to `b`. There exist several alternatives to this approach.

At the point where independent variables are assigned numerical values, the corresponding derivative values should also be initialized, unless they may default to zero.

Deinitializing Derivatives of Dependents

At the point where the dependent variable values are extracted, the corresponding derivative values can also be extracted.

Nothing else in the user-defined function evaluation code needs to change. In particular, all the rules for derivative propagation are concealed in the AD package nominated in a suitable header file.

A Small Example of Simple Forward

As a small example for the simple forward-mode implementation presented here, we consider the scalar-valued function

$$f : \mathbb{R}^2 \to \mathbb{R}\,, \qquad f(x) = \sum_{i=1}^{n} \left[\left(x_1 - \alpha_i\right)^2 + \left(x_2 - \beta_i\right)^2 \right] . \tag{6.1}$$

Here we assume that the real scalars $x = \left(x_1, x_2\right)^\top$ are the independent variables and the parameters α_i and β_i are constants. One may be interested in the directional derivative of f in direction $\dot{x} = (\dot{x}_1, \dot{x}_2)^\top$. For this purpose, first we present the code to evaluate $f(x)$. Note that `cin` and `cout` are the standard input/output stream operations used in C++ to read and write variable values.

```
double alphai, betai;
double x1, x2, y;
cin >> x1, cin >> x2;
y = 0.0;
for (int i=0; i<n; i++)
{ cin >> alphai; cin >> betai;
  y = y + (x1-alphai)*(x1-alphai) + (x2-betai)*(x2-betai);}
cout << y;
```

The augmented code to calculate the directional derivative is given below, where the vertical bar | denotes a modified or inserted line.

```
 1   double alphai, betai;
 2 | adouble x1, x2, y;
 3 | double dx1, dx2;
 4   cin >> x1;  cin >> x2;
 5 | cin >> dx1; cin >> dx2;
 6 | x1.setdotvalue(dx1);
 7 | x2.setdotvalue(dx2);
 8   y = 0.0;
 9   for (int i=0; i<n; i++)
10    { cin >> alphai;  cin >> betai;
11      y = y + (x1-alphai)*(x1-alphai) + (x2-betai)*(x2-betai);}
12 | cout << y;cout << y.dot;
```

The changes introduced into the augmented program correspond to the points enumerated at the beginning of this section, and the input/output stream operations have been overloaded to read and write `adoubles`, so that `cin` will assign input values to the `adoubles` `x` and `y`, together with a `dot` value of zero.

Lines 4 to 7 of the augmented code correspond to the initialization in the first loop of Table 3.4. The function evaluation in the second loop of Table 3.4 is performed in the lines 8 to 11. Line 12 represent the third loop of Table 3.4, namely, the extraction of the dependent variables.

Note, in particular, that the "main body" of the code, which actually calculates the function value, is unchanged. Of course, in this tiny example the main body is a relatively small proportion of the code, but in a more realistic application it might constitute 99% of the code, including nested loops and recursive procedure calls with parameter passing, as well as other programming structures. Note that our example already includes overwriting of `y`. Apart from redeclaring `doubles` as `adoubles`, no changes need be made to any of this code. However, a judicious choice of all variables that must be redeclared because they are active, as defined at the beginning of this chapter, may not be easy. However, performing this selection carefully is important for efficiency to avoid redundant calculations.

Vector Mode and Higher Derivatives

To evaluate a full Jacobian with a single pass of the forward mode, we can redefine the `adouble` type so that the `dot` field is an array of tangent components

```
class v_adouble                  with      class vector
{  double val;                             { double comp[p]; };
   vector dot; };
```

where `p` is the number of directional components sought. For efficiency it is preferable that `p` is a compile-time constant, but for flexibility it may also be defined as a static member of the class `vector`.

We also define a number of overloaded operations on the type `vector`, for example,

```
vector operator* (double a, vector b)
{  vector c;
   for (int i=0; i<p; i++)
      c.comp[i] = a * b.comp[i];
   return c;                           }
```

Once we have done so, the same code used to define the overloaded operations on scalars will also work correctly on vectors, allowing us to evaluate an entire Jacobian or, more generally, an arbitrary family of tangents, in a single forward sweep.

We can also extend the definition of an `adouble` to encompass higher-order derivatives. For example, we can define a type

```
class s_adouble
{  double val;
   double dot;
   double dotdot;  };
```

with appropriate overloaded operations acting on the various components to compute second-order derivative, or a more general type `taylor`.

Simple Reverse Implementation

In this subsection we describe a simple approach to implementing the basic reverse method introduced in section 3.2. This implementation evaluates a complete adjoint vector. This adjoint vector can be a normal vector corresponding to the gradient of one of the dependent variables, or it may be an arbitrary linear combination of such normal vectors, for example, a set of constraint normals scaled by Lagrange multipliers such as $\bar{x} = \sum_i \bar{y}_i \nabla_x F_i$.

During the execution of the evaluation procedure we build up an internal representation of the computation, which we will call *trace*. The trace is in effect a three-address code, consisting of an array of operation codes and a sequence of variable indices, both encoded as integers. In addition to these two symbolic data structures there is a third numerical record for storing the floating point numbers that represent the prevalues discussed in section 4.2. Both the symbolic information and the numerical data may be reused in various ways, so their distinct roles should be understood. Generally, the symbolic trace is specific to overloading tools, whereas the record of prevalues occurs also in source transformation based AD tools. The return sweep will be carried out by a simultaneous interpretation of the three records, which are illustrated in Fig. 6.3.

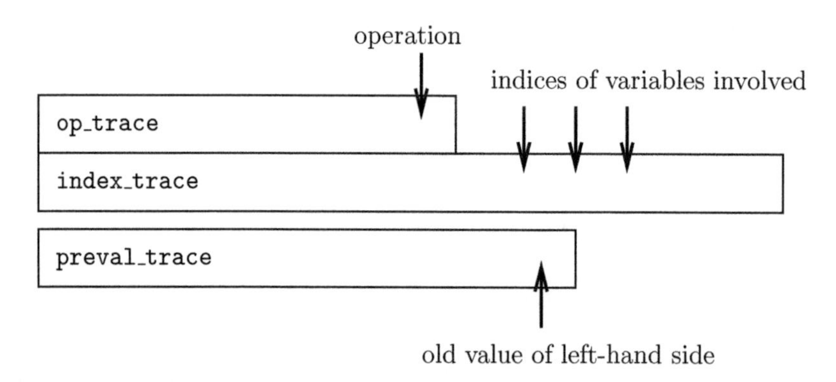

Figure 6.3: Structure of the Trace

Implementing a Reverse-Mode Tool

To build a simple reverse-mode AD package, we need to carry out six tasks.

- Define a new data type, here again called `adouble`, containing the numerical value of v_i and an identifier or index.

  ```
  class adouble
  {  double val;
     int index; };
  ```

  Additionally we need the trace consisting of the three arrays `op_trace`, `index_trace`, and `preval_trace` to store information about the performed operations, the involved variables, and the overwritten variables, respectively. Furthermore, a global variable `indexcount` is required to perform bookkeeping about the already used indices. This setup is illustrated in Fig. 6.3.

- Corresponding to the usual floating-point operations on scalars, define arithmetic operations of `adouble` type. These overloaded operations calculate the floating point value for v_i as usual, but as a side effect they also record themselves and their arguments in three arrays: `op_trace`, `index_trace`, and `preval_trace`. For this purpose we also need to define an integer `opcode` for each of the elemental operations, for example,

```
const int
emptyv = 0, constv = 1, indepv = 2, bplusv = 3,
bminusv = 4, bmultv = 5, recipv = 6,  ...
expv = 11, lnv = 12, sinv = 13, cosv = 14, ...
assignv = 61, ...
```

Then we may define typical operations as follows.

```
adouble sin (adouble a)
{  adouble b;
   indexcount += 1;
   b.index = indexcount;
   put_op(sinv, op_trace);
   put_ind(a.index, b.index, index_trace);
   b.val = sin(a.val);
   return b;               }

adouble operator* (adouble a, adouble b)
{  adouble c;
   indexcount += 1;
   c.index = indexcount;
   put_op(bmultv, op_trace);
   put_ind(a.index, b.index, c.index, index_trace);
   c.val = a.val*b.val;
   return c;               }
```

In the case of binary operations such as * we again need to provide "mixed-mode" versions for combining `adoubles` with constants or variables of type `double`. Assignments between `adoubles` and their initializations from `doubles` must also be recorded as operations. For example, the overloaded assignment operator to store the information required for the return sweep may be defined by

```
adouble&adouble::operator =  (adouble b)
 {  put_op(assignv, op_trace);
    put_val(val, preval_trace);
    put_ind(b.index, index_trace);
    val = b.val;
    index = b.index;          }
```

- We also need a way of initializing independent variables (and setting the corresponding adjoint values initially to zero). In the C++ package ADOL-C [GJU96] the binary shift operator was overloaded such that the statement `x<<=b` has the effect of specifying `x` an independent variable and initializing its value to `b`. There exist several alternatives to this approach.

- Define a `return_sweep` routine that reverses through the trace and calculates the adjoint variables $\bar{v}_i$ correctly, according to the adjoint evaluation procedure with tape and overwrites given in Table 4.4. The routine `return_sweep()` consists basically of a `case` statement inside a loop. The actual derivative calculation is performed by using the two additional arrays `value` containing intermediate results and `bar` containing the derivative information.

```
void return_sweep()
 { while (NOT_END_OF_OPTAPE)
    { opcode = get_op(op_trace);
      switch(opcode)
      { ...
        case assignv:
          res = get_ind(index_trace);
          value[res] = get_val(preval_trace);
        break;
        ...
        case sinv:
          res = get_ind(index_trace);
          arg = get_ind (index_trace);
          bar[arg] += bar[res]*cos(value[arg]);
        break;
        ...
        case bmultv:
          res  = get_ind(index_trace);
          arg1 = get_ind(index_trace);
          arg2 = get_ind(index_trace);
          bar[arg1] += bar[res]*value[arg2];
          bar[arg2] += bar[res]*value[arg1];
        break;
        ...                                   } } }
```

- Define some mechanism for extracting the current floating point value from an `adouble`, for initializing adjoint components to the correct values at the beginning of the return sweep and for extracting gradient component values when these become available at the end; in other words, define routines `value`, `setbarvalue`, and `getbarvalue`.

- Define a routine `resettrace()` to indicate that the adjoint calculation is finished such that the trace and the additional arrays can be reused for a new derivative evaluation.

Using the Reverse-Mode Tool

To apply our reverse-mode AD implementation to a particular numerical program, we need to make a number of modifications to the users code. More specifically we must modify the evaluation routine that defines the function to be differentiated and also the calling subprograms that invoke it to obtain function and derivative values.

Changing Type of Active Variables

Just like in the simple forward implementation, all floating point program variables that are active must be redeclared to be of type `adouble` rather than of type `double`.

Initializing Adjoints of Dependents

At some point after the dependent variables receive their final values, the corresponding adjoint values must also be initialized as in Table 3.5.

Invoking Reverse Sweep and Freeing the Traces

Before the gradient values can be extracted, the `return_sweep` routine must be invoked. This invocation can take place from within the gradient value extraction routine. We need to indicate when the adjoint calculation is finished so that we can reuse the storage for a subsequent derivative evaluation. Otherwise, even a function with a small computational graph will exhaust the available storage if it is evaluated many times.

Deinitializing Adjoints of Independents

When the adjoint calculation is completed, the corresponding derivative values can be extracted.

Remarks on Efficiency

While the reverse implementation sketched here incurs some overhead costs; it does bring out some salient features. Not only is the total temporal complexity a small multiple of the temporal complexity of the underlying evaluation code, but we observe that the potentially very large data structures represented by the three trace arrays are accessed strictly sequentially.

A Small Example of Simple Reverse

As a small example for the simple reverse-mode implementation discussed above, we consider again the scalar-valued function (6.1) and corresponding evaluation code given on page 114. The augmented code to calculate the gradient is as follows.

```
1    double alphai, betai;
2  | adouble x1, x2, y;
3  | double bx1, bx2, by;
4    cin >> x1;  cin >> x2;
```

```
 5   y = 0.0;
 6   for (int i=0; i<n; i++)
 7   { cin >> alphai;  cin >> betai;
 8     y = y + (x1-alphai)*(x1-alphai) + (x2-betai)*(x2-betai);}
 9 | cin >> by;
10 | setbarvalue(y, by);
11 | return_sweep();
12 | getbarvalue(x1,bx1);
13 | getbarvalue(x2,bx2);
14 | resettrace();
```

Lines 5 to 8 correspond to the recording sweep of Table 4.4. The initialization of the normal is done in lines 9 and 10. The second for-loop of the return sweep in Table 3.7 is invoked in line 11. Lines 12 and 13 represent the final for-loop of Table 3.7, namely, the extraction of the adjoint values.

6.2 Source Transformation

An AD tool based on operator overloading consists essentially of a library written in the same language as the program to be differentiated. Its size can be reasonably small. In contrast, source transformation is a more laborious approach to implementing an AD tool, which is similar in its complexity to developing a compiler. The source transformation description that we provide in this chapter should facilitate some understanding of what is going on inside sophisticated AD tools based on source transformation, i.e., an AD preprocessor.

In the earliest days of AD (see, e.g., [Con78] and [KK+86]), users were expected to completely rewrite their code, usually replacing all arithmetic operations with function calls. Later software designers tried to live up to the expectation generated by the label "automatic differentiation" by using preprocessors (e.g., Augment used by Kedem [Ked80]) or overloading (e.g., in Pascal-SC by Rall [Ral84]). Probably the first powerful general-purpose system was GRESS, developed at Oak Ridge National Laboratory [Obl83] in the 1980s and later endowed with the adjoint variant ADGEN [WO+87]. After the first international workshop on AD in 1991 in Breckenridge, three large Fortran 77 systems were developed: ADIFOR [BC+92] at Argonne National Laboratory and Rice University; Odyssée [R-S93] at INRIA, Sophia Antipolis; and TAMC as a one-man effort by Ralf Giering of the Meteorological Institute of Hamburg [GK98]. Currently, the tool Tapenade [HP04] is developed and maintained at INRIA, Sophia Antipolis, and TAF [GK98] as successor of TAMC by the company FastOpt. All these systems accept Fortran 77 codes that contain some constructs of Fortran 90 and generate derivative code in the same language. A similar one-man effort is PADRE2 [Kub96], which generates some scratch files in addition to the derived source code; this is also true of GRESS/ADGEN. While PADRE2 also calculates second-order adjoints, the other systems generate "only" first derivative codes. In principle, however, the source transformation may be applied repeatedly, yielding second and higher derivative codes if everything goes well. Naturally, the symmetry of Hessians and higher derivative tensors cannot be exploited by such repeated source transformations. Taylor coefficients and derivative tensors of arbitrary order have been a key aspect of the systems DAFOR [Ber90b] and COSY INFINITY [Ber95] by Martin Berz. The system PCOMP [DLS95], developed by Klaus Schittkowski and his coworkers, generates efficient derivative code from function specification in a restricted Fortran-like language.

AD preprocessors belong to a large family of static tools: they perform program analysis or source transformation at compile time without requiring input data or running the program. Static tools may have various objectives, among which one can find, for example, the compilation of source files into binary code, the detection of runtime errors, and the instrumentation of source code. These tools take as input a program written in a programming language (C, C++, Ada, Fortran, Perl, etc.), construct an internal representation of the program, perform some analysis, and possibly transform the internal representation using the result of the analysis. AD preprocessors are enhancers, that is, they generate a new program that computes supplementary values, namely, derivatives of the values of interest with respect to input values. A key distinction from other enhancers is that while efficiency of the generated program is of low priority in most tools, for instance, when debugging a program or finding runtime errors, it is highly important for AD. The fact that the generated code is meant to be used in the operational phase leads to specific concerns about the cost of the derivatives compared to the original function.

Enhancer tools are built from four components, as shown in Fig. 6.4. The parsing, printing, and optimization components are not specific to AD. Therefore they are not described here in detail. For the transformation task, AD preprocessors use general-purpose data structures to internally represent the program.

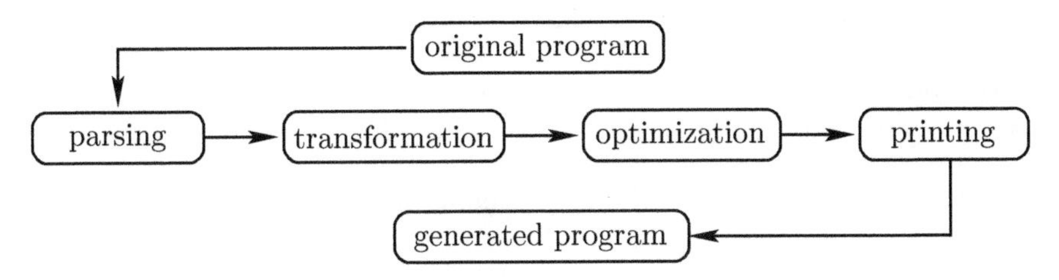

Figure 6.4: General Architecture of a Source Preprocessor

Among them one can find the following fundamental ones:

- **The call graph** represents the "call to" relation between procedures of the program as a graph whose nodes are the procedure names , say P and Q, which are connected by an arc if and only if P calls Q.

- **The abstract syntax tree** represents the procedures declarations and statements of each procedure in the program in a syntactical way.

- **The control flow graph** represents the "successor" relation between statements of a procedure as a graph the nodes of which are the statements indices. One arc relates statements s_i and s_j if and only if statement s_j must be executed directly after statement s_i.

- **The data flow graph** represents the "depends on" relation between variables of a procedure/program as a graph the nodes of which are the variables and an arc relates variables v_j and v_i if and only if the value of variable v_i is obtained directly from the value of variable v_j.

The differentiation of a program is a complex process involving the enhancement of the original program: one or more derivative statements are added in forward or reverse mode for each original active statement. This objective can be reached in two ways. One can apply differentiation to generate a completely differentiated code, then apply slicing [Tip95] with respect to dependent and independent variables to discard useless statements. This approach is called context-free differentiation. Alternatively, one can apply context-sensitive differentiation. The context-sensitive and context-free differentiation approaches differ in that in context-sensitive differentiation only active original statements with respect to dependent and independent variables are differentiated, whereas in context-free differentiation all (real) original statements are differentiated. Moreover, in context-sensitive differentiation only variables whose values are necessary for the evaluation of derivative values are recorded and retrieved, whereas in context-free differentiation the value of all modified variables is recorded or retrieved.

For clarity of the following description of an AD preprocessor, some hypotheses on the input code are imposed:

- Access paths to memory location are limited to scalar variable accesses (i.e., component of array, fields of structures, and the like are not considered here).

- Subprograms are limited to procedures (functions, function pointers, virtual functions, arrays of functions, and the like are not considered here).

- Input/output statements are considered as passive statements and are therefore not differentiated.

- Goto, jump, and exit statements are not dealt with.

With these restrictions, the description below is independent of the programming language and much simpler than a complete specification. At the same time, the description is general enough to give an overview of the method. Moreover, it allows straightforward extensions to overcome these restrictions.

In forward mode, the calculation of the derivative values is performed in the same order as for the original values. Consequently, the control flow (branches, loops) and the structure (procedure definitions and calls) of the program can be maintained.

Implementing a Forward-Mode Tool

The simplest version of a forward-mode tool first copies the original abstract syntax tree to obtain the derivative abstract syntax tree, which is then transformed by applying recursively the following forward differentiation rules:

- An assignment is differentiated by
  1. generating the derivative assignments according to section 3.1,
  2. associating new indices to the derivative statements,
  3. inserting the new statements before the original one,
  4. updating the control flow graph by adding the new arcs and modifying the old ones.
- A call to a procedure `eval(x, y)` is differentiated by
  1. generating a new call `eval_tang(x, dx, y, dy)`,
  2. associating the new call to a new index,
  3. replacing the original call by the new call,
  4. replacing the name of the original procedure by the name of the differentiated procedure in the call graph,
  5. updating the control flow graph by adding the new arcs and modifying the old ones.
- A control statement is differentiated by keeping the control as it is and applying the rules to the statements in the body.
- Any other statement is left unchanged.
- A body is differentiated by applying the differentiation rules to the statements from the body line by line and concatenating the results in the original statement order.
- A declaration is differentiated by
  1. generating the derivative declaration as a copy of the original declaration for the derivative variable,
  2. inserting the derivative declaration before the original one,
- A procedure `eval(x, y)` is differentiated by
  1. generating the derivative header `eval_tang(x, dx, y, dy)` from the original header `eval(x, y)`,
  2. generating the derivative declarations from the original ones,
  3. generating the derivative body from the original body,
  4. replacing each procedure declaration component by its derivative counterpart in the original declaration.
- A program is differentiated by applying the differentiation rules to the global declarations and each procedure in the program.

Note that the call- and control-flow graphs are updated during the transformation of the abstract syntax tree. This method is easily extendable to all forward-mode variants.

Using the Forward-Mode Tool

Using the forward-mode tool, the user gives as input the source files of the program that defines the function to be differentiated and simply runs the AD preprocessor to generate the derivative source program. Note that the derivative program must be modified to initialize the independent derivative variables to the required values. Then the derivative program may be used as any user-level program to obtain the derivative values.

We consider the same example as in section 6.1 on operator overloading. In context-free mode, the forward AD tool generates the following.

```
 1 | double dalphai, dbetai;
 2   double aplha1, beta1;
 3 | double dx1, dx2, dy;
 4   double x1, x2, y;
 5   cin >> dx1, cin >> dx2; /* Added by the user */
 6   cin >> x1, cin >> x2;
 7 | dy = 0.0;
 8   y = 0.0;
 9   for (int i=0; i<n; i++)
10   { cin >> alphai; cin >> betai;
11     cin >> dalphai; cin >> dbetai; /* Added by the user */
12 |   dy = dy + 2*(x1-alphai)*dx1 - 2*(x1-alphai)*dalphai +
13 |             2*(x2-beta2i)*dx2 - 2*(x2-beta2i)*dbetai;
14     y = y + (x1-alpha1)*(x1-alphai) + (x2-betai)*(x2-betai); }
15   cout << dy; /* Added by the user */
16   cout << y;
```

Here we have marked each extra line added by the AD preprocessor with a vertical bar. Lines 5 to 8 of the generated code correspond to the initialization in the first loop of Table 3.4. The function evaluation continued in lines 9 to 14 represents the second loop of Table 3.4. The extraction of the computed values, i.e., the third loop of Table 3.4, is performed in lines 15 and 16.

As we said before, the generated code is suboptimal: if only the derivative with respect to `x1`, `x2` is required, `alphai` and `betai` are passive, and the only nonzero contributions to the computation of `dy` are the first and the third terms of the right-hand side. In context-sensitive mode, the evaluation of `dy` will contain only these two nonzero terms.

Note that in this code, the initialization of the derivative values is performed by reading in the values from the standard stream, and the derivative values are printed on the standard stream. Because of the limitations described above, these statements are input/output and must be added by the user in the derivative program.

Implementing a Reverse-Mode Tool

In reverse mode, the evaluation of the derivative values is performed in the reverse order with respect to the original values. This makes the evaluation of the derivative statement more complex than in forward mode.

This section shows the simplest implementation of the reverse-mode: each derivative procedure executes the forward sweep (the original procedure and the storage of the prevalues of the assigned variables) and the return sweep (the values are restored and the derivatives are computed). This strategy is analyzed as *split reversal* in section 12.2 and is often used when writing adjoint code by hand because of its simplicity. Other strategies (including loop checkpointing) can be implemented by simply combining the forward/return sweeps described above in different manners, as discussed in Chapter 12.

A reverse-mode AD preprocessor may proceed in the same way as the forward-mode tool: it duplicates the original syntax tree to generate the forward sweep and inverts it to obtain the return sweep. This approach does not work for programs with `goto`, `jump`, or `exit` statements. To handle such constructs, the transformation must be applied on the control flow graph instead of the syntax tree.

The abstract syntax tree is transformed by applying recursively the following reverse differentiation rules that construct the forward and return sweeps simultaneously.

- An assignment is differentiated by
  1. generating the derivative assignments by using the rules from section 3.2,
  2. generating the record and retrieve statements for the modified variable,
  3. associating new indices to the derivative and record/retrieve statements,
  4. inserting the record and original statements in the forward sweep,
  5. inserting the retrieve and derivative statements in the return sweep,
  6. updating the control flow graph.
- A call to a procedure `eval(x, y)` is differentiated by
  1. generating a new call `eval_dual(bx, x, by, y)`,
  2. generating the record and retrieve statements for the modified parameters,
  3. associating the call of `eval_dual(bx, x, by, y)` and record/retrieve statements to new indices,
  4. inserting the record statements and the call of `eval(x, y)` in the forward sweep,
  5. inserting the retrieve and derivative statements in the return sweep,
  6. updating the call graph,
  7. updating the control flow graph.

- A branch statement is differentiated by
  1. applying the differentiation rules to the statements in the body and generating the forward and return sweep bodies,
  2. generating the forward branch statement by replicating the original test and inserting the forward sweep body,
  3. generating the return branch statement by replicating the original test and inserting the return sweep body,
  4. inserting the forward branch in the forward sweep,
  5. inserting the return branch in the return sweep,
  6. updating the call graph,
  7. updating the control flow graph.

- A loop statement is differentiated by
  1. applying the differentiation rules to the statements in the body and generating the forward and return sweep bodies,
  2. generating the forward loop statement by replicating the original header and inserting the forward sweep body,
  3. generating the return loop statement by reverting the original header and inserting the return sweep body,
  4. inserting the forward loop in the forward sweep,
  5. inserting the return loop in the return sweep,
  6. updating the call graph,
  7. updating the control flow graph.

- Any constant statement is added to the forward sweep with no counterpart in the return sweep.

- A procedure is differentiated by
  1. generating the derivative declaration as a copy of the original declaration for the derivative variable,
  2. inserting the derivative declaration before the original one.

- A body is differentiated by applying the differentiation rules to the statements from the sequence in order: each statement of the forward sweep is concatenated at the end of the forward sweep body and each differentiated statement is concatenated at the beginning of the return sweep body.

- A procedure with header `eval(x, y)` is differentiated by
  1. generating the derivative header `eval_dual(bx, x, by, y)`,
  2. replacing the procedure header by the derivative header,
  3. replacing the procedure declaration by the derivative declaration,
  4. replacing the procedure body by the derivative body.

- A program is differentiated by applying the differentiation rules to the global declarations and each procedure in the program.

Note that the call and control flow graphs are updated during the transformation of the abstract syntax tree.

The choice of the record/retrieve implementation is of great impact on the efficiency of the derivative program but is language, and machine-dependent and is therefore not described here.

Using the Reverse-Mode Tool

The reverse-mode AD preprocessor is used in the same way as a forward-mode AD preprocessor: the user gives as input all the source files of the program, simply runs the tool, and inserts the initializations of the derivative variables. Then the derivative program may be used as any other user-level program to obtain the derivative values.

For the same example as in section 6.1, the context-free reverse-mode tool generates the following.

```
 1 | double save_aplhai, save_betai;
 2 | double balphai, bbetai;
 3   double aplhal[n], betal[n];
 4 | double bx1, bx2, by;
 5 | double save_y[n];
 6   double x1, x2, y;
 7   cin >> x1; cin >> x2;
 8   y = 0.0;
 9   for(int i=0; i<n; i++)
10   { cin >> alphai; cin >> betai;
11 |   save_alphai[i] = alphai;
12 |   save_betai[i] = betai;
13 |   save_y[i] = y;
14     y = y + (x1-alphai)*(x1-alphai) + (x2-betai)*(x2-betai); }
15   cout << y;
16   cin >> by; /* Added by the user */
17 | bx1 = 0.0;
18 | bx2 = 0.0;
19 | balphai = 0.0;
20 | bbetai = 0.0;
21 | for (int i=n-1; i>=0; i--)
22 | { alphai=save_alphai[i];
23 |   betai=save_betai[i];
24 |   y=save_y[i];
25 |   bx1 += 2*(x1-alphai)*by;
26 |   balphai -=  2*(x1-alphai)*by;
27 |   bx2 += 2*(x2-beta2i)*by;
28 |   bbetai -=  2*(x2-beta2i)*by; }
29   cout << bx1; /* Added by the user */
30   cout << bx2; /* Added by the user */
31   cout << balphai; /* Added by the user */
32   cout << dbetai; /* Added by the user */
```

Note that as in forward mode, the initialization and retrieval of the derivative have been added manually but could have been generated automatically. Lines 7 to 15 represent the recording sweep of Table 3.7. The second for-loop of the return sweep in Table 3.7 can be found in lines 21 to 28. Lines 29 to 32 correspond to the final loop of the return sweep in Table 3.7. The generated code is suboptimal: if only the derivative with respect to `x1, x2` were required, `balphai` and `bbetai` should disappear, which happens in context-sensitive mode.

Context-Sensitive Transformation

If the source transformation methods presented above are applied, all (real) input variables are considered as independent, and all (real) output variables are considered as dependent variables. Hence all values of all modified variables are recorded and retrieved in reverse mode. This naive approach corresponds to the brute-force use of the operator overloading tool when all `double` variables are retyped as `adouble`.

To generate a program that computes the derivative of the dependent with respect to the independent variables, the dispensable derivative statements must be discarded by an optimization phase, or the active statements must be determined and the differentiation process must be applied only to them. Identifying the useless statements by program optimization is much more expensive than not generating them in the first place.

The general purpose AD preprocessors apply a static analysis generally called "activity analysis" to detect whether a particular variable, statement, or procedure is to be considered active with respect to the independent and dependent variables. The AD preprocessor optimizes the generated code by generating derivatives only for active program components (variable, statement, procedure). The activity analysis is performed on the original program before the forward- or reverse-mode differentiation and allows for the generation of a better derivative program by avoiding the differentiation of passive statements as much as possible.

In the same manner, recording and retrieving the values as it is performed in the naive reverse transformation are not efficient: the values of `y` are recorded all along the forward loop and restored all along the backward loop even though the value of `y` is not used. The "to-be-recorded analysis" described, for example, in [FN01] is also an a priori analysis that allows one to know for each occurrence of each variable whether it has to be recorded. However, this analysis has not yet been implemented in general-purpose AD tools.

Note that if the programming language allows the use of pointers, the previous analysis must be performed modulo aliases. Hence, for example, if a variable is active, all its aliases are also active. Alias analysis is a difficult subject on which a lot of work has been done (see, e.g., [Ste96, Deu94]), but it is not a problem specific to AD and is therefore not described here.

6.3 AD for Parallel Programs

The widespread use of large clusters and the advent of multicore processors have increased the push toward automatic differentiation of programs that are written by using libraries such as MPI [MPI] for message passing or pragma-based language extensions such as OpenMP [OMP]. The topic was first investigated in [Hov97]. We cannot hope to cover all the concepts by which parallel computing is supported, but we want to concentrate on the most important aspects of message passing as standardized by MPI and code parallelization enabled by OpenMP. The main problems arise in the context of source transformation AD, but a few are applicable to AD via operator overloading as well. In practice, source transformation AD tools so far have limited coverage of parallel programming constructs.

Extended Activity Analysis

The preceding section discussed activity analysis, which determines the program variable subset that needs to carry derivative information. The data flow analysis described in section 6.2 covers the constructs of the respective programming language but not dependencies that are established through library calls such as MPI's `send`, `recv`, or the collective communication operations. On the other hand, these hidden dependencies clearly do not originate only with MPI constructs. The same effect can easily be recreated in sequential programs, for instance, by transferring data from program variable `a` to `b` by writing to and reading from a file.

The default solution to this problem is the assumption that any program variable `b` occurring in a `recv` call potentially depends on any variable `a` occurring in a `send` call. In many practical applications this leads to a considerable overestimate of the active variable set. A recent attempt to reduce this overestimate is an activity analysis on the MPI control flow graph introduced in [SKH06]. A simple example of such a graph is shown in Fig. 6.5.

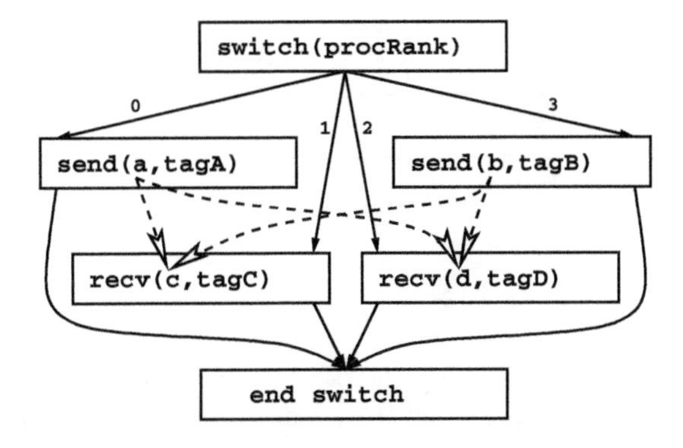

Figure 6.5: A Control Flow Graph Enhanced with Potential Communication Edges

We have a simple `switch` based on the MPI process rank (0-3) that determines the behavior for four processes executing in parallel. We start with the conservative assumption that all `recv`s are connected to all `send`s. In MPI the parameters that identify matching communication points are communicator, tag, and source/destination pairs. The communicator identifies a subset of the participating processes, the tag is an integer message identifier, and the source and destination are the process numbers within the respective communicator. The analysis on the MPI control flow graph uses constant propagation to establish guaranteed mismatches between `send/recv` pairs and to exclude certain dependencies. For simplicity we left out the communicator and source/destination parameters in Fig. 6.5 and indicate only the potential data-flow dependencies between the `send` and `recv` buffers.

If the code before the `switch` vertex contains assignments for the `tag`s, for example, `tagA=tagC=1; tagB=tagD=2`, then the analysis can propagate these constants, determine the mismatch, remove the two diagonal communication edges, and thereby exclude the dependence of `d` on `a` and of `c` on `b`. Collective communications are handled in a similar fashion. While this approach is fully automatic, its efficacy depends to some extent on the coding style of the original program. Complementing the appropriate use of the identifying parameters can be optional pragmas to identify communication channels. The idea was first introduced in [Fos95, Chapter 6], but no AD preprocessor has implemented this concept yet.

A different set of questions arises when one considers the implications for the data dependencies that can be inferred from OpenMP directives. Fig. 6.6 shows an example for a parallelizable loop where the actual dependencies are obfuscated by potential aliasing between arrays and by using a computed address that cannot be resolved with the typical induction variable mechanisms.

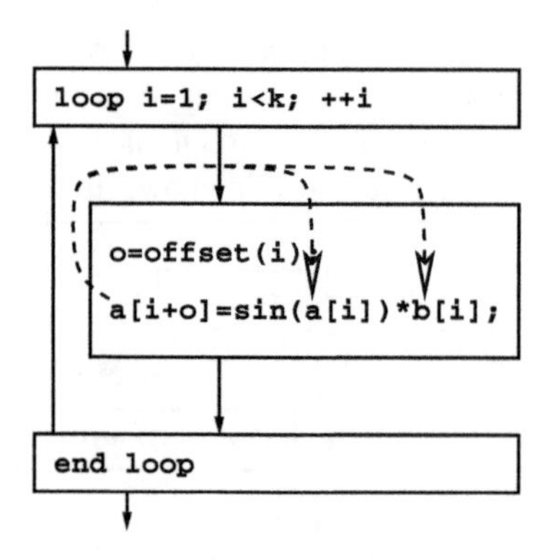

Figure 6.6: A Control Flow Graph with a Parallelizable Loop

Considering the main OpenMP workhorse `omp parallel do`, the need for this directive presumably is rooted in the inability of the automatic data flow and dependence analyses to remove the offending dependencies from the conservative overestimate. Otherwise an autoparallelizing compiler could verify that the loop in question is indeed free of loop-carried dependencies and could generate parallel instructions right away. That is, one would not require an OpenMP directive in the first place. In turn, an AD tool that is aware of such parallelization directives can use the implied dependency exclusions to enhance the analysis result. In our example in Fig. 6.6 an `omp parallel do` directive implies that `a[i+o]` does not overlap with `b[i]`, and therefore the data flow analysis does not need to propagate along the dashed edges. This situation might not be automatically detected because the code analysis has to prove that the computed offset always yields a value $\geq$`k` and the arrays `a` and `b` are not aliased. Conservatively the analysis assumes loop-carried dependencies (dashed edges) preventing an automatic parallelization. Further consequences of OpenMP directives on the reverse-mode code generation are explained at the end of this section.

Parallel Forward Mode

From a naive point of view the transformation of a given parallel program for forward mode would amount to merely mirroring the parallelization constructs for the derivative data. There are, however, a number of technical problems and the potential for amplifying a load imbalance of which the reader should be aware.

In a simple implementation, the forward mode adds derivative data and operations for derivative computations in a uniform fashion to the entire original program. Section 4.5 gives a theoretical overhead factor $\in [2, 5/2]$ for the forward mode. When one employs vector forward mode with p directions, this overhead factor grows with a problem-dependent fraction of p. Consequently, runtime differences caused by a load imbalance that were already present in the original program will be amplified by this factor; see also [RBB07]. An AD-specific source of load imbalance in forward mode is the propagation of dynamic sparse vectors [BK$^+$97], when the nonzero elements are distributed unevenly across the processes.

The correct association between program variables and their respective derivatives under MPI might be considered a negligible implementation issue but has been a practical problem for the application of AD in the past [HB98, CF96].

There is a noteworthy efficiency aspect to the differentiation of reduction operations. A good example is the product reduction, which is logically equivalent to the Speelpenning example discussed in section 3.3. Here, as well as in the other cases of arithmetic operations encapsulated within the MPI library calls, a direct AD transformation of the MPI implementation is not recommended. One has to determine a way to compute the product itself and the partial derivatives to propagate the directional derivatives. The computational complexity becomes easier to understand if one considers the reduction operations on a binary tree, as was done in [HB98]. The best forward-mode efficiency would be achieved with a special user-defined reduction operation that, for each node c in the tree with children a and b, simply implements the product rule $\dot{c} = b\dot{a} + a\dot{b}$ along with the product $c = ab$ itself. This approach allows the complete execution in just one sweep from the leaves to the root. Details on this topic can be found in [HB98]. MPI provides a user interface to define such specific reduction operations.

For OpenMP one could choose the safe route of reapplying all directives for the original program variables to their corresponding derivatives and stop there. When one considers the propagation of derivatives in vector mode, it can be beneficial to parallelize the propagation. When the derivative vectors are sufficiently long, a speedup of more than half the processor count was achieved for moderate processor counts [BL+01]. The approach made use of the OpenMP *orphaning* concept, which permits specifying parallelization directives outside the parallel region in which they are applied. The parallelization directives are applied to the routines that encapsulate the propagation without the overhead of parallel regions that are internal to these routines. In [BRW04] the concept of explicitly nesting the parallelization of the derivative vector propagation inside the given parallelization was explored, aimed at using a larger number of threads for the computation. The theoretical benefits have not yet been exploited in practice.

Parallel Reverse Mode

The code analysis and data association issues mentioned in the previous two subsections also pertain to the adjoint model for a parallel program.

One can consider a `send(`a`)` of data in a variable a and the corresponding `recv(`b`)` into a variable b to be equivalent to writing b=a. The respective adjoint statements are $\bar{a}$`+=`$\bar{b}$`;` $\bar{b}$`=0`. They can be expressed as `send(`$\bar{b}$`);` $\bar{b}$`=0` as the adjoint of the original `recv` call and `recv(t);` and $\bar{a}$`+=t` as the adjoint of the original `send` call, using a temporary variable `t` of matching shape. This has been repeatedly discovered and used in various contexts [FDF00, Che06].

Interpreting a `recv` call as an assignment $b = a$, it is clear that one has to replicate all the actions to record and restore the old values overwritten by the call to `recv(`b`)` when this is warranted, e.g., by TBR analysis. To keep the examples simple, we omit the recording code altogether and also leave out any statements in the return sweep that would restore recorded values in overwritten `recv` buffers.

A concern for parallel programs is the correctness of the communication patterns and, in particular, the avoidance of deadlocks. Proving that a given program is free of deadlocks in practice is possible only for relatively simple programs. A deadlock can occur if there is a cycle in the *communication graph*. The communication graph for a program (see, e.g., [SO98]) is similar to the MPI control flow graph shown in our example in Fig. 6.5; but instead of just adding edges for the communication flow, the communication graph also contains edges describing the dependencies between the communication endpoints. Often the noncommunication-related control flow is filtered out. The cycles relevant for deadlocks have to include communication edges—not just, for instance, loop control flow cycles. For the plain (frequently called "blocking") pairs of `send`/`recv` calls, the edges linking the vertices are bidirectional because the MPI standard allows a blocking implementation; that is, the `send/recv` call may return only after the control flow in the counterpart has reached the respective `recv/send` call. A cycle indicating a deadlock and the use of reordering and buffering to resolve it are shown in Fig. 6.7.

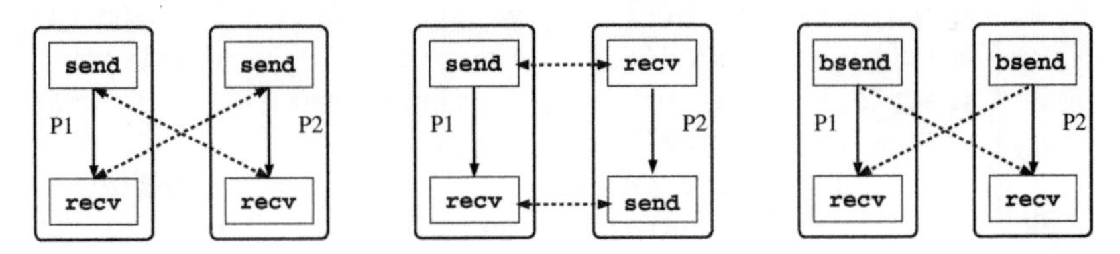

Figure 6.7: Deadlock, Reordered `send/recv`, Buffered Sends

In complicated programs the deadlock-free order may not always be apparent. For large data sets one may run out of buffer space, thereby introducing a deadlock caused by memory starvation. A third option to resolve the deadlock, shown in Fig. 6.8, uses the nonblocking `isend(`a`,r)`, which keeps the data in the program address space referenced by variable a and receives a request identifier `r`. The program can then advance to the subsequent `wait(r)` after whose return the data in the send buffer a is known to be transmitted to the receiving side. After the `wait` returns, the send buffer can be overwritten. The options for treating such programs have been explored in [UH+08].

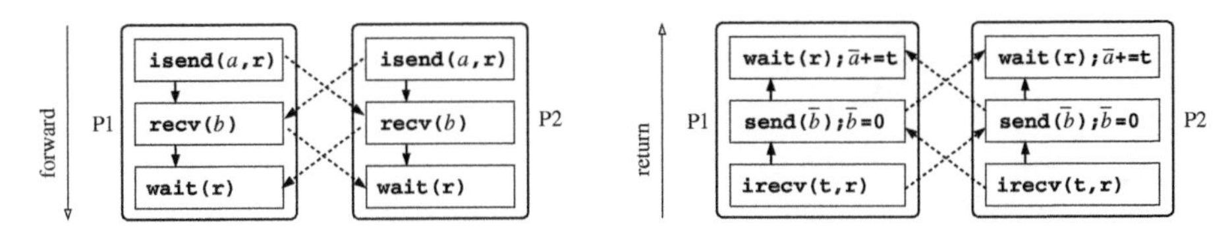

Figure 6.8: Nonblocking Send `isend` Followed by `wait` to Break Deadlock

For the adjoint of the parallel program and the corresponding adjoint communication graph, the direction of the communication edges needs to be reversed. This requirement imposes rules on the choice of the `send` call as the adjoint for a given `recv` and the treatment of `wait` calls. We can determine a set of patterns where simple rules suffice for the adjoint generation. To limit the number of distinct cases, we assume that `send(`a`)` is equivalent to `isend(`a`,r);` `wait(r)`, and similarly for `recv`.

The straightforward edge direction reversal as shown in Fig. 6.8 is implied when the original program contains only calls fitting the adjoining rules listed in Table 6.1. We omit all parameters except the buffers a, b, and a temporary buffer `t` and the request parameter `r` for nonblocking calls.

The combinations of nonblocking, synchronous, and buffered send and receive modes not listed in the table can be easily derived. As evident from the table entries, the proper adjoint for a given call depends on the context in the original code. One has to facilitate the proper pairing of the `isend/irecv` calls

Table 6.1: Rules for Adjoining a Restricted Set of MPI `send/recv` Patterns

	Forward Sweep		Return Sweep	
Rule	Call	Paired with	Call	Paired with
1	isend(a,r)	wait(r)	wait(r); $\bar{a}$+=t	irecv(t,r)
2	wait(r)	isend(a,r)	irecv(t,r)	wait(r)
3	irecv(b,r)	wait(r)	wait(r); $\bar{b}$=0	isend($\bar{b}$,r)
4	wait(r)	irecv(b,r)	isend($\bar{b}$,r)	wait(r)
5	bsend(a)	recv(b)	recv(t); $\bar{a}$+=t	bsend($\bar{b}$)
6	recv(b)	bsend(a)	bsend($\bar{b}$); $\bar{b}$=0	recv(t)
7	ssend(a)	recv(b)	recv(t); $\bar{a}$+=t	ssend($\bar{b}$)
8	recv(b)	ssend(a)	ssend($\bar{b}$); $\bar{b}$=0	recv(t)

with their respective individual `waits` for rules 1–4 (intra process) and the pairing of the `send` mode for a given `recv` for rules 5–8 (inter process). An automatic code analysis may not be able to determine the exact pairs and could either use the notion of communication channels identified by pragmas or wrap the MPI calls into a separate layer. This layer essentially encapsulates the required context information. It has distinct `wait` variants and passes the respective user space buffer as an additional argument, for example, swait(r,a) paired with isend(a,r). Likewise the layer would introduce distinct `recv` variants; for instance, `brecv` would be paired with `bsend`.

Multiple Sources and Targets

In the examples considered so far, we have had only cases where the communication vertices in the communication graph had single in- and out-edges. This is a critical ingredient for inverting the communication. There are three common scenarios where the single in/out edge property is lost.

1. Use of wildcard for the tag or the source parameter.
2. Use of collective communication (reductions, broadcasts, etc.).
3. Use of the collective variant of `wait` called `waitall`.

The use of the MPI wildcard values for parameters `source` or `tag` implies that a given `recv` might be paired with any `send` from a particular set; that is, the `recv` call has multiple communication in-edges. Inverting the edge direction for the adjoint means that we need to be able to determine the destination. In a single threated setup a simple solution is to store the values of the actual tag and source during the recording sweep; they can be retrieved through MPI calls. Conceptually this means that we pick at runtime an incarnation of the communication graph in which the single source/target property is satisfied and that therefore can be inverted by replacing the wildcard parameters in the return sweep with the previously recorded actual values.

For collective communications the transformation of the respective MPI calls is essentially uniform across the participating calls. To illustrate the effect, we consider a product reduction followed by a broadcast of the result, which could be accomplished by calling `allreduce`, but here we want to do it explicitly. Essentially we compute the product $p = \prod a_i$ and broadcast the value to all processes i so that $b_i = p$ for all i. The apparent adjoint is the summation reduction $\bar{p} = \sum \bar{b}_i$ followed by a broadcast of $\bar{p}$ and subsequent increment $\bar{a}_i$ += $\frac{p}{a_i}\bar{p}$ assuming $a_i \neq 0$. The respective communication graphs are shown in Fig. 6.9. Note that we assume here that the processes retain their rank between the forward and the reverse sweep.

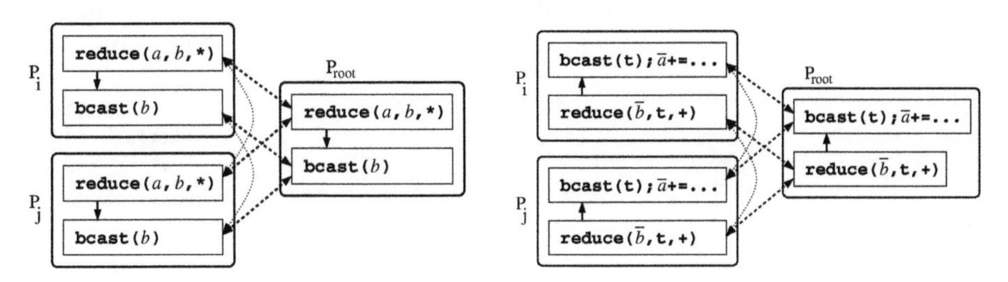

Figure 6.9: Adjoining of Collective Reduction and Broadcast

The calculation of $\partial p/\partial a_i$ as p/a_i does not work when $a_i = 0$ and may be inaccurate when $a_i \approx 0$. In section 3.3, we recommend a division-free way of differentiating Speelpenning's product, which has a similar operations count but less parallelism.

Similar to treating the reduction in forward mode, there is again an efficiency concern that is best visualized by considering an execution of the reduction on a tree. In principle, the partials could be computed explicitly by using prefix and postfix reduction operations during the recording sweep. Alternatively, one could record the a_i and then in the return sweep first compute all the intermediate products from the leaves to the root in the reduction tree followed by propagating the adjoints from the root to the leaves. This approach requires only two passes over the tree and is less costly than any approach using the explicit computation of the partials. Unlike the custom reduction operation for the forward case or the explicit partials computation using pre- and postfix reductions, MPI does not provide interfaces facilitating the two-pass approach. Consequently, one would have to implement it from scratch.

The use of `waitall` as a collective completion point poses the most complex problem of adjoining MPI routines that AD, at least in theory, can handle at the moment. This is a commonly used MPI idiom and occurs, for instance, in the logic of the MIT general circulation model [MIT]. There the adjoint MPI logic for the exchange of boundary layers of the grid partitions has been hand-coded because currently no source transformation tool can properly handle nonblocking MPI calls. Any change to the grid implementation necessitates a change to the respective hand-coded adjoint. This and the desire to provide choices for the grid to be used illustrate the practical demand for such capabilities.

Both the use of nonblocking point-to-point communications and the associated collective completion aim at reducing the order in processing the messages imposed on the message-passing system by the program that uses it. Often such a program-imposed order is artificial and has been shown to degrade the efficiency on processors with multiple communication links. While one can imagine many different orders of calls to `isend`, `irecv`, and `wait`, without loss of generality we consider a sequence of `isend` calls, followed by a sequence of `irecv` calls, followed by a `waitall` for all the request identifiers returned by the `isend`s and `irecv`s. For simplicity we assume all processes have the same behavior. We distinguish the buffers by an index, denote `r` as the vector of all request identifiers ($r_1, r_2, \ldots$), and show in the communication graph only placeholder communication edges that refer to nodes in the representer process; see Fig. 6.10.

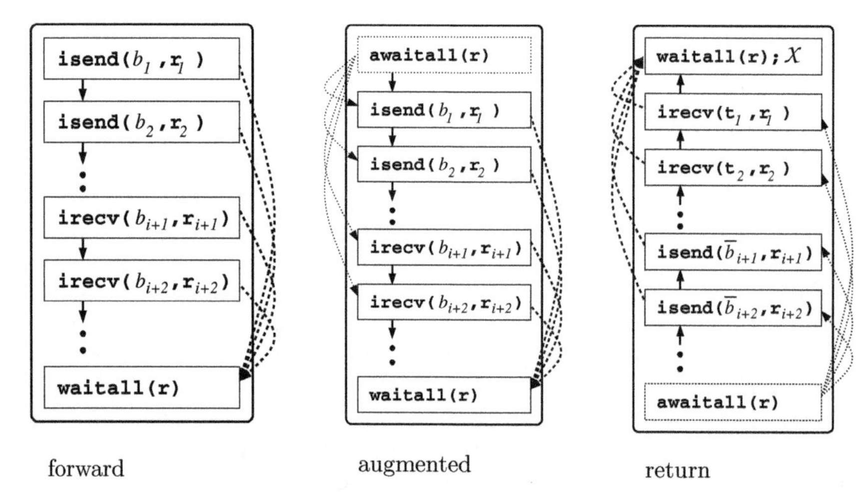

Figure 6.10: Original Code with `waitall` (left), Manually Augmented with `awaitall` (center), and Corresponding Return Sweep (right)

One could, of course, separate out all requests, introduce individual `wait` calls, and follow the recipe in Table 6.1. That approach, however, imposes an artificial order on the internal message-passing system, which the use of `waitall` tries to avoid. As shown in Fig. 6.10, the completion `waitall` has multiple communication in-edges which makes a simple vertex-based adjoint transformation impossible. We introduce a symmetric nonoperational counterpart *anti wait* denoted as `awaitall` in the augmented forward sweep and form the adjoint by turning the `awaitall` into a `waitall` and the original `waitall` into the nonoperational `awaitall` in the return sweep. The nonoperational communcation edges emanating from `awaitall` do not affect the forward sweep, and the transformation of the original `waitall` into an `awaitall` for the return sweep renders its problematic edges nonoperational.

With the transformations performed in the fashion suggested, we are able to generate the adjoint communication graph by reversing the edge direction. Consequently, if the original communication graph is free of cycles, then the adjoint communication graph also will be free of cycles, and we can be certain that the adjoint transformations do not introduce deadlocks.

The final $\mathcal{X}$ in the rightmost graph in Fig. 6.10 denotes the buffer updates $\bar{b}_j$+=$\mathtt{t}_j, j = 1, \ldots, i$, and $\bar{b}_j = 0, j = i+1, \ldots$, that have to wait for completion of the nonblocking calls. Here, similar to the recipes in Table 6.1, the additional context information that is required to accomplish $\mathcal{X}$ could be avoided if one wrapped the MPI calls and performed the bookkeeping on the buffers to be incremented and nullified inside the wrapper.

A frequently used MPI call is `barrier`, for instance, in the context of `rsend`; see Fig. 6.11. The standard requires that a `recv` has to be posted by the time `rsend` is called, which typically necessitates a `barrier` in the forward sweep.

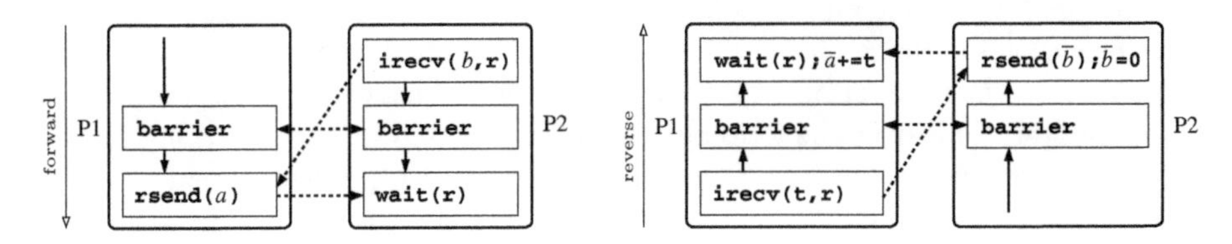

Figure 6.11: Handling of the `barrier` Routine

For the adjoint the `barrier` call stays in place; however, a vertex transformation recipe requires context information or a nonoperational counterpart to the `rsend` similar to the treatment of `waitall` in Fig. 6.10. In a logically correct program we can leave the `barrier` call in place for the adjoint transformation. We point out that a logically correct use of `barrier` to demarcate a critical section always requires synchronization on entry or exit in the original program whenever such synchronization is required for the respective exit and entry of the adjoint of that critical section. MPI one-sided communication routines that resemble a shared-memory programming style require explicit library calls on entry and exit of the section performing remote memory accesses.

A naive view of the adjoint transformation of an OpenMP parallel loop assumes that dependencies in the adjoint loop could arise only from antidependencies (i.e., overwrites) in the original loop, and vice versa. While any such dependency would prevent the original loop from being parallelizable, the absence of such dependencies asserted by the OpenMP directive does not imply that the adjoint loop is parallelizable. We illustrate the problem in Fig. 6.12. The original loop (left) can be parallelized. A standard transformation inverting the loop direction and generating the adjoint of the loop body (right) exhibits various increment statements of `bx`.

Here the problem lies with the increment operations of the adjoints because there can be a race condition between the reads and the writes of the `bx[i-1]`, `bx[i]`, and `bx[i+1]` when the adjoint loop is executed in parallel, leading to erroneous computations. If the increment operation were atomic, which is not guaranteed in practice, then this problem would disappear.

While it may not be possible to parallelize the adjoint loop, an important benefit is the ability to cheaply recompute the values needed by the adjoint loop body. In general the adjoint loop needs to be executed in the order reverse to

```
loop i=2; i<n; ++i
     a[i]=x[i-1]-2*x[i]+x[i+1]
     b[i]=a[i]+sqrt(x[i])
end loop
```

```
loop i=n-1; i>=2;--i
     ba[i]     += bb[i]
     bx[i]     += bb[i]*1./(2*sqrt(x[i])
     bb[i]     =  0
     bx[i-1]   += ba[i]
     bx[i]     += (-2)*ba[i]
     bx[i+1]   += ba[i]
     ba[i]     =  0
enddo
```

Figure 6.12: Introduction of a Race Condition by Adjoining

the original loop. Recomputing values (with loop-carried dependencies) needed by the adjoint implies either recording them or recomputing them at a cost quadratic in the number of loop iterations. A parallelizable forward loop implies that the iterations over the loop body can be executed in any order, e.g., in the reverse order that would be required in the return sweep when the adjoint loop is not parallelizable, as in our example. The values needed for the adjoint of the loop body can then be recomputed immediately prior to the return sweep of that loop body iteration by running the recording sweep over that single iteration. Therefore, the quadratic recomputation complexity vanishes.

6.4 Summary and Outlook

While the principles and basic techniques of algorithmic differentiation are quite simple and may even appear trivial to some people, their effective application to larger problems poses quite a few challenges. Here the size and difficulty of a problem depends not so much on the sheer runtime of an evaluation procedure as on complexity of its coding. Heterogeneous calculations pieced together from various software components and possibly run concurrently on several platforms can realistically not be differentiated with current AD implementations. Perspectively, each software component provided by a public or private vendor should have calling modes for propagating sensitivity information forward or backward in addition to its normal simulation functionality. This would of course require an agreement upon a standard for direct and adjoint derivative matrices in a suitably compressed format. For the time being such a prospect seem a fair way off and for the most part we have to be content with differentiating suites of source programs in a common language, typically from the Fortran or C family. As described in section 6.3 concurrency encoded in MPI or OpenMP can be largely preserved through the differentiation process, through some care must be taken to ensure correctness and maintain load balancing. There is even some prospect of gaining parallelism, for example, through stripmining of Jacobians or concurrent recomputations as briefly described at the end of Chapter 12. Naturally these kinds of techniques will remain in the domain of expert users, who develop a large-scale application where savings in wall-clock time are critical.

For user with small- or medium-sized AD applications the following considerations for writing (or if necessary rewriting) code are useful. Unless they are really completely separate, the evaluation of all problem functions whose derivatives are needed (e.g., optimization objectives and constraints) should be invoked by one *top-level* function call. In doing so one should make as much use as possible of common subexpressions, which then can also be exploited by the differentiation tool. For some tools the top-level call can be merely conceptual, that is, represent a certain *active section* of the user program where function evaluations take place.

Within the subroutines called by the top-level function, one should avoid extensive calculations that are passive, in that they do not depend on the actual values of variables (e.g., the setting up of a grid). Even if the corresponding code segments are conditioned on a flag and may, for example, be executed only at an initial call, the AD tool may be unable to predict this runtime behavior and wind up allocating and propagating a large number of zero derivatives. For the same reason one should not share work arrays or local variables between passive parts of the code and the top-level call that is to be differentiated. Generally, one should separate code and data structures as much as possible into an active and a passive part. If one knows that some calculations within an active subroutine have no or only a negligible influence on the derivative values, one might go through the trouble of deactivating them by suppressing the dependence on the independent variables.

These coding style recommendations are not really specific to AD. They would similarly assist parallelizing or ("merely") optimizing compilers. Anything that simplifies compile-time dependence analysis by making the control and data flow more transparent is a good idea. Whenever possible, one should provide fixed upper bounds on array sizes and iteration counters.

On the general assumption that deferring choices until runtime almost inevitably produces code that runs more slowly than when the decision is already made at compile time, we can formulate the following sweeping recommendation:

RULE 12

DO AS MUCH AS POSSIBLE AT COMPILE-TIME,
AS MUCH AS NECESSARY AT RUNTIME.

6.5 Examples and Exercises

Exercise 6.1 (*Scalar Forward by Overloading*)
Code the `adouble` proposal for the forward mode as outlined in section 6.1, including the four basic arithmetic operations $+, -, *$, and $/$ and the intrinsic functions sin, cos, exp, and sqrt. For flexibility you may add mixed-mode versions where one argument is a `double` for the four binary operations, though this is not necessary for the examples suggested.

a. Check and debug your implementation on the trivial test functions

$$y = x - x, \quad x/x, \quad \sin^2(x) + \cos^2(x), \quad \text{sqrt}(x * x), \quad \text{and} \quad \exp(x)$$

and reproduce the results listed in Table 1.3 for the baby example.

b. For the norm problem discussed in Exercise 2.2, compare the accuracy obtained by your implementation with that of difference quotient approximations.

c. Overload the incremental (C++) operation += for adoubles, and use it to simplify the code for the norm problem.

Exercise 6.2 (*Scalar Reverse by Overloading*)
Code the adouble proposal for the reverse mode as outlined in section 6.1 with the same elemental operations as for your adoubles in Exercise 6.1. Check and debug your code on the same trivial test functions.

a. Reproduce the results in Table 1.4 for the baby example.

b. Apply your implementation to Speelpenning's product as discussed in Exercise 3.6. At the argument $x_i = i/(1+i)$, compare the results and runtimes to a hand-coded adjoint procedure and also to n runs of your forward implementation from Exercise 6.1.

c. If one of the source transformation tools listed in section 6.2 is or can be installed on your system, verify the consistency of its results with yours and compare compilation times and runtimes.

Exercise 6.3 (*Second-Order Adjoints*)
Combine the forward and the reverse mode by modifying the class adouble from your reverse-mode implementation. Verify the correctness of this extension of the scalar reverse mode on the coordinate transformation example discussed in Exercise 5.4.

Exercise 6.4 (*Second Derivatives Forward*)
In order to propagate first and second directional derivatives, implement a version of the class adouble that has the same structure as the adouble for the forward mode, but whose field dot is of type adouble. For this purpose, you may simply replicate the code overloading the restricted set of elementals for adoubles. Check the correctness of your implementation on the trivial identities, and compute the acceleration of the light point in the lighthouse example of section 2.1. Observe that the new adouble could also be defined as an adouble<adouble<double>>, namely, by a recursive application of templates. Such techniques are at the heart of the expression templates techniques [Cés99] used to handle complicated right-hand sides more efficiently.

Exercise 6.5 (*Vector Forward by Overloading*)
As suggested in subsection 6.1 on page 113, code a class vector providing additions between them and multiplication by a real scalar.

a. Replicate the source generated for the adouble class in Exercise 6.1 above with dot now a vector rather than a double. As in the previous exercises you may define adoubles as a class template parametrized by the type of dot.

b. Test the new vector implementation of the forward mode on the trivial identities, the norm example, and Speelpenning's product. Compare the runtimes with that of the scalar forward multiplied by n.

Exercise 6.6 (***Taylor Forward***)
To propagate higher derivatives forward, write an active class `taylor` whose data member is a vector of d coefficients. Implement arithmetic operations and intrinsic functions according to the formulas given in Tables 13.1 and 13.2, respectively. Test your implementation on the trivial test functions and the lighthouse example to compute time derivatives of arbitrary order.

Exercise 6.7 (***Overloading Assignments***)
All exercises above can be performed in a language that does not allow the overloading of assignments. However, this restriction may severely impair efficiency for the following reason.

In the vector-mode implementation according to the previous exercise, the results from each arithmetic operation and other elemental function are first placed into a temporary return variable and then typically copied to a named variable on the left-hand side of an assignment. To avoid the often repeated copying of the vector part of an `adouble` one may redefine its field `dot` to be merely a pointer to a `vector` of derivatives, whose length need not be limited at runtime. These objects can be dynamically allocated within each overloaded elemental and by the initialization function `makeindepvar` for the independent variables.

However, as one can easily see, for example, on Speelpenning's product, each multiplication would allocate an additional vector and none of those would ever be released again so that the storage demand would grow by a factor of about n compared to the base implementation according to Exercise 6.5. The simplest way to avoid this effect is to overload the assignment so that the old vector pointed to by the left-hand side is deallocated, for example, by the statement `free(dot)`.

a. Modify your vector forward mode as sketched above, and verify on Speelpenning's example that the results are correct and the storage is only of order n. Check whether the elimination of vector copying actually reduces the runtime.

b. Construct an example where an intermediate variable is first assigned to another named variable and then receives a new value itself. Show that the modified vector forward cannot work properly and correct it by adding a reference counter to the data structure `vector`. Now deallocation happens only when the reference counter has been reduced to zero.

Exercise 6.8 (***User-Defined Constructors/Destructors***)
Even the last implementation sketched in Exercise 6.7 can work properly only if the `dot` is automatically initialized to a null pointer upon variable construction. This is ensured by most compilers, even if not strictly enforced by all language standards. Otherwise looking up and modifying the reference counter may already cause segmentation faults. In C++ the user can be sure of proper initialization by writing his or her own constructor function for variables of type `adouble` and such. Similarly, he or she can write a corresponding destructor routine that is called by the compiler whenever a variable of type `adouble` goes out of scope. In our context, the destructor can decrement the reference counter

of the `vector` structure being pointed to by the variable being destructed. Otherwise, the reference counter mechanism introduced in part **b** of Exercise 6.7 cannot really work properly on multilayered programs, though the derivative values obtained would be correct.

Upgrade the last version of vector forward once more by adding a constructor and a destructor for `adouble` as described above. Test the efficacy of the new version, for example, on the norm problem recoded such that the squaring of the x_i happens in a function with at least one local variable.

Part II

Jacobians and Hessians

Chapter 7

Sparse Forward and Reverse

In this second part of the book we consider some extensions of the techniques described in the first part to calculate entire Jacobian and Hessian matrices. As we shall see, the answer to the frequently asked question, "How much will a Jacobian or Hessian cost?" is a resounding, "It depends!" From a user's point of view this situation is not very satisfactory, but as we will show in Chapter 11, the need to investigate and exploit the structure of the functions at hand opens up numerous opportunities for research and development. The elimination procedures discussed in Chapter 9 elevate AD from an exercise in implementing the forward and reverse modes efficiently to a research topic centered on an NP-hard combinatorial problem akin to sparse matrix factorization. In this chapter and the next, we will concentrate on the more mundane methods of evaluating derivative matrices in the forward or reverse mode.

In Chapter 3 we showed that the product of the Jacobian $F'(x)$ or the transpose $F'(x)^\top$ with a given vector can be computed as a *tangent* $F'(x)\dot{x}$ and *gradient* $F'(x)^\top \bar{y}$, respectively. In both cases the computational complexity is a small multiple of the complexity of evaluating F itself. We also showed that by replacing $\dot{x}$ and $\bar{y}$ by matrices $\dot{X} \in \mathbb{R}^{n\times p}$ and $\bar{Y} \in \mathbb{R}^{m\times q}$ with p columns and q columns, respectively, we obtain $F'(x)\,\dot{X} \in \mathbb{R}^{m\times p}$ and $\bar{Y}^\top F'(x) \in \mathbb{R}^{q\times n}$ at a complexity that grows linearly with p and q, respectively. This result seems reasonable because the number of derivative components obtained grows accordingly. On the other hand, we found that the number of components in a gradient did not influence the total cost of its evaluation (relative to the cost of evaluating the function by itself). Hence, one might hope that a similarly small growth in computational cost might be encountered in going from vector functions to Jacobians or from gradients to Hessians. Perhaps if a gradient can be obtained for 5 evaluations, a complete Hessian could be obtained for 25, independent of the number of variables. As we will see in a simple example in section 11.1, this wish cannot always be fulfilled by either the techniques advocated here or any other differentiation method conceivable.

Hence, we consider first applying the vector-tangent or vector-gradient procedures to calculate

$$\dot{Y} = F'(x)\,\dot{X} \quad \text{with} \quad \dot{X} = I_n \quad \text{or} \quad \bar{X}^\top = \bar{Y}^\top F'(x) \quad \text{with} \quad \bar{Y} = I_m \tag{7.1}$$

in order to obtain the whole Jacobian $F'(x)$ in one sweep. Should the storage growth by the factor $p = n$ or $q = m$ be a problem, we can *strip-mine*, i.e., calculate subgroups of columns or rows at a time by partitioning $\dot{X}$ or $\bar{Y}$ appropriately. Depending on the relative size of m and n, one would prefer the vector-forward mode if $n \leq m$ and would otherwise use the vector-reverse mode. Hence the complexity growth for calculating the Jacobian can be limited, in general, to the factor $\min(n, m)$. On some problems this method is optimal up to a small factor. On other problems, however, substantial savings can be achieved by applying the chain rule with a little more imagination. Such *cross-country* elimination methods will be introduced and analyzed in Chapter 9. According to a result of Herley [Her93], finding the elimination order with the lowest fill-in is an NP-hard combinatorial optimization problem. The same is true for the closely related task of minimizing the number of arithmetic operations [Nau06]. This fact may explain why, despite its great potential savings, cross-country elimination has not been used much in its full generality. The sparse forward and reverse methods to be discussed in the remainder of this chapter can be viewed as special cases of cross-country elimination.

Apart from using sparsity of the Jacobians and Hessians, differentiation techniques can also exploit other structural properties of a given F. Here as before F is not viewed simply as a mathematical mapping between Euclidean spaces, because one must take into account the structure of its evaluation procedure. In this chapter we consider two scalars, the *average domain* and *range size*, to characterize specific evaluation procedures at hand. Also, the maximal numbers of nonzeros per rows or columns of the Jacobian and the chromatic numbers of the row- and column-incidence graphs will be important to gauge the efficacy of sparse differentiation procedures.

Fortunately, many practical problems are sparse or otherwise partially separable (sparsity implies partial separability; see Lemma 11.1). Substantial savings can be achieved by exploiting function structure either dynamically, pseudostatically, or even fully statically. By "dynamically" we mean the use of dynamic data structures at runtime. "Statically" refers to the detection of some sparsity at compile-time, which has been explored, for example, in the Ph.D. thesis [Tad99] and is the subject of current research. The techniques and difficulties of this fully static approach are similar to those occurring in parallelizing compilers. They require in particular the careful analysis and description of array indices, which is in general a rather difficult task. A simpler way of avoiding the runtime penalty of fully dynamic sparsity propagation is the matrix compression technique described in Chapter 8. It is related to the inspector-executor paradigm in parallel computing, where a first preliminary execution on a representative data set provides the dependence information for many subsequent "production runs." Therefore the label "pseudostatic" seems appropriate.

Even nonsparse problems may have a structural property called *scarcity*, which we introduce in section 10.3, as a generalization of sparsity. It can be exploited for calculating products of the Jacobian or its transpose with a minimal number of operations. It is as yet unclear whether there is a canonical scarcity-preserving Jacobians representation.

7.1 Quantifying Structure and Sparsity

In this section we consider two numbers: the average domain size $\bar{n} \leq n$ and the average range size $\bar{m} \leq m$. They determine the complexity of various methods for evaluating or estimating Jacobians.

Smart Difference Quotients and Index Domains

Suppose we estimate $F'(x)$ by columns according to

$$F'(x)\, e_j \approx [\, F(x + \varepsilon e_j) - F(x)] \,/ \varepsilon \quad \text{for} \quad j = 1 \ldots n \,. \tag{7.2}$$

Assume the evaluation procedure for F involves no overwriting, so that all intermediate quantities $v_k = v_k(x)$ evaluated at the point x can be kept in store. Going through again after changing just one independent variable, say the jth, we have to recompute only those elemental relations $v_k = \varphi_k(u_k)$ that depend on x_j in a nontrivial way. In other words, only those v_k need to be reevaluated for which j belongs to the index set

$$\mathcal{X}_k \equiv \{j \leq n : j - n \prec^* k\} \quad \text{for} \quad k = 1 - n \ldots l \,. \tag{7.3}$$

Here $\prec^*$ is the transitive closure of the data dependence relation $\prec$ discussed in section 2.2. The *index domains* $\mathcal{X}_k$ may be computed according to the forward recurrence

$$\mathcal{X}_k = \bigcup_{j \prec k} \mathcal{X}_j \quad \text{from} \quad \mathcal{X}_{j-n} = \{\, j \,\} \quad \text{for} \quad j \leq n \,. \tag{7.4}$$

Obviously we must have the inclusion

$$\{j \leq n \, : \, \partial v_k / \partial x_j \not\equiv 0\} \subseteq \mathcal{X}_k$$

and, barring degeneracy, the two index sets actually will be equal. A proper subset relation may, for example, arise through statement sequences such as

$$v_1 = \sin(v_0)\,; \quad v_2 = \cos(v_0)\,; \quad v_3 = v_1 * v_1 + v_2 * v_2 \,.$$

Here $\partial v_3 / \partial v_0 \equiv \partial v_3 / \partial x_1$ would vanish identically even though we have clearly $\mathcal{X}_3 = \mathcal{X}_1 = \mathcal{X}_2 = \{1\}$. However, such degeneracies are unlikely to occur frequently in well-designed codes, so that the use of the index domain $\mathcal{X}_k$ defined by (7.3) is probably not too much of an overestimate. Fischer and Flanders [FF99] established the following result which can be used to obtain simple lower complexity bounds for evaluation procedures of a given function.

Lemma 7.1 (Fischer and Flanders)
Provided all elemental functions $\varphi \in \Psi$ are unary or binary, we have

$$|\mathcal{X}_i| \le i + 1 \quad \textit{for all} \quad 1 \le i \le l\,,$$

where $|\cdot|$ denotes cardinality for set arguments.

Proof. The proof is left as Exercise 7.1. ∎

Denoting the task of incrementally evaluating F at $x + \varepsilon e_j$ in the way described above by $eval_{+j}(F)$, we obtain the complexity estimate

$$WORK\{eval_{+j}(F)\} = \sum_{k:\mathcal{X}_k \ni j} WORK\{eval(\varphi_k)\}\,.$$

We denote by *desti* the task of estimating the Jacobian by differencing. Summation of the previous equality over j yields

$$\begin{aligned} WORK\{desti(F)\} &= WORK\{eval(F)\} + \sum_{j=1}^{n} WORK\{eval_{+j}(F)\} \\ &= \sum_{k=1}^{l} (1 + |\mathcal{X}_k|\,) WORK\{eval(\varphi_k)\}\,. \end{aligned} \tag{7.5}$$

Here the subtractions and multiplications (by $1/\varepsilon$) for finally forming the divided differences according to (7.2) have been neglected. For the runtime functional introduced in (4.8) we obtain

$$TIME\{desti(F)\} = \sum_{k=1}^{l} (1 + |\mathcal{X}_k|\,)\, TIME\{eval(\varphi_k)\}\,. \tag{7.6}$$

When the component functions $F_i(x)$ are additively separable in the usual sense, all $\mathcal{X}_k$ associated with nonlinear φ_k must be singletons. In that case the finite difference estimation of the Jacobian takes only twice the time of evaluating the function F itself.

Index Range and Path Connectedness

In analogy to the index domain $\mathcal{X}_k$ we define the *index ranges*

$$\mathcal{Y}_k \equiv \{i \le m \,:\, k \prec^* l - m + i\} \quad \text{for} \quad k = l, l-1, \ldots, 1-n\,, \tag{7.7}$$

which contain the indices of all those dependents y_i that are impacted by the intermediate v_k in a nontrivial fashion. They can be computed according to the backward recurrence

$$\mathcal{Y}_k \equiv \bigcup_{j \succ k} \mathcal{Y}_j \quad \text{from} \quad \mathcal{Y}_{l-i} \equiv \{m - i\} \quad \text{for} \quad i < m\,. \tag{7.8}$$

To relate some of our complexity estimates directly to Jacobian and Hessian sparsity, we formulate the following condition.

Assumption (PC): PATH CONNECTEDNESS

All intermediate variables v_k depend on some independent x_j and impact some dependent y_i in that

$$\mathcal{X}_k \neq \emptyset \neq \mathcal{Y}_k \quad \text{for} \quad 1-n \leq k \leq l\,.$$

Since by the definition of $\mathcal{X}_k$ and $\mathcal{Y}_k$

$$j \prec^* i \quad \Longrightarrow \quad \mathcal{X}_j \subset \mathcal{X}_i \quad \text{and} \quad \mathcal{Y}_j \supset \mathcal{Y}_i\,,$$

it follows immediately from Assumption (PC) that all $\mathcal{X}_k$ are contained in some $\mathcal{X}_{l-m+i}$ and all $\mathcal{Y}_k$ are contained in some $\mathcal{Y}_{j-n}$. Hence we may define the width of the Jacobian $F'(x)$ and its transpose $F'(x)^\top$, respectively.

Definition (MW): MATRIX WIDTH OR MAXIMAL DOMAIN AND RANGE SIZE

Abbreviating $p_i = |\mathcal{X}_{\hat{\imath}}|$ *with* $\hat{\imath} = l - m + i$ *and* $q_j = |\mathcal{Y}_{j-n}|$, *we call*

$$\hat{n} \equiv \hat{n}\left(F'(x)\right) = \max_{1\leq i\leq m} p_i = \max_{1-n\leq i\leq l} |\mathcal{X}_i| \tag{7.9}$$

the maximal domain size *of* F *or the* width *of* $F'(x)$, *and we call*

$$\hat{m} \equiv \hat{m}\left(F'(x)\right) = \max_{1\leq j\leq n} q_j = \max_{1-n\leq j\leq l} |\mathcal{Y}_j| \tag{7.10}$$

the maximal range size *of* F *or the* width *of* $F'(x)^\top$.

As a consequence we will observe in Chapter 8 that the matrix compression approach does not depend on the internal structure of the evaluation procedure but only on the resulting sparsity structure of the Jacobian.

The Average Domain Size $\bar{n}$

In general, we may quantify the separability of F by taking the average of the domain sizes $|\mathcal{X}_k|$ weighted by the elemental runtimes $TIME\{eval(\varphi_k)\}$. More specifically, we set

$$\bar{n} \equiv \frac{\sum_{k=1}^{l} |\mathcal{X}_k|\, TIME\{eval(\varphi_k)\}}{\sum_{k=1}^{l} TIME\{eval(\varphi_k)\}} \equiv \frac{\sum_{j=1}^{n} TIME\{eval_{+j}(F)\}}{TIME\{eval(F)\}} \leq \hat{n}\,. \tag{7.11}$$

The three function-specific numbers $\bar{n} \leq \hat{n} \leq n$ and the analogous transposed quantities $\bar{m} \leq \hat{m} \leq m$ defined below will occur in various complexity bounds. In contrast to $\hat{n}$ and n, the average domain size $\bar{n}$ is, in general, not an integer, but it may be rounded up if that property seems desirable. When $\bar{n}$ reaches its minimal value 1, the function F is completely separable in the traditional sense. Thus, $\bar{n}$ can be interpreted as an inverse measure of what we shall call *value separability* in section 11.2. Substituting the definition of $\bar{n}$ into the right-hand side of (7.6), one obtains the estimate

$$\boxed{TIME\{desti(F)\} = (1+\bar{n})\,TIME\{eval(F)\}\,.} \tag{7.12}$$

As we will see in the following section, a similar bound holds for a sparse implementation of the forward mode. Thus, the average domain size $\bar{n}$ does indeed seem to be a crucial quantity regarding the complexity of Jacobians.

Two Illustrative Examples

Let $F = (F_i)_{i=1\ldots m}$ be of the form

$$F_i(x) = \sum_{j=1}^{n} F_{ij}(x_j) \quad \text{for} \quad i = 1\ldots m\,,$$

with computationally expensive functions $F_{ij} : \mathbb{R} \longmapsto \mathbb{R}$. Then all nonlinear elementals have index domains $\mathcal{X}_k$ that are singletons, so that $\bar{n}$ must be close to its minimal value 1. Consider, on the other hand,

$$F_i(x) = \tilde{F}_i\left(\prod\nolimits_{j=1}^{n}(x_j - \alpha_{ij})\right) \quad \text{for} \quad i = 1\ldots m$$

with some matrix $(\alpha_{ij}) \in \mathbb{R}^{m\times n}$ of constant coefficients and nonlinear functions $\tilde{F}_i : \mathbb{R} \longmapsto \mathbb{R}$. By making the $\tilde{F}_i$ very expensive compared with the n subtractions and multiplications needed to compute their arguments, the average domain size $\bar{n}$ can be pushed arbitrarily close to n.

These two examples confirm that $\bar{n}$ is a reciprocal measure of value separability in that a small $\bar{n}$ means high separability, and vice versa. In the first example, smart differencing and the forward mode yield the Jacobian cheaply. In the second example, this is definitely not the case because the arguments x_i are so thoroughly intertwined. However, the second example has a high degree of argument separability in that its average range size as defined below is small.

The Average Range Size $\bar{m}$

In perfect analogy to $\bar{n}$ we may define the average range size

$$\bar{m} \equiv \frac{\sum_{k=1}^{l} |\mathcal{Y}_k|\, TIME\{eval(\varphi_k)\}}{\sum_{k=1}^{l} TIME\{eval(\varphi_k)\}} = \frac{\sum_{i=1}^{m} TIME\{eval(F_i)\}}{TIME\{eval(F)\}} \le \hat{m}\,. \tag{7.13}$$

We may consider $\bar{m}$ as an inverse measure of *argument separability*, as defined in section 11.2. For the two examples discussed above, all index ranges are singletons so we have $\bar{m} = 1$ exactly. As we shall see in the next section, the average sizes $\bar{n}$ and $\bar{m}$ determine the complexity of the sparse forward and reverse modes of AD, respectively. More specifically, there exists a small constant γ such that for all $F \in \mathcal{F}$

$$\boxed{WORK\{eval(F')\} \leq \gamma \min\{\bar{n}, \bar{m}\}\, WORK\{eval(F)\}\,.} \qquad (7.14)$$

We shall demonstrate in section 11.1 that in some cases no differentiation method can evaluate F' significantly more cheaply than indicated by this bound.

7.2 Sparse Derivative Propagation

The realization of the ambitiously low complexity bound (7.12) by differencing could be based on a *remember* option [SMB97] for elemental evaluations. Such facilities are customary in computer algebra systems, but completely alien to the procedural languages used in most numerical computations.

Alternatively, one might simultaneously evaluate F at all $n+1$ points x and $x + \varepsilon e_j$ for $j = 1 \dots n$, with the intermediates v_k being kept in a sparse representation. Here "sparse" means that in addition to $v_k(x)$, only those values $v_k(x + \varepsilon e_j)$ are kept that differ from $v_k(x)$ and each other. This requires knowledge of the index domains $\mathcal{X}_k$. These can be (re)computed at runtime or possibly determined at (pre)compile-time and then "hard-wired" into an expanded evaluation procedure. However, if one goes through the trouble of getting the $\mathcal{X}_k$, much better use of them can be made, namely, for evaluating derivatives *analytically*. Thus, instead of carrying along function values at n neighboring points, we might as well propagate partial derivatives in the corresponding directions.

Let us evaluate $\dot{Y} = \dot{F}(x, \dot{X}) = F'(x)\dot{X}$ with $\dot{X} = I_n$ by the forward procedure of Table 3.2 with the scalars $\dot{v}_i$ replaced by n-vectors $\dot{V}_i$. Then the jth component $\dot{V}_{i,j}$ of $\dot{V}_i = \nabla_x v_i$ can be nonzero only when $j \in \mathcal{X}_i$, which is the union of all $\mathcal{X}_k$ for $k \prec i$. Hence we obtain the sparse Jacobian procedure listed in Table 7.1 with $c_{ij} = \partial\varphi_i/\partial v_j$, as in (2.8).

Table 7.1: Sparse Forward Jacobian Procedure

v_{i-n}	$\equiv$	x_i	
$\mathcal{X}_{i-n}$	$=$	$\{i\}$	$i = 1 \dots n$
$\dot{V}_{i-n,i}$	$=$	1	
v_i	$=$	$\varphi_i(v_k)_{k\prec i}$	
$\mathcal{X}_i$	$=$	$\bigcup_{k\prec i} \mathcal{X}_k$	$i = 1 \dots l$
$\dot{V}_{i,j}$	$=$	$\sum_{k\prec i} c_{ik}\, \dot{V}_{k,j}$ for $j \in \mathcal{X}_i$	
y_{m-i}	$\equiv$	v_{l-i}	$i = m-1 \dots 0$
$\dot{Y}_{m-i,j}$	$\equiv$	$\dot{V}_{l-i,j}$ for $j \in \mathcal{X}_{l-i}$	

Note that the sparse forward Jacobian calculation is not an additive task. The number of nonzero components of the column vector $\dot{V}_i$ is at most $|\mathcal{X}_i|$, which also equals the number of nonvanishing columns in $\dot{U}_i = (\dot{V}_k^\top)_{k \prec i}$. This number varies with i. Hence, the task to be performed for the ith elemental function is the propagation of $|\mathcal{X}_i|$ tangents through φ_i. Consequently, the complexity ratio for evaluating the elementals themselves is anything but constant. It varies between 1 and the width $\hat{n}$, the maximal number of nonzeros in any row of $F'(x)$ defined in (7.9). Using the results of section 4.5, we obtain from (4.18) the estimate

$$\boxed{TIME\left\{\overrightarrow{eval}(F')\right\} \leq (1 + 1.5\,\bar{n})\,TIME\{eval(F)\}\ .} \tag{7.15}$$

Here $\overrightarrow{eval}$ and $\overleftarrow{eval}$ denote the tasks of evaluating some derivative object in the forward or reverse mode, respectively. It should be appreciated that this bound may be much tighter than the trivial one obtained by replacing the average domain size $\bar{n}$ with n, even when the Jacobian $F'(x)$ is not sparse at all.

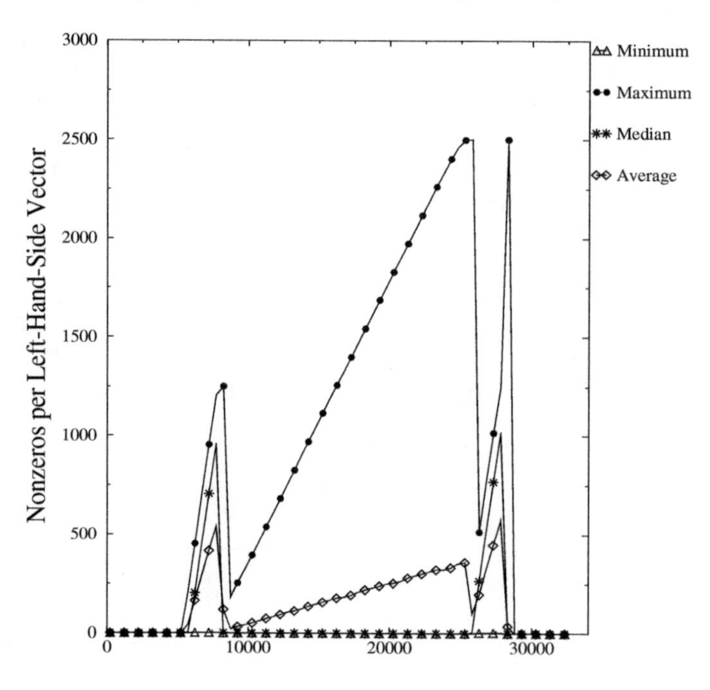

Figure 7.1: Saxpys Performed during Execution [BK+97]

Fig. 7.1, from the study [BK+97], displays the frequency with which sparse gradients $\dot{V}_k$ of a certain length $|\mathcal{X}_k|$ occur on a particular example problem differentiated with ADIFOR. The horizontal axis counts the number of Fortran assignments, which are effectively treated as elemental functions by virtue of ADIFOR's statement-level reversal approach (see section 10.2). As one can see most derivatives $\dot{V}_i$ are quite short, especially during the early part of the calculation. Only toward the end are some dense vectors of size 2 500, the number of independents in this problem, reached. Assuming that all assignments are of comparable complexity, we have on this example $\bar{n} \approx 600 \ll 2\,500 = n = \hat{n}$. Since the incompressible elastic rod (IER) problem in question is value separable in the sense given in section 11.2, the large gap between $\hat{n}$ and $\bar{n}$ can be significantly reduced by reformulating the function, as discussed in Chapter 11.

When variables are overwritten according to an allocation function &, the corresponding indices $\mathcal{X}_k$ and values $\dot{V}_k$ may also be deallocated and reused. The practical difficulty for an efficient implementation is that $|\mathcal{X}_k|$ may greatly vary in size in a way that is, in general, unpredictable.

We may also use the reverse mode to evaluate complete Jacobians $F'(x)$ in a sparse fashion. Starting with $\bar{Y} = I_m$ we obtain $\bar{X}^\top = \bar{F}(x, \bar{Y}) = \bar{Y}^\top F'(x)$ by applying Table 4.4, with $\bar{v}_i$ replaced by m-vectors $\bar{V}_i$. The component $\bar{V}_{i,j}$ of $\bar{V}_i$ can be nonzero only when $j \in \mathcal{Y}_i$. Thus we obtain the sparse procedure listed in Table 7.2.

Table 7.2: Sparse Reverse Jacobian Procedure

$\mathcal{Y}_i$	$\equiv$	$\emptyset$; $\bar{V}_i \equiv 0$	$1-n \le i \le l$
v_{i-n}	$=$	x_i	$i = 1 \ldots n$
v_i	$\rightarrowtail$	◉○	$i = 1 \ldots l$
v_i	$=$	$\varphi_i(v_j)_{j \prec i}$	
y_{m-i}	$=$	v_{l-i}	$i = m-1 \ldots 0$
$\mathcal{Y}_{l-i}$	$=$	$\{m-i\}$	$i = 0 \ldots m-1$
$\bar{V}_{l-i,m-i}$	$=$	1	
v_i	$\leftarrowtail$	○◉	$i = l \ldots 1$
$\mathcal{Y}_j$	$=$	$\mathcal{Y}_j \cup \mathcal{Y}_i$; $j \prec i$	
$\bar{V}_{j,k}$	$\mathrel{+}=$	$\bar{V}_{i,k}\, c_{ij}$ for $k \in \mathcal{Y}_i$; $j \prec i$	
$\mathcal{Y}_i$	$=$	$\emptyset$; $\bar{V}_i \equiv 0$	
$\bar{X}_{i,k}$	$=$	$\bar{V}_{i-n,k}$ for $k \in \mathcal{Y}_{i-n}$	$i = 1 \ldots n$

The task performed for each elemental function φ_i is the backward propagation of $|\mathcal{Y}_i|$ gradients. Thus, we derive from (4.23) for the sparse reverse mode

$$\boxed{TIME\{\overleftarrow{eval}\,(F')\} \le (1.5 + 2.5\,\bar{m})\, TIME\{eval(F)\}} \tag{7.16}$$

where $\bar{m}$ is the average range size defined in (7.13). Both runtime estimates (7.15) and (7.16) are very optimistic: they fail to account for the effort to manipulate the index sets $\mathcal{X}_i$ or $\mathcal{Y}_i$. The storage and modification of sparse vectors are a research area in itself [DER89], and it would be impossible to make any general statements. For a thorough implementation in the context of AD, see SparsLinC [BC+96].

When an evaluation procedure represents a subroutine with fixed dimensions and without program branches, the index sets $\mathcal{X}_i$ and $\mathcal{Y}_i$ can be determined at compile-time. One may then implement the sparse procedures of Tables 7.1 and 7.2 with much less runtime overhead. We will return to this option in Chapter 9 in the context of more general edge-elimination techniques. The compiled sparse approach offers a great potential for right-hand sides arising in ODEs or finite-element calculations. Typically these derivative matrices are of moderate size and need to be evaluated at many points in space or time, so that an extensive preprocessing effort may be worthwhile. In the following chapter we consider another way of exploiting sparsity in AD, namely, compression.

7.3 Sparse Second Derivatives

Provided the differentiability Assumption (ED) holds with $d \geq 2$, each intermediate v_i has a Hessian

$$\ddot{V}_i = \left(\frac{\partial^2 v_i}{\partial x_j \, \partial x_k}\right)_{j=1\ldots n}^{k=1\ldots n} \in \mathbb{R}^{n\times n} .$$

Since $v_i(x)$ depends only nontrivially on the variables x_j with $j \in \mathcal{X}_i$, we know that the nonzero entries $\ddot{V}_{ijk} = e_j^\top \ddot{V}_i \, e_k$ must have indices $(j,k) \in \mathcal{X}_i \times \mathcal{X}_i$. In fact, we can be more precise. Just as $\mathcal{X}_i$ contain all indices j with $\dot{V}_{ij} \not\equiv 0$, we are looking for a set of index pairs

$$\dot{\mathcal{X}}_i \supseteq \left\{(j,\, k) \in \mathcal{X}_i \times \mathcal{X}_i \;\middle|\; \ddot{V}_{ijk} \neq 0\right\} .$$

By differentiating the vector relation $\dot{V}_i = \sum_{j\prec i} c_{ij} \, \dot{V}_j$ once more, we obtain the matrix recursion

$$\ddot{V}_i = \sum_{j\prec\, i}\left[c_{ij}\, \ddot{V}_j + \dot{V}_j \sum_{k\prec\, i} c_{ijk}\, \dot{V}_k^\top\right] ,$$

where the c_{ijk} are as defined in (5.11).

Since the first derivatives c_{ij} are almost certainly nonzero, we must always require that

$$\dot{\mathcal{X}}_i \supseteq \bigcup_{j\prec\, i} \dot{\mathcal{X}}_j \,. \tag{7.17}$$

On the other hand, the second derivatives c_{ijk} vanish for all linear elemental functions φ_i, that is, additions, subtractions, and multiplication by constants. For univariate nonlinear functions we have $c_{ijj} \neq 0$ and require therefore that

$$\dot{\mathcal{X}}_i \supseteq \mathcal{X}_j \times \mathcal{X}_j \,. \tag{7.18}$$

If we restrict ourselves to the basic library of elementals Ψ, that leaves only multiplications $v_i = v_j * v_k$, where $c_{ijk} = 1 = c_{ikj}$ and $c_{ijj} = 0 = c_{ikk}$. Then we require that

$$\dot{\mathcal{X}}_i \supseteq \mathcal{X}_j \times \mathcal{X}_k \cup \mathcal{X}_k \times \mathcal{X}_j \; . \tag{7.19}$$

The Cartesian product $\dot{\mathcal{X}}_i = \mathcal{X}_i \times \mathcal{X}_i$, which neglects *internal* sparsity of the $\ddot{V}_i$, would be a valid choice for an upper bound satisfying (7.17), (7.18), and (7.19). The addition to Table 7.1 listed in Table 7.3 tracks the sparsity of the $\ddot{V}_i$ starting from $\ddot{V}_{i-n} = 0$, and hence $\dot{\mathcal{X}}_{i-n} = \emptyset$. In the assignments and increments for the $\ddot{V}_i$, it is understood that only the elements corresponding to indices in $\dot{\mathcal{X}}_i$ need to be computed. Writing the relations componentwise would have made the expressions too complicated.

Table 7.3: Additions to Table 7.1 for Evaluating Hessians Forward

$\dot{\mathcal{X}}_{i-n}$	$=$	$\emptyset; \quad \ddot{V}_{i-n} = 0$	$i = 1 \ldots n$
$\dot{\mathcal{X}}_i$	$=$	$\bigcup_{j \prec i} \dot{\mathcal{X}}_j$	
$\ddot{V}_i$	$=$	$\sum_{j \prec i} c_{ij} \ddot{V}_j$	
if (v_i	$\equiv$	$\psi(v_j)$ with $\psi'' \not\equiv 0$)	
$\dot{\mathcal{X}}_i$	$=$	$\dot{\mathcal{X}}_i \cup (\mathcal{X}_j \times \mathcal{X}_j)$	$i = 1 \ldots l$
$\ddot{V}_i$	$+=$	$\dot{V}_j \, c_{ijj} \, \dot{V}_j^\top$	
if (v_i	$\equiv$	$v_j * v_k$)	
$\dot{\mathcal{X}}_i$	$=$	$\dot{\mathcal{X}}_i \cup (\mathcal{X}_j \times \mathcal{X}_k) \cup (\mathcal{X}_k \times \mathcal{X}_j)$	
$\ddot{V}_i$	$+=$	$\dot{V}_j \, \dot{V}_k^\top + \dot{V}_k \, \dot{V}_j^\top$	
$\ddot{Y}_{m-i}$	$=$	$\ddot{V}_{l-i}$	$i = m-1 \ldots 0$

Since $|\dot{\mathcal{X}}_i| \le |\mathcal{X}_i|^2$ the computational effort in Table 7.3 is of order $|\mathcal{X}_i|^2$ for each $i \le l$ and will quite likely dominate the cost of the underlying Table 7.1. Summing over i and replacing one factor $|\mathcal{X}_i|$ by its upper bound, the Jacobian width $\hat{n}$, we obtain the complexity bound

$$\boxed{TIME\{\overrightarrow{eval}(F'')\} \le c\,\hat{n}\,\bar{n}\; TIME\{eval(F)\}} \tag{7.20}$$

where c is a constant of about 10. If partial separability (see section 11.2) has been used to reduce the maximal row length of the Jacobian, we might hope that all domain sizes $|\mathcal{X}_i|$ are reasonably small. Otherwise, we may reduce storage and computation by keeping track of the sets $\dot{\mathcal{X}}_i \subseteq \mathcal{X}_i \times \mathcal{X}_i$ and thus using internal sparsity. Naturally, this procedure entails significant overhead, which one generally tries to avoid.

Alternatively, we may apply the reverse mode. By differentiating the adjoint procedure Table 3.5 for evaluating the row vector $\bar{x}^\top = \bar{y}^\top F'(x)$, we find that the second-order adjoints

$$\dot{\bar{V}}_i \equiv \left(\frac{\partial \bar{v}_i}{\partial x_j}\right)_{j=1\ldots n} \equiv \left(\dot{\bar{V}}_{ij}\right)_{j=1\ldots n} \in \mathbb{R}^n \quad \text{for} \quad i = 1-n\ldots l$$

need to be propagated. The derivative of $\bar{x}$ with respect to x is the symmetric Hessian matrix

$$H(x,\bar{y}) \equiv \sum_{i=1}^{m} \bar{y}_i\, F_i''(x) = \nabla_x^2\, \bar{y}^\top y \in \mathbb{R}^{n\times n}\,.$$

The ith row and column of the Hessian $H(x,\bar{y})$ is given by $\dot{\bar{V}}_{\hat{\imath}}$ with $\hat{\imath} \equiv l-m+i$ as before. $H(x,\bar{y})$ is likely to be sparse too, especially when F is partially separable. The relation between the sparsity pattern of $H(x,\bar{y})$ for dense $\bar{y}$ and that of the Jacobian $F'(x)$ as described by the index sets $\mathcal{X}_{\hat{\imath}}$ and $\mathcal{Y}_{j-n}$ will be examined in Proposition 7.1 at the end of this subsection.

Before that, we will discuss a dynamically sparse procedure for computing $H(x,\bar{y})$, which consists of a forward sweep identical to Table 7.1 and the subsequent reverse sweep listed in Table 7.4. The index domains $\bar{\mathcal{X}}_i$ are defined to satisfy

$$\bar{\mathcal{X}}_i \supseteq \left\{ j \leq n \mid \dot{\bar{V}}_{ij} \neq 0 \right\}\,. \tag{7.21}$$

Hence the $\bar{\mathcal{X}}_i$ (over)estimate the nonzero pattern of the ith Hessian row. Differentiating the incremental statement in Table 3.5, we obtain the relation

$$\dot{\bar{V}}_j \mathrel{+}= \dot{\bar{V}}_i\, c_{ij} + \bar{v}_i \sum_{k\prec i} c_{ijk}\, \dot{V}_k\,,$$

where $\dot{V}_k \equiv \nabla_x v_k$ with index pattern $\mathcal{X}_k$, as in Table 7.1. Now we again make a case distinction regarding the nature of φ_i. Since generally $c_{ij} \neq 0$, we must require that

$$\bar{\mathcal{X}}_j \supseteq \bigcup_{i\succ j} \bar{\mathcal{X}}_i\,, \tag{7.22}$$

which corresponds nicely to (7.17). For nonlinear univariate φ_i we also need

$$\bar{\mathcal{X}}_j \supset \mathcal{X}_j\,, \tag{7.23}$$

which corresponds to (7.18). Finally, for $\varphi_i = v_j * v_k$ we need

$$\bar{\mathcal{X}}_j \supset \mathcal{X}_k \quad \text{and} \quad \bar{\mathcal{X}}_k \supset \mathcal{X}_j\,, \tag{7.24}$$

which corresponds to (7.19).

These relations are built into the addition listed in Table 7.4 to the forward procedure of Table 7.1. For simplicity we have assumed that there is no overwriting, which means in particular that all $\mathcal{X}_i$ and $\dot{V}_i$ are known from the

Table 7.4: Additions to Table 7.1 for Evaluating Hessians, No Overwrite!

Statement	Range
$\bar{\mathcal{X}}_i \equiv \emptyset\,; \quad \bar{v}_i \equiv 0\,; \quad \dot{\bar{V}}_i = 0$	$i = 1-n \ldots l$
$\bar{v}_{l-i} = \bar{y}_{m-i}$	$i = 0 \ldots m-1$
$\bar{v}_j \mathrel{+}= \bar{v}_i\, c_{ij}\,; \quad j \prec i$	
$\bar{\mathcal{X}}_j = \bar{\mathcal{X}}_j \cup \bar{\mathcal{X}}_i\,; \quad j \prec i$	
$\dot{\bar{V}}_j \mathrel{+}= \dot{\bar{V}}_i\, c_{ij}\,; \quad j \prec i$	
if $(v_i \equiv \psi(v_j)$ with $\psi'' \not\equiv 0)$	
$\quad \bar{\mathcal{X}}_j = \bar{\mathcal{X}}_j \cup \mathcal{X}_j$	
$\quad \dot{\bar{V}}_j \mathrel{+}= \bar{v}_i\, c_{ij}\, \dot{V}_j$	$i = l \ldots 1$
if $(v_i \equiv v_j * v_k)$	
$\quad \bar{\mathcal{X}}_j = \bar{\mathcal{X}}_j \cup \mathcal{X}_k$	
$\quad \dot{\bar{V}}_j \mathrel{+}= \bar{v}_i\, \dot{V}_k$	
$\quad \bar{\mathcal{X}}_k = \bar{\mathcal{X}}_k \cup \mathcal{X}_j$	
$\quad \dot{\bar{V}}_k \mathrel{+}= \bar{v}_i\, \dot{V}_j$	
$\dot{\bar{X}}_i = \dot{\bar{V}}_{i-n}$	$i = 1 \ldots n$

forward sweep Table 7.1 when we are on the way back according to Table 7.4. The combination of Tables 7.1 and 7.4 can also be run with overwriting, provided all v_i, $\mathcal{X}_i$, and $\dot{V}_i$ are saved on the way forward and recovered from the tape ⌂⌂ on the way back. The complexity is governed by the size of the index domains $\bar{\mathcal{X}}_i$, which can be bounded as follows.

Proposition 7.1 (Sparsity of Lagrangian Hessians)
The sets $\bar{\mathcal{X}}_i$ defined recursively by Table 7.4 satisfy

$$\bar{\mathcal{X}}_i \subseteq \bigcup_{k \in \mathcal{Y}_i} \mathcal{X}_{l-m+k}\,, \qquad |\bar{\mathcal{X}}_i| \le \hat{n}\,|\mathcal{Y}_i|$$

and

$$\max_{1 \le i \le l} |\bar{\mathcal{X}}_i| = \hat{n}\,(\nabla_x^2 \bar{y}^\top F(x)) \le \hat{m}\,(F'(x))\,\hat{n}(F'(x))$$

where $\hat{n}(A)$ denotes the width of a matrix A and $\hat{m}(A) \equiv \hat{n}(A^\top)$ according to Definition (MW).

Proof. As a result of the initialization $\bar{\mathcal{X}}_i = \emptyset$ for all $1-n \le i \le l$, the asserted inclusion holds trivially at the very beginning. Suppose it also holds until i has reached a certain value and we begin to execute the main loop of Table 7.4. The first expansion of $\bar{\mathcal{X}}_j$ for some $j \prec i$ means that

$$\bar{\mathcal{X}}_j = \bar{\mathcal{X}}_j \cup \bar{\mathcal{X}}_i \subset \bigcup_{k \in \mathcal{Y}_j} \mathcal{X}_{\hat{k}} \cup \bigcup_{k \in \mathcal{Y}_i} \mathcal{X}_{\hat{k}} = \bigcup_{k \in \mathcal{Y}_j} \mathcal{X}_{\hat{k}}\,,$$

since $\mathcal{Y}_i \subset \mathcal{Y}_j$ because of the path connectedness Assumption (PC). When $\varphi_i = \psi$ with $\psi'' \neq 0$, we have $j \prec i \prec \hat{k}$ for at least one $k = \hat{k} + m - l$, so that $\mathcal{X}_j \subset \mathcal{X}_{\hat{k}}$ and $k \in \mathcal{Y}_j$, which again confirms the assertion. When $\varphi_i(v_j, v_k) = v_j * v_k$, the same argument applies, which completes the proof of the asserted set inclusion. The inequality follows from

$$|\bar{\mathcal{X}}_i| \leq \Big| \bigcup_{k \in \mathcal{Y}_i} \mathcal{X}_{\hat{k}} \Big| \leq |\mathcal{Y}_i| \max_{l-m<k\leq l} |\mathcal{X}_k|$$

by definition $\hat{m}$ and $\hat{n}$ in (7.10) and (7.9). ■

Proposition 7.1 asserts that $\bar{v}_i$ can depend only on those x_j that impact some dependent variable $y_k = v_{l-m+k}$ that also depends on v_i. In graph terminology the maximal node $\hat{k} \equiv l - m + k$ is the common endpoint of two directed paths emanating from $x_j = v_{j-n}$ and v_i, respectively. In other words, there exists a walk between v_i and x_j. Since the computational effort associated with intermediate i in Table 7.4 is of order $|\bar{\mathcal{X}}_i| \leq |\mathcal{Y}_i|\hat{n}$, it follows that

$$\boxed{TIME\{\overrightarrow{eval}(H)\} \leq c\, \hat{n}\, \bar{m}\; TIME\{eval(F)\}} \tag{7.25}$$

where c is a constant of about 10. By comparison with (7.20) we note that the average range size $\bar{m}$ has taken the place of the average domain size $\bar{n}$, which may entail a significant reduction of the complexity bound. On the other hand, it should be noted that one only obtains one Hessian $H = \bar{y}^\top F''$ and not the full second derivative tensor F'' calculated by the purely forward calculation of second derivatives in section 7.3.

Summary

The forward and reverse modes of differentiation can be used to propagate derivative objects in sparse storage mode. This approach greatly diminishes operations counts, with n and m being effectively reduced to the average domain and range size $\bar{n}$ and $\bar{m}$, respectively. However, it is not clear to what extent this theoretical gain can be turned into runtime savings. The problem is that the required index manipulations are of the same order as the number of the arithmetic operations. For the problem of calculating Hessians in the forward mode, one may overestimate $\dot{\mathcal{X}}_i \equiv \mathcal{X}_i \times \mathcal{X}_i$, thus making the indexing effort linear but the arithmetic count quadratic with respect to $|\mathcal{X}_i| \leq \hat{n}$.

As an alternative to the dynamic sparsity exploitation examined in this chapter, we discuss in the next chapter pseudostatic schemes based on matrix compression. However, these must rely on either the user or a preliminary, dynamically sparse forward or reverse sweep providing the index sets $\mathcal{X}_{\hat{\imath}}$, $\mathcal{Y}_{j-n}$, $\dot{\mathcal{X}}_i$, or $\bar{\mathcal{X}}_i$ and thus the sparsity pattern of the Jacobian, or the Hessian, respectively.

7.4 Examples and Exercises

Exercise 7.1 (*Proof of Fischer and Flanders* [FF99])
Establish Lemma 7.1 by proving the implication

$$1 \leq i_1 \leq \cdots \leq i_r \leq k \implies \Big| \bigcup_{1\leq j\leq r} \mathcal{X}_{i_j} \Big| \leq r + k$$

by induction on $k \geq 1$.

Exercise 7.2 (*One-Dimensional Control*)
For some constant stepsize h consider the evaluation procedure displayed in the box on the right. The independents $x_k = v_{k-n}$ can be thought of as accelerations added to the velocities v_{k-1} by $h * v_{k-n}$ to yield v_k. The positions w_{k-1} are updated by $h * v_k$ to w_k accordingly. Both position and velocity start from 1.

for $k = 1 \ldots n$
  $v_{k-n} = x_k$
$v_1 = 1 + h * v_{1-n}$
$w_1 = 1 + h * v_1$
for $k = 2 \ldots n$
  $v_k = v_{k-1} + h * v_{k-n}$
  $w_k = w_{k-1} + h * v_k$
$y = w_n$

a. Draw the computational graph and determine for each intermediate v_i the index domain $\mathcal{X}_i$ and the index range $\mathcal{Y}_i$. Compute the average domain size considering the scalar *saxpy* $c = a + h * b$ as one temporal complexity unit.
b. Verify that the gradient component $\frac{\partial y}{\partial x_k}$ can be computed by smart differencing at a cost of $2(n-k)+1$ scalar *saxpies*. Discuss the validity of (7.12).

Exercise 7.3 (*Partially Separable Energy*)
With $b = (b_1, b_2, \ldots, b_n)^\top \in \mathbb{R}^n$ a vector of positive constants, consider the energy function

$$y = f(x) = \sum_{i=1}^{n} x_i \ln[x_i/(1 - b^\top x)]$$

which is well defined for all $x \in \mathbb{R}^n$ with $b^\top x < 1$ and $x > 0$ componentwise. This kind of function is used for modeling the energy density of reservoir oils and other mixed fluids in phase equilibrium calculations [Mic82].
a. Write two distinct procedures for evaluating $f(x)$, one involving divisions and the other involving only additions, subtractions, multiplications, and logarithms. Sketch the computational graph for the *original* as well as the *division-free* formulation for $n = 4$.
b. Determine the index domains $\mathcal{X}_i$ and index ranges $\mathcal{Y}_i$ for each intermediate variable v_i in either formulation. Express the average domain size $\bar{n}$ for both procedures as a function of n and $|\ln|$. Here $|\ln|$ denotes the temporal complexity of the logarithm relative to the arithmetic operations, including divisions,

which are all assumed to require one time unit each. Find the limit of $\bar{n}$ when $|\ln|$ is very large.
c. Demonstrate how the dense gradient ∇f can be obtained more cheaply by smart differencing or sparse forward differentiation (see Table 7.1) applied to the division-free formulation. Observe, in contrast, that for large $|\ln|$, even smart differencing applied to the original formulation incurs the full cost equivalent to $(n+1)$ function evaluations for the Jacobian approximation.

Chapter 8

Exploiting Sparsity by Compression

As it turned out in Chapter 7 the straightforward dynamic exploitation of sparsity generally does not pan out, due to overhead. Therefore we shall develop in this chapter an alternative based on matrix compression.

It is well known that sparse Jacobians can be determined on the basis of directional derivatives,

$$F'(x)\,s = [F(x + \varepsilon\, s) - F(x)]\,/\varepsilon + O(\varepsilon)\;, \tag{8.1}$$

where the directions $s \in \mathbb{R}^n$ are usually chosen as 0-1 vectors. Combining p such directions into a *seed matrix* $S \in \mathbb{R}^{n\times p}$, we obtain the matrix equation

$$B = F'(x)\,S \in \mathbb{R}^{m\times p}\;.$$

In the notation of section 3.1 we may interpret $\dot{X} \equiv S$ as the input to the vector tangent procedure and then obtain $B = \dot{Y}$ as the result. In contrast to the difference quotient approximation (8.1), the tangent procedure yields entries for B that are correct up to working accuracy.

The ith row of B is given by

$$b_i^\top \equiv e_i^\top B \equiv e_i^\top F'(x)^\top S = \nabla F_i(x)^\top S\;, \tag{8.2}$$

where $e_i \in \mathbb{R}^m$ is the ith Cartesian basis vector. This means that $b_i \in \mathbb{R}^p$ is uniquely determined by the entries in the ith row of the Jacobian $F'(x)$, which represent the gradient of the ith component function $F_i(x) \equiv e_i^\top F(x)$. The following question arises: Under which condition is this linear relation reversible, so that the entries of $\nabla F_i(x)$ can be reconstructed from the elements of b_i? Clearly, this can be done only if p, the number of computed values, is not smaller than the number of unknown nonzeros in the ith row, namely,

$$p_i \equiv |\mathcal{X}_{l-m+i}| \quad \text{for} \quad 1 \le i \le m\;.$$

Since we want to obtain all rows of B using the same seed matrix S, we need

$$cols(S) = p \ge \max_{1\le i\le m} p_i = \hat{n}\;,$$

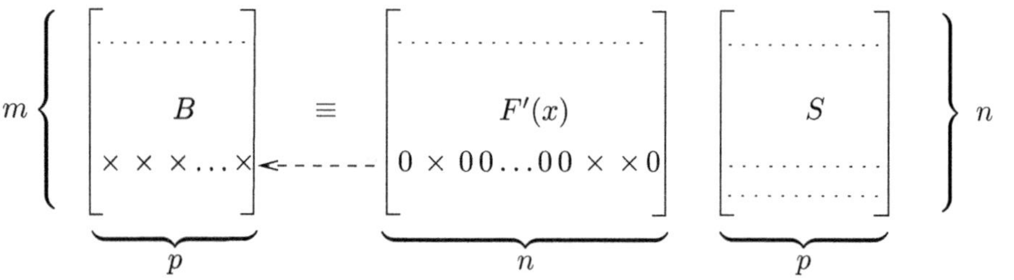

$$m\left\{\underbrace{\begin{bmatrix} \cdots\cdots \\ B \\ \times\ \times\ \times\ldots\times \\ \ \end{bmatrix}}_{p}\right. \equiv \underbrace{\begin{bmatrix} \cdots\cdots \\ F'(x) \\ 0\times 00\ldots 00\times\times 0 \\ \ \end{bmatrix}}_{n} \left.\underbrace{\begin{bmatrix} \cdots\cdots \\ S \\ \cdots\cdots \\ \cdots\cdots \end{bmatrix}}_{p}\right\} n$$

Figure 8.1: Row Compression/Reconstruction

where $\hat{n}$ is the Jacobian width defined in (7.9). The relation among $S, F'(x)$, and B is illustrated in Fig. 8.1.

The sparsity of the ith row of $F'(x)$ is determined by the index set $\mathcal{X}_{\hat{\imath}}$, where $\hat{\imath} = l - m + i$. Abbreviating $A \equiv F'(x)$, we have by (8.2) that for any $i \leq m$

$$b_i^\top = e_i^\top A\, S = \sum_{j \in \mathcal{X}_{\hat{\imath}}} e_i^\top A\, e_j\, e_j^\top S = a_i^\top S_i \ ,$$

where

$$a_i \equiv (e_i^\top A\, e_j)_{j \in \mathcal{X}_{\hat{\imath}}} \in \mathbb{R}^{p_i} \quad \text{and} \quad S_i \equiv (e_j^\top S)_{j \in \mathcal{X}_{\hat{\imath}}} \in \mathbb{R}^{p_i \times p} \ . \tag{8.3}$$

All we have done here is to leave out terms that vanish because of the sparsity of A. Since $p \geq p_i = |\mathcal{X}_{\hat{\imath}}|$ by definition, the linear system

$$S_i^\top a_i = b_i \tag{8.4}$$

is generally overdetermined. We have $p \geq p_i$ equations to solve for the p_i unknown components of a_i. However, since (8.2) holds by the chain rule, this linear system is always mathematically consistent. Because of roundoff errors it is unlikely that the right-hand side b_i obtained from the forward mode lies exactly in the range of $S_i^\top$. Nevertheless, a_i can be computed accurately provided the p_i columns of $S_i^\top$ are linearly independent.

Another important consideration is the computational effort, since m such linear systems need to be solved in order to reconstruct the Jacobian $A = F'(x)$ from its compression $B = A\,S$. For choosing S we later discuss two basic approaches and a combination of them. Partitioning

$$S_i = [\ \underbrace{\hat{S}_i}_{p_i}\ ,\ \underbrace{\tilde{S}_i}_{p - p_i}\ ] \in \mathbb{R}^{p_i \times p} \ ,$$

we observe that $\delta_i(S) \equiv \det(\hat{S}_i)$ is a polynomial in the entries of S, which clearly does not vanish identically. Hence the same is true for the product

$$\delta(S) \equiv \prod_{i=1}^{m} \delta_i(S) \ ,$$

whose roots are nowhere dense in $\mathbb{R}^{n \times p}$. For all $S \in \mathbb{R}^{n \times p}$ with $\delta(S) \neq 0$ the linear system (8.4) has for each i a unique solution a_i. This can then be obtained by solving the square system

$$\hat{S}_i^\top a_i = \hat{b}_i \equiv \left(e_k^\top b_i\right)_{k=1\ldots p_i} . \tag{8.5}$$

Thus we have established the following generalization of a result from [NR83].

Proposition 8.1 (Newsam–Ramsdell Principle)
Suppose F is once continuously differentiable on the whole of $\mathbb{R}^n$.

(i) *If no component function F_i depends nontrivially on more than $\hat{n}$ of the n variables x_i, then the restriction $F|_{\mathcal{S}}$ uniquely determines $F(x)$ and its Jacobian $F'(x)$ for almost all subspaces $\mathcal{S} \subset \mathbb{R}^n$ of dimension $\hat{n}$ or greater.*

(ii) *If no variable x_j impacts more than $\hat{m}$ of the m component functions F_i nontrivially, then the projection PF uniquely determines the Jacobian $F'(x)$ for almost all projections $P = PP$ of rank $\hat{m}$ or greater.*

Proof. Leading up to the proposition we have already shown that $F'(x)$ is uniquely determined by $F'(x)S$ for almost all matrices $S \in \mathbb{R}^{n\times\hat{n}}$. By the mean value theorem this implies that $F(x)$ is also uniquely determined for all $x \in \mathbb{R}^n$, provided $\mathcal{S}$ is spanned by the columns of S. The second assertion follows similarly. Any projection P of rank $\hat{m}$ can be written as $W^\dagger W^\top = P$ with $W^\dagger$ a generalized inverse in that $W^\top W^\dagger = I_{\hat{m}}$. The Jacobian of $PF(x)$ premultiplied with $W^\top$ yields the $\hat{m} \times n$ matrix $W^\top F'(x)$ and, equivalently, its transpose $F'(x)^\top W \in \mathbb{R}^{n\times\hat{m}}$. To that we may apply the same arguments as already given with rows and columns of the Jacobian interchanged to reconstruct $F'(x)$ uniquely for almost all $W \in \mathbb{R}^{m\times\hat{m}}$. ∎

The proposition is basically a degree-of-freedom argument. It also shows that row compressions of the Jacobian by a seed matrix S can be interpreted as restricting F to a suitable subspace $range(S) \subset \mathbb{R}^n$. Similarly, column compressions can be interpreted as projections of F onto a certain subspace $range(P) \subset \mathbb{R}^m$. In the CPR approach discussed below, the subspaces $range(S)$ or $range(P)$ are required to have a set of basis vectors that represent sums of Cartesian basis vectors. Generally, this extra condition will require that $rank(S) > \hat{n}$ or $rank(P) > \hat{m}$.

The fact that the linear systems are solvable for most S is not entirely satisfactory from a computational point of view. Powell and Toint [PT79] distinguished among *direct*, *substitution*, and *elimination* methods for estimating Jacobians. In our notation these three cases correspond to all m square matrices $\hat{S}_i$ being permutations of

1. the identity matrix,
2. a unitary triangular 0-1 matrix,
3. a general nonsingular square matrix.

Corresponding to these three cases, the reconstruction of $A = F'(x)$ involves no arithmetic operations (1), only subtractions (2), or all four arithmetic operations (3). In the remainder of this section we will meet examples of all three kinds of reconstruction methods. First we consider a *direct* estimation technique.

8.1 Curtis–Powell–Reid Seeding

Since the columns of B are directional derivatives, they can be approximated by difference quotients according to (8.1). This well-established technique for approximating Jacobians can be made much more efficient by finding a mapping

$$c : \{1 \dots n\} \longmapsto \{1 \dots p\}$$

such that for any pair $j, k \leq n$

$$\mathcal{Y}_{j-n} \cap \mathcal{Y}_{k-n} \neq \emptyset \Longrightarrow c(j) \neq c(k) \, . \tag{8.6}$$

The inverse images $c^{-1}(i)$ for $i \leq p$ are usually called "column groups" associated with the "color" i. The implication (8.6) requires that two column indices j and k can belong to the same group only when the corresponding independent variables $x_j = v_{j-n}$ and $x_k = v_{k-n}$ do not jointly impact any dependent variable. This condition implies that none of the rows of $F'(x) = A$ can have a nonzero entry in both the jth and kth columns.

Coloring Techniques for Sparse Jacobians

As observed by Coleman and Moré [CM83], the task of finding a suitable function $c(j)$ is equivalent to coloring the column incidence graph

$$\mathcal{G}_c = (\mathcal{V}_c, \mathcal{E}_c) \quad \text{with} \quad \mathcal{V}_c \equiv \{1, 2, \dots, n\}$$

and

$$(j, k) \in \mathcal{E}_c \quad \text{if} \quad \mathcal{Y}_{j-n} \cap \mathcal{Y}_{k-n} \neq \emptyset \, .$$

In this context the condition (8.6) means that two vertices j and k cannot receive the same color when they are adjacent, i.e., connected by an edge. This corresponds to a so-called distance-1 coloring. Alternatively, one can also perform a distance-2 coloring of an appropriate bipartite graph as proposed by [GMP05].

To keep the computational effort low, we would like to keep the number p of different colors as small as possible. The absolute minimum for any given graph $\mathcal{G}$ is called its *chromatic number* $\chi(\mathcal{G})$, and its determination is well known to be NP-hard. Fortunately, various heuristic algorithms have been developed for both coloring approaches mentioned above. They generally yield values p that are only slightly larger than the chromatic number itself. A comprehensive introduction of corresponding coloring algorithms together with a comparison of their performance can be found in [GMP05]. All this methodology can be immediately transferred to AD. Moreover, it can also be applied by grouping, or coloring, Jacobian rows rather than columns.

Reconstructing the Jacobian

The Curtis–Powell–Reid (CPR) approach [CPR74] amounts to using the seed matrix

$$S = \left[e_{c(j)}^{\top} \right]_{j=1\ldots n} \in \mathbb{R}^{n \times p} , \tag{8.7}$$

where e_i denotes the ith Cartesian basic vector in $\mathbb{R}^p$. Then the submatrices $S_i \subset S$ defined in (8.3) take the form

$$S_i = \left[e_{c(j)}^{\top} \right]_{j \in \mathcal{X}_i} \in \mathbb{R}^{p_i \times p} . \tag{8.8}$$

In other words, each row of S_i contains a single 1 and is otherwise zero. Moreover, each column contains no more than one 1, so that S_i actually consists of a permuted identity matrix of order p_i appended with $p - p_i$ zero columns. To see this, suppose that in the same column of S_i there is a 1 in the jth and the kth rows. Then we must have $c(j) = c(k)$ and $j \in \mathcal{X}_{\hat{\imath}} \ni k$. However, that implies $i = \hat{\imath} + m - l \in \mathcal{Y}_{j-n} \cap \mathcal{Y}_{k-n}$, which requires by (8.6) that $c(j) \neq c(k)$, thus leading to a contradiction.

In terms of the transpose $S_i^{\top} \in \mathbb{R}^{p \times p_i}$ we have shown that each column contains a single 1, which belongs to a row that has no other nonzero entries. Therefore, the linear system (8.4) has the unique solution

$$a_{ij} = e_i^{\top} A \, e_j = e_i^{\top} B \, e_{c(j)} = b_{i\,c(j)} \, .$$

In other words, all nonzero elements of the Jacobian $A = F'(x)$ can be identified among the nonzero entries of the compressed Jacobian B. In fact, we may interpret B simply as a sparse data structure for A.

As an example, we may consider the IER problem from the Minpack-2 test suite [AC+92]. Its sparsity pattern in case of an extremely coarse grid with only 93 variables is displayed on the left-hand side of Fig. 8.2. Probably due to the regular structure of this discretization, the chromatic number $\mathcal{X}$ equals $\hat{n} = 17$, the maximal number of nonzeros per row. The corresponding compressed Jacobian structure is displayed on the right-hand side of Fig. 8.2. Clearly, there are not many zeros left, so matrix compression with CPR must be considered fairly optimal. The sparse matrix in the middle is the seed.

The following runtimes were measured on an Fedora Linux system with an AMD Athlon XP 1666 MHz processor and 512 MB RAM for the IER problem with a much finer grid involving 15,003 variables. The sparsity pattern, the calculation of the seed matrix, and the evaluation of the compressed Jacobian were computed using ADOL-C version 1.10.2. The function evaluation needed 0.0013 seconds. The sparsity pattern of the Jacobian was computed within 0.0671 seconds, yielding the runtime ratio of 51.6. This can be seen as an initial investment that becomes negligible when the Jacobian needs to be evaluated more than a handful of times. The generation of the seed matrix was based on the `GreedyPartialD2Coloring` algorithm of [GMP05] and required 0.4901 seconds, yielding the runtime ratio of 377. Once again 17 colors were needed such that the seed matrix S has 17 columns. Finally, the computation of the

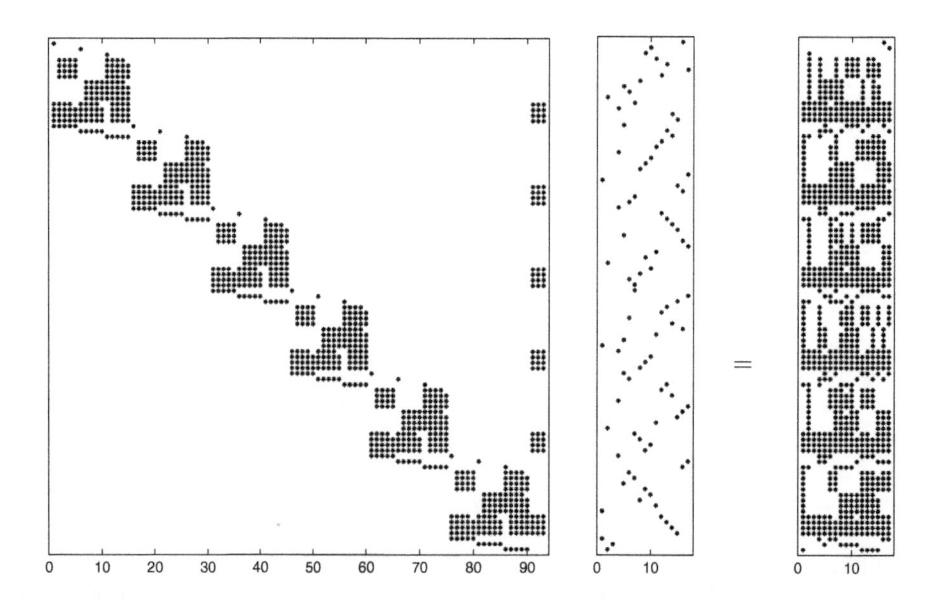

Figure 8.2: Jacobian Sparsity (left) and Compressed Sparsity (right) on IER Problem

Jacobian-matrix product JS took 0.0372 seconds and is therefore only slightly larger than the theoretical bound of $(1 + 1.5 * 17)0.00013 = 0.03445$ seconds derived from (4.18). The evaluation of the hand-coded Jacobian provided for all Minpack-2 test problems required 29.62 seconds and agreed with the ADOL-C values up to the last digit. Unfortunately, row compression with CPR seeding does not always work very well, which leads us to investigate alternatives in section 8.2.

The Binary Interaction Example

On some problems the chromatic number $\chi(\mathcal{G}_c) = \chi(F')$ can be much larger than its lower bound, the Jacobian width $\hat{n}$. One particular example due to Steihaug and Hossain [SH98] is a set of binary interactions

$$F_k(x_1, \ldots, x_n) = \varphi_k(x_i, x_j)$$

with $i < j$ and $k = i + (j-1)(j-2)/2 \leq m \equiv n(n-1)/2$. Here φ_k may be any elementary function in two variables, which can be thought of as interaction energy between particles i and j at positions x_i and x_j, respectively. We have for any pair $i < j$ exactly

$$\mathcal{Y}_{i-n} \cap \mathcal{Y}_{j-n} = \{i + (j-1)(j-2)/2\} \neq \emptyset \,,$$

so that the column incidence graph $\mathcal{G}_c$ is complete in that any two vertices are adjacent. Consequently, the chromatic number is n, and we have

$$\chi(F') = \chi(\mathcal{G}_c) = n \gg \hat{n} = 2 \, ,$$

where the last equation follows from $p_i = |\mathcal{X}_{\hat{\imath}}| = 2$ for all $i = 1 \dots m$. Thus, the CPR approach allows no compression of the Jacobian at all, and we face an n-fold increase in computational cost, irrespective of whether difference quotients or the forward mode of AD is employed. One way to estimate $F'(x)$ at a lower cost is to split it vertically, as advocated by Steihaug and Hossain [SH98] and others.

In the binary interaction example, each function $F_k(x) = \varphi_k(x_i, x_j)$ could be evaluated separately at the three points $(x_i + \varepsilon, x_j)$, $(x_i, x_j + \varepsilon)$, and (x_i, x_j). Similarly, one could apply the forward mode of AD to obtain the gradient $\nabla\varphi_k(x_i, x_j)$ representing the kth row of $F'(x) \in \mathbb{R}^{m \times n}$ quite cheaply. However, this vertical splitting is no longer such a good idea when

$$F_k(x) = \varphi_k(\psi_i(x_i), \psi_j(x_j)) \, ,$$

with the univariate transformations $\psi_i : \mathbb{R} \longmapsto \mathbb{R}$ being much more costly to evaluate than the φ_k. Then treating each $F_k(x)$ separately means reevaluating each $\psi_i(x_i)$ exactly $n - 1$ times. In terms of the average domain and range sizes $\bar{n}$ and $\bar{m}$, and abbreviating $|\varphi_k| \equiv TIME\{eval(\varphi_k)\}$ and $|\psi_i| \equiv TIME\{eval(\psi_i)\}$, we find that

$$\bar{n} = \frac{\sum_{i=1}^{n} |\psi_i| + \sum_{k=1}^{m} 2|\varphi_k|}{\sum_{i=1}^{n} |\psi_i| + \sum_{k=1}^{m} |\varphi_k|} \in [1, 2]$$

but

$$\bar{m} = \frac{\sum_{i=1}^{n} (n-1)|\psi_i| + \sum_{k=1}^{m} |\varphi_k|}{\sum_{i=1}^{n} |\psi_i| + \sum_{k=1}^{m} |\varphi_k|} \in [1, n-1] \, .$$

Here we have used the fact that the ψ_i depend on one independent variable and impact all dependents, whereas the φ_k depend on two and impact one. In terms of the partial separability concepts to be introduced in section 11.2, the binary interaction example is always quite value separable but is only argument separable when $\sum_i |\psi_i|$ is small compared to $\sum_k |\varphi_k|$. When the former dominates the latter, we have $\bar{m} \approx n - 1$, so that F is not at all argument separable. In other words, evaluating all F_k separately is almost n times as time consuming as evaluating them jointly. Then, even the smart differencing scheme and sparse forward discussed in sections 7.1 and 7.2 would be quite cheap, but matrix compression by the CPR approach is much more expensive.

8.2 Newsam–Ramsdell Seeding

Newsam and Ramsdell [NR83] originally looked for a class of matrices S for which the subsystems (8.4) could be solved efficiently and sufficiently accurately. Their suggestions, as well as the seeds matrices to be considered in this subsection, are of the generalized Vandermonde form

$$S = \left[P_k(\hat{\lambda}_j)\right]_{k=1\ldots p}^{j=1\ldots n} \in \mathbb{R}^{n\times p}.$$

Here the $P_k(\lambda)$ for $k = 1\ldots p$ denote linearly independent polynomials of degree less than p and the n real abscissas $\hat{\lambda}_j$ may be restricted to the interval $[-1, 1]$ without loss of generality. In the classical Vandermonde case the P_k are the monomials $P_k(\lambda) = \lambda^{k-1}$. If all n abscissa values are distinct, it follows immediately that all square submatrices of S are nonsingular, which ensures in particular that each one of the m (possibly overdetermined but always consistent) linear relations $b_i^\top = a_i^\top S_i$ is invertible. At least this is true in a mathematical sense without regard to the conditioning of these submatrices and thus the numerical stability of the whole reconstruction procedure.

The effort for solving an unstructured $p \times p_i$ system of linear equations via a matrix factorization grows cubically with the smaller dimension p_i. By exploiting the generalized Vandermonde structure as proposed, for example, in [RO91] one can reduce this effort to grow only quadratically in p_i, which has long been known for the classical Vandermonde matrices [GV96]. The trouble with Vandermonde matrices is that their condition number grows exponentially with the order p_i. This drawback loomed very large in the context of difference quotient approximations, where the right-hand sides b_i are significantly perturbed by truncation errors. However, it was shown by Björck and Pereyra [BP97] that the algorithm just mentioned yields much better accuracy than could be expected from the general estimates based on conditioning. Moreover, in the context of AD the b_i are usually obtained with working accuracy. Nevertheless, the conditioning of the S_i is a reason for concern.

Mathematically, the matrices S_i have full rank if and only if the following implication holds:

$$j, k \in \mathcal{X}_{\hat{\imath}} \Longrightarrow \lambda_j \neq \lambda_k \, .$$

In other words, for each i, the p_i values λ_j with $j \in \mathcal{X}_{\hat{\imath}}$ must all be distinct. A simple way to achieve this is the setting

$$\lambda_j = 2\,\frac{(j-1)}{(n-1)} - 1 \quad \text{for} \quad j = 1\ldots n\,,$$

which distributes the λ_j uniformly between -1 and $+1$. The symmetry of this distribution with respect to the origin is known to be beneficial for the conditioning. However, for large n, the λ_j are rather close together, and if two neighbors occur in some linear system (8.5), the conditioning cannot be very good.

Coloring for Conditioning

Usually the situation can be improved considerably by setting $\lambda_j = \mu_{c(j)}$, with μ_k for $k = 1 \ldots p$ a collection of distinct values and $c : [1 \ldots n] \to [1 \ldots \chi]$ the coloring function discussed in the preceding section. The resulting matrices

$$\hat{S}_i = \left[P_k\left(\mu_{c(j)}\right)\right]_{j \in \mathcal{X}_i}^{k=1\ldots p_i} \tag{8.9}$$

are still nonsingular since

$$\{j,k\} \subset \mathcal{X}_{\hat{\imath}} \Longrightarrow \mathcal{Y}_{j-n} \cap \mathcal{Y}_{k-n} \ni i \Longrightarrow c(j) \neq c(k) \Longrightarrow \mu_{c(j)} \neq \mu_{c(k)} \,,$$

where the central implication holds because of the condition (8.6) on $c(j)$.

We may interpret this approach as two-stage seeding, where $S = GC$ with $G \in \mathbb{R}^{n \times \chi}$ a grouping matrix and $C \in \mathbb{R}^{\chi \times p}$ a combination matrix, as displayed in Fig. 8.3. From now on we may assume without loss of generality that precompression by coloring has been performed so that $n = \chi$, $\mu_j = \lambda_j$, $G = I_\chi$, and $C = S$. In general we would expect the gap $\chi - p$ to be only a small multiple of p. However, as we have seen on the binary interaction example we may have $\chi = n$ and $p = 2$, so that $(\chi - p)/p = \frac{1}{2}n - 1$ can be arbitrary large.

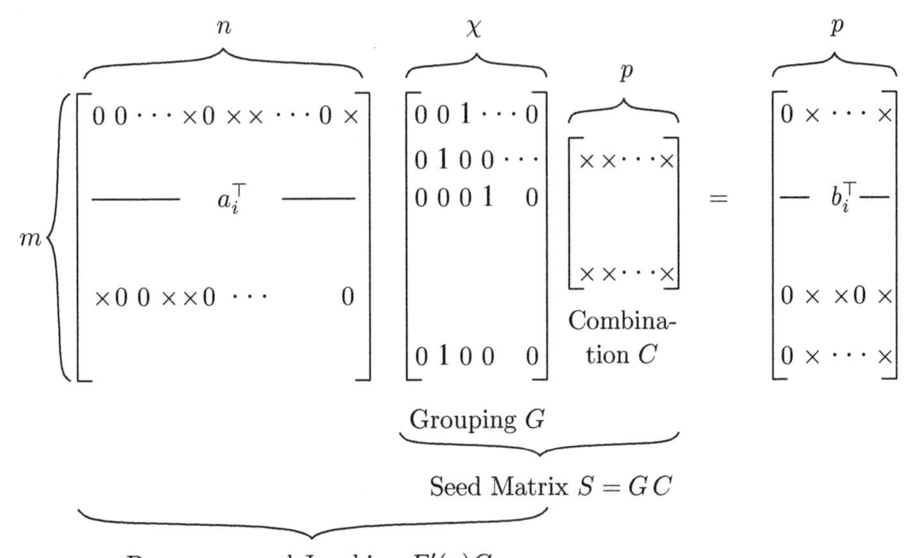

Figure 8.3: Two-Stage Seeding

Hossain and Steihaug have proposed to set $\lambda_j = j$ and to define the polynomials $P_k(\lambda_j)$ for $k = 1 \ldots p$ such that

$$P_k(\lambda_j) \equiv \binom{\chi - p}{j - k} = P_1(j - k + 1) \quad \text{for} \quad j = 1 \ldots \chi \,.$$

The resulting matrix $[P_k(\lambda_j)]_{k=1\ldots\varrho}^{j=1\ldots\lambda} \in \mathbb{R}^{\chi\times\varrho}$ has only $(\chi - \varrho + 1)$ nontrivial diagonals so that it can be factorized cheaply if no pivoting is required. It should be noted that since the degree of the P_k is $\chi > p$ the special fast solution methods mentioned above are not applicable. One can easily see that the matrices S_i obtained for $J_i \equiv \{1\ldots p\}$ and $J_i \equiv \{\chi - p + 1\ldots\chi\}$ are triangular. However, when $\chi - p$ is of significant size the subdiagonale for which $j - k \approx \frac{1}{2}(\chi - p)$ dominate all other entries and pivoting becomes inevitable. As a result the low complexity count for bounded matrices may be lost completely. Moreover, it has been found that the L_1 norm of the triangular inverses defined above is exactly

$$\|S_i^{-1}\| = \begin{pmatrix} \chi - 1 \\ p \end{pmatrix} \left(\frac{\chi}{p} - 1\right) \approx \begin{pmatrix} \chi \\ \varrho \end{pmatrix},$$

which may grow exponentially. Nevertheless, since the binomial coefficient on the right-hand side represents exactly the number of ways one may select a $p \times p$ submatrix from the compression matrix $C = S$, very poor conditioning occurs only when the number of square subsystems to be solved is just a minute fraction of the possible combinations. Hence one can hope to avoid the poorly conditioned subsystems by reordering the rows of C appropriately. The same idea is applicable for other choices of the μ_i and P_k. A very promising approach is to define the μ_i for $j = 1\ldots\chi$ as Chebyshev points in $[-1, 1]$ and to select the P_k as Lagrange interpolation polynomials on a more or less equally spaced subset of p of the abscissas μ_j. Then the computational complexity per subsystem resolution is reduced to the order of $(\chi - p)p$ and the worst-case conditioning is about the same as for the Steihaug–Hossain choice.

Checking Pattern Consistency

Throughout this chapter we assume that the sparsity pattern of the Jacobian $F'(x)$ is correctly known so that it can be reconstructed from its compression $F'(x)S$. If this assumption happens to be wrong the resulting Jacobian values are very likely to be wrong too. To detect this highly undesirable situation we may append an extra column $s \in \mathbb{R}^n$ to the seed matrix S and check whether the resulting last column b of $F'(x)[S, s] = [B, b]$ really satisfies $F'(x)s = b$, where $F'(x)$ is reconstructed using the alleged sparsity pattern. This test is implemented in the sparse AD library SparsLinC [BC+96].

Even without appending an extra column one may check for all rows with $p_i < p$ whether the overdetermined linear system (8.4) is really consistent. In the case of CPR seeding this test may be applicable for all rows, and consistency is likely to occur in the form of exact zeros. These must arise when some column of S is structurally orthogonal to some row of $F'(x)$. In the case of Newsam–Ramsdell (NR) seeding S is likely to be dense with or without the extra column s, so a consistency check requires the selection of a numerical tolerance.

8.3 Column Compression

As discussed in section 3.3, the vector reverse mode yields products

$$C^\top \equiv W^\top F'(x) \in \mathbb{R}^{q\times n} \quad \text{for} \quad W \in \mathbb{R}^{m\times q}$$

at a cost proportional to q. Hence we may also apply the compression and reconstruction approaches discussed above to $F'(x)^\top = A^\top \in \mathbb{R}^{n\times m}$. The column incidence graph $\mathcal{G}_c$ is now replaced by the row incidence graph

$$\mathcal{G}_r = (\mathcal{V}_r, \mathcal{E}_r) \quad \text{with} \quad \mathcal{V}_r \equiv \{1, 2, \ldots, m\}$$

and

$$(i,k) \in \mathcal{E}_r \quad \text{if} \quad \mathcal{X}_{\hat{\imath}} \cap \mathcal{X}_{\hat{k}} \neq \emptyset\,.$$

We use the abbreviation $\hat{\imath} \equiv l - m + i$ and $\hat{k} \equiv l - m + k$ as before. Let

$$d : \{1 \ldots m\} \longmapsto \{1 \ldots q\}$$

denote a coloring of $\mathcal{G}_r$ so that

$$\mathcal{X}_{\hat{\imath}} \cap \mathcal{X}_{\hat{k}} \neq \emptyset \Longrightarrow d(i) \neq d(k)\,.$$

Then we may use the CPR approach

$$W \equiv \left[e_{d(i)}^\top\right]_{i=1\ldots m} \in \mathbb{R}^{m\times q}$$

with e_k now denoting the kth Cartesian basis vector in $\mathbb{R}^q$. Similarly, we can implement the NR approach with

$$W \equiv \left[\mu_{d(i)}^{k-1}\right]_{i=1\ldots m}^{k=1\ldots q} \in \mathbb{R}^{m\times q} \quad \text{or} \quad W \equiv \left[P_k\left(\mu_{c(j)}\right)\right]_{i=1\ldots m}^{k=1\ldots q}\,.$$

The square linear systems to be solved now take the form

$$W_j^\top a_j = c_j \equiv C^\top e_j \in \mathbb{R}^q\,,$$

where

$$a_j \equiv \left(e_k^\top A\, e_j\right)_{k\in\mathcal{Y}_{j-n}} \quad \text{and} \quad W_j = \left[e_k^\top W\, e_i\right]_{k\in\mathcal{Y}_{j-n}}^{i=1\ldots q} \in \mathbb{R}^{q_i\times q}\,.$$

Again we may partition $W_j = [\hat{W}_j, \tilde{W}_j]$ and solve the square system

$$\hat{W}_j^\top a_j = \hat{c}_j \equiv \left(e_k^\top c_j\right)_{k=1\ldots q_j}\,.$$

The conditions and methods of solvability are exactly the same as for row compression.

Comparison of CPR and NR on an Example

Let us consider the following small example with $n = 4 = m$. Suppose the Jacobian has the structure

$$A = F' \equiv \begin{bmatrix} a & b & 0 & 0 \\ c & 0 & d & 0 \\ e & 0 & 0 & f \\ 0 & 0 & g & h \end{bmatrix}.$$

Then the column and row incidence graphs $\mathcal{G}_c$ and $\mathcal{G}_r$ take the following forms.

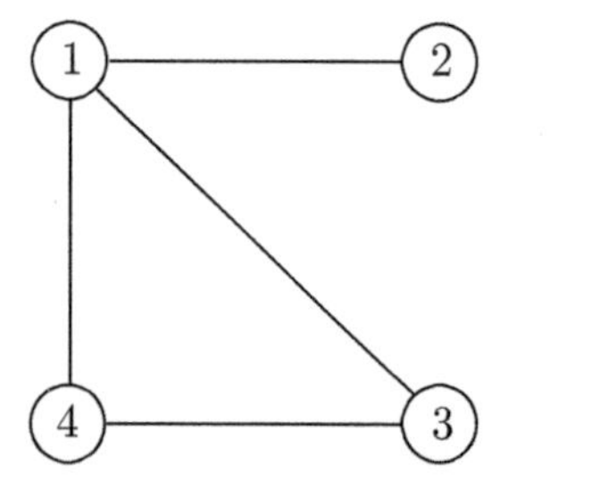

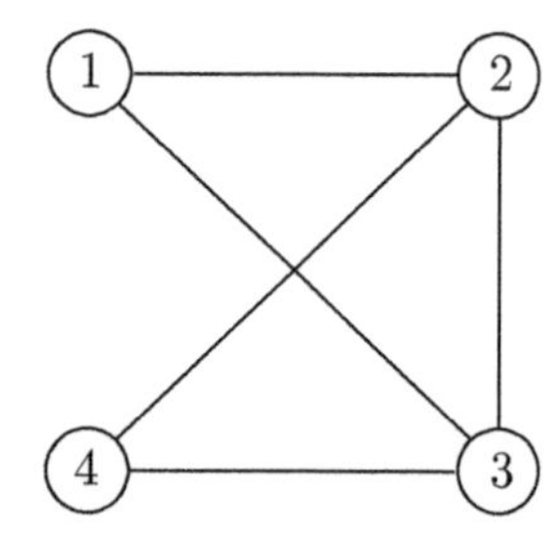

Both $\mathcal{G}_c$ and $\mathcal{G}_r$ consist of a complete subgraph of three vertices and only one extra vertex so that their chromatic numbers satisfy $p = 3 = q$. In the graph $\mathcal{G}_c$ columns 2 and 3 can receive the same color, and in the graph $\mathcal{G}_r$ this applies to rows 1 and 4. Hence the CPR approach can be implemented with the seed matrices

$$S = \begin{bmatrix} 1 & 0 & 0 \\ 0 & 1 & 0 \\ 0 & 1 & 0 \\ 0 & 0 & 1 \end{bmatrix} \quad \text{and} \quad W = \begin{bmatrix} 1 & 0 & 0 \\ 0 & 1 & 0 \\ 0 & 0 & 1 \\ 1 & 0 & 0 \end{bmatrix}$$

yielding the compressed Jacobians

$$B = \begin{bmatrix} a & b & 0 \\ c & d & 0 \\ e & 0 & f \\ 0 & g & h \end{bmatrix} \quad \text{and} \quad C = \begin{bmatrix} a & c & e \\ b & 0 & 0 \\ g & d & 0 \\ h & 0 & f \end{bmatrix}.$$

Thus we see that each nonzero entry of the Jacobian A can be identified and recovered from the compressed version $B = AS$ or $C = A^\top W$. We also note that both B and C each contain four exact zeros, whose absence would indicate deviations from the assumed sparsity pattern. Here there is no difference in complexity between column and row compression based on the forward and reverse modes, respectively.

By inspection of $A = F'(x)$ we note that the maximal number of nonzeros per row and column is given by $\hat{n} = 2$ and $\hat{m} = 3$, respectively. Clearly the general relations $n \geq p \geq \hat{n}$ and $m \geq q \geq \hat{m}$ with $p = \mathcal{X}(\mathcal{G}_c)$ and $q = \mathcal{X}(\mathcal{G}_r)$ are satisfied. Now we can pick the seed matrix according to the NR approach with monomial basis and coloring as

$$S = \begin{bmatrix} 1 & \mu_1 \\ 1 & \mu_2 \\ 1 & \mu_2 \\ 1 & \mu_3 \end{bmatrix} \quad \text{with} \quad \mu_1 < \mu_2 < \mu_3 \,, \quad \text{e.g.,} \quad \begin{bmatrix} \mu_1 \\ \mu_2 \\ \mu_3 \end{bmatrix} = \begin{bmatrix} -1 \\ 0 \\ 1 \end{bmatrix} .$$

In this case we get

$$B = A\,S = \begin{bmatrix} a+b\,, & \mu_1\, a + \mu_2\, b \\ c+d\,, & \mu_1\, c + \mu_2\, d \\ e+f\,, & \mu_1\, e + \mu_3\, f \\ g+h\,, & \mu_2\, g + \mu_3\, h \end{bmatrix} .$$

Having obtained $B \in \mathbb{R}^{4\times 2}$ by the forward mode, we can solve the linear system

$$(a, b) \begin{bmatrix} 1 & \mu_1 \\ 1 & \mu_2 \end{bmatrix} = b_1^\top \equiv e_1^\top B \in \mathbb{R}^2$$

in order to compute the entries (a, b) of the Jacobian $F'(x) = A$. The regularity follows from the assumption that all the μ-values are distinct. Here there is no redundant information available to check consistency, as we have evaluated exactly eight compressed Jacobian elements to reconstruct the same number of original elements.

Obviously, this little example is not very convincing in terms of computational complexity. However, it was found in [Gei95] that the effort for solving the linear Vandermonde systems for the problems from the Minpack test suite is negligible compared with the differentiation effort. Hence, the reduction in the number of derivative vectors needed must be considered a significant advantage. Here the number has been reduced from three to two. The same reduction can be achieved by the two-sided reduction discussed in the following subsection.

8.4 Combined Column and Row Compression

Neither approach discussed above yields any advantages when the Jacobian has the classical inverted arrowhead shape

$$F'(x) \;=\; A \;\equiv\; \begin{bmatrix} \delta_1 & \alpha_2 & & \alpha_{n-1} & \alpha_n \\ \beta_2 & \delta_2 & & 0 & 0 \\ & & \ddots & & \\ \beta_{n-1} & 0 & & \delta_{n-1} & 0 \\ \beta_n & 0 & & 0 & \delta_n \end{bmatrix} .$$

The lines separating and splitting the first row will be explained below. Then both $\mathcal{G}_c$ and $\mathcal{G}_r$ are complete graphs and we have $n = \hat{n} = p = q = \hat{m} = n$. However, one can easily see that the three derivative vectors

$$A \begin{bmatrix} 0 & 1 \\ 1 & 0 \\ 1 & 0 \\ \vdots & \vdots \\ 1 & 0 \end{bmatrix} = \begin{bmatrix} \sum_{i=2}^{n} \alpha_i & \delta_1 \\ \delta_2 & \beta_2 \\ \vdots & \vdots \\ \delta_{n-1} & \beta_{n-1} \\ \delta_n & \beta_n \end{bmatrix} \quad \text{and} \quad A^\top \begin{bmatrix} 1 \\ 0 \\ \vdots \\ 0 \\ 0 \end{bmatrix} = \begin{bmatrix} \delta_1 \\ \alpha_2 \\ \vdots \\ \alpha_{n-1} \\ \alpha_n \end{bmatrix}$$

contain all nonzero elements of A. In other words, a combination of two *forward* Jacobian-vector products and one *reverse* vector-Jacobian product yields the complete Jacobian, whose reconstruction would require n derivative vectors if either pure column or pure row compression were applied.

More generally the question is, "How can suitable matrices $S \in \mathbb{R}^{n\times p}$ and $W \in \mathbb{R}^{m\times q}$ with small $p+q$ be found such that $B = AS \in \mathbb{R}^{m\times p}$ together with $C = A^\top W \in \mathbb{R}^{n\times q}$ contain enough information to reconstruct the nonzero entries of A?" Again one may follow the CPR or the NR approach, yielding direct substitution or elimination methods. Coleman and Verma [CV96] suggested partitioning a permutation $A = QF'(x)P$ of the Jacobian as shown in Fig. 8.4.

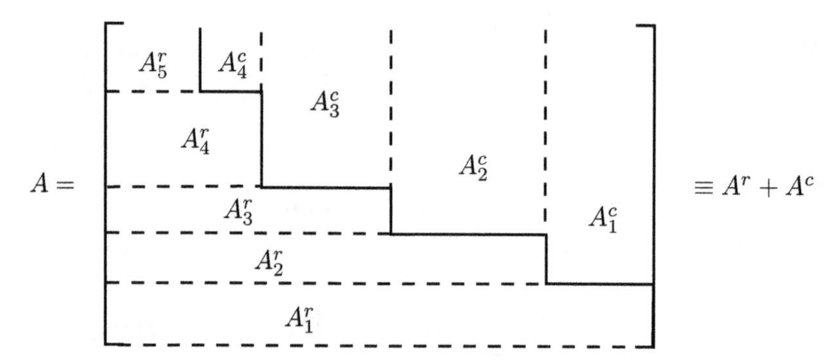

Figure 8.4: Coleman–Verma Partition of Rectangular A

Here the matrix A^r formed by the blocks A_s^r and padded by zeros in place of the A_s^c is block lower triangular. Its complement $A^c = A - A^r$ is block upper triangular. It should be noted that none of the blocks needs to be square; and since the number of row blocks and column blocks may differ by one, it is not even clear which blocks belong to the diagonal. Fortunately, none of this matters for the following reconstruction procedure.

First, we can construct $S \in \mathbb{R}^{n\times p}$ such that it is suitable for column compression of the lower left part A^r formed by the A_s^r. Similarly we can define $W \in \mathbb{R}^{m\times q}$ such that it compresses the upper right part A^c, again using either the CPR or the NR approach. On the inverted arrowhead example above we have chosen $p = 2$ and $q = 1$, but due to the symmetry one could of course also use $p = 1$ and $q = 2$. However, here neither parameter p nor q can be zero without the other jumping to n, which is the whole point of the combined procedure. The forward and reverse modes yield

$$B \equiv A\,S = A^r\,S + A^c\,S \quad \text{and} \quad C \equiv A^\top W = (A^c)^\top W + (A^r)^\top W\,.$$

By inspection of the partitioning Fig. 8.4 we note that the last rows of $B = AS$ do not depend at all on the upper right part A^c, so they can be used to compute A_1^r by row reconstruction as though A^c were zero altogether. We can correct C by subtracting the contributions of A_1^r and obtain

$$\tilde{C} \equiv C - (\bar{A}_1^r)^\top W\,,$$

where $\bar{A}_1^r \in \mathbb{R}^{m\times n}$ has the same format as A and is obtained from A_1^r by padding it with zeros on top. Now we have effectively eliminated the last rows from the system and can determine A_1^c by column reconstruction as though the other A_s^r for $s > 1$ were all zero. Clearly this process can be continued until all of A has been obtained.

Computing a Feasible Partition

The evaluation cost for the combined compression procedure sketched above is proportional to the sum $p{+}q$. Depending on which approach is used to construct S, the integer p represents either a coloring number of the column incidence graph of A^r (for CPR) or the maximal number of nonzeros in any row of A^r (for NR). Similarly, q represents a coloring number for the row incidence graph of A^c or its maximal column length. We will call an integer pair (p, q) *feasible* if a suitable decomposition $A = A^r + A^c$ can be found with p relating to A^r and q to A^c, as just specified. Clearly, any pair $(\tilde{p}, \tilde{q})$ with $\tilde{p} \geq p$ and $\tilde{q} \geq q$ must be feasible if (p, q) has this property. Hence we may look for marginally feasible pairs (p, q), where reducing either component leads to infeasibility. When $\hat{p}$ is the chromatic number of the column incidence graph of $\mathcal{G}(A)$ or the maximal row length in A, the pair $(\hat{p}, 0)$ is always marginally feasible. The same is true for $(0, \hat{q})$ with $\hat{q}$ analogously defined in terms of $A^\top$. Hence there must always be a unique profile of marginally feasible pairs (p, q). For the NR approach the procedure listed in Table 8.1 determines for each $p \in [0, \hat{p}]$ the corresponding minimal $q \in [0, \hat{q}]$ and a suitable partition of the form Fig. 8.4.

The one-pass procedure listed in Table 8.1 yields a minimal q only if its initial value is chosen sufficiently small. For example, $q = 0$ will always work. However, when q needs to be incremented repeatedly, the final number of stages s is likely to be unnecessary large. Hence, it makes sense to run the procedure twice: first to determine the minimal q for the given p, and second to actually set up the matrix A^r with a minimal number of stages. Naturally, one may also want to search for better combinations of p and q, for example, by minimizing $p + q$. It appears that in general the whole profile of marginal pairs (p, q) needs to be traced out in order to find the global minimum of any weighted sum

$\alpha p + (1-\alpha)q$ with $\alpha \in (0,1)$. Fortunately, this is still a polynomial search. In any case the same procedure must be rerun with A replaced by $A^\top$ and (p,q) by (q,p) to make sure the value of p cannot be reduced.

Table 8.1: Computation of Coleman–Verma Partition for Given p

Read p and initialize q

For $s = 1, 2, \ldots$

1. Combine some rows of A with no more than p nonzero entries into the matrix A_s^r and remove it from A.
2. Stop if A is empty.
3. If necessary, increment q until at least one column of A has no more than q entries.
4. Combine some columns of A with no more than q nonzero entries into the matrix A_s^c and remove it from A.

When the Jacobian $F'(x) = A$ is actually symmetric because $F(x) = \nabla f(x)$ is the gradient of some scalar function $f(x)$, we may set $p = q$. The search procedure then reduces to the one proposed originally by Powell and Toint [PT79]. In that case the Hessian-matrix products $F'(x)s = \nabla^2 f(x)\, s$ can be computed as second-order adjoints with $s = \dot{x}$, by the procedure listed in Table 5.6. We may also use a vector version to obtain $F'(x)S = \nabla^2 f(x)\, S$, where $\dot{X} = S \in \mathbb{R}^{n\times p}$ is a matrix.

In the nonsymmetric case our search procedure is essentially due to Coleman and Verma, who used it to find a block-triangular partition $A = A^r + A^c$. They then applied the CPR approach to compress A^r and $(A^c)^\top$ even though the row and column lengths p and q of A^r and A^c are only lower bounds of the relevant chromatic numbers $\chi(A^r)$ and $\chi(A^{cT})$.

In Exercise 8.3 we construct a set of matrices of variable dimension n on which compression works with $p + q = 5$, but pure column or row compression requires $p \geq \frac{n}{2}$ or $q \geq \frac{n}{2}$. However, on the limited set of test problems examined in [SH98], [CC86], and [Keh96], the gap between simple and combined compression was found to be mostly rather small.

8.5 Second Derivatives, Two-Sided Compression

For scalar-valued functions $f : \mathbb{R}^n \to \mathbb{R}$, a compressed representation of the sparse Hessian $\nabla^2 f$ can be computed as

$$H = \nabla^2 f(x) S$$

for a given seed matrix $S \in \mathbb{R}^{n\times p}$ by p Hessian-vector products or even only one Hessian-matrix product as discussed in section 5.4. So far, the development of corresponding compression techniques is restricted to direct and substitution methods, where the entries of the sparse Hessian $\nabla^2 f$ can be recovered either directly or by the solution of simple triangular systems from the compressed representation H. For the direct recovery of the Hessian matrix, one may use a so-called star coloring. For a recovery of the Hessian entries based on substitution, an acyclic coloring yielding a collection of trees may be applied. These colorings represent again heuristic algorithms since once more the computation of the minimal numbers of colors in known to be NP-hard [CC86, CM83]. Efficient coloring algorithms have been developed [GT$^+$07] to provide a seed matrix such that a compressed representation of the corresponding Hessian can be evaluated. This paper contains also a detailed analysis and numerical results for different coloring techniques in the context of sparse Hessian computations.

For vector-valued functions $F : \mathbb{R}^n \to \mathbb{R}^m$, the sparsity of the derivative tensor $F'' \in \mathbb{R}^{m\times n\times n}$ is closely related to that of the Jacobian $F'(x) \in \mathbb{R}^{m\times n}$. Therefore, it is perhaps not too surprising that the seed and weight matrices S and W can also be used for computing and reconstructing F''. In this section we first consider compression of second derivatives in the forward mode based solely on S and then compression in the reverse mode, where both S and W come into play.

Each component function $F_i(x)$ depends only nontrivially on the x_j with $j \in \mathcal{X}_i$. Hence, all nonzeros in its Hessian F_i'' must belong to the square matrix

$$\hat{F}_i'' = \left[\frac{\partial^2}{\partial x_j \partial x_k} F(x)\right]_{k\in\mathcal{X}_i}^{j\in\mathcal{X}_i} \in \mathbb{R}^{p_i\times p_i} .$$

With $S_i \equiv \left[\hat{S}_i, \tilde{S}_i\right] \in \mathbb{R}^{p_i\times p}$, as at the beginning of this chapter, we have

$$S^\top F_i'' S = S_i^\top \hat{F}_i'' S_i = \begin{bmatrix} \hat{S}_i^\top \hat{F}_i'' \hat{S}_i \ , \ \hat{S}_i^\top \hat{F}_i'' \tilde{S}_i \\ \tilde{S}_i^\top \hat{F}_i'' \hat{S}_i \ , \ \tilde{S}_i^\top \hat{F}_i'' \tilde{S}_i \end{bmatrix} \in \mathbb{R}^{p\times p} .$$

Denoting the leading $p_i \times p_i$ minor of $S_i^\top F_i'' S_i$ by H_i, we obtain (using nonsingularity of $\hat{S}_i$)

$$\boxed{\hat{F}_i'' = \hat{S}_i^{-T} H_i \hat{S}_i^{-1} \in \mathbb{R}^{p_i\times p_i} .} \tag{8.10}$$

Thus $\hat{F}_i''$, and equivalently the nonzero entries of F_i'', can be reconstructed from the compressed Hessian H_i.

When S is defined on the basis of the CPR approach $\hat{S}_i$ is a permutation matrix, so all nonzero entries of F_i'' can be directly identified in $S^\top F_i'' S$. When the NR approach is used, one needs to solve $2n$ linear systems in the nonsingular matrix $\hat{S}_i$ and its transpose. It is somewhat surprising that this straightforward extension of row compression has not yet been used much for calculating the Hessian of partially separable functions as defined in Lemma 11.1 with moderately sized $\hat{n}$. The compressed Hessians $S_i^\top \ F_i'' \ S_i$ can be obtained by the procedures of Tables 7.1 and 7.3 with the initialization

$$\dot{V}_{i-n} = e_i^\top S \in \mathbb{R}^p \,, \quad \ddot{V}_{i-n} = 0 \in \mathbb{R}^{p\times p} \,, \quad \text{and} \quad \dot{\mathcal{X}}_{i-n} = \mathcal{X}_{i-n} = \{1\ldots p\} \,.$$

The last setting makes it look as though there are only p independent variables. Hence, no dynamic sparsity management is required. The complexity bound (7.20) for the dynamically sparse forward propagation of the second derivative applies with the average domain size $\bar{n}$ replaced by the larger factor $\hat{n}$.

Simultaneous Row Compression and Column Compression

Alternatively, we can also employ a simultaneous row compression and column compression in order to determine the full second derivative tensor $F''(x)$. Suppose we have, in addition to S, a weighting matrix $W \in \mathbb{R}^{m\times q}$ so that $F'(x)$ can be reconstructed from the compression $C^\top = W^\top F'(x)$. With

$$F^{(j)}(x) \equiv \frac{\partial}{\partial x_j} F'(x) \in \mathbb{R}^{m\times n}$$

denoting a *slice* of the tensor $F''(x)$, we may define the *compressed tensor*

$$W^\top F''(x) S \equiv \left(W^\top F^{(j)}(x) S\right)_{j=1}^{n} = \left(w_i^\top F''(x) s_k\right)_{k=1\ldots p}^{i=1\ldots q} \in \mathbb{R}^{q\times p\times n} \,.$$

Here the $w_i \in \mathbb{R}^m$ and $s_k \in \mathbb{R}^n$ range over the columns of W and S, respectively. Thus, each $w_i^\top F''(x) s_k$ can be calculated as a second-order adjoint vector according to Table 5.6. Alternatively, the slices $w_i^\top F''(x) S \in \mathbb{R}^{n\times p}$ or the whole of $W^\top F''(x) S$ can be calculated by an adaption of Tables 7.1 and 7.4. Before discussing this, let us examine how $F''(x)$ can be reconstructed from its compression $W^\top F''(x) S$.

The key observation for this purpose is that all columns of the slice $F^{(j)}(x) \in \mathbb{R}^{m\times n}$ are derivatives of the jth Jacobian column $F'(x) e_j$. Consequently, all its nonzero entries must have row indices in $\mathcal{Y}_{j-n}$, and the same is true for arbitrary linear combinations of its columns. Hence we deduce that

$$e_i^\top F^{(j)}(x) s \neq 0 \Longrightarrow i \in \mathcal{Y}_{j-n} \,.$$

In other words, the columns of $B^{(j)} \equiv F^{(j)}(x) S$ have at least the sparsity of $F'(x) e_j$ and can therefore be reconstructed from the matrix

$$H^{(j)} = W^\top F^{(j)}(x) S = W^\top B^{(j)} \in \mathbb{R}^{q\times p} \,.$$

This task requires solving up to $\hat{n}$ linear systems in the nonsingular matrix $\hat{W}_j \in \mathbb{R}^{q_j \times q_j}$. As a result of this first reconstruction stage, we have recovered the one-sided compression

$$F_i''(x)\, S \equiv \left(e_i^\top F_i^{(j)}(x)\, S\right)_{j=1}^{n} \in \mathbb{R}^{n\times p} \,.$$

There are at most p_i such nontrivial rows, each of which cost $O(p_i^2)$ operations to solve for. Each row in the Hessian F_i'' is at least as sparse as the Jacobian row $e_i^\top F'(x) = \nabla F_i^\top$ and can therefore be reconstructed from its product with S.

Again, both reconstruction stages involve only entry identification when W and S are chosen according to the CPR approach. Otherwise, a large number of linear systems must be solved, a task that probably makes prefactoring all square matrices $\hat{W}_j \in \mathbb{R}^{q_j \times q_j}$ and $\hat{S}_i \in \mathbb{R}^{p_i \times p_i}$ worthwhile. In any case the total reconstruction effort has an operations count of order

$$\sum_{j=1}^{n} q_j^2\, \hat{n} + \sum_{i=1}^{m} p_i^3 \le n\, \hat{n}\, \hat{m}^2 + m\, \hat{n}^3 = \hat{n}(n\, \hat{m}^2 + m\, \hat{n}^2)\,.$$

Provided the distribution of all nonvanishing elements of $F'(x)$ among rows and columns is approximately even, we may assume $n\hat{m} \approx m\hat{n}$; hence the bound on the right-hand side is roughly the number of nonzeros in $F'(x)$ times $\hat{n}(\hat{m} + \hat{n})$.

It was found in [Keh96] that this effort for reconstructing the second derivative tensor $F_i''(x)$ was usually small compared with the effort for evaluating its compression. This experimental evidence was based on evaluating the matrix $W^\top F^{(j)}(x) s_k \in \mathbb{R}^{q \times n}$ for each $k = 1 \dots p$ in one reverse sweep of ADOL-C. This mode corresponds to the second-order adjoint procedure listed in Table 5.6, with the adjoints $\bar{v}_i$ and their derivatives $\dot{\bar{v}}_i$ being vectors of length q. Each one of these p sweeps has a temporal complexity of order $(4+6q)TIME\{eval(F)\}$ and a RAM requirement of order $qRAM\{eval(F)\}$. The SAM requirement is independent of q because only the scalars v_i and $\dot{v}_i$ need to be saved on the tape. An even better amortization of the bookkeeping overhead might be achieved if the p return sweeps were executed simultaneously with the $\dot{v}_i$ replaced by p-vectors $\dot{V}_i$ on the way forward and the $\dot{\bar{V}}_i$ extended to $q \times p$ matrices on the way back. However, this would increase both the RAM and the SAM requirements by the factor p, which may or may not be a problem. If the reconstruction effort can be neglected, the complexity bound (7.24) for the dynamically sparse reverse propagation of second derivatives applies here, with the average range size $\bar{m}$ replaced by the larger factor $\hat{m}$.

Exploiting Symmetry

In a subtle way the two-stage reconstruction process discussed in the preceding subsection relies on the symmetry of the second derivative tensor. A more direct exploitation of symmetry in the way proposed by Powell and Toint [PT79] is also possible. Suppose we want to calculate the Lagrangian Hessian

$$H(x, \bar{y}) \equiv \sum_{i=1}^{m} \bar{y}_i\, F_i''(x) = \nabla_x^2\, \bar{y}^\top F(x)\,,$$

where the weights $\bar{y}_j \in \mathbb{R}$ may be thought of as Lagrange multipliers.

According to Proposition 7.1 the width of the Hessian is bounded by the product of the maximal row length $\hat{n}$ and $\hat{m}$ of the Jacobian $F'(x)^\top$. Hence, disregarding symmetry, one might evaluate H at the cost of $O(\hat{n}\hat{m}TIME\{eval(F)\})$

by applying the NR method to the vector function $\bar{y}F'(x) : \mathbb{R}^n \longmapsto \mathbb{R}^n$ with $\bar{y}$ fixed. This is not, however, a good strategy because for essentially the same effort we can obtain the whole tensor $F''(x)$ using simultaneous row compression and column compression, as described in the preceding section.

To obtain the Hessian $H(x, \bar{y})$ more cheaply, one may exploit its symmetry in the way proposed by Powell and Toint [PT79] and later analyzed by Coleman and Cai [CC86], both in the context of differencing. Powell and Toint proposed a specialization of the algorithm shown in Table 8.1 to the symmetric case, which we give in Table 8.2.

Table 8.2: Computing the Reduced Width $\tilde{n}(H) = \max p$ of $H = H^\top$

Initialize $p = 0$
For $s = 1, 2, \ldots, n-1$
1. Increment p until a row of H has no more than p nonzero entries.
2. Remove all rows and columns with no more than p nonzero entries.

We refer to the number p obtained by the algorithm in Table 8.2 as the *reduced width* $\tilde{n}(H)$. Clearly, it cannot be smaller than the minimal number of nonzeros in any row of H. It is shown in Exercise 8.4 that not even this lower bound is always optimal in that a more general substitution or elimination method may determine H on the basis of fewer Hessian-vector products.

We may also use symmetry to further reduce the effort for computing the full tensor $F''(x)$ by two-sided compression. The number

$$\tilde{n}(F''(x)) \equiv \max_{1 \leq i \leq m} \tilde{n}(F_i''(x)) \leq \hat{n}(F'(x))$$

may be significantly smaller than $\hat{n}(F'(x))$, especially when partial value separability (which we will discuss in section 11.2) has not been fully exploited to reduce $\hat{n}(F'(x))$ toward its lower bound $\bar{n}(F'(x))$. For generic $S \in \mathbb{R}^{n \times \tilde{n}}$ we may then reconstruct all F'' from the one-sided compression $(F''S) = (F_i''S)_{i=1\ldots m}$ derived from the two-sided compression

$$W^\top F''S = \left(H^{(j)}\right)_{j=1\ldots n} \in \mathbb{R}^{q \times n \times \tilde{n}} .$$

In other words the row seeding in simultaneous row and column compression remains unchanged, since the slices $F^{(j)}(x)S$ are nonsymmetric but the column seeding exploits symmetry.

Hence, the total reconstruction effort is now $n\hat{n}\hat{m}^2 + m\,\tilde{n}^3$, where $\tilde{n} \leq \hat{n}$ is the maximal reduced width of any component Hessian F_i'' for $i = 1 \ldots m$.

Table 8.3: Cost Ratio for Tensors and Hessians

	$\longrightarrow$	$\longleftrightarrow$
$F''(x)$	$\hat{n}^2$	$\hat{m}\,\tilde{n}(F'')$
$H(x,\bar{y})$	$\hat{n}^2$	$\tilde{n}(H)$

Assuming that the evaluation effort dominates the reconstruction effort, we obtain the approximate complexities in $TIME\{eval(F)\}$ units shown in Table 8.3.

In calculating $\hat{n}$ and $\hat{m}$ as the width of the Jacobian $F'(x)$ and its transposed $F'(x)^\top$, one may disregard constant entries because they have no impact on $F''(x)$ and $H(x, \bar{y})$. Then $\hat{n}$ becomes a lower bound for $\tilde{n}(H)$.

8.6 Summary

In this chapter we have considered various possibilities for compressing the $m \times n$ Jacobians A by multiplying them from the right or left with suitably chosen rectangular matrices $S \in \mathbb{R}^{n \times p}$ and $W \in \mathbb{R}^{m \times q}$. We call these alternative methods row compression and column compression. The resulting rectangular matrices $B \equiv AS \in \mathbb{R}^{m \times p}$ and $C = A^\top W \in \mathbb{R}^{n \times q}$ are directly obtained without truncation error by the forward and reverse vector modes described in sections 3.1 and 3.2.

If S or W are chosen as 0-1 matrices following the CPR approach, all entries in A can be directly identified in B or C, respectively. Then p and q cannot be smaller than the chromatic numbers $\chi(\mathcal{G}_c) = \chi(A)$ and $\chi(\mathcal{G}_r) = \chi(A^\top)$ of the column and row incidence graph of A, respectively. In turn, these chromatic numbers are bounded below by $\hat{n}$ and $\hat{m}$, which represent the maximal number of nonzeros in any row or column of A, respectively.

Matrices S and W with $p = \hat{n}, q = \hat{m}$ can be constructed following the NR approach. This situation is optimal in that $p < \hat{n}$ or $q < \hat{m}$ is impossible by a simple degree-of-freedom argument. The drawback of the NR approach is that linear systems of order up to p or q need to be solved to reconstruct A from B or C, respectively. The choice of S and W as Vandermonde matrices leads to low operations counts of $O(mp^2)$ or $O(nq^2)$. However, numerical conditioning is a problem, though not quite as debilitating as in the traditional difference quotient scenario.

Finally, row compression and column compression can be combined to further reduce the number of derivative vectors needed to reconstruct A. Hence we can compile Table 8.4 estimating the runtime ratio of various methods for computing Jacobians.

Table 8.4: Jacobian with $\bar{n} \le \hat{n} \le \chi(\mathcal{G}_c) \le n$ and $\bar{m} \le \hat{m} \le \chi(\mathcal{G}_r) \le m$

Sparse		NR compressed			CPR compressed		
$\longrightarrow$	$\longleftarrow$	$\longrightarrow$	$\longleftarrow$	$\rightleftarrows$	$\longrightarrow$	$\longleftarrow$	$\rightleftarrows$
$\bar{n}$	$\bar{m}$	$\hat{n}$	$\hat{m}$	$\le \min(\hat{n}, \hat{m})$	$\chi(A)$	$\chi(A^\top)$	??

The entries in the last row are correct only up to a (one hopes) small constant. There does not seem to be a natural analog of the combined compression idea in the context of dynamically sparse derivative calculations. However, as we will see in Chapter 9, there is a much larger range of cross-country elimination procedures that include sparse forward and reverse as special cases.

Since according to (7.11) and (7.13) $\bar{n} \leq \hat{n}$ and $\bar{m} \leq \hat{m}$, the (dynamically) sparse forward or reverse methods look (on paper) better than the corresponding column compression or row compression scheme. Unfortunately, the overhead in dealing with dynamic sparsity is typically so large that the gaps $\hat{n} - \bar{n}$ or $\hat{m} - \bar{m}$ must be quite sizable to make it worthwhile. Since the maximal row and column lengths $\hat{n}$ and $\hat{m}$ are much easier to determine than the related chromatic numbers $\chi(\mathcal{G}_c) \geq \hat{n}$ and $\chi(\mathcal{G}_r) \geq \hat{m}$, we will in the remainder of this book always discuss the cost of compression in terms of $\hat{n}$ and $\hat{m}$. On many problems the chromatic numbers are the same or only slightly larger, in which case no harm is done by replacing them with $\hat{n}$ and $\hat{m}$ in complexity estimates. On the other hand, when there is a substantial gap, we may assume that NR with coloring for stability is the method of choice.

In section 8.5 the compression approach was extended to the calculation of sparse Hessians and general second-order tensors. One can apply two-sided compression based on the forward mode alone or its combination with reverse mode differentiation. Finally, one may utilize the inherent symmetry of second derivatives to reduce the cost a little more.

8.7 Examples and Exercises

Exercise 8.1 (***Column Compression or Row Compression***)
Consider the Jacobian sparsity pattern

$$A = \begin{bmatrix} \times & \times & \times & \times & 0 & 0 & 0 & 0 & 0 & 0 \\ \times & \times & 0 & 0 & \times & \times & 0 & 0 & 0 & 0 \\ \times & \times & 0 & 0 & 0 & 0 & \times & \times & 0 & 0 \\ 0 & 0 & \times & \times & \times & \times & 0 & 0 & 0 & 0 \\ 0 & 0 & \times & \times & 0 & 0 & \times & \times & 0 & 0 \\ 0 & 0 & \times & \times & 0 & 0 & 0 & 0 & \times & \times \end{bmatrix} \in \mathbb{R}^{6\times 10} .$$

a. Draw the column and row incidence graphs. Determine optimal colorings for both and write down the corresponding CPR matrices S and W.
b. Show that S but not W can be made smaller in the NR approach, and write down the resulting matrix S according to (8.9).

Exercise 8.2 (***Combined Row and Column Compression***)
Consider the Jacobian sparsity pattern

$$A = \begin{bmatrix} \times & \times & \times & 0 & \times & 0 \\ \times & 0 & 0 & \times & 0 & \times \\ \times & 0 & \times & 0 & 0 & 0 \\ 0 & \times & 0 & \times & 0 & 0 \\ 0 & \times & 0 & 0 & \times & 0 \\ 0 & \times & 0 & 0 & 0 & \times \end{bmatrix} \in \mathbb{R}^{6\times 6} .$$

Note that the maximal row length is $\hat{n} = 4$ and the maximal column length is also $\hat{m} = 4$.

a. Determine for each $p \in [1,4]$ the corresponding minimal $q \in [1,4]$ and vice versa using the algorithm of Table 8.1. Draw all marginally feasible combinations in a (p,q) diagram.
b. Consider the Coleman–Verma partition $A = A^r + A^c$ for $p = 1$ and $q = 2$. Show that the chromatic numbers of A^r and A^c are equal to the lower bounds $p = 1$ and $q = 2$, respectively. Determine suitable CPR matrices $S \in \mathbb{R}^{6\times 1}$, $W \in \mathbb{R}^{6\times 2}$, and write down the resulting products AS and WA in terms of the individual entries

$$A_{ij} = e_i^\top A e_j \quad \text{for} \quad 1 \le i,j \le 6\,.$$

Successively circle all entries that can be computed from the values of AS and WA.

Exercise 8.3 (*Multistage Partitioning*)
Starting with a dense $J^{(0)} \in \mathbb{R}^{2\times 2}$ recursively define the matrix patterns

$$J^{(k+1)} = \begin{bmatrix} J^{(k)} & D_{2^{(k+1)}} \\ D_{2^{(k+1)}} & P_{2^{(k+1)}} \end{bmatrix} \in \mathbb{R}^{2^{(k+1)}\times 2^{(k+1)}}$$

where

$$P_i = D_i + e_i\, v_i^\top + v_i\, e_i^\top \in \mathbb{R}^{i\times i}$$

with D_i diagonal, $v_i \in \mathbb{R}^i$ dense, and e_i a Cartesian basis vector. Draw the sparsity pattern $J^{(3)}$, and determine its Coleman–Verma partition with $p = 3$ and $q = 2$. Show that the same values of p and q are sufficient for $J^{(k+1)}$ with arbitrary k.

Exercise 8.4 (*Nontriangular Substitution*)
Consider

$$H = \begin{bmatrix} \times & \times & \times & 0 & \times & \times \\ \times & \times & \times & \times & 0 & \times \\ \times & \times & \times & \times & \times & 0 \\ 0 & \times & \times & \times & \times & \times \\ \times & 0 & \times & \times & \times & \times \\ \times & \times & 0 & \times & \times & \times \end{bmatrix} \in \mathbb{R}^{6\times 6} \quad \text{and} \quad S = \begin{bmatrix} 0 & 0 & 1 & 1 \\ 1 & 0 & 0 & 0 \\ 0 & 1 & 0 & 0 \\ 0 & 0 & 1 & 0 \\ 0 & 0 & 0 & 1 \\ 1 & 1 & 0 & 0 \end{bmatrix} \in \mathbb{R}^{6\times 4}\,.$$

The symmetric sparsity of H is a circulant version of an example proposed by Powell and Toint [PT79]. Show that the seed matrix S allows the reconstruction H from HS by substitution, whereas simple compression without symmetry requires five Hessian-vector products.

Exercise 8.5 (*Compression of Second Derivatives*)
Consider the Jacobian sparsity pattern

$$A = \begin{bmatrix} \times & 0 & \times & 0 & 0 & \times \\ \times & \times & 0 & \times & 0 & 0 \\ 0 & 0 & \times & 0 & \times & \times \end{bmatrix}^\top = F'(x) \in \mathbb{R}^{6\times 3}\,.$$

a. Determine CPR matrices $S \in \mathbb{R}^{3\times 2}$ and $W \in \mathbb{R}^{6\times 3}$ so that the compression AS or WA yields enough information to reconstruct A.
b. Abbreviating

$$H_{ijk} \equiv \partial^2 F_i(x)/\partial x_j \partial x_k \quad \text{for} \quad i = 1\ldots 6 \quad \text{and} \quad 1 \leq j, k \leq 3\,,$$

write down the compression $S^\top \nabla^2 F_i S$ and check that each H_{ijk} can be identified.
c. Determine the sparsity pattern of the slices $F^{(j)} \equiv \partial F'(x)/\partial x_j \in \mathbb{R}^{6\times 3}$, and write down the compressions $B^{(j)} \equiv W^\top F^{(j)}(x) S$. Observe as in part **b** that each H_{ijk} can be identified in some $B^{(j)}$.
d. Note that the calculation of the full second derivative tensor

$$F''(x) = (H_{ijk}) \in \mathbb{R}^{6\times 3\times 3}$$

according to part **c** requires the evaluation of six second-order adjoints of the form $w^\top F''(x)s$, with w and s ranging over the columns of W and S, respectively. Determine for fixed $w = \bar{y}$ the sparsity of the Hessian

$$H(x, \bar{y}) \equiv \sum_{i=1}^{6} \bar{y}_i \nabla^2 F_i\,,$$

and apply the Powell–Toint algorithm. Show that only two second-order adjoints are needed to determine $H(x, \bar{y})$, exploiting its symmetry.

Chapter 9

Going beyond Forward and Reverse

So far we have considered two ways of calculating the Jacobian entries $\partial y_i/\partial x_j$ from the elemental partials $c_{ij} = \partial\varphi_i/\partial v_j$, namely, the forward and reverse modes. In both cases only multiplications and additions of floating point numbers are needed once the elemental partials c_{ij} have been evaluated at the current argument $x \in \mathbb{R}^n$. From now on we will refer to this process as derivative *accumulation*. As will be shown in this chapter, the "global" partials $\partial y_i/\partial x_j$ can be accumulated from the "local" partials c_{ij} in very many different ways. The forward and reverse modes turn out to be "pure" choices in a huge variety of strategies for accumulating derivatives by eliminating edges or vertices on the linearized computational graph.

The natural goal of accumulating Jacobians with a minimal number of arithmetic operations is on larger problems elusive because it has been proven by Naumann [Nau06] to be NP-hard, as will be discussed in section 9.5. Nevertheless, the freedom to apply the chain rule in varying orders opens up avenues for various preaccumulations, for example, the statement-level reverse mode employed by the tangent mode of ADIFOR or the dirty-vector propagation suggested by Christianson, Dixon, and Brown [CDB96]. Another benefit is that the complexity of Jacobian calculations seems to become a little bit less dependent on the particular way in which the underlying function has been written down. Finally, the associativity of the chain rule can also be used to perform accumulation in parallel on independent sections of the graph, thus introducing concurrency along the (evaluation) "time-axis," which is typically not present in the underlying forward simulation.

A Small Example

Before we approach the general problem in a rather abstract way let us consider a small problem in two variables, namely, a function given by the formulas

$$F(x_1, x_2) = \big(x_1 * x_2 * \exp(\sin(x_1 * x_2)),\ \cos(\exp(\sin(x_1 * x_2)))\big)^\top .$$

Obviously the evaluation should use the many common subexpressions and could, for example, be effected by the program

$$v_1 = x_1 \cdot x_2 \, , \qquad v_2 = \sin(v_1) \, , \qquad v_3 = \exp(v_2) \, ,$$
$$y_1 = v_1 \cdot v_3 \, , \qquad y_2 = \cos(v_3) \, .$$

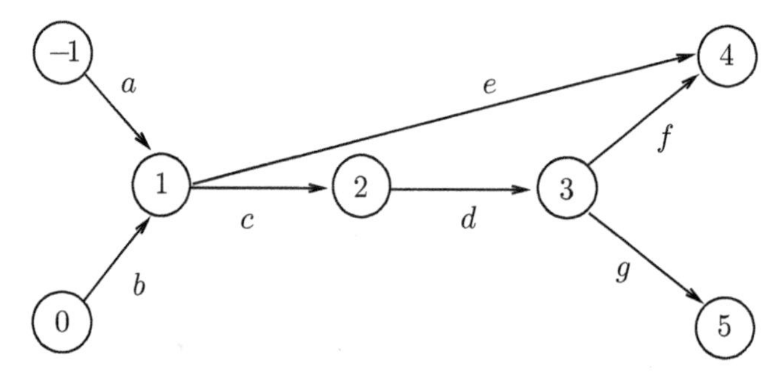

Figure 9.1: Naumann's First Example

The corresponding computational graph is illustrated by Fig. 9.1 where the arcs are annotated by the local partials c_{ij}:

$$a = x_2 \, , \; b = x_1 \, , \; c = \cos(v_1) \, , \; d = v_3 \, , \; e = v_3 \, , \; f = v_1 \, , \; g = -\sin(v_3) \, .$$

Accumulating the Jacobian in a symbolic fashion gives us

$$F'(x_1, x_2) = \underbrace{\begin{bmatrix} ae + acdf \, , & be + bcdf \\ acdg \, , & bcdg \end{bmatrix}}_{OPS \; = \; 14 \, mults + 2 \, adds} \, .$$

By identifying the many common partial products we may arrive at the following much more economical way of applying the chain rule to compute the Jacobian:

$$F'(x_1, x_2) = \underbrace{\begin{bmatrix} au & bu \\ at & bt \end{bmatrix} \quad \text{with} \quad \begin{matrix} r = cd \\ s = rf \end{matrix} \quad \text{and} \quad \begin{matrix} t = rg \\ u = e + s \end{matrix}}_{OPS \; = \; 7 \, mults + 1 \, add} \, .$$

It should be noted that the very first multiplication $r = cd$ concerns two arcs in the interior of the graph, so we are definitely not performing the accumulation process forward or reverse through the computational graph, but the number of arithmetic operations is reduced considerably. Hence, we face the question of generalizing the accumulation procedure to decrease the overall number of operations performed.

Matrix Chaining and Vertex-Elimination

Accumulation strategies that generalize the forward and reverse modes can be introduced in various ways. An obvious approach is to reconsider the product representation

$$F'(x) = Q_m A_l A_{l-1} \ldots A_2 A_1 P_n^\top \in \mathbb{R}^{m \times n}$$

that was used in section 3.2 to derive the reverse mode in the first place. Rather than multiplying the elemental Jacobians A_i in the original order $i = 1 \ldots l$ or exactly in the opposite order $i = l \ldots 1$, we may allow any parenthesizing scheme to accumulate the Jacobian $F'(x)$. For example, one might recursively multiply pairs of neighbors A_i and A_{i+1}, an approach that looks especially promising on a parallel system. Then only $\log_2 l$ recursion levels would be required and the Jacobian could be accumulated in logarithmic time, provided $l/2$ processors were available initially.

The task of parenthesizing the A_i to minimize the serial operations count is known as the matrix chaining problem and has attracted considerable attention in the dynamic programming literature [AHU74, RT78]. There it is usually assumed that the factors A_i are dense and rectangular of compatible dimensions $n_i \times n_{i-1}$. Then the operations count for multiplying two neighboring factors A_{i+1} and A_i is simply $n_{i+1}\, n_i\, n_{i-1}$ and the overall effort can be globally minimized by dynamic programming.

In our context this approach is not directly applicable since the factors A_i are extremely sparse. More importantly, unless $i \prec j$ or $j \prec i$, the factors A_i and A_j commute with each other and can therefore be interchanged. This extra flexibility for reducing the computational effort cannot be naturally incorporated into the matrix chaining approach, but it fits very well into the vertex-elimination framework of accumulation that was proposed in [GR91] and has been discussed in a few other papers [BH96] since then. Both possibilities can be interpreted as edge-elimination procedures on the computational graph, which is a more general technique for performing Jacobian accumulation. Unfortunately, we have not yet learned to utilize the resulting freedom very well and face instead a difficult combinatorial optimization problem. Greedy heuristics, for example, those that always eliminate vertices of minimal Markowitz degree, often work well but can be shown to fail on some large classes of problems that appear not at all contrived, as will be discussed in section 10.1. While matrix chaining can be performed in polynomial time, general vertex-elimination is NP-hard with respect to fill-in.

This chapter is organized as follows. In section 9.1 we use numerical linear algebra techniques to derive Jacobian accumulation procedures from the implicit function theorem. When intermediate columns or rows of the extended Jacobian E', as defined in (2.9), are eliminated in their natural or reversed order we obtain again the forward and reverse modes of differentiation, respectively. In section 9.2 we rederive the same class of accumulation procedures as edge-eliminations on the linearized computational graph that keep its input/output characteristic unchanged. Although opinions are divided, this approach for deriving general accumulation procedures and designing suitable heuristics seems the more natural and economical to us. Section 9.3 introduces the concept of vertex-elimination. In section 9.4 we describe a generalization of edge-elimination called face-elimination. It must be performed on the line-graph associated with the original linearized computational graph. Face-elimination and edge-elimination all reduce to vertex-elimination when the graph is diagonal in that every interior vertex has exactly one predecessor and one successor. In section 9.5 it is shown that even on this extremely restricted class of problems, Jacobian accumulation with a minimal operations count is in some sense NP-hard. The chapter concludes with a summary and outlook in section 9.6.

9.1 Jacobians as Schur Complements

As we have stated in section 2.2 functions defined by an evaluation procedure of the form of Table 2.2 can be equivalently characterized as solutions of the triangular system of nonlinear equations $E(x; v) = 0$ defined by (2.5) with the normalization (2.8). Applying the implicit function theorem to the equivalent system $-E(x; v) = 0$ we immediately obtain the derivative

$$\frac{\partial v}{\partial x} = -\begin{bmatrix} -I & 0 & 0 \\ B & L-I & 0 \\ R & T & -I \end{bmatrix}^{-1} \begin{bmatrix} I \\ 0 \\ 0 \end{bmatrix} = \begin{bmatrix} I \\ (I-L)^{-1}B \\ R+T(I-L)^{-1}B \end{bmatrix} \in \mathbb{R}^{(n+l)\times n}.$$

The $(n+i)$th row of $\frac{\partial v}{\partial x}$ represents exactly the gradient $\dot{V}_i = \nabla_x v_i$ of the ith intermediate value v_i with respect to all independents $x_j = v_{j-n}$ for $j = 1 \ldots n$. Thus we see that the Jacobian

$$\begin{aligned} F'(x) \;&\equiv A = R+T(I-L)^{-1}B \\ &= R+T[(I-L)^{-1}B] \;=\; R+[T(I-L)^{-1}]B \end{aligned} \tag{9.1}$$

is the Schur complement of R, representing the reduced system after the intermediate variables v_i with $i = 1 \ldots l-m$ have been eliminated. In the equation above we have indicated two different bracketings that could be obeyed in computing the Jacobian and lead to the following familiar faces.

Once Again: Forward and Reverse

First, since

$$L \equiv \left(c_{ij}\right)_{j=1\ldots l-m}^{i=1\ldots l-m} \in \mathbb{R}^{(l-m)\times(l-m)}$$

is strictly lower triangular, the systems of n linear equations

$$\left(I-L\right)\dot{Z} = B \quad \text{with} \quad \dot{Z} \equiv \left(\dot{V}_i^\top\right)_{i=1}^{l-m} \in \mathbb{R}^{(l-m)\times n}$$

can be solved quite naturally by the process of forward substitution—familiar from numerical linear algebra. Subsequently, one may simply multiply $\dot{Z}$ by T to obtain the Jacobian $A(x) = R + T\dot{Z}$. By comparison with Table 7.1 we observe that this amounts exactly to the vector forward mode considered in Chapter 7, where we may of course make use of sparsity. As a by-product we obtain $\dot{Z}$, whose rows represent the gradients of all intermediate quantities with respect to the independents.

Second, one may prefer (especially if $m < n$ so that T has fewer rows than R columns) the alternative of solving the m linear systems

$$\left(I - L^\top\right) \bar{Z}^\top = T^\top \quad \text{with} \quad \bar{Z} \equiv \left(\bar{V}_i\right)_{i=1}^{l-m} \in \mathbb{R}^{m \times (l-m)}$$

and subsequently multiplying the solution $\bar{Z}^\top$ by $B^\top$. By comparison with Table 7.2, we observe that this amounts exactly to the vector reverse mode, where we may also make use of sparsity. This time the by-product $\bar{Z}$ contains the sensitivities $\bar{V}_i = \partial y / \partial v_i$ of the dependent vector y with respect to all intermediate values. It may be used for estimating evaluation errors as discussed already in section 3.4.

The forward and reverse modes require the solution of n or m linear systems, respectively. Since the triangular matrices involved are transposed to each other, they have exactly the same degree of sparsity. Therefore, one might expect that the ratio of n to m gives a reliable indication of the relative cost of the forward and reverse modes. As we have already intimidated in section 7.1, this notion is very inaccurate and one should instead compare the average domain size $\bar{n}$ and the average range size $\bar{m}$. To reinforce this claim in terms of linear algebra observations, let us consider a situation where L is block diagonal, which means that the evaluation proceeds in several mutually independent subprocedures. Then the extended Jacobian can be partitioned into the form

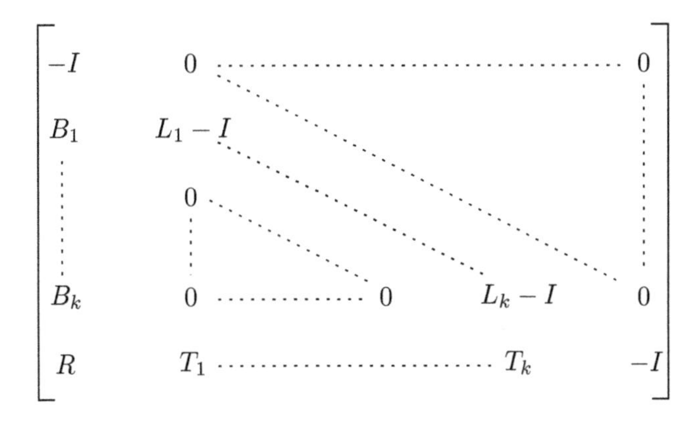

The Schur complement is given by the sum

$$A = R + \sum_{j=1}^{k} T_j \left(I - L_j\right)^{-1} B_j \, .$$

Then the k additive terms may again be computed as either $T_j \left[(I - L_j)^{-1} B_j\right]$ or as $\left[T_j (I - L_j)^{-1}\right] B_j$, which agrees exactly with the forward and reverse modes. Now suppose for simplicity that all B_j and T_j are square matrices of the same size $m = n$. Moreover, let us assume that each one of the k blocks computes exactly one of the m dependent variables so that we must have $m = k$ and $T_j = e_j\, a_j^\top$, where e_j is the jth Cartesian basis vector and a_j some other vector in $\mathbb{R}^n$. Then the computation of the matrix

$$\left(I - L_j^\top\right)^{-1} T_j^\top = \left[\left(I - L_j^\top\right)^{-1} a_j\right] e_j^\top$$

requires only one linear system solve. On the other hand all B_j may well have full rank so that computing $(I - L_j)^{-1} B_j$ does require n linear solves. Hence we find that in this situation the reverse mode is orders of magnitude faster than the forward mode even though $m = n$. In the terminology of section 7.1 the problem constructed has an average range size $\bar{m} = 1$, whereas the average domain size $\bar{n}$ is significantly larger than 1. While the choice between forward and reverse seems difficult enough, we open the door even further.

Going Cross-Country

There are many more ways to compute the Schur complement representing the Jacobian. Consider, for example, the linear problem

$$\begin{array}{|l|}\hline
v_{-1} = x_1 \\
v_0 \;\;= x_2 \\ \hline
v_1 \;\;= c_{1,-1} * v_{-1} + c_{1,0} * v_0 \\
v_2 \;\;= c_{2,1} * v_1 \\
v_3 \;\;= c_{3,2} * v_2 \\
v_4 \;\;= c_{4,3} * v_3 \\
v_5 \;\;= c_{5,-1} * v_{-1} + c_{5,4} * v_4 \\
v_6 \;\;= c_{6,0} * v_0 + c_{6,4} * v_4 \\ \hline
y_1 \;\;= v_5 \\
y_2 \;\;= v_6 \\ \hline
\end{array} \qquad (9.2)$$

with $n = 2$ independent, $l - m = 4$ intermediate, and $m = 2$ dependent variables. The corresponding extended Jacobian has the form

$$\begin{bmatrix} -I & 0 & 0 \\ B & L - I & 0 \\ R & T & -I \end{bmatrix} = \left[\begin{array}{cc:cccc:cc}
-1 & 0 & 0 & 0 & 0 & 0 & 0 & 0 \\
0 & -1 & 0 & 0 & 0 & 0 & 0 & 0 \\ \hdashline
c_{1,-1} & c_{1,0} & -1 & 0 & 0 & 0 & 0 & 0 \\
0 & 0 & c_{2,1} & -1 & 0 & 0 & 0 & 0 \\
0 & 0 & 0 & c_{3,2} & -1 & 0 & 0 & 0 \\
0 & 0 & 0 & 0 & c_{4,3} & -1 & 0 & 0 \\ \hdashline
c_{5,-1} & 0 & 0 & 0 & 0 & c_{5,4} & -1 & 0 \\
0 & c_{6,0} & 0 & 0 & 0 & c_{6,4} & 0 & -1
\end{array}\right] \qquad (9.3)$$

The structure of this matrix is persymmetric, which leads to the forward and reverse modes requiring exactly the same kind and number of arithmetic operations. In the forward mode listed in Table 9.1 we compute the successive

Table 9.1: Jacobian Accumulation by Forward Mode: $10\,mults$, $2\,adds$

$$
\begin{array}{l}
\dot{V}_2 \equiv (c_{2,-1}, c_{2,0}) = (c_{2,1} * c_{1,-1} \quad , c_{2,1} * c_{1,0} \quad) \\
\dot{V}_3 \equiv (c_{3,-1}, c_{3,0}) = (c_{3,2} * c_{2,-1} \quad , c_{3,2} * c_{2,0} \quad) \\
\dot{V}_4 \equiv (c_{4,-1}, c_{4,0}) = (c_{4,3} * c_{3,-1} \quad , c_{4,3} * c_{3,0} \quad) \\
\dot{V}_5 \equiv (c_{5,-1}, c_{5,0}) = (c_{5,-1} + c_{5,4} * c_{4,-1} \, , c_{5,4} * c_{4,0} \quad) \\
\dot{V}_6 \equiv (c_{6,-1}, c_{6,0}) = (c_{6,4} * c_{4,-1} \quad , c_{6,0} + c_{6,4} * c_{4,0})
\end{array}
$$

gradients which involves 10 multiplications and two additions. The corresponding reverse-mode calculation is left to the reader, as Exercise 9.1.

The five lines in Table 9.1 can be interpreted as successive elimination of the subdiagonal entries $c_{2,1}$, $c_{3,2}$, $c_{4,3}$, $c_{5,4}$, and the entry $c_{6,4}$ in the second subdiagonal of (9.3) by elementary row operations. The four diagonal elements -1 in the middle box are used as pivots to reduce all entries of L and T to zero. At each stage the submatrices remaining after removing first the third, then the fourth, and finally the fifth row-column pair are in fact the Schur complements after the intermediates v_1, v_2, v_3, and finally v_4 have been removed from the system, respectively. Exactly the same sequence of reduced systems would be obtained if the same entries -1 were used to eliminate the nonzeros to their left in the same order. As shown in Exercise 9.1, the reverse mode corresponds to eliminating the intermediates in the opposite order v_4, v_3, v_2, and v_1.

Now it appears tempting to try other orders for eliminating the intermediate variables, for example, first v_2, v_3, and v_4, and then v_1 last. In contrast to the previous situation, the first two elementary row operations cost only one multiplication and generate only one new nonzero entry. To be specific we obtain with $c_{3,1} = c_{3,2} * c_{2,1}$ and $c_{4,1} = c_{4,3} * c_{3,1}$ the top matrix of Table 9.2.

Now we can eliminate the two subdiagonal elements in the last column of the middle block, generating two new elements $c_{5,1} = c_{5,4} * c_{4,1}$ and $c_{6,1} = c_{6,4} * c_{4,1}$ at the cost of two more multiplications. Finally, we may eliminate these newly created entries in the third column using the -1 above them and obtain the matrix on the bottom of Table 9.2 with $\tilde{c}_{5,-1} = c_{5,-1} + c_{5,1} * c_{1,-1}$, $c_{5,0} = c_{5,1} * c_{1,0}$, $c_{6,-1} = c_{6,1} * c_{1,-1}$, and $\tilde{c}_{6,0} = c_{6,0} + c_{6,1} * c_{1,0}$. This time the total operations count for accumulating the Jacobian in the bottom left corner as Schur complement is only eight multiplications and two additions. Hence we have saved two multiplications, compared to the forward- or reverse-elimination of the intermediates.

One might object that we have not eliminated the first three subdiagonal entries in the third column, which would cost another six multiplications. This cleaning up is not necessary since it would not alter the values in the Schur complement, and the entries $c_{2,-1} = c_{2,1} * c_{1,-1}$, $c_{2,0} = c_{2,1} * c_{1,0}$ represent gradients of intermediates that are not of primary interest to us. Alternatively, we could also have eliminated $c_{4,1}$ first (introducing $c_{4,-1} = c_{4,1} * c_{1,-1}$, $c_{4,0} = c_{4,1} * c_{1,0}$) and then finally eliminating the two subdiagonal zeros in the last column of the middle block using again four more multiplications and two additions.

Table 9.2: Accumulation by Cross-Country Elimination: $8\,mults$, $2\,adds$

$$\left[\begin{array}{cc:cccc:cc}
-1 & 0 & 0 & 0 & 0 & 0 & 0 & 0 \\
0 & -1 & 0 & 0 & 0 & 0 & 0 & 0 \\
\hdashline
c_{1,-1} & c_{1,0} & -1 & 0 & 0 & 0 & 0 & 0 \\
0 & 0 & c_{2,1} & -1 & 0 & 0 & 0 & 0 \\
0 & 0 & c_{3,1} & 0 & -1 & 0 & 0 & 0 \\
0 & 0 & c_{4,1} & 0 & 0 & -1 & 0 & 0 \\
\hdashline
c_{5,-1} & 0 & 0 & 0 & 0 & c_{5,4} & -1 & 0 \\
0 & c_{6,0} & 0 & 0 & 0 & c_{6,4} & 0 & -1
\end{array}\right]$$

$$\left[\begin{array}{cc:cccc:cc}
-1 & 0 & 0 & 0 & 0 & 0 & 0 & 0 \\
0 & -1 & 0 & 0 & 0 & 0 & 0 & 0 \\
\hdashline
c_{1,-1} & c_{1,0} & -1 & 0 & 0 & 0 & 0 & 0 \\
0 & 0 & c_{2,1} & -1 & 0 & 0 & 0 & 0 \\
0 & 0 & c_{3,1} & 0 & -1 & 0 & 0 & 0 \\
0 & 0 & c_{4,1} & 0 & 0 & -1 & 0 & 0 \\
\hdashline
\tilde{c}_{5,-1} & c_{5,0} & 0 & 0 & 0 & 0 & -1 & 0 \\
c_{6,-1} & \tilde{c}_{6,0} & 0 & 0 & 0 & 0 & 0 & -1
\end{array}\right]$$

The actual arithmetic updates for eliminating c_{ij} by an elementary row or column operation are

$$c_{ik} \mathrel{-}= c_{ij} * c_{jk} \quad \text{for all} \quad k < j \quad \text{or} \quad c_{kj} \mathrel{-}= c_{ki} * c_{ij} \quad \text{for all} \quad k > i \,, \tag{9.4}$$

respectively. It should be noted that due to the -1's in the diagonal, there appears to be little problem with numerical stability so that good pivoting strategies can be selected once and for all, without regard for actual numerical values of the elemental partials. Hence we may formulate our cross-country method for accumulating Jacobians as Schur complements as shown in Table 9.3.

Since eliminating one intermediate entry is likely to create fill-in at other intermediate positions, the following result may come as a surprise.

Proposition 9.1 (Finite Termination of Jacobian Accumulation)
The procedure specified in Table 9.3 *always reduces the extended Jacobian $E'(x;v)$ to the compact $A = F'(x)$ after finitely many elimination steps.*

Proof. See Exercise 9.2. ∎

By now the reader will have noted that our Jacobian accumulation problem closely resembles the classical linear algebra problem of sparse Gaussian elimination [RT78]. There, off-diagonal elements are always eliminated in complete

Table 9.3: General Jacobian Accumulation Procedure

- In the extended Jacobian, label all rows and columns, except for the first n and last m, as *intermediate*. Their entries are also called intermediate.
- Using intermediate diagonal elements as pivots, perform elementary row or column operations to eliminate any one off-diagonal intermediate entry.
- Whenever an intermediate diagonal element is the only remaining nonzero entry in its row or column, label it as eliminated.
- When all intermediate diagonal elements are eliminated, the $m \times n$ matrix in the bottom left corner is the desired Jacobian A.

rows or columns so that the corresponding pivot variable and equation is eliminated at once. Going after individual off-diagonal entries would certainly be possible, but one might wind up cycling, which is impossible here. Moreover, it is not clear whether any saving can be achieved this way in linear equation solving. However, as we will see in the next section, that is possible in Jacobian accumulation.

In many applications the primary purpose for evaluating or approximating Jacobians is the computation of Newton steps. As we will discuss briefly in section 10.3, this goal can be achieved directly and sometimes at a drastically reduced cost without first accumulating the Jacobian at all; see [Gri90, Utk96a, Utk96b, Hos97].

The finite termination result stated above can be derived from the graph theoretical point of view developed in the next section, which also provides more intuitive insight, especially for the design of heuristics to select promising elimination orderings. Naturally, the number of individual elimination steps, though always finite, may vary dramatically for various strategies of picking the subdiagonal entries to be eliminated.

9.2 Accumulation by Edge-Elimination

As we first mentioned in section 2.2 the properties of evaluation procedures are often visualized by computational graph(s). The set of vertices is always the same, namely, the $n + l$ indices $\mathcal{V} \equiv \{1 - n, \ldots, 0, 1, \ldots, l\}$ or equivalently the corresponding variables v_i. Similarly the set $\mathcal{E}$ of directed edges is always defined by the transitive dependence relation $\prec$ introduced in section 2.2, so that

$$(j, i) \in \mathcal{E} \Leftrightarrow j \prec i \, .$$

Depending on the context, we may consider vertices and arcs as labeled by various symbols, especially vertices by elemental functions φ_i and arcs by elemental partials c_{ij}. When function names are left off and partials are evaluated as real numbers at the current argument x, we will refer to the computational graph as being *linearized.* Other than that, there is no need to distinguish versions of the graph depending on the labeling.

The linearized computational graph is simply the incidence graph that would be associated with the extended Jacobian $E'(x;v)$ in numerical linear algebra. For the little example used in section 9.1 to illustrate cross-country elimination, the graph is drawn from left to right in Fig. 9.2.

The dashed lines represent the intermediate gradient components that are computed in the forward mode. Two such *fill-in* arcs occur for nodes ②, ③, and ④ each.

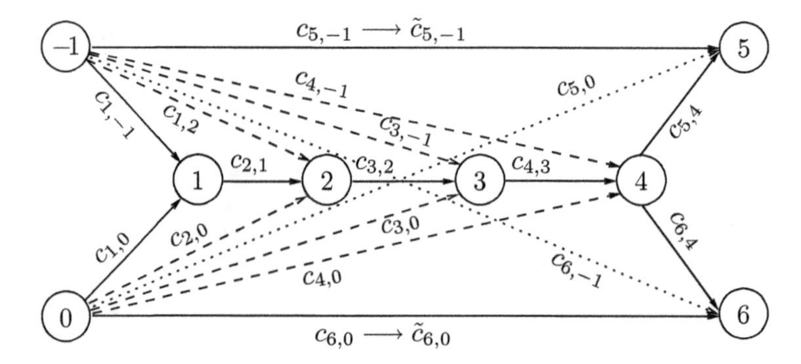

Figure 9.2: Graph for Equation (9.2)

Because the original graph represented by the solid lines is clearly symmetric, the reverse mode would have the analogous effect of introducing pairs of sensitivities $(c_{5,3}, c_{6,3})$, $(c_{5,2}, c_{6,2})$, and $(c_{5,1}, c_{6,1})$. In either "unidirectional" mode one finally obtains the same Jacobian represented by the dotted new arcs $\{c_{5,0}, c_{6,-1}\}$ and the updated values $\{\tilde{c}_{5,-1}, \tilde{c}_{6,0}\}$.

In the more efficient cross-country mode that was introduced above in terms of elementary matrix operations the vertices ② and ③ are eliminated first, resulting in an arc with value $c_{4,1} = c_{4,3} * c_{3,2} * c_{2,1}$ that directly connects nodes ① and ④. Subsequently vertices ① and ④ can be eliminated in either order yielding the unique graph representing the Jacobian. After restating our assumptions we may describe this reduction procedure in general as follows.

Assumptions Revisited

Let $C = \left(c_{ij}\right)_{j=1-n\ldots l}^{i=1-n\ldots l}$ be any square matrix of order $n + l$ satisfying

$$c_{ij} = 0 \quad \text{if} \quad i \leq j \quad \text{or} \quad i < 1 \quad \text{or} \quad j > l - m\,.$$

Then the corresponding incidence graph is acyclic and the corresponding partial order $\prec$ strict. Without loss of generality we make the connectedness Assumption (PC) (see p. 149), which requires that the first n rows and the last m columns of C are the only ones that vanish identically. If the connectedness assumption is not satisfied initially, one may enlarge n or m appropriately and then renumber the entries of C accordingly. Hence we have a one-to-one correspondence between linear computational graphs and arbitrary, strictly lower triangular matrices. The first n vertices $\textcircled{j}$ for $j = 1 - n \ldots 0$ are the only minimal elements, or roots, and the last m vertices $\textcircled{j}$ for $j = l - m + i$ with $i = 1 \ldots m$ are the only maximal elements, or leaves. The remaining $(l - m)$ vertices are called intermediate, and if there are none, the graph is called *bipartite*. Given any input vector $x \in \mathbb{R}^n$, the evaluation procedure in Table 9.4 defines an image vector $y \in \mathbb{R}^m$ that depends linearly on x. Hence we may write $y = F(x) \equiv Ax$ for some uniquely determined matrix $A \in \mathbb{R}^{m \times n}$. When the graph is bipartite we must have

$$a_{ij} \equiv e_i^\top A\, e_j = c_{ij} \; ,$$

where the e_i and e_j denote the Cartesian basis vectors in $\mathbb{R}^m$ and $\mathbb{R}^n$, respectively.

Table 9.4: Linearized Evaluation Procedure

$v_{i-n} = x_i$	for	$i = 1 \ldots n$
$v_i = \sum_{j \prec i} c_{ij}\, v_j$	for	$i = 1 \ldots l$
$y_{m-i} = v_{l-i}$	for	$i = m - 1 \ldots 0$

An arbitrary $m \times n$ matrix A may always be embedded into a strictly lower triangular matrix of size $n + m$ by appending it with n zero rows on top and m zero columns on the right. The corresponding graph is bipartite and has been used extensively for the analysis of linear systems solvers [DER89]. Hence we may identify arbitrary rectangular matrices with bipartite graphs.

For any computational graph C there exists a corresponding bipartite graph A, which we will call its Jacobian. Any two graphs C and $\tilde{C}$ with the same Jacobian will be called functionally equivalent, because they define the same mapping from x to y. Conversely, each equivalence class $[C]$ contains exactly one bipartite element, which we wish to compute with minimal effort.

Pathvalues and Pathlengths

Let $\mathcal{P} = \langle (k_0, k_1), (k_1, k_2), \ldots, (k_{r-1}, k_r) \rangle$ with $k_s \prec k_{s+1}$ for $s = 0 \ldots r - 1$ denote some directed path of length $|\mathcal{P}| \equiv r$ that connects its origin k_0 and its destination k_r. The set of paths with the same origin j and destination i will be denoted by $\left[j \mapsto i\right]$. For subsets of vertices $\mathcal{U}, \mathcal{W} \subset \mathcal{V}$, we denote the set of all paths connecting them by

$$\left[\mathcal{U} \mapsto \mathcal{W}\right] \equiv \bigcup_{j \in \mathcal{U},\, i \in \mathcal{W}} \left[j \mapsto i\right] . \tag{9.5}$$

Since the sets on the right are disjoint we find for their cardinality denoted by $|\cdot|$ that

$$\left|\mathcal{U} \mapsto \mathcal{W}\right| = \sum_{j \in \mathcal{U}, i \in \mathcal{W}} \left|j \mapsto i\right| . \tag{9.6}$$

The last quantity of interest is the sum over all the length of all paths in $[\mathcal{U} \mapsto \mathcal{W}]$, which we will represent by the "norm"

$$\|\mathcal{U} \mapsto \mathcal{W}\| \equiv \sum_{\mathcal{P} \in [\mathcal{U} \mapsto \mathcal{W}]} |\mathcal{P}| \,. \tag{9.7}$$

For any path $\mathcal{P}$ we define its value as the product of its arc values:

$$c_{\mathcal{P}} \equiv \prod_{(j,i) \in \mathcal{P}} c_{ij} \; \in \mathbb{R} \,. \tag{9.8}$$

Summing over all pathvalues between an origin j and a destination i we obtain the mutual sensitivity

$$c_{[j \mapsto i]} \equiv \sum_{\mathcal{P} \in [j \mapsto i]} c_{\mathcal{P}} \,. \tag{9.9}$$

For $j - n$ and $l - m + i$ denoting an independent and a dependent vertex, respectively, we may now write the chain rule in the following deceptively simple form.

Proposition 9.2 (Explicit Jacobian Accumulation)
The Jacobian entry a_{ij} is given by the sum over the products of all the elemental partials along all paths connecting x_j and y_i. Formally,

$$a_{ij} = c_{[j-n \mapsto l-m+i]} \tag{9.10}$$

where the sensitivity on the right is defined by (9.9) *and* (9.8).

Proof. The proof follows by induction on the intermediate variables v_i, as sketched in Exercise 9.3. ∎

Equation (9.10) has been known for a long time (see, for example, [Bau74]). In terms of computational complexity the expression on the right resembles the explicit expansion of a determinant, which also consists of a large number of additive terms that have many common multiplicative subexpressions. Hence, the challenge is now to organize the computation such that a large number of these subexpressions is identified and reused not only for a single pair (j, i) but across all nonzero partials a_{ij}. This is exactly what our Jacobian accumulation procedures attempt to do.

Computing the value of a path $\mathcal{P}$ requires $|\mathcal{P}|-1$ multiplications, or possibly a few less since some of the arc values c_{ij} may be equal to 1. Summing these contributions requires one extra addition for each path so that $\|j \mapsto i\|$ is a fairly sharp upper bound on the number of arithmetic operations needed to calculate $c_{[j \mapsto i]}$ by evaluating each pathvalue separately. Abbreviating $\mathcal{X} = \{1-n \ldots 0\}$ and $\mathcal{Y} = \{l-m+1 \ldots l\}$ we conclude that $\|\mathcal{X} \mapsto \mathcal{Y}\|$ is an upper bound on the multiplications and additions needed to accumulate the Jacobian A from C. While we certainly do not advocate using this "naive" approach, one may use $\|\mathcal{X} \mapsto \mathcal{Y}\|$ in designing heuristics for more efficient Jacobian evaluations.

The number $|\mathcal{X} \mapsto \mathcal{Y}|$ is also a lower bound on the length of the symbolic expression for the Jacobian. Here we mean by symbolic expression a closed formula in terms of the independent variables x_j for $j = 1 \ldots n$ without any named intermediate variables. If there are any subexpressions that are common to several components of the Jacobian, they will be replicated in their full length. There is a one-to-one correspondence between each occurrence of a variable symbol x_j in the expression for the ith function component $y_i = F_i$ and a path connecting nodes $j - n$ and $l - m + i$ in the computational graph.

The computation and updates of the number of paths and their lengths can be performed using the following recursions:

$$\begin{aligned}
|\mathcal{X} \mapsto i| &= \sum_{j \prec i} |\mathcal{X} \mapsto j| \,, \\
|j \mapsto \mathcal{Y}| &= \sum_{i \succ j} |i \mapsto \mathcal{Y}| \,, \\
\|\mathcal{X} \mapsto i\| &= |\mathcal{X} \mapsto i| + \sum_{j \prec i} \|\mathcal{X} \mapsto j\| \,, \\
\|j \mapsto \mathcal{Y}\| &= |j \mapsto \mathcal{Y}| + \sum_{i \succ j} \|i \mapsto \mathcal{Y}\| \,.
\end{aligned} \tag{9.11}$$

Here we may start with the initial conditions

$$\begin{aligned}
& |\mathcal{X} \mapsto j - n| = 1 = |l - m + i \mapsto \mathcal{Y}| \\
\text{and} \quad & \hspace{16em} \text{for} \quad j \le n,\, i > 0\,. \\
& \|\mathcal{X} \mapsto j - n\| = 0 = \|l - m + i \mapsto \mathcal{Y}\|
\end{aligned}$$

Then we must have at the end

$$\begin{aligned}
\sum_{1 \le j \le n} \|j - n \mapsto \mathcal{Y}\| = \|\mathcal{X} \mapsto \mathcal{Y}\| = \sum_{1 \le i \le m} \|\mathcal{X} \mapsto l - m + i\| \,, \\
\sum_{1 \le j \le n} |j - n \mapsto \mathcal{Y}| = |\mathcal{X} \mapsto \mathcal{Y}| = \sum_{1 \le i \le m} |\mathcal{X} \mapsto l - m + i| \,.
\end{aligned}$$

The effort for computing the four integers on the left of (9.11) for all vertices is obviously a small multiple of the number of arcs in the graph. However, it is not so clear how much effort must be put into updating these numbers when individual arcs or whole vertices are eliminated, as described in subsequent sections.

Back- and Front-Elimination of Edges

Suppose we want to simplify a given linear graph by eliminating a certain arc (j, i). Unless its value c_{ij} happens to be zero, the arc signifies a certain dependence of v_i and its successors on v_j and also of v_i on the predecessors v_k of v_j. These dependencies must somehow be accounted for if the relation between the input vector x and the output vector y is to be maintained. In other words we want to keep the corresponding Jacobian A invariant as the graph C is successively simplified. To this end we have to make sure that the values for the Jacobian entries a_{ij} as given in Proposition 9.2 remain unchanged, while the complexity of the expressions is reduced. In terms of the linear algebra interpretation given in section 9.1 this amounts again to eliminating off-diagonal entries using either one of the corresponding diagonal entries with value -1. Graphically we may visualize this process as back- or front-elimination of edges and can formulate the following consequence of Proposition 9.2.

Corollary 9.1 (Edge-Elimination Rules)
The input/output relation of a computational graph C, and hence its Jacobian A, remain unaltered if an arc (j,i) is back-eliminated according to

$$c_{ik} \mathrel{+}= c_{ij} * c_{jk} \quad \textit{for all} \quad k \prec j \quad \textit{and then} \quad c_{ij} = 0$$

or front-eliminated according to

$$c_{hj} \mathrel{+}= c_{hi} * c_{ij} \quad \textit{for all} \quad h \succ i \quad \textit{and then} \quad c_{ij} = 0\,.$$

Here $a \mathrel{+}= b$ means $a = a + b$ and it is assumed that $j > 0$ for back-elimination and $i \leq l - m$ for front-elimination.

Proof. To prove the invariance of pathvalues let us consider any path $\mathcal{P} \in [\mathcal{X} \mapsto \mathcal{Y}]$ that contains the arc (j,i). For back-elimination it must then also contain one of the arcs (k,j) with $k \prec j$. If k is not already a predecessor of i, the new arc (k,i) with value $c_{ik} = c_{ij} * c_{jk}$ introduces a new path with exactly the same endpoints and the same value as the original one, which is subsequently disconnected by the setting $c_{ij} = 0$. Conversely, any other new path containing (k,i) must correspond to an old path containing (k,j) and (j,i) with the same value and endpoints. When the arc (k,i) did already exist the arguments apply just the same, except that now pairs of old paths with the same endpoints are merged to a single new one with the pathvalues being added. This completes the proof. ∎

By comparison with (9.4) we see that this update of the c_{ik} to their new values describes exactly the elementary row operation on the extended Jacobian that eliminates the off-diagonal element c_{ij}. Since the arc (j,i) has been combined with its predecessors (k,j) we will call this process *back-elimination* of the arc (j,i). Alternatively, we may also *front-eliminate* the arc (j,i) by merging it with its successors (i,h), which corresponds to elementary column operations in the matrix setting. Thus we have certainly not discovered anything new, but the graph description appears somewhat more intuitive, as one may glean from Fig. 9.3. The arc (j,i) of the original computational subgraph displayed in the top left corner is eliminated first. Back-elimination leads to the situation displayed in the top right corner requiring only two multiplications for the new arc values $c_{ik} = c_{ij} * c_{jk}$ and $c_{ik'} = c_{ij} * c_{jk'}$. Front-elimination leads to the situation at the bottom left corner requiring two multiplications and one addition for the new arc values $\tilde{c}_{hj} = c_{hj} + c_{hi} * c_{ij}$ and $\tilde{c}_{h'j} = c_{h'i} * c_{ij}$. Since (j,i) was the only incoming arc of ⓘ and has been eliminated, the arcs (i,h) and (i,h') simply can be deleted, which formally corresponds to their back-elimination.

Comparing the top right and bottom left corners of Fig. 9.3 we see that the effects of back-eliminating and front-eliminating the same arc can be quite different. Unfortunately, it is usually very hard to tell which alternative is more advantageous in the long run, and indeed which arc should be eliminated in the first place. Suppose for the moment that $\textcircled{k}$, $\textcircled{k'}$ represent the independent and $\textcircled{h}$, $\textcircled{h'}$ the dependent variables in Fig. 9.3. Then the elimination of (j, i) leaves only arcs that either emanate from an independent or reach a dependent vertex.

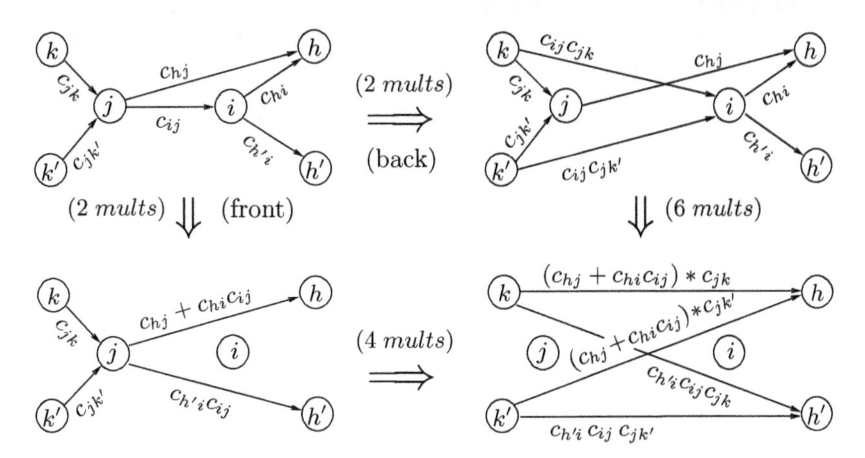

Figure 9.3: Eliminations on Computational Subgraph

In the former case they can only be front-eliminated, and in the latter they can only be back-eliminated. One can easily verify that starting from the top right or bottom left corner the order of elimination does not matter as one always obtains eventually the bipartite graph in the bottom right corner. However, coming from above involves 6 multiplications and 2 additions, while coming from the left requires only 4 multiplications. Thus we see that beginning with the front-elimination of (j, i) yields the local Jacobian for a total of 6 multiplications, whereas starting with the back-elimination of (j, i) requires 2 more.

In both cases some arcs are merely updated and some others are newly created, thus representing fill-in. In fact, by looking at the top right graph, one might at first not be convinced that there is any simplification at all, because the number of arcs has actually gone up. However, we already know from Proposition 9.2 that the process of successively back- or front-eliminating arcs must terminate eventually. The same result follows also from the next observation.

Proposition 9.3 (Finite Termination of Edge-Elimination)
When the arc (j, i) is either back- or front-eliminated the total length $\|\mathcal{X} \mapsto \mathcal{Y}\|$ of all paths in the graph is reduced by at least the product of $|\mathcal{X} \mapsto j|$, the number of paths from the independents $\mathcal{X}$ to $\textcircled{j}$, and $|i \mapsto \mathcal{Y}|$, the number of paths from $\textcircled{i}$ to the dependents $\mathcal{Y}$. The number of multiplications needed to perform the elimination equals either $|k \prec j|$, the number of immediate predecessors of $\textcircled{j}$, or $|h \succ i|$, the number of successors of $\textcircled{i}$, respectively.

Proof. Originally, there are exactly $|\mathcal{X} \mapsto j|\ |i \mapsto \mathcal{Y}|$ paths running through the arc (j, i). As a consequence of the elimination each of them is either reduced in length by 1 or merged with another path altogether. Hence the assertion indeed establishes a lower bound but the actual reduction is likely to be much larger. ∎

Rather than eliminating single arcs across the graph, one mostly tends to eliminate groups with the same source or target vertex ⓘ. If this is done for i in ascending or descending order one obtains some familiar methods, as characterized in the following list:

Nonincremental Forward:
Back-elimination of all incoming edges of ⓙ for $j = 1 \ldots l$
Nonincremental Reverse:
Front-elimination of all outgoing edges of ⓙ for $j = l - m \ldots 1 - n$
Incremental Reverse:
Front-elimination of all incoming edges of ⓙ for $j = l - m \ldots 1$
Incremental Forward:
Back-elimination of all outgoing edges of ⓙ for $j = 1 \ldots l - m$

Actually the last method, called incremental forward, has not yet been considered and was included only to achieve full duality with the two reverse modes.

9.3 Accumulation by Vertex-Elimination

Even in the tiny example displayed in Fig. 9.3, discussing the elimination process in terms of back- and front-elimination of arcs is rather tedious. At least in that particular case we can describe the two alternatives much more concisely in terms of vertex-eliminations. As we have noted above, the front-elimination of arc (j, i) effectively takes its destination ⓘ completely out of the picture, because there are no other incoming arcs. To take out the node ⓙ we would have to front-eliminate both of its incoming arcs or back-eliminate both of its outgoing arcs. The result is the same, as we show in the following corollary.

Corollary 9.2 (Vertex-Elimination Rule)
The Jacobian of a given graph C remains unaltered if any one of its intermediate vertices $j \in \{1 \ldots l - m\}$ is eliminated by updating for all pairs $i \succ j$, $k \prec j$

$$c_{i\,k} \mathrel{+}= c_{i\,j} * c_{j\,k}$$

and subsequently setting $c_{i\,j} = c_{j\,k} = 0$ for all $i \succ j$ and $k \prec j$. The number of multiplications is bounded by the Markowitz degree [Mar57]

$$mark(j) = |i \succ j||k \prec j|\,, \tag{9.12}$$

which is also a lower bound on the reduction in the total pathlength $\|\mathcal{X} \mapsto \mathcal{Y}\|$. Here $|\cdot|$ denotes the cardinality of the predecessor set $\{k \prec j\}$ and the successor set $\{i \succ j\}$ and $\|X \mapsto Y\|$ is defined in (9.7).

It is very important to realize that the Markowitz degree of a particular vertex is likely to change whenever a neighboring vertex is eliminated. Hence $mark(v_j)$ must be understood to be defined at the time when v_j is eliminated, which makes the minimization of the total multiplication count a difficult combinatorial problem. In fact, there are exactly $(l - m)!$ different orderings in which the intermediate nodes $j \in \{1 \ldots l - m\}$ can be eliminated. This huge variety is already a reduction from the even larger number of possibilities for eliminating individual edges, which we implicitly considered before. The trivial elimination orders $(1 \ldots l - m)$ and $(l - m, l - m - 1, \ldots, 1)$ correspond to the incremental forward and the incremental reverse modes, since the elimination of vertex j is equivalent to the front-elimination of all its incoming edges or the back-elimination of all its outgoing edges. Thus the usual nonincremental forward mode is, strictly speaking, not a vertex-elimination procedure. However, the difference concerns only the order in which certain edges are eliminated. In either case, for each intermediate v_i all gradient components $c_{[j-n \to i]}$ are calculated before they are eventually front-eliminated.

An obvious question is whether the restriction to vertex-eliminations may increase the minimal operations count achievable. Unfortunately, the answer is yes and it is not yet clear how large the gap may be. The example shown in Fig. 9.4 was given by Naumann in his Ph.D. thesis [Nau99].

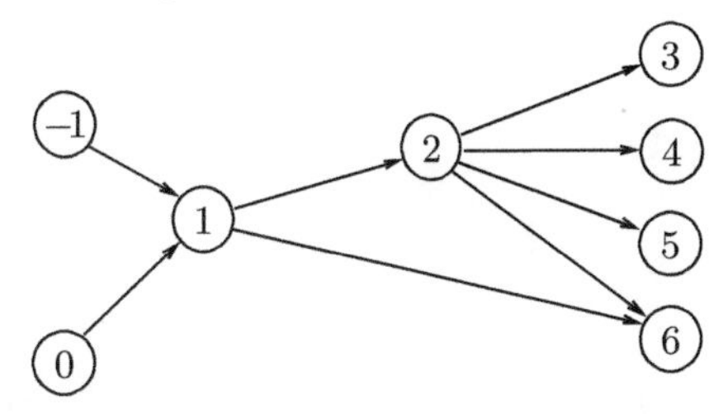

Figure 9.4: Naumann's Lion Example

There are two independents, four dependents, and two intermediate nodes. Eliminating the vertex ① first requires 4 multiplications leading to a graph with two direct arcs from the independents to ⑥ and connections from ② to all other remaining vertices. Then eliminating ② requires 8 more multiplications and 2 additions, yielding a total count of 12 multiplications and 2 additions. Similarly, eliminating first the vertex ② and then ① requires 12 multiplications but only 1 addition. However, one can do better than either of these vertex-elimination procedures by first back-eliminating the arc $(2, 6)$, which requires 1 multiplication and 1 addition. Subsequently, one may eliminate ① at a cost of 4 multiplications and finally ② at a cost of 6 multiplications, yielding a total of 11 multiplications and 1 addition.

On the other hand it is quite clear that general vertex-elimination can be orders of magnitude more efficient than the forward or reverse mode. For example, one may consider generalizations of Fig. 9.2 with $n > 2$ independents, $m > 2$ dependents, and a central chain of $1 + k \gg 4$ intermediate nodes. Then the multiplication counts are $(k+m)n$ for the forward mode and $(k+n)m$ for the reverse mode, but only $k + mn$ for the cross-country mode, where nodes 1 and $1+k$ are eliminated last. In some cases the freedom of cross-country elimination may alleviate unforeseen consequences of problem reformulation. Consider, for example, the Helmholtz energy already examined in Exercise 7.3. The graphs corresponding to the original and the division-free formulation for dimension $n = 4$ are displayed in Figs. 9.5 and 9.6.

In both versions ① represents the inner product $v_1 = b^\top x$ and ② the shift $v_2 = 1 - v_1$. The next vertex ③ represents the reciprocal $v_3 = 1/v_2$ in the original version and the logarithm $v_3 = \ln(v_2)$ in the alternative, division-free version. Subsequently we have for $i = 1 \ldots 4$ the quotients $v_{3+i} = v_{i-4}/v_3$ in the original formulation and the logarithms $v_{3+i} = \ln(v_{i-4})$ in the alternative formulation. The next layer consists of the four vertices $v_{7+i} = \ln(v_{3+i})$ or, alternatively, $v_{7+i} = v_{3+i} - v_3$, which already represent identical values. The final vertices $v_{11+i} = v_{7+i} * v_{i-4}$ for $i = 1 \ldots 4$ are completely identical. All edges except for $c_{2,1} = -1$ have real values that vary with the independent variables. In fact, we could have interpreted v_3 in both versions directly as an elemental function of v_1, namely, as $v_3 = 1/(1 - v_1)$ or $v_3 = \ln(1 - v_1)$, respectively. However, for the purpose of this discussion we will consider $c_{2,1}$ as a generic real number as well. The subgraph formed by the first $4 + 7 = 11$ vertices has the same structure as our earlier example displayed in Fig. 9.2, where eliminating the vertex ① and ② first turned out to be advantageous. This is not the point we are trying to make here though. As we have noted in Exercise 7.3**b**, the average domain size $\bar{n}$ is smaller in the division-free version because the vertices ④, ⑤, ⑥, and ⑦ depend only on one, rather than all, independent variables. While this makes the division-free version more advantageous for the sparse forward mode, there happens to be no difference in the reverse mode. More specifically, we have, according to Exercise 9.5, the following multiplication counts.

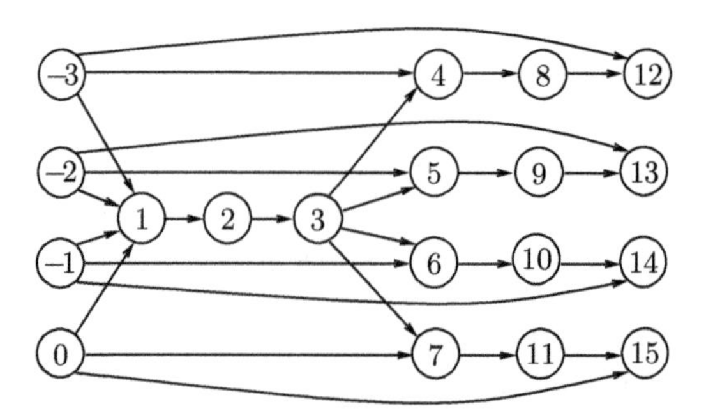

Figure 9.5: Original Graph of Separated Energy

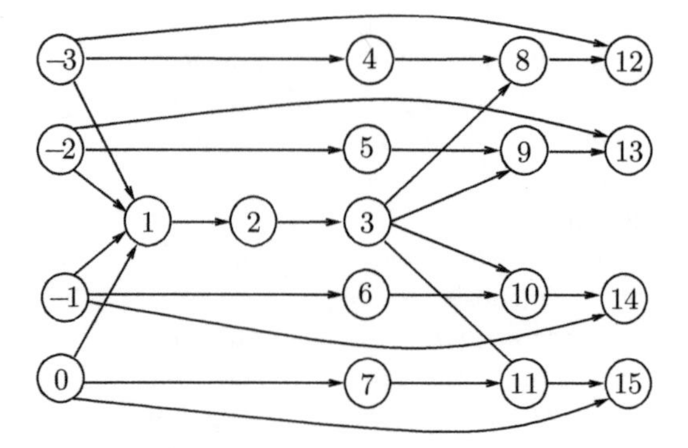

Figure 9.6: Alternate Graph of Separated Energy

	Forward	Reverse	Cross-country
Original	56	36	33
Alternative	44	36	33

Here the elimination order chosen for cross-country was ②, ④–⑦, ⑧–⑪, ③, and finally ①. This ordering is consistent with the greedy Markowitz criterion to be discussed later.

There are two lessons we draw from this. First, we confirm our observation that cross-country elimination can be more efficient than either the forward or the reverse mode. Second, the cost of the forward (and similarly the reverse) mode can be affected by seemingly unimportant reformulations of the evaluation procedure. In contrast, cross-country elimination provides enough flexibility to make up for such incidental variations. Unfortunately, we have currently no polynomial criterion to decide whether the vertex-elimination order given above is in fact optimal, let alone whether edge-elimination might be better still. Naumann conjectures that the minimal operations count achievable by vertex-elimination is never more than twice that achievable by edge-elimination. Moreover, as we will see in the next section, there is an even more general elimination procedure called face-elimination, which yields in general lower operation counts. This techniques was introduced by Naumann in [Nau99].

As a final remark let us observe that even though Jacobian accumulation involves no divisions, certain vertex-elimination orderings may lead to numerical instabilities. John Reid gave an example similar to the one depicted in Fig. 9.7, where the arcs are annotated by constant partials u and h, whose values are assumed to be rather close to 2 and -1, respectively.

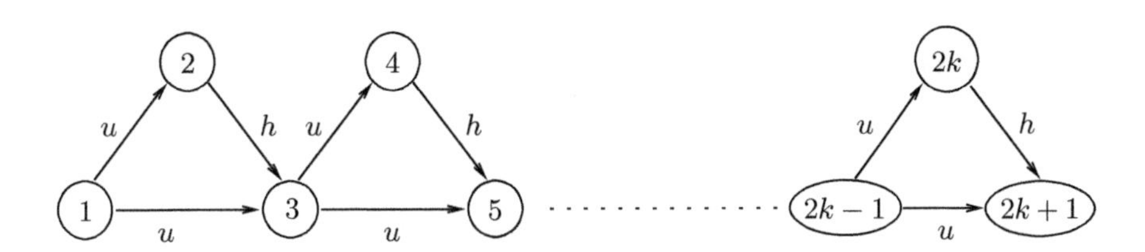

Figure 9.7: Reid's Example of Potential Accumulation Instability

When the intermediate vertices $2, 3, 4, \ldots, 2k-1, 2k$ are eliminated forward or backward all newly calculated partial derivatives are close to zero, as should be the final result $c_{2k+1,1}$. However, if we first eliminate only the odd notes $3, 5, \ldots, 2k-1$ the arc between vertex 1 and vertex $2k+1$ reaches temporarily the value $u^k \approx 2^k$. The subsequent elimination of the even intermediate vertices $2, 4, 6, \ldots, 2k$ theoretically balances this enormous value out to nearly zero. However, in floating point arithmetic, errors are certain to be amplified enormously. Hence, it is clear that numerical stability must be a concern in the study of suitable elimination orderings. In this particular case the numerically unstable, prior elimination of the odd vertices is also the one that makes the least sense in terms of minimizing fill-in and operation count. More specifically, the number of newly allocated and computed arcs grows quadratically with the depth k of the computational graph. Practically any other elimination order will result in a temporal and spatial complexity that is linear in k.

9.4 Face-Elimination on the Line-Graph

When edges of the computational graph are back- or front-eliminated as specified in Corollary 9.1 we are forced to simultaneously merge them with all their predecessor or successor edges, respectively. Sometimes we may wish to do that only with one of them. However, that desire cannot be realized by a simple modification of the linearized computational graph. Instead we have to consider the so-called line-graph of the computational graph with a slight extension. Specifically, we append $(\mathcal{V}, \mathcal{E})$ by a source vertex $\textcircled{-\infty}$ and a sink vertex $\textcircled{+\infty}$. Hence we have

$$\begin{aligned} \bar{\mathcal{V}} &\equiv \mathcal{V} \cup \{-\infty, +\infty\} \qquad \text{and} \\ \bar{\mathcal{E}} &\equiv \mathcal{E} \cup \{(-\infty, j-n)\}_{j=1\ldots n} \cup \{(l-m+i, \infty)\}_{i=1\ldots m} \,. \end{aligned}$$

By calling one edge a predecessor of another when its destination is the origin of the latter we have already established an acyclic precedence relationship between the edges. Now we consider the edges (j,i) of the extended graph $(\bar{\mathcal{V}}, \bar{\mathcal{E}})$ as vertices of the line-graph and connect them by a new edge exactly if one precedes the other. Formally we have a new edge set

$$\mathfrak{F} \equiv \{(j,k,i) \; : \; (j,k) \in \mathcal{E} \ni (k,i)\} \,.$$

Geometrically, we may interpret the edges of the line-graph $(\bar{\mathcal{E}}, \mathfrak{F})$ as faces of the extended graph $(\bar{\mathcal{V}}, \bar{\mathcal{E}})$. Therefore, we will refer to them as faces. The line-graph corresponding to Fig. 9.4 is displayed in Fig. 9.8.

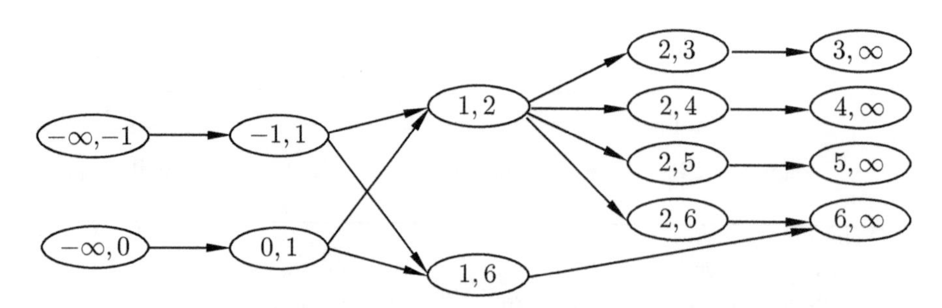

Figure 9.8: Line-Graph for Naumann's First Example

The minimal and maximal vertices of the line-graph are exactly the extra edges introduced in the original graph to connect its minimal and maximal vertices with the sink $\textcircled{-\infty}$ and the source $\textcircled{+\infty}$. Furthermore, there is an obvious one-to-one correspondence between maximal paths in the original graph $(\mathcal{V}, \mathcal{E})$ and maximal paths in the line-graph $(\bar{\mathcal{E}}, \mathfrak{F})$, which we will refer to as edge- and vertex-paths, respectively. The number of edges in an edge-path of $(\mathcal{V}, \mathcal{E})$ is exactly the same as the number of interior vertices in $(\bar{\mathcal{E}}, \mathfrak{F})$. Here the qualification *interior* excludes the first and last elements of any maximal vertex chain in $(\bar{\mathcal{E}}, \mathfrak{F})$, which are always elements of $\bar{\mathcal{E}} \backslash \mathcal{E}$. All interior vertices

$e \equiv (i,j) \in \mathcal{E}$ may be labeled by their value $c_e \equiv c_{ij}$, whereas the faces in $\mathfrak{F}$ have no numerical value associated with them. For the sake of brevity we write $o_j \equiv (-\infty, j-n)$ and $d_i \equiv (l-m+i, +\infty)$ to denote the origins and destinations in $(\bar{\mathcal{E}}, \mathfrak{F})$. After these notational preparations we may rewrite Bauer's formula (9.10) as

$$a_{ij} \equiv \sum_{o_j \overset{\mathcal{P}}{\mapsto} d_i} \prod_{e \in \mathcal{P} \cap \mathcal{E}} c_e \tag{9.13}$$

where the path $\mathcal{P} \cap \mathcal{E}$ is interpreted as a sequence of vertices in $(\bar{\mathcal{E}}, \mathfrak{F})$ rather than a sequence of edges in $(\mathcal{V}, \mathcal{E})$. Now we observe that the vertex version (9.13) of Bauer's formula can be evaluated for any directed acyclic graph $(\bar{\mathcal{E}}, \mathfrak{F})$ with minimal elements o_j for $j = 1 \dots n$ and maximal elements d_i for $i = 1 \dots m$ whose interior vertices $e \in \mathcal{E}$ have certain values $c_e \in \mathbb{R}$.

Face-Elimination

We can change the graph $(\bar{\mathcal{E}}, \mathfrak{F})$ by the following modifications without altering the accumulated values a_{ij}, but hopefully simplifying their computation.

Interior Face-Elimination
Replace any interior face (e, f) by a new vertex g with value $c_g = c_e \cdot c_f$ and connect g to all successors of f as well as all predecessors of e to g.

Interior Vertex Merge
Merge any two interior vertices with identical predecessor and successor sets by adding their values.

Interior Vertex Split
Make a copy of any interior vertex by assigning the same value and the same predecessor or successor set while splitting the other set, i.e., reattaching some of the incoming or some of the outgoing edges to the new copy.

One can easily check that all three modifications leave the accumulated values a_{ij} unaltered and do not affect the extreme vertices in $\bar{\mathcal{E}} \setminus \mathcal{E}$. Moreover the first two modifications also reduce the sum of the length of all maximal paths, whereas the third one leaves that integer constant. Hence the first two operations can only be applied finitely often, resulting in a tripartite graph without any interior edges with all interior vertices having distinct neighborhoods. Then we can split and merge until every interior vertex e is simply connected in that it has exactly one incoming and one outgoing edge connecting it to the vertices o_j and d_i for which then $c_e = a_{ij}$.

Applying the modifications described above, one may generate new directed graphs that are no longer line-graphs of an underlying computational graph. More specifically, one may lose the following characteristic property of line-graphs.

Definition (BP): BICLIQUE PROPERTY
Each edge belongs to a unique biclique, i.e., a maximal complete bipartite subgraph. Each interior node is a minimal element in exactly one biclique and a maximal element in another.

Given such a decomposition into bicliques one can easily reconstruct the underlying directed graph by associating one original vertex with each biclique. Obviously the property of being a line-graph will be maintained by modifications that can be interpreted as the elimination of an edge, say (j,i), on the original graph. It can be checked quite easily that the back-elimination of (j,i) is equivalent to the elimination of all faces arriving at (j,i) on the line-graph, including mergers of newly created with existing vertices if they have the same set of neighbors. Analogously, front-elimination of the edge (j,i) corresponds to the elimination of all faces emanating from (j,i) in the line-graph. For example the back-elimination of $(1,6)$ in the original graph corresponds to the simultaneous elimination of $(-1,1,6)$ and $(0,1,6)$ in the line-graph yielding the new structure displayed in Fig. 9.9 with $(1,6)$ and $\widetilde{(1,6)}$ representing the newly introduced edges, one with an old label.

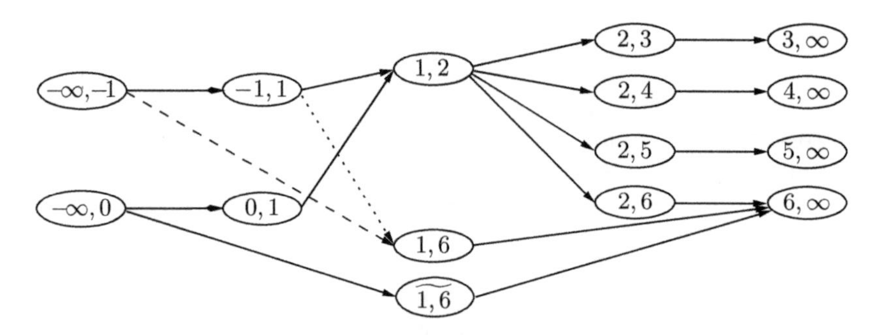

Figure 9.9: Result of Eliminating the Faces $(-1,1,6)$ and $(0,1,6)$

After the elimination of the faces $(-1,1,6)$ and $(0,1,6)$ on the line-graph one can check by inspection of Fig. 9.9 without the dotted line that a biclique decomposition is still possible. This corresponds of course to the computational graph obtained by back-eliminating the edge $(1,6)$. However, suppose we only eliminate the face $(0,1,6)$ but not $(-1,1,6)$. Then Fig. 9.9 applies without the dashed but with the dotted line and we see that the vertex $(1,2)$ is now a maximal element in two distinct maximal bicliques, namely, $\{(-1,1),(0,1),(1,2)\}$ and $\{(-1,1),(1,6),(1,2)\}$. Hence there is no underlying computational graph. Nevertheless, the vertex version of Bauer's formula, namely, (9.13), remains valid and we can continue eliminating interior faces one by one in arbitrary order until eventually none are left. Naturally, we will merge vertices whenever this is possible as work is otherwise duplicated. At least at the end we will also have to split vertices to arrive at a tripartite graph, whose interior is simply connected and thus represents the line-graph of a bipartite graph with the source and sink extension. In [Nau04] it is demonstrated that the minimal operations count achievable by face-elimination is sometimes indeed lower than the optimal result over all possible edge-eliminations.

RULE 13

FACE-ELIMINATION ALLOWS LOWER OPERATIONS COUNTS THAN EDGE-ELIMINATION, WHICH IS IN TURN CHEAPER THAN VERTEX-ELIMINATION.

9.5 NP-hardness via Ensemble Computation

Like many other computational tasks on graphs our various elimination procedures have a combinatorial flavor and are therefore suspect of being NP-hard in some sense. We give here a brief description of the reduction from ensemble computation, a classical NP-complete problem to Jacobian accumulation. The Jacobians that arise are of a very simple diagonal nature so that face-, edge-, and vertex-elimination are all equivalent. Suppose we have a collection of real factors

$$c_j \in \mathbb{R}, \quad j = 1 \dots k$$

and index subsets

$$J_i \subset \{1, 2 \dots k\}, \quad i = 1 \dots n\,.$$

Then the task of ensemble computation is to compute the family of products

$$a_i = \prod_{j \in J_i} c_j\,, \quad i = 1 \dots n$$

using a minimal number of binary multiplications. It is known to be NP-complete [GJ79]. Obviously it does not really matter whether the c_j are reals or belong to some other commutative ring of scalars. The intrinsic combinatorial problem consists here in identifying common subproducts.

A corresponding Jacobian accumulation problem is obtained for the function $y = F(x) : \mathbb{R}^n \mapsto \mathbb{R}^n$ defined by the nested evaluation procedure

for $i = 1 \dots n$
  $y_{i,0} = x_i$
  for $j \in J_i$
    $y_{i,j} = \psi(c_j * y_{i,j-})$

where $j-$ denotes the predecessor of j in J_i and $\psi(v)$ is some univariate function possibly also depending on the index pair (i, j). Naumann originally considered the case where $\psi(v) = v$ is the identity. With y_i representing the last value of the $y_{i,j}$ computed in the inner loop, we have directly

$$\frac{\partial y_i}{\partial x_i} = \prod_{j \in J_i} c_j$$

so that Jacobian accumulation with a minimal number of binary operations is indeed equivalent to ensemble computation. Obviously, the effort of rewriting the problem is linear in its size. Because all intermediate quantities have exactly one predecessor and one successor, both edge- and even face-elimination reduce to vertex-elimination. Hence we have shown that every ensemble computation problem can be interpreted as a special Jacobian accumulation problem.

There are two main objections to accepting this result as general proof that accumulating Jacobians with minimal complexity is a difficult problem. The first criticism can be partly removed by a nontrivial choice of $\psi(v)$. As we have done throughout this book one normally relates the effort of evaluating derivatives to that of evaluating the underlying scalar or vector function. Only if the former is in some sense much greater than the latter is one inclined to say that differentiation is hard. When $\psi(v) = v$ the vector function $F(x)$ is linear and its evaluation is virtually equivalent to that of computing its Jacobian. Any savings we can make by sharing common subproducts in evaluating the Jacobian can be realized in the same way for the function evaluation itself. So life is difficult with or without differentiation and the cost of evaluating the function and Jacobian together is practically the same as evaluating the function.

However, if we choose $\psi(v) = \sin(v)$ or set it to any other nonlinear intrinsic function then the evaluation of F itself cannot be simplified, because for a general argument vector $x \in \mathbb{R}^n$ there are no algebraic relations between the intermediate quantities $y_{i,j}$ for $j \in J_i$ generated by the inner loop in the evolution procedure above. On the other hand on the level of the derivatives we have

$$\frac{\partial y_i}{\partial x_i} = \prod_{j \in J_i} \psi'(c_j \cdot y_{i,j-}) \cdot c_j = \prod_{j \in J_i} \psi'(c_j \cdot y_{i,j-}) \cdot \prod_{j \in J_i} c_j$$

where we have used the commutativity of reals. So, if we phrase our task now as follows: "Evaluate the function or evaluate the Jacobian with a minimal arithmetic complexity, i.e., number of intrinsic function evaluations and binary multiplications;" then the second task is indeed much more difficult. But that is only true on a meta-level because of our self-imposed requirement to evaluate the Jacobian with absolutely minimal arithmetic cost, which entails the combinatorial effort of solving an ensemble computation problem. If we simply are content with evaluating some subexpressions repeatedly, we find that the evaluation of F' together with F requires at most three times the effort of evaluating F by itself in its given, nonimprovable form. Moreover, the reductions that we may achieve reduce this cost increase at best from 3 to 2. Here we have again assumed that evaluating $\psi'(v)$ is no more expensive than evaluating $\psi(v)$, which is very realistic for $\sin(v)$ and all other usual intrinsics. A similar observation applies in general, namely, the naive forward or reverse modes always yield Jacobians with the polynomial effort $\min(n, m)OPS(F) \leq OPS(F)^2$, so we may get into trouble only if we try to be smart. This is exactly the same situation as in sparse linear equation solving which can always be done by dense Gauss with a polynomial effort of $O(n^3)$ operations but becomes NP-complete if one tries to absolutely minimize the fill-in [RT78].

The second doubt concerns the fact that identical multiplier values c_j occur at different parts of the calculation, whereas arc values in computational parts are usually considered independent. This objection could be removed by introducing the factors c_j for $j = 1 \ldots k$ as additional independents, so that every intermediate $c_j * y_{i,j-}$ represents a vertex with two incoming arcs. However, then the Jacobian being computed is no longer the full Jacobian so that we have the interpretation that the task of evaluating partial Jacobians with minimal multiplication count is NP-hard.

9.6 Summary and Outlook

In this chapter we have shown that there is more to AD than the forward and reverse modes. By applying the chain rule "cross-country" and allowing the elimination of individual edges in the linearized computational graph, we obtain a huge variety of ways to accumulate all nonzero Jacobian entries from the elemental partials c_{ij}. The most general procedure for such division-free accumulation appears to be face-elimination on the associated line-graph. Selecting an elimination ordering that minimizes the total number of multiplications or any other natural measure of computational complexity turns out to be a hard combinatorial problem.

Edge- and vertex-elimination can be interpreted as elementary row or column operations on the extended Jacobian $E'(x; v)$, as demonstrated in section 9.1. According to Proposition 9.1 the Jacobian itself is then always obtained as a Schur complement after a finite, though possibly large, number of elimination steps. The accumulation procedure does not involve any divisions, but merely additions and multiplications. Nevertheless, a more careful analysis of the numerical properties of the accumulation process is desired, especially in the light of Reid's example.

9.7 Examples and Exercises

Exercise 9.1 (*Reverse as Backward Substitution*)
Consider again the example discussed in section 9.1 and write down the transposed linear system

$$(I - L^\top)\,\bar{Z}^\top = T^\top\,,$$

where L and T are given by (9.3) and

$$\bar{Z} = \left(\bar{V}_1^\top, \bar{V}_2^\top, \bar{V}_3^\top, \bar{V}_4^\top\right) \in \mathbb{R}^{2\times 4}\,.$$

a. Solve this upper triangular system by backward substitution to obtain a sequence of expressions for the $\bar{V}_i$ with $i = 4, 3, 2, 1$. Finally multiply $\bar{Z}$ by $B^\top$ as given in (9.3) to obtain $F'(x) = \left(\bar{V}_{-1}^\top, \bar{V}_0^\top\right)^\top$.

b. Compare this columnwise expression for $F'(x)$ with $F'(x) = \left(\dot{V}_5, \dot{V}_6\right)^\top$ as obtained in (9.2). Show that the numbers of multiplications and additions in the forward and reverse modes are exactly the same.

Exercise 9.2 (*Finite Termination*)
To prove Proposition 9.1, show that any newly created element, say c_{ik} or c_{kj}, must lie on a subdiagonal below the one to which the element being eliminated, c_{ij}, belongs. Hence, the number of nonzeros in the $(i–j)$th subdiagonal is reduced by 1. Then prove by induction on $k = 1 \ldots l + \min(m, n) - 1$ that the number of nonzero elements in the kth subdiagonal must reach a limit, which implies that the process comes to a halt after finitely many steps. Finally, argue that all off-diagonal entries outside the bottom left $m \times n$ submatrix must be zero.

Exercise 9.3 (*Explicit Jacobian Accumulation*)
To prove Proposition 9.2 by induction assume that it is true for a particular linearized function $y = Ax$, with $A \in \mathbb{R}^{m \times n}$ the accumulation of a given C. Now define the scalar value function

$$\tilde{y} = c^\top y = c^\top A x \quad \text{for} \quad c^\top \in \mathbb{R}^m .$$

The computational graph of $\tilde{y} = \tilde{y}(x)$ and the corresponding matrix $\tilde{c}$ have the form

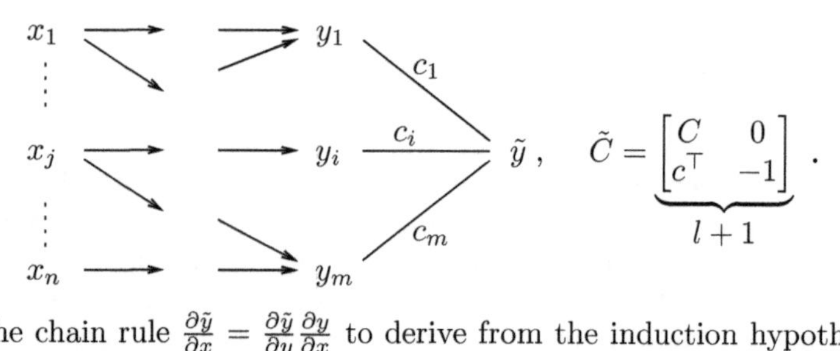

Apply the chain rule $\frac{\partial \tilde{y}}{\partial x} = \frac{\partial \tilde{y}}{\partial y}\frac{\partial y}{\partial x}$ to derive from the induction hypothesis that the elements of $\tilde{A} \equiv \frac{\partial \tilde{y}}{\partial x} \in \mathbb{R}^n$ satisfy Proposition 9.2 with $\tilde{C}$ as above.

Finally, argue that every linear computational graph can be build up from its independent nodes x_i for $i = 1 \ldots n$ by a sequence of l single node augmentations described above. This completes the proof.

Exercise 9.4 (*Naumann's Lion Example, Fig.* 9.4)
Work through both vertex-elimination sequences from Naumann's lion example and find one cheaper edge-elimination sequence.

Exercise 9.5 (*Alternative Proof of Corollary* 9.1)
The given proof of Corollary 9.1 is based on the invariance of sums of pathvalues under edge-eliminations. The same result, and in particular the update formulas for the c_{ik} or c_{hj}, can also be obtained directly as follows.

a. For a fixed pair $j \prec i$ with $j > 0$ consider the unique sum in Table 9.4 that depends on c_{ij}. Substitute for v_j the corresponding sum of terms $c_{ik} v_k$ with $k \prec j$ and combine the resulting new terms in the sum for v_i with old ones corresponding to common antecedents $k \prec i$ and $k \prec j$. Compare the total coefficients with the update formula for c_{ik} given in Corollary 9.1.

b. Examine whether the alternative proof for back-elimination of edges given in part **a** can also be used to prove the validity of front-elimination more directly.

Chapter 10

Jacobian and Hessian Accumulation

Given the combinatorial nature of the face-, edge-, and vertex-elimination problems, one has to try one's luck with heuristics that hopefully yield nearly optimal solutions most of the time. For simplicity we will consider only vertex-elimination. In section 10.1 we sketch some heuristics for selecting suitable elimination orderings and discuss preliminary observations on their performance. In section 10.2 we consider local preaccumulation techniques that promise significant reductions without the need for extensive global analysis and optimization. Section 10.3 is dedicated to the efficient evaluation of numerous Jacobian-vector and vector-Jacobian products, respectively. In section 10.4 we discuss the special case where the Jacobian is in fact a Hessian and the computational graph can be kept in some sense symmetric throughout the elimination process.

10.1 Greedy and Other Heuristics

The number of multiplications needed for eliminating an intermediate vertex $\textcircled{j}$ is bounded by its Markowitz degree defined in (9.12). This bound is attained, unless some of the arc values are 1 or -1, representing additions or subtractions in the underlying evaluation procedure. The number of new fill-in edges generated by this elimination equals $mark(j)$ minus the number of already existing arcs. Our task is to find an elimination order that achieves acceptably low values for the sums of the Markowitz degrees and the corresponding fill-in at the time of elimination.

With the aim of minimizing the total number of multiplications and fill-ins during the Jacobian elimination process, one may adopt the customary greedy approach of always selecting a vertex for elimination that has a minimal Markowitz degree or causes minimal fill-in right now [DD$^+$90]. As a first illustration of how greedy Markowitz works let us look at the graph of the lighthouse example with its five intermediate nodes. Of the $5! = 120$ possible elimination orderings, in Fig. 10.1 we have displayed three: on the left, forward, in the middle, reverse, and on the right, Greedy Markowitz.

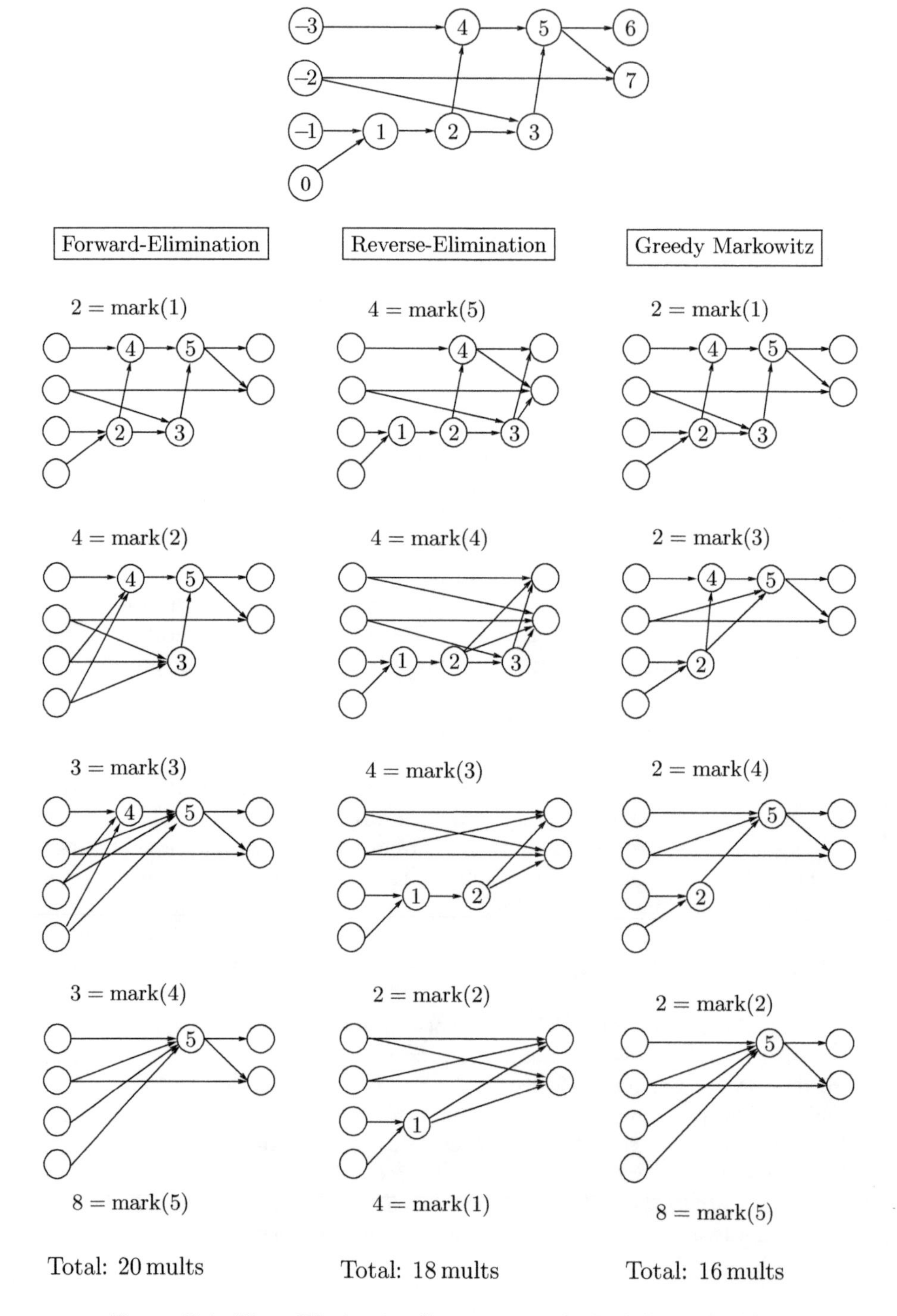

Figure 10.1: Three Elimination Sequences on the Lighthouse Problem

The numbers between the diagrams denote the Markowitz degree of the node just being eliminated, representing the cost of that transition. As one can see in Fig. 10.1, greedy Markowitz works in that it is 4 multiplications cheaper than forward and still 2 multiplications cheaper than reverse.

Safe Preeliminations

One may ask whether certain preeliminations can be performed a priori to simplify the subsequent combinatorial problem without deteriorating the theoretically optimal operations count.

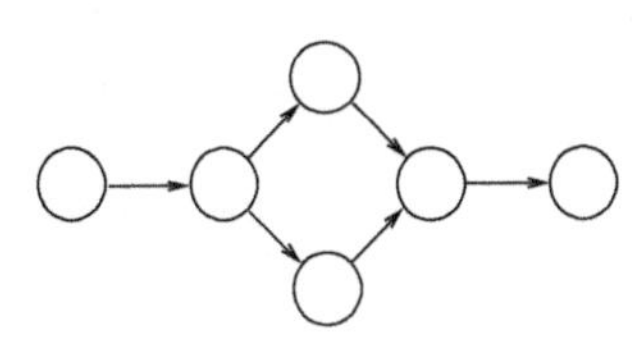

Figure 10.2: Diamond Example

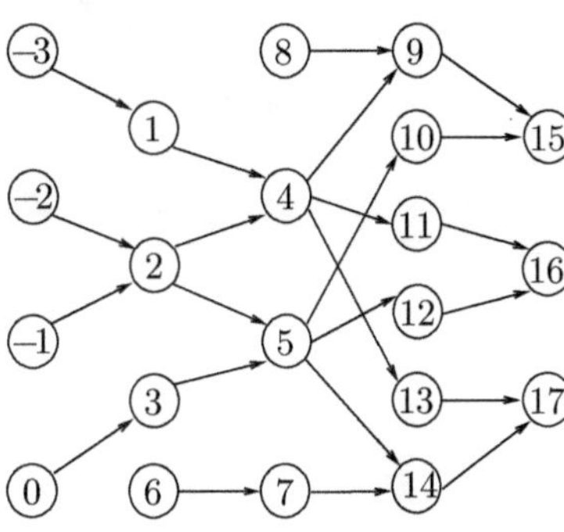

Figure 10.3: Graph for Table 10.1

Nodes with Markowitz degree 0 have either no predecessor or no successor. They represent dead branches of the computational graph and can therefore be deleted at no cost in terms of multiplications.

Nodes with Markowitz degree 1 have exactly one predecessor and one successor, representing a chain of at least two univariate dependencies. They can always be eliminated without negatively affecting the overall operations count. At first one may expect the same to be true for other vertices with only one predecessor or one successor. However, this is not the case, as one may easily see from the "diamond" example displayed in Fig. 10.2. Assuming that all arcs have nonunitary values, one finds that eliminating the first and last intermediate vertices first leads to an overall multiplication count of 6, compared to which eliminating the intermediate layer first leads to a saving of 2 multiplications. Either way, a single addition is performed, which is unavoidable since two paths connect the independent with the dependent vertices. The "diamond" effect occurs similarly for the data-fitting example examined in Exercise 3.5.

Table 10.1: Another Example

$$
\begin{array}{rcl}
v_{-3} &=& x_1 \\
v_{-2} &=& x_2 \\
v_{-1} &=& x_3 \\
v_0 &=& x_4 \\
\hline
v_1 &=& \tan(v_{-3}) \\
v_2 &=& v_{-2} * v_{-1} \\
v_3 &=& \sinh(v_0) \\
v_4 &=& v_1/v_2 \\
v_5 &=& v_2 * v_3 \\
v_6 &=& \arctan(1.) \\
v_7 &=& 4. * v_6 \\
v_8 &=& \log(4.3) \\
v_9 &=& v_8 * v_4 \\
v_{10} &=& \sin(v_5) \\
v_{11} &=& \cos(v_4) \\
v_{12} &=& \exp(v_5) \\
v_{13} &=& \sqrt{v_4} \\
v_{14} &=& v_5 * v_7 \\
v_{15} &=& v_9 * v_{10} \\
v_{16} &=& v_{11}/v_{12} \\
v_{17} &=& v_{13} * v_{14} \\
\hline
y_1 &=& v_{15} \\
y_2 &=& v_{16} \\
y_3 &=& v_{17}
\end{array}
$$

Even though the range of safe preeliminations appears very limited, they typically achieve a significant reduction in the number of vertices. Consider for example the single-assignment procedure listed in Table 10.1 and the computational graph sketched in Fig. 10.3.

By eliminating all nodes with Markowitz degree zero or one we can reduce the total number of vertices from 21 to 10 and obtain the graph sketched in the top left corner of Fig. 10.4. It is not at all surprising that instant gratification à la greedy Markowitz may cause trouble further down the road, as can be seen in Fig. 10.4. Here we face the alternative of eliminating either the central node on the left or one of the vertex pairs on the right. The left one seems a natural choice as it has the minimal Markowitz degree 4; its elimination results in Stage b, where the degree of the remaining two vertices has gone up from 6 to 9. Their subsequent elimination yields first Stage c and then the final bipartite Stage d at a total cost of $4 + 2 * 9 = 22$ multiplications. If instead we eliminate both of the symmetric nodes before the left one as shown in Stages b' and c', the total count is only $3 * 6 = 18$. Minimizing fill-in rather than operations makes no difference; here the greedy heuristic fails also with respect to that criterion.

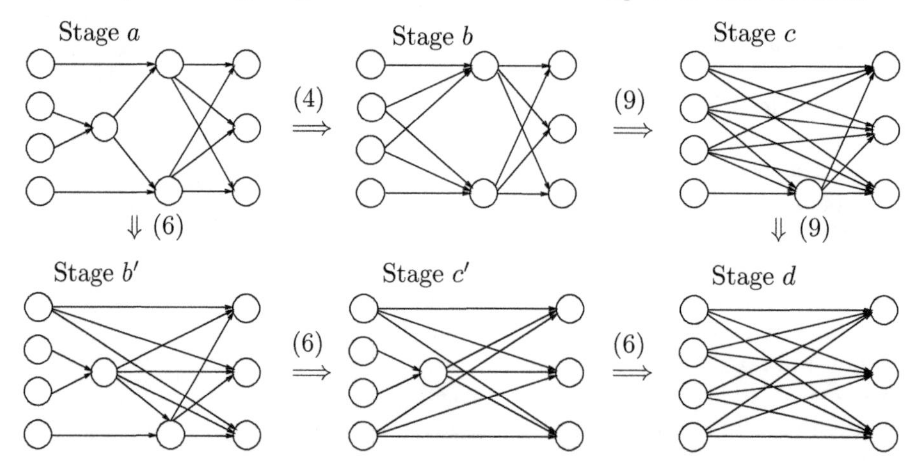

Figure 10.4: Example where Only Relatively Greedy Markowitz is Optimal

Relatively Greedy Markowitz

To save greedy Markowitz from failing on our extremely simple example, Naumann suggested the following modification in his thesis [Nau99]. In an attempt to look a little further down the road, we may ask ourselves for each node, "How much will it cost to eliminate this vertex later rather than now?" This question of course cannot be answered since the Markowitz degree of any particular vertex may oscillate up and down as its neighbors are eliminated. However, we can be quite sure how much it would cost if we kept the vertex in question around until the very end and eliminated it last. Then we must have $mark(i) = |\mathcal{X}_i||\mathcal{Y}_i|$, where $\mathcal{X}_i$ and $\mathcal{Y}_i$ are the index domain and index range defined in section 7.1.

They can be computed initially for all vertices using the recurrences (7.4) and (7.8) and are invariant with respect to subsequent vertex-eliminations. Subtracting this ultimate Markowitz degree of the current value, we obtain the relative Markowitz degree

$$relmark(i) \equiv mark(i) - |\mathcal{X}_i||\mathcal{Y}_i| \,. \tag{10.1}$$

These integers as well as their sums over $i = 1 \dots l$ at any stage of the elimination process may be positive or negative. In the example above, the left vertex initially has the relative Markowitz degree $4 - 6 = -2$, whereas the other two have only $6-9 = -3$ each. Eliminating one of them leaves the relative Markowitz degree of its symmetric partner unaffected but increases that of the left vertex to $8 - 6 = 2$, as can be seen in Fig. 10.4, Stage b'. Hence relative Markowitz does indeed yield the optimal elimination order on that particular example.

Breakdown of Greedy Markowitz on Evolutions

In sparse linear equation solving by direct methods, the greedy Markowitz heuristic is widely considered to be the pivoting criterion of choice despite faring quite poorly on some more or less contrived examples. Many schemes are derived by relaxing the Markowitz criterion to reduce the cost of its implementation while still obtaining essentially the same elimination sequences. With regards to Jacobian elimination it appears that the greedy Markowitz heuristic leaves a lot to be desired not only in its original but also in its relative form.

Suppose some combustion process is taking place in a thin annulus, i.e., a circular tube. Let $u(t, x)$ represent the temperature as a function of the temporal variable t and the spatial variable x. Assuming that the circumference is normalized to 1, we impose the harmonic boundary condition

$$u(t,\, 1) = u(t,\, 0) \quad \text{for all} \quad t \geq 0\,.$$

Assuming that the supply of combustibles and all other parameters are constant in time, one may describe the evolution of the temperature by a PDE of the form

$$u_t = R(u,\, u_x,\, u_{xx}) \quad \text{for} \quad t \geq 0; \quad 0 \leq x \leq 1\,.$$

Here

$$u_t = u_t(t,\, x), \quad u_x = u_x(t,\, x), \quad \text{and} \quad u_{xx} = u_{xx}(t,\, x)$$

represent first and second partial derivatives with respect to t and x, respectively.

Using central differences on a grid of width $h = 1/n$ and explicit Euler in time, one obtains a discrete dynamical system of the form

$$U_{i+1,j} = T_{i+1,j}\,(U_{i,j-1},\, U_{i,j},\, U_{i,j+1}) \quad \text{for} \quad j = 1 \dots n$$

where on the right-hand side $U_{i,0} \equiv U_{i,n}$ and $U_{i,n+1} \equiv U_{i,1}$. In the following we will consider the ternary functions $T_{i,j}(\cdot,\cdot,\cdot)$ as elemental. The resulting computational graph is sketched in Fig. 10.5.

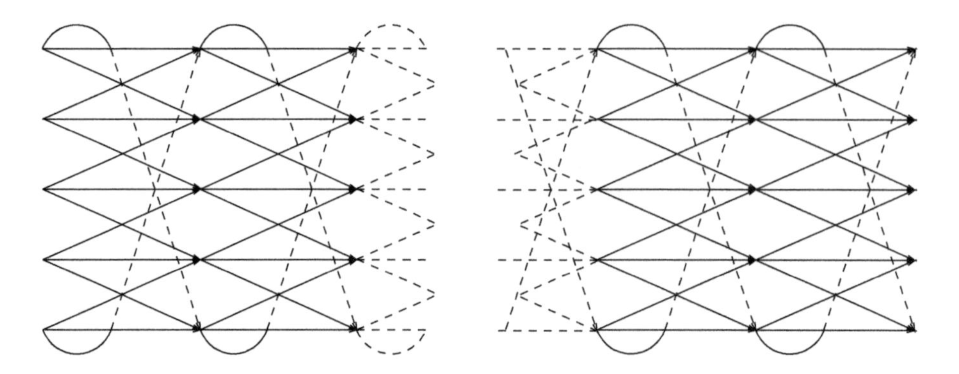

Figure 10.5: Graph of Evolution with Harmonic Boundary Conditions

The algebraic vector functions $T_i \equiv (T_{i,j})_{1 \leq j \leq n}$ map the state space $\mathbb{R}^n$ into itself and have tridiagonal Jacobians T_i', provided they are sufficiently smooth, as we will assume. In fact the T_i' are circulant since the $(1, n)$ and $(n, 1)$ elements are also nonzero due to the harmonic boundary conditions. The rows of the Jacobians T_i' are the gradients of the elemental functions $T_{i,j}$. In view of the inherent symmetry we will only consider elimination orderings that eliminate all vertices in one layer together. This happens automatically if Markowitz and other heuristics are combined with forward or reverse as a tiebreaking criterion. In other words we always pick the first or last vertex that ties with others in satisfying the primary selection criterion. Here first and last are defined according to the lexical order of the indices (i, j). As a result all methods begin by multiplying two successive tridiagonal matrices T_{i-1}' and T_i', yielding a pentadiagonal matrix. This matrix may then be multiplied by its neighbor, and so on.

Under these a priori restrictions we now consider the task of computing

$$\frac{\partial U_l}{\partial U_0} = T_l' T_{l-1}' \ldots T_2' T_1' \,,$$

with a minimal number of operations. The forward and reverse methods would calculate according to the bracketing

$$T_l'(T_{l-1}' \ldots (T_3'(T_2'T_1')) \ldots) \quad \text{and} \quad (\ldots((T_l'T_{l-1}')T_{l-2}') \ldots T_2')T_1' \,,$$

respectively. The product of two circulant band matrices with half bandwidths (number of superdiagonals) α and β has half bandwidth $\min(n-1, \alpha+\beta)$. Since the full bandwidths are $2\alpha+1$ and $2\beta+1$, the number of multiplications required for the product is $n(2\alpha + 1)(2\beta + 1)$ multiplications. The partial product of l circulant tridiagonal matrices has a half bandwidth of l, so that it takes only $l \geq (n - 1)/2$ factors to build up a dense product. In the following we will assume that the number of time steps l remains lower than $n/2$. However, the resulting conclusions remain valid for arbitrarily large l, as will be shown in Exercise 10.3.

With $2\,l < n$ we obtain for the forward and reverse modes the multiplications count

$$n\left[3*3+3*5+\cdots+3\,(2\,l-1)\right]=3n\sum_{i=1}^{l-1}(2i+1)=3n(l^2-1)\approx 3n\,l^2\,. \quad (10.2)$$

Now let us examine the question of whether $\frac{\partial U_l}{\partial U_0}$ can be obtained cheaper by other methods, i.e., bracketing schemes. A good contender would seem to be the Markowitz compatible scheme

$$((T'_l\,T'_{l-1})\,(T'_{l-2}\,T'_{l-3}))\ldots((T'_4\,T'_3)\,(T'_2\,T'_1))\,. \quad (10.3)$$

Here neighboring tridiagonal matrices are pairwise multiplied, yielding pentadiagonal matrices, which are then multiplied to yield circulant matrices of bandwidth $9=2^3{+}1$, and so on. In other words, the half bandwidth doubles at each stage, while the number of products is halved every time. Assuming that $l=2^k<n/2$ as an explicit count for (10.3) Markowitz yields the complexity

$$\begin{aligned} &n\,\left[3^2\,2^{k-1}+5^2\,2^{k-2}+9^2\,2^{k-3}+\cdots+(2^k\,+\,1)^2 2^0\right] \\ =\;& n[2\,l^2+2l\,\log_2 l\,-\,l\,-\,1]\,, \end{aligned} \quad (10.4)$$

a result that was obtained using Maple. Hence we see that the leading coefficient in the operations count for Markowitz is 33% lower than that for the unidirectional mode. In fact, Markowitz is optimal up to leading order, as can be seen from the following result.

Proposition 10.1 (Complexity Bracket on One-Dimensional Evolution) *Provided $n>2l$, the number of multiplications needed to accumulate the Jacobian $\frac{\partial U_l}{\partial U_0}$ by multiplying the factors T'_k in any arbitrary order is bounded below by $2nl(l-1)$ and above by $3nl(l-1)$.*

Proof. See Exercise 10.1. ∎

Unfortunately, the superiority of the Markowitz-compatible bracketing scheme on the one-dimensional evolution with harmonic boundary conditions is by no means a typical effect. To see this we only have to increase the dimension by 1, i.e., consider an evolution on the unit square with harmonic boundary conditions and a mesh width of $h=1/n$ in each spatial direction. For a discretization based on the nine-point stencil, the products of α Jacobian factors have a generalized bandwidth of $(1+2\alpha)^2$. They have n^2 rows and columns. Consequently, the cost for the product of two such matrices is n^2 times the product of their bandwidths. Thus we obtain in analogy to (10.2) for forward and reverse the total multiplications count

$$3^2\,n^2[3^2+5^2+\cdots+(2i+1)^2+\cdots+(2l-1)^2]=n^2(12\,l^3-3\,l-9)\approx 12\,n^2\,l^3\,,$$

which means the complexity grows like l^3 as long as $n>2l$.

In contrast, we obtain for Markowitz with $l = 2^k$ in analogy to (10.4) the lower bound

$$n^2 \left[3^4 \, 2^{k-1} + 5^4 \, 2^{k-2} + \cdots + (1 + 2^k)^4 2^0\right] \geq n^2 \, 16^k = n^2 \, l^4 \, .$$

Maple yields the exact leading term of the right-hand side as $n^2(1+1/7)l^4$, which is only slightly larger than our lower bound that takes into account only the cost for the last matrix-matrix multiply. Thus we see that the Markowitz bracketing causes a rapid fill-in in both factors of all multiplies. In contrast, forward, and reverse always keep one factor rather sparse, so that the overall complexity is one order of magnitude smaller than that of Markowitz with respect to l as long as $l < n/2$.

In three dimensions, i.e., an evolution on the unit cube, this effect is even more pronounced (see Exercise 10.4), as greedy Markowitz is two powers of l more costly than forward or reverse. When we allow l to become larger than a given n, the total complexity eventually grows linearly with respect to l irrespective of the domain dimension and the elimination strategy. However, on squares and cubes the slope is again much larger for Markowitz (see Exercise 10.3) than for forward and reverse. Moreover, in all these cases the relative Markowitz criterion described above does not fare much better, and alternative tiebreakers cannot improve the situation either. While the assumption of harmonic boundary conditions and uniform grids greatly simplified keeping track of operation counts, there is no reason to believe that variations of Markowitz would fare any better for other boundary conditions or on irregular grids. Since this class of problems can hardly be dismissed as contrived counterexamples without practical importance, we may formulate the following rule.

RULE 14

DO NOT USE GREEDY HEURISTICS BASED ON THE MARKOWITZ DEGREE FOR COMPUTING THE JACOBIANS OF FINAL STATES WITH RESPECT TO INITIAL STATES IN EVOLUTIONS.

One may even argue that such Jacobians should never be accumulated explicitly at all, because they are in the following sense redundant. When we have exactly $1 + 2l = n$ then the Jacobian $\frac{\partial U_l}{\partial U_0}$ is already dense with $n^4 = (n^2)^2$ and $n^6 = (n^3)^2$ elements in two or three dimensions, respectively. Yet the total number of nonzero elements in the l factors T_j' for $j = 1 \ldots l$ is only $9n^2 l \approx 9n^3/2$ or $27n^3 \approx 27n^4/2$ and thus orders of magnitude smaller than n^4 and n^6, respectively. Hence, these Jacobians belong to a submanifold of large codimension in the space of all real square matrices with the appropriate size. This intrinsic information is lost when they are accumulated rather than kept in factorized form (which also requires much less storage).

Fortunately, the explicit accumulation of such Jacobians can frequently be avoided altogether, for example, in shooting methods to satisfy nonlinear boundary conditions. In the remainder of this section we will consider elimination methods that do work better on the evolution example above.

Optimal Pathlength Reduction

In section 9.2 we have observed that the number of multiplications for accumulating a given Jacobian is bounded above by the total pathlength $\|\mathcal{X} \mapsto \mathcal{Y}\|$ of its computational graph. Hence it seems natural to look for those simplifications in the graph that reduce this upper bound most, at least relative to the cost caused by this simplification itself. We have seen in Corollary 9.2 that a lower bound of the reduction in $\|\mathcal{X} \mapsto \mathcal{Y}\|$ caused by the vertex-elimination of v_i is $|\mathcal{X} \mapsto i||i \mapsto \mathcal{Y}|$, the total number of paths running through ⓘ. The computational effort in terms of multiplications is given by the Markowitz degree. The elimination of ⓘ alters the values of $|\mathcal{X} \mapsto j|$ and $\|\mathcal{X} \mapsto j\|$ or $|j \mapsto \mathcal{Y}|$ and $\|j \mapsto \mathcal{Y}\|$ for all j with $j \succ^* i$ or $j \prec^* i$, respectively. In contrast, the Markowitz degree changes only for the immediate neighbors. Hence, a greedy heuristic maximizing the reduction of $\|\mathcal{X} \mapsto \mathcal{Y}\|$ at each step is more expensive to implement than the simple Markowitz criterion considered earlier. Nevertheless, this effort may have its rewards, as one can see in Table 10.2.

Table 10.2: Multiplication Count for Accumulation on Planar Evolution

Forward	Abs Mark	Rel Mark	Pathred
944 960	2 150 144	2 019 584	944 960

The results were obtained over $l = 50$ steps on a grid with $h = 1/8$ and thus $8^2 = 64$ variables. The resulting graph has 3,136 intermediate nodes, whose elimination involves quite a lot of work. The pathlength reduction heuristic effectively reduces to the forward mode and fares a lot better than the two Markowitz variations. However, on other problems the relative performance between the Markowitz and pathlength reduction heuristic is reversed so a lot remains to be explored.

Matrix Chaining and Quotient Graphs

On closer examinations one finds that by using forward or reverse as tiebreakers, we have effectively restricted the vertex-elimination problem on the evolution equations to a matrix chaining problem. Then dynamic programming would yield the optimal solution at an acceptable cost. On less structured problems this transition from the computational graph to a matrix chain is not as natural and certainly not unique. As we noted before we limit our options a priori when we write the Jacobian as the product of elemental Jacobians in a particular order, usually the one in which the elemental statements were executed originally. Apart from this arbitrariness, there is the problem that the number of elementals l is normally very large and their Jacobians are extremely sparse, so that solving the resulting dynamic programming problem might be quite difficult. To simplify the situation we may merge successive elemental Jacobians at zero cost if the result variable of one elemental is not an argument of the other, and vice versa.

Suppose we have a mapping $h : \{1-n, \ldots, 0, 1, \ldots, l\} \longmapsto \mathbb{R}$ that is monotonic in that

$$j \prec i \implies h(j) < h(i) \quad \text{for} \quad 1-n \leq j < i \leq l\,.$$

The nonempty inverse images $h^{-1}(\alpha)$ for $\alpha \in \mathbb{R}$ form a disjoint partition of the vertex set $\{1-n \ldots l\}$ and any two elements of the same subset $h^{-1}(\alpha)$ must be mutually independent. The minimal number of subsets in such a partition is the height of the computational graph, i.e., the maximal length of any directed path. To achieve this number we may define, for example,

$$h(k) \equiv \max\{|\mathcal{P}| : \mathcal{P} \in [\mathcal{X} \mapsto k]\} \quad \text{or} \quad h(k) \equiv -\max\{|\mathcal{P}| : \mathcal{P} \in [k \mapsto \mathcal{Y}]\}$$

or any integer combination thereof. The quotient graph with the supernodes $h^{-1}(\alpha)$ is a chain with some additional arcs. Nevertheless optimal matrix chaining can be performed as described in [Nau99]. Another interesting aspect is that, provided there are enough processors, all elemental functions φ_i with $i \in h^{-1}(\alpha)$ for some α can be evaluated concurrently.

10.2 Local Preaccumulations

Given the enormous number of nodes in a typical computational graph, it seems highly unlikely that global searches and cost function minimizations can ever be performed with a reasonable effort. In any case there is currently no AD tool that can exploit the chain rule in a true cross-country manner on a larger scale. However, some of the tools may well achieve some peephole optimizations by local analysis of the derivative code. A classic example is the statement-level reverse mode employed by ADIFOR, and a more or less dual technique is the dirty-vector approach suggested by Christianson, Dixon, and Brown [CDB96]. We will refer to these techniques as local preaccumulation. It may also be interpreted as *interface narrowing*, or *contraction*.

Suppose we have a computationally intensive function

$$(w_1, w_2, \ldots, w_{\tilde{m}}) = G(u_1, u_2, \ldots, u_{\tilde{n}})$$

embedded in the larger context of a function F with n independents $x_1, \ldots, x_n$ and m dependents $y_1, \ldots, y_m$. A situation like this is depicted in Fig. 10.6 with $n = 5$, $m = 4$, $\tilde{n} = 2$, and $\tilde{m} = 3$. The unlabeled nodes outside G represent dependencies between the global independents x_j and dependents y_i that bypass G, of which there are usually very many. The key property of G is that it involves a large number of internal nodes compared to the size of its interface $\tilde{n} + \tilde{m}$, here equal to 5. One may view Fig. 10.6 as a generalization of the chain example of Fig. 9.2.

Now imagine that the vector forward mode is used to calculate the Jacobian $F'(x)$ in one sweep. Then $\tilde{n}$ gradient vectors $\nabla_x u_i$ for $i = 1 \ldots \tilde{n}$ are associated with the independents of G. Consequently the gradients $\nabla_x v \in \mathbb{R}^n$ of all internal nodes v of G are linear combinations of the $\tilde{n} < n$ input gradients $\nabla_x u_i$.

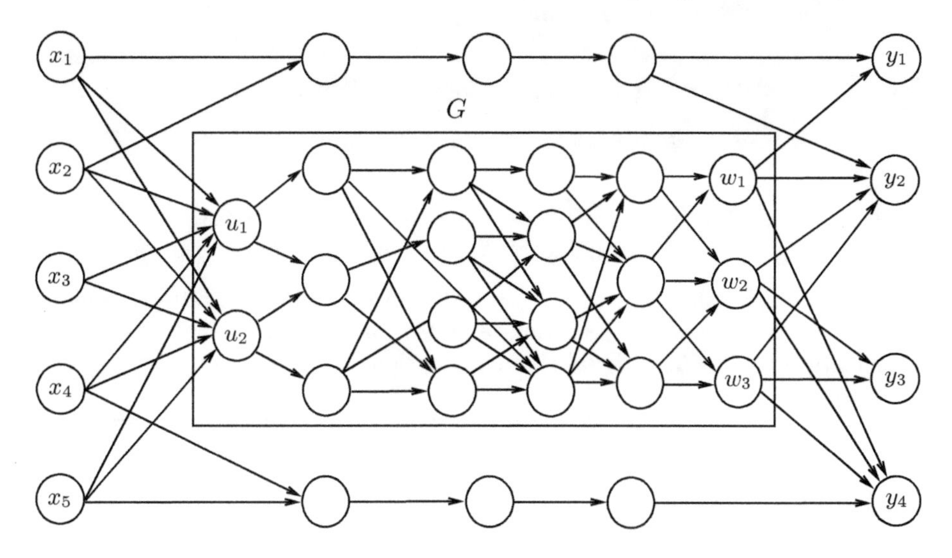

Figure 10.6: Local Evaluation Procedure G in Context of Large Function F

Obviously, propagating these long vectors involves redundancy that should be avoided. For example, we could first compute the Jacobian $G' = \partial w/\partial u$ in the forward mode, which involves the propagation of derivative vectors $\nabla_u v \in \mathbb{R}^{\tilde{n}}$ rather than $\nabla_x v \in \mathbb{R}^n$. Provided $\tilde{n}$ is significantly smaller than n, that entails a significant saving. Subsequently we may apply the chain rule in the form

$$\begin{pmatrix} \nabla_x w_1^\top \\ \vdots \\ \nabla_x w_{\tilde{m}}^\top \end{pmatrix} = \begin{pmatrix} \frac{\partial w_1}{\partial u_1} & \cdots & \frac{\partial w_1}{\partial u_{\tilde{n}}} \\ \vdots & & \vdots \\ \frac{\partial w_{\tilde{m}}}{\partial u_1} & \cdots & \frac{\partial w_{\tilde{m}}}{\partial u_{\tilde{m}}} \end{pmatrix} \begin{pmatrix} \nabla_x u_1 \\ \vdots \\ \nabla_x u_{\tilde{n}} \end{pmatrix} \tag{10.5}$$

to obtain the gradients $\nabla_x w_i$ associated with the local dependents. From then on, the forward sweep on the computational graph can continue as before and must yield up to roundoff the same Jacobian $F'(x)$ as without preaccumulation of the local Jacobian $G'(x)$. If F consists largely of small functions like G with up to $\tilde{n}$ local independents, one can expect that calculating $F'(x)$ with preaccumulation as described requires roughly $\tilde{n}/n$ times the computational effort as the standard forward mode. If Jacobian compression is used, n must be replaced by p, the number of columns in the seed matrix.

In terms of the computational graph we may perform preaccumulation for all subgraphs of the following form.

Definition (LP): LOCAL PROCEDURE AND JACOBIAN
Given the computational graph of an evaluation procedure F, a subgraph G is called a local evaluation procedure *if*

(i) *the $\tilde{n}$ minimal vertices $\{u_j\}_{j=1\ldots\tilde{n}}$ of G are mutually independent and the $\tilde{m}$ maximal vertices $\{w_j\}_{j=1\ldots\tilde{m}}$ are mutually independent;*

(ii) *all vertices v with $v \succ u_j$ for some $j \leq \tilde{n}$ or $v \prec w_i$ for some $i \leq \tilde{m}$ belong to G.*

The Jacobian $(\partial w_i/\partial u_j) \in \mathbb{R}^{\tilde{m}\times\tilde{n}}$ of the mapping from the $(u_1, \ldots, u_{\tilde{n}})$ to the $(w_1, \ldots, w_{\tilde{m}})$ is called a local Jacobian, *and* $\min(\tilde{m}, \tilde{n})$ *its generic rank.*

To emphasize the contrast, $F'(x)$ itself will sometimes be called the global Jacobian. Now, recalling the estimate (7.14) for Jacobian complexity, we can formulate the following general recommendation.

RULE 15

LOCAL JACOBIANS SHOULD BE PREACCUMULATED IF THEIR GENERIC RANK IS SIGNIFICANTLY SMALLER THAN THE NUMBER OF DERIVATIVE OR ADJOINT COMPONENTS BEING PROPAGATED GLOBALLY.

Of course the "should" must be taken with a grain of salt. Suppose the generic rank $\min\{\tilde{n}, \tilde{m}\}$ is just slightly smaller than p, the number of derivative components, or q, the number of adjoint components, being propagated forward or backward, respectively. Then treating the local evaluation graph differently from the rest may not be worth the trouble. Moreover, if $\tilde{n}$, $\tilde{m}$, p, and q are not fixed at compile-time, then one has to provide alternative derivative procedures for the various size combinations that may occur at runtime.

When the local Jacobian is sparse or the local procedure is otherwise structured, one may also employ any one of the economizations discussed in Chapters 7 and 8 or even the more general elimination techniques discussed in sections 9.1, 9.2, and 10.1. This effort is likely to be worthwhile when hundreds or thousands of the local subgraphs have the same fixed structure as is likely to occur in finite-element codes and similar applications. There, a few prototype model routines are called repeatedly with varying parameter values, and optimized Jacobian routines should pay off handsomely. First attempts in this direction are described in [BGP06] and [TR$^+$02]. For general applications, the user still has to generate code for the local Jacobians (hopefully using an AD tool) and then put them together to compute the global Jacobian by hand.

Whatever the value of $p \geq 1$ and $q \geq 1$, things are relatively clear in the cases $\tilde{n} = 1$ or $\tilde{m} = 1$, where we cannot lose by preaccumulating in the forward or reverse mode, respectively.

Statement-Level Reverse

A single-assignment statement with a scalar on the left-hand side represents a local evaluation procedure no matter how complex the right-hand side is.

Consider, for example, the assignment

$$w = -10 * u_2 * \exp(u_3) + \ln(u_1) - 3 * 10^7 * u_3 * (u_2 - 1) * \sqrt{u_1}$$

whose corresponding computational subgraph is depicted in Fig. 10.7. In contrast to our usual habits, we have labeled the intermediate nodes not with a variable name for the quantity being calculated but with the arithmetic operation or special function employed.

The intermediate values may remain anonymous because they have only one way to go in that they are only used once as an argument. In other words, the computational subgraph corresponding to an assignment is almost a tree in that only the minimal vertices and thus locally independent nodes are allowed to have more than one outgoing arc. Hence it is clear that the gradient vector $(\partial w/\partial u_j)_{j=1}^{\tilde{n}}$ is best calculated in the reverse mode, where each elemental partial enters exactly once into the calculation. The same observation is still true for more general *funnels*, which were defined by Speelpenning [Spe80] as any computational subgraph with one local dependent but arbitrarily many independents. A funnel occurs, for example, when one calls a program function with a scalar return value provided it is guaranteed not to have any side effects.

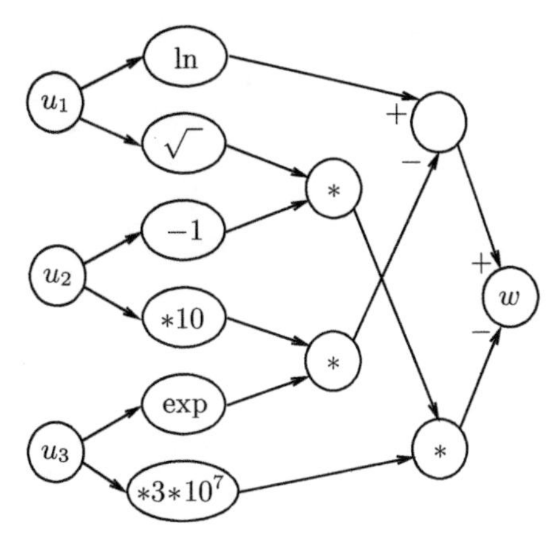

Figure 10.7: Scalar Assignment Tree

On the somewhat contrived right-hand side displayed in Fig. 10.7, the savings on a global forward sweep with $p = 20$ derivative components would be rather dramatic. Calculating a derivative vector of length 20 for each of the 10 intermediate nodes and w would require $6 \cdot 20 + 3 \cdot 60 + 2 \cdot 20 = 340$ arithmetic operations. Here the cost factor 60 applies to the derivative $\nabla v = u\nabla w + w\nabla u$ of a multiplication, whereas the univariate functions and the two subtractions require only 20 operations each for their derivatives. In contrast, preaccumulating the gradient $(\partial w/\partial u_1, \partial w/\partial u_2, \partial w/\partial u_3)^\top$ requires only the backward elimination of its 16 arcs, and the final linear combination

$$\nabla w = \frac{\partial w}{\partial u_1}\nabla u_1 + \frac{\partial w}{\partial u_2}\nabla u_2 + \frac{\partial w}{\partial u_3}\nabla u_3$$

costs 100 more arithmetic operations. The resulting total of 116 operations for the statement-level reversal approach compares very well with the corresponding number of 340 for the standard forward mode.

Of course, such extremely complicated right-hand sides are rather exceptional in practice, where most assignment statements tend to involve only one or two arithmetic operations. Hence, it has been suggested [BK$^+$97] that one considers preaccumulations as basic blocks and whole subroutines of evaluation programs. A case study of the driven cavity problem from Minpack yielded the results displayed in Fig. 10.8.

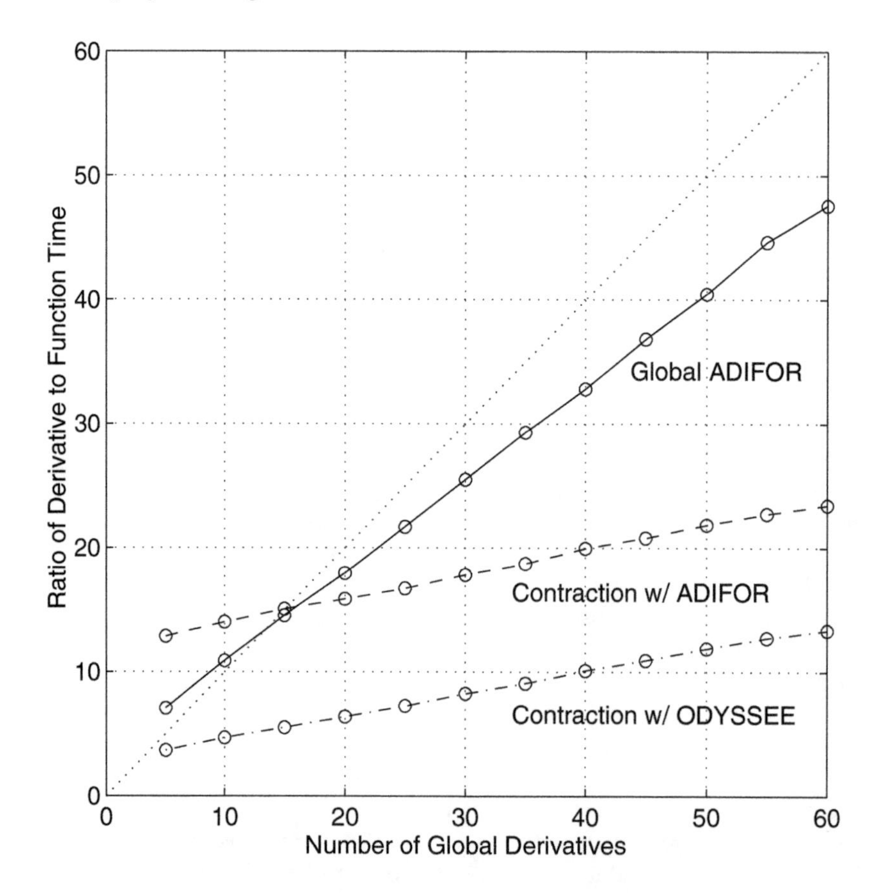

Figure 10.8: Runtime Reduction by Preaccumulation from [BK+97]

The driven cavity code in question uses a grid stencil involving $\tilde{n} = 13$ local dependent variables. Hence the local Jacobians are in fact gradients with 13 components. The original Fortran source was transformed using the tool ADIFOR for propagating p global derivatives in the forward mode. The resulting runtime ratios as a function of p are labeled "global ADIFOR." The preaccumulation of the local gradients also generated with ADIFOR and then hand-spliced into the overall code yielded the line labeled "contraction with ADIFOR."

Since there is very little going on in the code outside the loop over the stencil, the crossover point occurs just to the right of $p = 13$, where $\tilde{n}/p = 1$. The third straight line was obtained by preaccumulating the local gradients in the reverse mode using adjoint code generated by Odyssée. As one can see, this hybrid approach (globally forward, locally reverse) is consistently more efficient than the two purely forward variants. The two variants with preaccumulation have exactly the same slope, since only the size of the common linear combination (10.5) changes with p.

The opposite of a funnel might be called a trumpet, where $\tilde{m} > 1$ local dependents are calculated from one local independent. In that case the local Jacobian is in fact a tangent of size $\tilde{m}$, which should naturally be preaccumulated in the forward mode no matter which method is employed globally. Unfortunately, this dependence structure seems more rarely detectable at compile-time. Therefore Christianson, Dixon, and Brown [CDB96] have suggested a runtime version, which can also be employed in the reverse mode, i.e., for funnels.

10.3 Scarcity and Vector Products

So far we have always assumed that our natural goal is the complete accumulation of the Jacobian, presumably for a subsequent factorization to solve linear systems or estimate eigenvalues. For many large-scale problems this effort may be unjustified because one can only use iterative solution procedures based on Jacobian-vector or vector-Jacobian products. The primary computational task is then to evaluate a long succession of these derivative vectors accurately and with minimal complexity. We will attempt to minimize the number of multiplications while being fully aware that this theoretical measure provides at best an indication of actual computing times. The practical goal of minimizing the multiplications count may be intuitively related to the aesthetic ambition of maintaining structure. This may be motivated and intuitively explained as follows. Disregarding structure of a given problem means embedding it into a larger class of problems, whose representation requires more data and is thus for the more specific class of problems redundant. Hence, reducing the amount of data is also likely to reduce the computational cost of resolving the problems.

To see how accumulating can destroy structure we look again at Naumann's lion example displayed in Fig. 9.4. Here accumulation yields a dense 4×2 matrix with 8 nonzero entries. Hence we have exactly as many global partial derivatives as local nonzero partials including the two unitary ones. However, the accumulated representation of the Jacobian no longer reveals an essential feature ingrained in the linearized computational graph, namely, that the rank of $F'(x)$ can nowhere be more than 1. The reason for this is that the generic rank of the Jacobian $F'(x)$ is always equal to the minimal size of a vertex cut of the underlying computational graph. For a proof of this rather elementary fact, see, for example, [Gri03]. On the other hand, it is quite clear that all rank-one matrices in $\mathbb{R}^{4\times 2}$ can arise as Jacobian of the lion example if we let the eight local partials c_{1-1}, c_{10}, c_{21}, c_{32}, c_{42}, c_{52}, c_{62}, and c_{61} roam freely. Moreover, one can see quite easily that this set of Jacobians is still the same if we impose the artificial conditions $c_{21} = 1$ and $c_{62} = 0$ so that only 6 degrees of freedom are left.

In some sense that is nothing special, since, for sufficiently smooth F the set of reachable Jacobians

$$\left\{F'(x) \in \mathbb{R}^{m\times n} : x \in \mathbb{R}^n\right\}$$

forms by its very definition an n-dimensional manifold embedded in the $m \times n$ dimensional linear space $\mathbb{R}^{m \times n}$. Some of that structure may be directly visible as *sparsity*, where certain entries are zero or otherwise constant. Then the set of reachable Jacobians is in fact contained in an affine subspace of $\mathbb{R}^{m \times n}$, which often has a dimension of order $O(n+m)$. It is well understood that such sparsity structure can be exploited by storing, factoring, and otherwise manipulating these matrices economically, especially when m and n are large. A similar effect can occur if the computational graph rather than the accumulated Jacobian is sparse in a certain sense. This concept of *scarcity* is developed below.

Let $\mathcal{X} \subset \mathbb{R}^n$ denote some open neighborhood in the domain $\mathcal{D} = \mathrm{dom}(F)$, which we may restrict further as required. Then the Hessenberg matrix C defined in (2.9) for the extended system formulation may be interpreted as the matrix-valued function

$$C : \mathcal{X} \quad \longrightarrow \quad \mathbb{R}^{(l+n)\times(l+n)} .$$

Assuming that C is real-analytic we find that all its entries are either constant or may be assumed to have open ranges $\{c_{ij}(x) \,:\, x \in \mathcal{X}\}$ on the possibly restricted open neighborhood $\mathcal{X} \subset \mathbb{R}^n$. Once the nonconstant coefficients are evaluated at the current global argument x we have lost all information about their correlation. This will be particularly true when they are subject to roundoff in finite precision arithmetic. Also, the very nature of an evaluation procedure composed of elementary functions suggests that neither we nor anybody else has global information that restricts the values of the φ_i and their partials c_{ij}. Therefore we will suppose that whatever computational procedure uses the values c_{ij} it can only be based on the assumption that

$$C \in \mathcal{C} \equiv \left(\{c_{ij}(x)\}_{x\in\mathcal{X}}\right)^{i=1-n\ldots l}_{j=1-n\ldots l} \supset C(\mathcal{X}) .$$

In other words $\mathcal{C}$ is the relatively open interval enclosure of the actual range $C(\mathcal{X})$. Consequently $\mathcal{C}$ can be made arbitrary small by restricting $\mathcal{X}$ accordingly. The number of nonconstant entries, i.e., proper interval components of $\mathcal{C}$, will be denoted by $\dim(\mathcal{C})$, so that certainly

$$\dim(C(\mathcal{X})) \le \dim(\mathcal{C}) \le l\,(l-m+n) .$$

The accumulation of the local partials yields the Jacobian matrix $A = \left(a_{ij}\right)^{i=1\ldots m}_{j=1\ldots n}$, which can be expressed as in (9.1)

$$A \equiv A(C) \equiv R + T\left(I - L\right)^{-1} B .$$

Hence we can view A as a matrix-valued function

$$A : \mathcal{C} \quad \longrightarrow \quad \mathbb{R}^{m\times n}$$

whose components are polynomial since $\det(I - L) \equiv 1$ due to the strict triangularity of L. The matrices $A(C)$ for $C \in \mathcal{C}$ may have a certain sparsity pattern, which can be represented by the set

$$\mathcal{A} = \left(\{a_{ij}(C)\}_{C \in \mathcal{C}} \right)_{j=1\ldots n}^{i=1\ldots m} \supset A(\mathcal{C})\ .$$

Then $\dim(\mathcal{A})$ gives the number of entries in the Jacobian $A(C)$ for $C \in \mathcal{C}$ that are nonzero and nonunitary. Now we can define the key concepts of this section as follows.

Definition (SC): JACOBIAN DIMENSION AND SCARCITY
For F given by a certain evaluation procedure we call $A(\mathcal{C})$ the Jacobian set, $\dim(A(\mathcal{C}))$ *the* Jacobian dimension, *and refer to the corresponding computational graph $\mathcal{G}$ and equivalently $\mathcal{C}$ as* scarce *if there is a positive codimension*

$$\text{scarce}(\mathcal{G}) \equiv \dim(\mathcal{A}) - \dim(A(\mathcal{C}))\ .$$

Finally we call $\mathcal{G}$ and $\mathcal{C}$ injective *if* $\dim(A(\mathcal{C})) = \dim(\mathcal{C})$.

In other words scarcity means that the Jacobian set of matrices that can be accumulated from elements $C \in \mathcal{C}$ forms a lower-dimensional subset of the sparse matrix set $\mathcal{A} \subset \mathbb{R}^{m\times n}$. Injectivity means that at least locally all elements of the Jacobian set $A(\mathcal{C})$ have a unique inverse image C.

While sparsity is a structure that meets the eye directly, scarcity is much more subtle. Naumann's lion is scarce but not sparse, and it is also not injective as $\dim(A(\mathcal{C})) = 5 < 6 = \dim(\mathcal{C})$. Hence $\mathcal{C}$ is in some sense a redundant representation of the Jacobian set $A(\mathcal{C})$ on that example.

Quest for a Minimal Representation

This section is the only part of the book in which we will take special note of unitary values. Rather than worrying about the cost of the presented transformation methods we will target here exclusively the quality of the end result, i.e., the number of nonzero and nonunitary edges.

Since multiplications by 1 or -1 are free, the number of costly multiplications needed for the calculation of $\dot{y}$ and $\bar{x}$ is exactly equal to the nonunitary edges in the linearized computational graph. Since we wish to evaluate many such products we consider first simplifying the graph without altering the corresponding accumulated Jacobian. In case of the lion example illustrated in Fig. 9.4 which has originally 8 arcs we may simplify the graph to the one displayed in Fig. 10.9. The transformation from the edge-valued graph $\mathcal{G}$ in Fig. 9.4 to $\tilde{\mathcal{G}}$ in Fig. 10.9 can be performed in two stages by first eliminating the vertex ②, and then normalizing the edge $(-1, 1)$ to the value 1. Each of these modification requires a certain adjustment of the edge values and incurs by itself a computational cost.

In the example above we have computed from $C \in \mathbb{R}^{6\times 4}$ another lower Hessenberg matrix $\tilde{C} = P(C) \in \mathbb{R}^{5\times 3}$, which is *linearly equivalent* in that

$$A(C) = A(\tilde{C}) = A(P(C))\ .$$

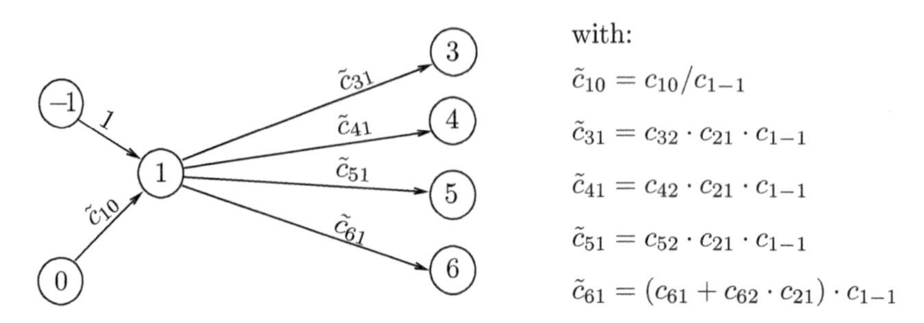

Figure 10.9: Modified Computational Graph $\tilde{\mathcal{G}}$

We will refer to such transformations $P : C \to \tilde{C}$ as preaccumulations of the given Hessenberg structure C. In the example the transformation P is rational and thus not globally defined. Naturally, the big question is how to find a transformation P to a set of Hessenberg matrices $\tilde{\mathcal{C}} = P(\mathcal{C})$ that has a low dimension $\dim(\tilde{\mathcal{C}})$ hopefully as low as $\dim(A(\mathcal{C}))$. It should be noted that any sequence of additions and multiplications, i.e., (homogeneous) linear transformations on scalars, can be written as a lower Hessenberg matrix. This is in particular true for all realizations of the linear transformation from $\dot{x}$ to $\dot{y}$ or equivalently from $\bar{y}$ to $\bar{x}$.

A priori, there is an infinite variety of possible lower Hessenberg structures $\tilde{\mathcal{C}}$, but we can restrict ourselves to a finite number of them by limiting the number $l - m$ of intermediate vertices as follows. We are interested only in Hessenberg structures whose number of free, i.e., nonconstant, entries is at most equal to the original $\dim(\mathcal{C})$. Suppose the ith row and the $(n+i)$th column of the structure contain only constants, which corresponds to all edges incident to vertex ⓘ being constant in the computational graph. Then this vertex may be eliminated by multiplying all incoming and outgoing edge values yielding again constant edges. This process can be continued until each intermediate vertex has at least one free edge. Then the number $l - m$ of these vertices equals at most twice the number of free edges $\leq 2\dim(\mathcal{C})$. Consequently, for a given computational graph $\mathcal{G}$ and the corresponding *natural* matrix representation $\mathcal{C}$ we have to consider only a finite number of related Hessenberg structures $\tilde{\mathcal{C}}$, which may contain linearly equivalent elements $\tilde{C}$ to a given $C \in \mathcal{C}$. We will consider their union as the domain of the accumulation function A. For certain special values of $C \in \mathcal{C}$ there may exist linearly equivalent representations $\tilde{C} = P(C)$ with a very small number of nonzero and nonunitary edges. An extreme case would be the particular linearized graph where all nonunitary edges have the value zero. In other cases we may have fortuitous cancellations during accumulation so that A considered as a special Hessenberg structure with $l = m$ contains very few nonunitary elements. Obviously such extreme cases are of little practical interest. Instead we look for a Hessenberg structure $\tilde{\mathcal{C}}$ and a transformation P defined on a relatively open subset $\mathcal{C}' \subset \mathcal{C}$ so that

$$A(C) = A(P(C)) \quad \text{for all} \quad C \in \mathcal{C}' .$$

For simplicity we identify $\mathcal{C}'$ with $\mathcal{C}$ and assume that on this possibly further restricted domain P is analytic. Hence all entries of $\tilde{C} = P(C)$ may be assumed constant or open with respect to C. Then we find immediately that for $\tilde{\mathcal{C}} \equiv P(\mathcal{C})$

$$\dim\big(A(\mathcal{C})\big) = \dim\big(A(P(\mathcal{C}))\big) \leq \dim\big(P(\mathcal{C})\big) = \dim\big(\tilde{\mathcal{C}}\big) \ .$$

In other words the number of free entries in $\tilde{\mathcal{C}}$ cannot be lower than the Jacobian dimension, which is therefore a lower bound on the number of floating point values needed to represent the Jacobian $F'(x)$ as well as the number of multiplications needed to calculate tangent and adjoint vectors $\dot{y} = F'(x)\dot{x}$ and $\bar{x}^\top = \bar{y}^\top F'(x)$.

In theory it is quite easy to construct a transformation P such that $\dim(P(\mathcal{C})) = \dim(A(\mathcal{C}))$ as follows. Since the mapping $A : \mathcal{C} \to \mathbb{R}^{m\times n}$ is polynomial one can deduce that its Jacobian has a maximal rank $r \leq \dim(\mathcal{A}) \leq m\,n$. Moreover, after suitably restricting $\mathcal{X}$ and thus $\mathcal{C}$, we have the following generalization of the implicit function theorem [GGJ90].

Proposition 10.2 (Implication of the Rank Theorem)

(i) *$A(\mathcal{C}) \subset \mathbb{R}^{m\times n}$ is a (regular) smooth manifold of dimension r.*

(ii) *There exists a transformation $P : \mathcal{C} \to \mathcal{C}$ with $P(C) = C$ for some $C \in \mathcal{C}$ such that all but r components of $P(C)$ are constant and*

$$\dim\big(A(\mathcal{C})\big) = \dim\big(A(P(\mathcal{C}))\big) = \dim\big(P(\mathcal{C})\big) \ .$$

The second assertion means that locally all Jacobians in $A(\mathcal{C})$ can be traced out by varying only r of the edge values and keeping the others constant. Unfortunately, we cannot necessarily choose these constants to be zero or unitary because the proposition above is purely local. The multilinearity of the accumulation function A might possibly help in strengthening the result. Nevertheless, it would seem unlikely that one could always reduce the number of nonzero and nonunitary edges in $\mathcal{C}$ to $\dim(A(\mathcal{C}))$ without changing the structure of the graph, for example, by introducing some new edges. Still, we have that on some neighborhood of every $C \in \mathcal{C}$ there exists a transformation $P : \mathcal{C} \to \tilde{\mathcal{C}}$ such that $\tilde{\mathcal{C}}$ has only $\dim(A(\mathcal{C}))$ nonzero and nonunitary edges and $A(\mathcal{C}) = A(P(\mathcal{C}))$ locally. If such a transformation P can be found, the result $P(C)$ would be an optimal representation of the Jacobian $F'(x)$ in terms of both floating point operations and storage.

Scarcity-Preserving Simplifications

In our quest for the optimal transformation P we consider some local implementations that are guaranteed to preserve the Jacobian set $A(\mathcal{C})$ while reducing the number of free edges. The danger in such local transformations is that structure is lost in that $A(P(\mathcal{C}))$ might become a proper superset of $A(\mathcal{C})$. For example, this would be the case in the lion example if we were to eliminate all intermediate edges to arrive at $P(C) \equiv A(C)$, which contains $m\,n = 8$ nonzero and nonunitary edges compared to $\dim(\mathcal{C}) = 6$ and $\dim(A(\mathcal{C})) = 5$. To avoid these redundancies we must be more cautious in transforming the graph. Table 10.3 is a list of six modifications of a particular edge $(j, i) \in \mathcal{E}$. They are displayed graphically in Fig. 10.10.

Table 10.3: Modifications of a Particular Edge $(j, i) \in \mathcal{E}$

Edge-Elimination at front

Delete (j, i) from $\mathcal{E}$ after incrementation
$c_{hj} \mathrel{+}= c_{hi} \cdot c_{ij}$ for $h \succ i$.

Edge-Elimination at back

Delete (j, i) from $\mathcal{E}$ after incrementation
$c_{ik} \mathrel{+}= c_{ij} \cdot c_{jk}$ for $k \prec j$.

Edge-Normalization forward

With $\gamma = c_{ij} \neq 0$ adjust
$c_{hi} \mathrel{*}= \gamma$ for $h \succ i$
$c_{ik} \mathrel{/}= \gamma$ for $k \prec i$.

Edge-Normalization backward

With $\gamma = c_{ij} \neq 0$ adjust
$c_{jk} \mathrel{*}= \gamma$ for $k \prec j$
$c_{hj} \mathrel{/}= \gamma$ for $h \succ j$.

Edge-Prerouting

Delete (j, i) from $\mathcal{E}$ after setting for some pivot edge (k, i) with $c_{ik} \neq 0$
$c_{kj} \mathrel{+}= \gamma \equiv c_{ij}/c_{ik}$ and
$c_{hj} \mathrel{-}= c_{hk} \cdot \gamma$ for $h \succ k,\ h \neq i$.

Edge-Postrouting

Delete (j, i) from $\mathcal{E}$ after setting for some pivot edge (j, h) with $c_{hi} \neq 0$
$c_{ih} \mathrel{+}= \gamma \equiv c_{ij}/c_{hj}$ and
$c_{ik} \mathrel{-}= c_{hk} \cdot \gamma$ for $k \prec h,\ k \neq j$.

As was shown for edge-elimination at front and at back in Corollary 9.1 the local modifications of a computational graph $\mathcal{G}$ and the associated Hessenberg matrix C, to a new pair $\tilde{\mathcal{G}}$ and $\tilde{C}$, ensure that $A(C) = A(\tilde{C})$. All of them make at least one edge attain a special value in $\{-1, 0, 1\}$, but the modification may result in other edges becoming nonspecial even when they were zero before. Hence the graph $\tilde{\mathcal{G}}$ need not necessarily be simpler than the original $\mathcal{G}$ in a natural sense. The dashed lines represent edges that may or may not have been present beforehand and are thus newly introduced or merely altered in value, respectively. The resulting edges are labeled with $+$ and $-$ depending on whether they were incremented or decremented. Correspondingly the edges that are multiplied or divided by c_{ij} during the normalization are labeled by $*$ and $\div$, respectively. In our discussion of Jacobian accumulation strategies in Chapter 8 no attention was paid to the issue of scarcity, i.e., whether or not the Jacobian set was enlarged through these modifications. At first one might think that the Jacobian set is maintained whenever the number of free, i.e., nonspecial, edges does not grow. This need not be the case but we have the following result.

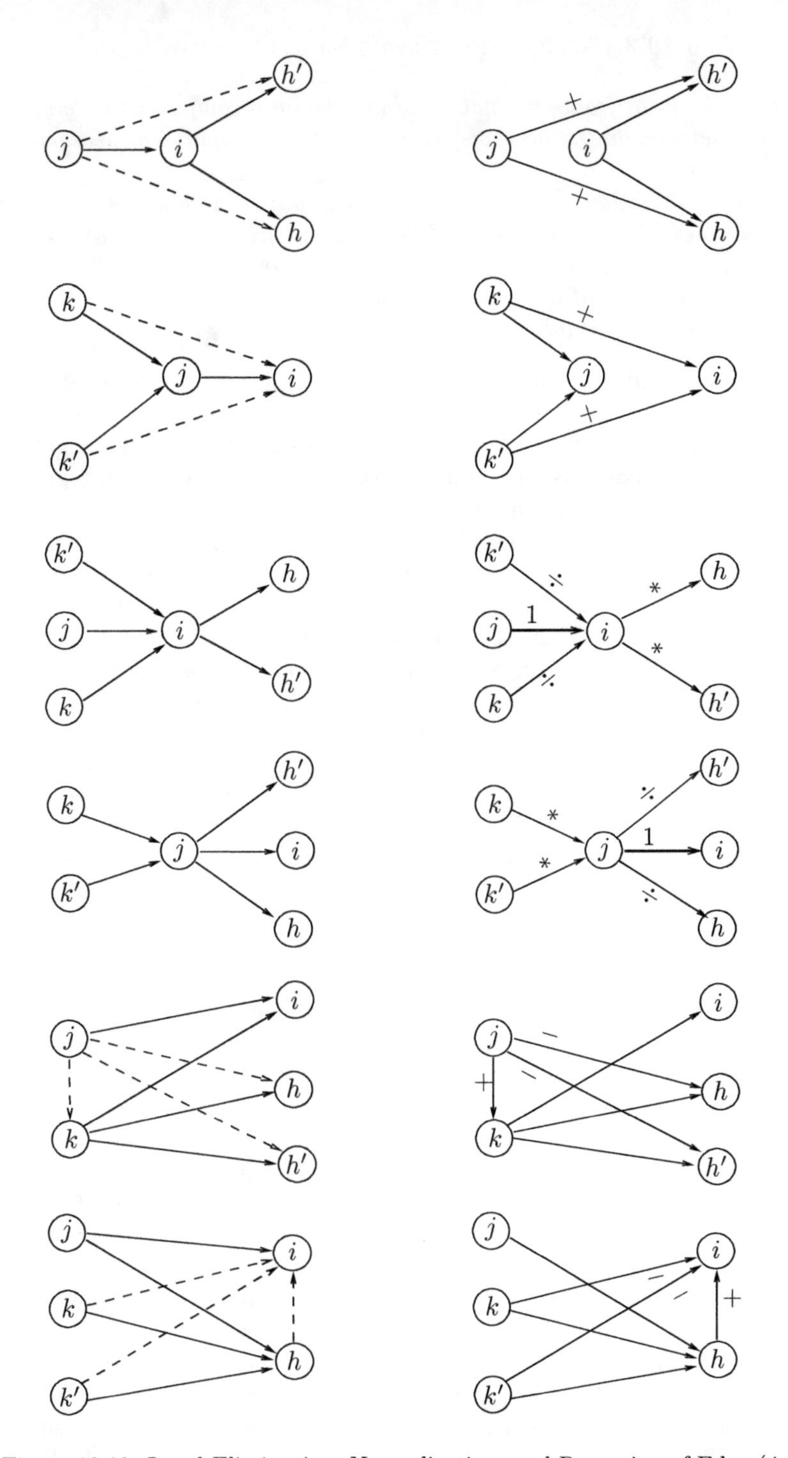

Figure 10.10: Local Elimination, Normalization, and Rerouting of Edge (j, i)

Proposition 10.3 (Scarcity-Preserving Modifications)

(i) *If the back- or front-elimination of an edge in a computational graph does not increase the number of free edges, the Jacobian dimension remains constant.*

(ii) *If the elimination of a vertex would lead to a reduction in the number of free edges, then at least one of the incident free edges can be eliminated via* (i) *without an increase in the Jacobian dimension.*

(iii) *If the rerouting of an edge via a pivot does not increase the number of nonunitary edges, the Jacobian dimension remains constant.*

Proof. The key idea in this proof is based on reconstructing suitable arc values of the graph prior to elimination from the edge value of the graph obtained afterwards. In that way we can ensure that the Jacobian set of the graph does not grow. By hypothesis the number of degrees of freedom in the graph, i.e., the number of free arcs, cannot go up. To prove (i) let us first suppose that the edge (j, i) is free and that it is being back-eliminated. Then there will afterwards be free edges connecting all predecessors $k \prec j$ of its origin j to its destination i. Suppose that j has another successor so that the edges (k, j) cannot be eliminated. Then at most one new free edge may be generated as fill-in replacing (j, i) in the total count. Given such an a posteriori value $\tilde{c}_{ik}$ we can reconstruct the eliminated arc value $c_{i\,j} = \tilde{c}_{i\,k}/\tilde{c}_{j\,k}$ assuming that $\tilde{c}_{j\,k} \equiv c_{j\,k} \neq 0$. Then all other arc values can be reconstructed as $c_{i\,k} = \tilde{c}_{i\,k} - c_{i\,j} * c_{j\,k}$. If on the other hand j has no other successor we can simply set $c_{i\,j} = 1$ and $c_{j\,k} = \tilde{c}_{j\,k}$. Finally if (j, i) is not free and thus has the value ± 1, there may be no fill-in and we can set either $c_{i\,k} = \tilde{c}_{i\,k} \mp c_{j\,k}$ or $c_{j\,k} = \pm\tilde{c}_{j\,k}$ depending on whether j has another successor or not. The proof of (ii) and (iii) is left as an exercise. ∎

Obviously normalization does not change the Jacobian set, and it was therefore not mentioned in the proposition. The application of Proposition 10.3 (iii) to the 3×3 subgraph depicted in Fig. 10.11 shows that there are $6 + 5$ edges in the original graph on the left which would be reduced to 9 by the elimination of the central vertex. However, this modification would destroy the property that the leading 2×2 matrix is singular, a scarcity feature that is maintained by the elimination of the two co-incident edges (j, i_2) and (k_2, j).

On the lion example the transformation from $\mathcal{G}$ to $\tilde{\mathcal{G}}$ as displayed in Figs. 9.4 and 10.9 can be interpreted in terms of two scarcity-preserving transformations, namely, the front-elimination of (1,2) and thus the elimination of vertex ②. The final normalization of $(-1, 1)$ or any other nonzero-valued edge yields the minimal representation $\tilde{\mathcal{G}}$ of Fig. 10.9.

As another example for a successful reduction to a minimal representation we consider the computational graph depicted on the left side of Fig. 10.12. On the right side of Fig. 10.12 we see the result of postrouting $(-2, 2)$ via $(-2, 1)$ assuming $c_{1\,-2} \neq 0$. Whereas the number of free edges stays constant during this modification the subsequent front-elimination of the newly inserted edge $(1, 2)$ yields no fill-in and thus a reduction of the number of free arcs by 1. The result is displayed on the left-hand side of Fig. 10.13. On the right side of Fig. 10.13 we

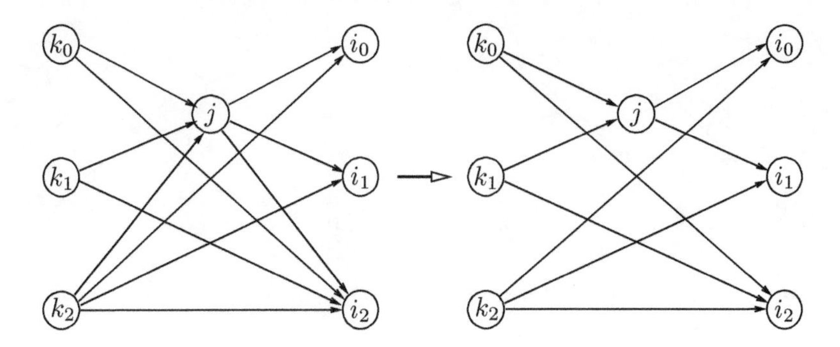

Figure 10.11: Scarcity-Preserving Elimination of (j, i_2) and (k_2, j)

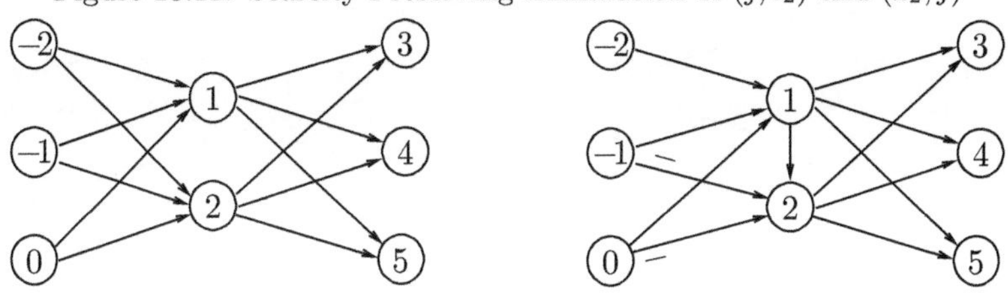

Figure 10.12: Postrouting of $(-2, 2)$ via $(-2, 1)$ to be Followed by Absorption of $(1, 2)$

see the result of prerouting $(2, 3)$ via $(1, 3)$, and its subsequent back-elimination leads to the graph depicted on the left side of Fig. 10.14.

There we still have 10 free edges, a number that can be reduced to 8 by normalizing $(-2, 1)$ and $(2, 4)$ or some other suitable pair of edges. This representation is minimal because the Jacobian set consists of all rank-one matrices in $\mathbb{R}^{3\times3}$, whose Jacobian dimension is clearly 8. What we have computed in effect is some kind of LU factorization for a rank deficient matrix. From the above examples one might gain the impression that the structural property of scarcity always manifests itself as a singularity of the Jacobian or some of its submatrices. This plausible notion is wrong, as can be seen from the following *upwinding* example.

Consider the time evolution of a function $v(t, u, w)$ with (u, w) restricted to the unit square and t to the unit interval. Suppose t, u, and w are discretized with a common increment $h = 1/\tilde{n}$ for some $\tilde{n} > 0$ so that

$$t_k = k\,h\,, \quad u_j = j\,h\,, \quad w_i = i\,h \quad \text{for} \quad 0 \le i, j, k \le \tilde{n}\,.$$

Furthermore we impose periodic boundary conditions in space, i.e.,

$$v(t, u_n, w) = v(t, 1, w) = v(t, 0, w) = v(t, u_0, w)$$

and

$$v(t, u, w_n) = v(t, u, 1) = v(t, u, 0) = v(t, u, w_0)\,.$$

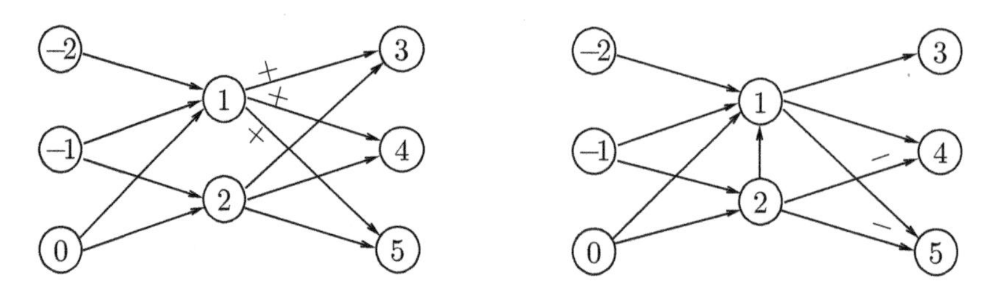

Figure 10.13: Prerouting of $(2, 3)$ via $(1, 4)$ to be Followed by Absorption of $(2, 1)$

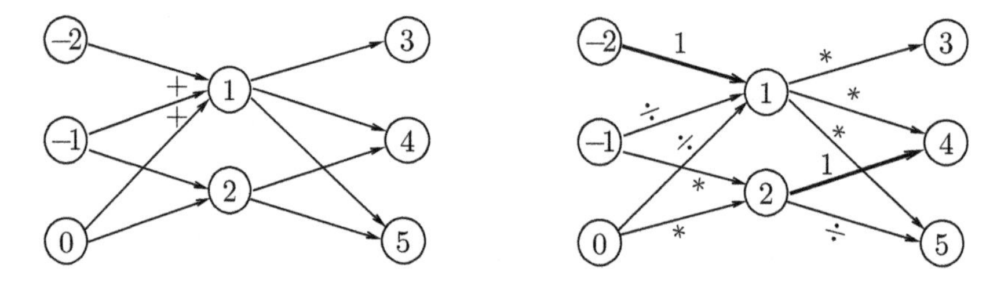

Figure 10.14: Normalization of Edges $(-2, 1)$ and $(2, 4)$

Now suppose the underlying evaluation equation can be discretized such that the approximations

$$v_{k,j,i} \approx v(t_k, u_j, w_i)$$

satisfy a difference equation of the form

$$u_{k+1,j,i} = f_h(t_k, u_j, w_i, v_{k,j,i}, v_{k,j-1,i}, v_{k,j,i-1}) \; .$$

Here $v_{k-1,i} \equiv v_{k,\tilde{n}-1,i}$ and $v_{k,j-1} = v_{k,j,\tilde{n}-1}$. In other words the new value $v_{k+1,j,i}$ depends on the old values of v at the same grid point as well as its immediate neighbors to the west (left), south (underneath), and southwest. The dependence between new and old values at position $p = (i-1)*3+j$ is depicted in Fig. 10.15 for $\tilde{n} = 3$.

Considering the initial values $\left(v_{0,j,i}\right)_{j=1\ldots\tilde{n},i=1\ldots\tilde{n}}$ as independents and the final $m = \tilde{n}^2$ values $\left(v_{\tilde{n}-1,j,i}\right)_{j=1\ldots\tilde{n},i=1\ldots\tilde{n}}$ as dependent variables, we obtain a directed acyclic graph with $\tilde{n}^3$ vertices and $\dim(\mathcal{C}) = 4\tilde{n}^2(\tilde{n}-1)$ edges. Since each independent vertex is connected by a path to each dependent vertex, all $\tilde{n}^4$ entries of the Jacobian are nonzero. Hence there is no sparsity and we have $\dim(\mathcal{A}) = \tilde{n}^4$. However, this number is $\frac{1}{4}\tilde{n}^2/(\tilde{n}-1)$ times larger than $\dim(\mathcal{C}) \geq \dim(A(\mathcal{C}))$, so that we have

$$\text{scarce}(\mathcal{G}) \geq \tilde{n}^4 - 4\tilde{n}^2(\tilde{n}-1) = \tilde{n}^2\left[\tilde{n}-2\right]^2 \quad .$$

Therefore, the Jacobian is scarce for all $\tilde{n} \geq 3$. On closer examination one finds that none of its minors has a vanishing determinant.

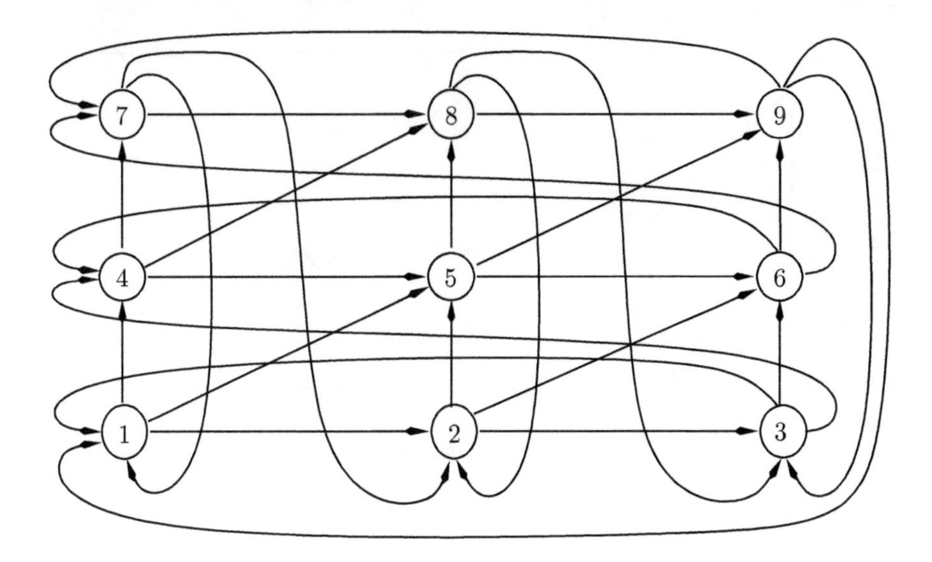

Figure 10.15: Update Dependences on the Upwinding Example

Moreover, it is quite easy to see that in general one incoming edge per vertex can be normalized to 1. For the considered example, this can be done quite naturally with the quarter of the edges that connect the new to the old value of v at the same location (u_j, w_i). Thus a minimal Jacobian representation contains at most $3\tilde{n}^2(\tilde{n}-1)$ floating point numbers compared to the count of $\tilde{n}^2$ for the accumulated matrix representation.

It was shown in the study [LU08] that the benefits of scarcity-preserving elimination strategies are not purely theoretical. Fig. 10.16 illustrates the number of edges in a Jacobian graph from a fluid dynamics kernel. The initial number of edges is 419 and the number of nonzero elements in the fully accumulated Jacobian is 271. Up to the dashed vertical line only scarcity-preserving edge-eliminations are applied, which lead to a minimum of 192 edges after 326 elimination steps. Hence on average each elimination reduces the total number of edges by less than one. After the minimum is reached, further non-scarcity-preserving elimination steps lead to a temporary increase in the number of edges, then their number drops down again and finally reaches 271. The circles represent *refills* where edges that had been eliminated already reoccur through the elimination of neighbors.

If we keep the partially accumulated graph structure obtained after 326 steps, each Jacobian vector or transposed Jacobian vector product will require only about 70 percent of the number of multiplications needed for multiplying them with the fully accumulated Jacobian.

On other test problems Lyons and Utke observed a ratio of 3 between the final and minimal edge count, which represents a savings of 66 percent for matrix-vector products. In terms of nonunitary edges the reduction was almost 4. Interestingly a small number of reroutings appeared to be instrumental

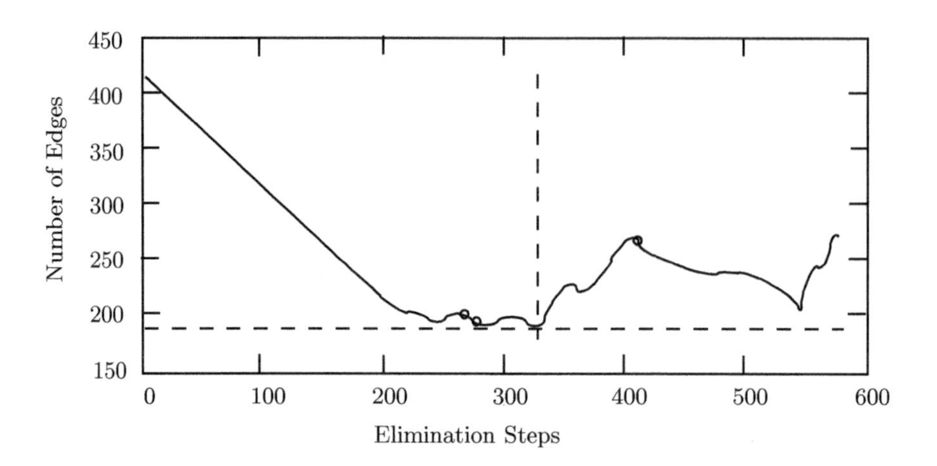

Figure 10.16: Number of Edges during Graph Simplification

in achieving these very encouraging results. Whether or not these numbers are actually the best obtainable cannot be determined as of now as we have no theoretical results of how to obtain or verify an absolutely minimal Jacobian graph representation.

10.4 Hessians and Their Computational Graph

It seems obvious that special considerations should apply when the vector function F in question is in fact the gradient ∇f of a scalar-valued function $f : \mathbb{R}^n \longmapsto \mathbb{R}$. Then we must have $F \equiv \nabla f : \mathbb{R}^n \longmapsto \mathbb{R}^n$ and the Jacobian $F'(x)$ is the symmetric Hessian $\nabla^2 f(x)$, provided $f(x)$ is at least twice continuously differentiable. On simply connected components of the domain, the converse is also true in that symmetry of the Jacobian $F'(x)$ implies the existence of an $f(x)$ with $F(x) = \nabla f(x)$. Some of the observations made below are due to Dixon [Dix91].

Suppose $f(x)$ is given as an evaluation procedure and we wish to compute its Hessian in a cross-country manner to exploit structure as much as possible. Since all methods for computing the gradients are derived from the evaluation procedure for f, we should not need to augment the computational graph of f with additional structural information. Also, during the elimination of intermediate nodes the resulting graphs should maintain a certain symmetry property that shows them representing symmetric Hessians rather than general Jacobians. All these desirable goals can be achieved, as we demonstrate by example.

Consider the simple scalar function

$$y = f(x) \equiv \exp(x_1) * \sin(x_1 + x_2) , \tag{10.6}$$

Table 10.4: Nonincremental Reverse for (10.6)

v_{-1}	$=$	x_1
v_0	$=$	x_2
v_1	$=$	$v_{-1} + v_0$
v_2	$=$	$\exp(v_{-1})$
v_3	$=$	$\sin(v_1)$
v_4	$=$	$v_3 * v_2$
y	$=$	v_4
$\bar{v}_4$	$=$	$\bar{y}$
$\bar{v}_3$	$=$	$\bar{v}_4 * v_2$
$\bar{v}_2$	$=$	$\bar{v}_4 * v_3$
$\bar{v}_1$	$=$	$\bar{v}_3 * \cos(v_1)$
$\bar{v}_0$	$=$	$\bar{v}_1$
$\bar{v}_{-1}$	$=$	$\bar{v}_1 + \bar{v}_2 * \exp(v_{-1})$
$\bar{x}_2$	$=$	$\bar{v}_0$
$\bar{x}_1$	$=$	$\bar{v}_{-1}$

which can be evaluated by the upper half of the procedure listed in Table 10.4. The lower half of that table represents the nonincremental reverse mode, whose general form was listed in Table 3.6.

One might be tempted to replace the factor $\exp(v_{-1})$ in the last nontrivial assignment of Table 10.4 by v_2. However, that algebraic simplification would destroy the symmetry of the corresponding computational graph displayed in Fig. 10.17. Here we have labeled the nodes representing the intermediate adjoints $\bar{v}_i$ simply by $\bar{\imath}$ so that $v_{\bar{\imath}} \equiv \bar{v}_i$. All arcs have been labeled by the corresponding elementary partials $\partial v_i/\partial v_j$, $\partial v_{\bar{\imath}}/\partial v_{\bar{\jmath}}$, or $\partial v_{\bar{\imath}}/\partial v_j$. Only the last represents genuine second derivatives, as we will see below.

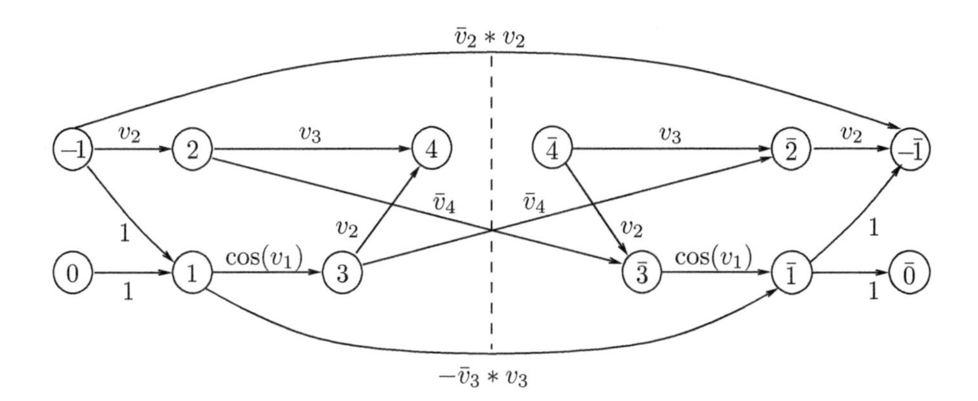

Figure 10.17: Symmetric Graph Corresponding to Table 10.4

The left half of Fig. 10.17 represents the upper half of Table 10.4 and is therefore simply the computational graph of f itself with all "internal" arc values $\partial v_i/\partial v_j$ being elemental partials of first order. The corresponding adjoint nodes form a mirror image in the right half of Fig. 10.17, with all "internal" arcs having the same values

$$\partial v_{\bar{\jmath}}/\partial v_{\bar{\imath}} \;=\; \partial v_i/\partial v_j \;=\; c_{ij} \tag{10.7}$$

if $j \prec i$ but with the opposite orientation. Finally, there are some arcs crossing the central dashed line whose values are of the form $\partial v_{\bar{k}}/\partial v_i$. Their exact values can be expressed as follows. In the nonincremental form each adjoint value $\bar{v}_k$ is calculated as

$$v_{\bar{k}} \equiv \bar{v}_k \equiv \sum_{j \succ k} \bar{v}_j \, c_{jk} = \sum_{j \succ k} \bar{v}_j \, \frac{\partial}{\partial v_k} \varphi_j \big(v_i\big)_{i \prec j} \, . \tag{10.8}$$

In drawing the symmetric computational graph we must consider the complicated expression on the right-hand side of (10.8) as a single elemental function. Consequently, we have the data dependence relations

$$\bar{j} \prec \bar{k} \iff j \succ k$$

and

$$i \prec \bar{k} \iff \frac{\partial^2 \varphi_j}{\partial v_k \partial v_i} \neq 0 \quad \text{for some} \quad j \, .$$

The last relation is again symmetric in that clearly

$$i \prec \bar{k} \iff k \prec \bar{\imath} \, .$$

The corresponding arc values are equal to the sums

$$\frac{\partial \bar{v}_k}{\partial v_i} = \sum_{k \prec j \succ i} \bar{v}_j \, \frac{\partial^2 \varphi_j}{\partial v_k \partial v_i} = \frac{\partial \bar{v}_i}{\partial v_k} \, .$$

Hence we have shown that the computational graph generated by the nonincremental reverse mode has the following property.

Definition (SG): SYMMETRIC COMPUTATIONAL GRAPH
A linearized computational graph is called symmetric if there exists on its vertex set an involution $i \to \bar{\imath}$, with $\bar{\imath} \neq i = \bar{\bar{\imath}}$ and

$$j \prec i \quad \iff \quad \bar{j} \succ \bar{\imath} \quad \Longrightarrow \quad c_{ij} \; = \; c_{\bar{j}\bar{\imath}} \, ,$$

where c_{ij} is the value of the arc (j, i).

This property is sufficient but not necessary for symmetry of the resulting Jacobian, as stated below.

Lemma 10.1 (GRAPH SYMMETRY IMPLIES MATRIX SYMMETRY)
Between the independent (minimal) and dependent (maximal) nodes of a symmetric computational graph, one can find a one-to-one correspondence that makes the resulting Jacobian symmetric.

Proof. See Exercise 10.6. ∎

In the remainder of this section we will assume for simplicity that in the original code for f, multiplication is the only nonlinear binary operation. This convention requires in particular that division be split into a reciprocal followed by a multiplication. Then all arcs $(i, \bar{k})$ that cross the symmetry line of the Hessian graph are made up of the contributions

$$
\begin{array}{lll}
\bullet \quad c_{\bar{k}i} \mathrel{+}= \bar{v}_j & \text{when } v_j = v_k * v_i & \text{for some } j \succ k, i \\
\text{or} & & \\
\bullet \quad c_{\bar{k}k} \mathrel{+}= \bar{v}_j \, \varphi_j''(v_k) & \text{when } v_j = \varphi_j(v_k) & \text{for some } j \succ k\,.
\end{array}
\tag{10.9}
$$

Here the $+=$ indicates that several such contributions may be added to each other. For example, some v_k may occur as factors in a multiplication and as arguments of a nonlinear elemental φ_j. In the example graph of Fig. 10.17, such superpositions do not occur, and the crossing arcs are one pair $c_{\bar{2},3} = c_{\bar{3},2} = \bar{v}_4$ generated by the multiplication $v_4 = v_3 * v_2$ and two separate arcs $c_{-\bar{1},-1} = \bar{v}_2 * v_2$ and $c_{\bar{1},1} = -\bar{v}_3 * v_3$ stemming from the univariate elementals $v_2 = \exp(v_{-1})$ and $v_3 = \sin(v_1)$, respectively. Like all linear operations the addition $v_1 = v_0 + v_{-1}$ does not generate any crossing arcs.

By replicating intermediates if necessary one can always rewrite f such that each v_k occurs at most once as an argument of a multiplication or nonlinear univariate function. Moreover, if v_k occurs as the operand of a multiplication we may assume that the other operand is either v_{k-1} or v_{k+1}, which can be achieved by renumbering or making more replicas. Now let the $(l-m+n) \times n$ matrix $\dot{Z} \equiv (\dot{V}_i^\top)_{i=1-n}^{l-m}$ denote again, as in section 9.1, the matrix formed by the gradients $\dot{V}_i^\top = \nabla_x v_i$ of all intermediates v_i with respect to the independents $x \in \mathbb{R}^n$. It can be computed during a forward sweep to calculate the gradient $\nabla f = \nabla_x f$. Then the Hessian takes the product form

$$
\nabla^2 f \;=\; \dot{Z}^\top D \dot{Z} \in \mathbb{R}^{n \times n}\,, \tag{10.10}
$$

where $D \in \mathbb{R}^{(l-m+n)\times(l-m+n)}$ is zero except for diagonal elements of the form $\bar{v}_j \varphi_j''$ and diagonal blocks of the form $\left(\begin{smallmatrix} 0 & \bar{v}_j \\ \bar{v}_j & 0 \end{smallmatrix}\right)$. For our example problem we have $l - m + n = 4 - 1 + 2 = 5$ and

$$
D = \begin{bmatrix} \bar{v}_2 v_2 & 0 & 0 & 0 & 0 \\ 0 & 0 & 0 & 0 & 0 \\ 0 & 0 & -\bar{v}_3 v_3 & 0 & 0 \\ 0 & 0 & 0 & 0 & \bar{v}_4 \\ 0 & 0 & 0 & \bar{v}_4 & 0 \end{bmatrix}, \quad \dot{Z} = \begin{bmatrix} 1 & 0 \\ 0 & 1 \\ 1 & 1 \\ v_2 & 0 \\ \cos(v_1) & \cos(v_1) \end{bmatrix}.
$$

Multiplying the factors together, we obtain the sum of outer products

$$
\nabla^2 f = \begin{pmatrix} 1 \\ 0 \end{pmatrix} \bar{v}_2 v_2 (1,0) - \begin{pmatrix} 1 \\ 1 \end{pmatrix} \bar{v}_3 v_3 (1,1) + \bar{v}_4 \cos(v_1) v_2 \left[\begin{pmatrix} 1 \\ 0 \end{pmatrix} (1,1) + \begin{pmatrix} 1 \\ 1 \end{pmatrix} (1,0) \right].
$$

This "dyadic" representation was used extensively by Jackson and McCormic [JM88], who referred to functions defined by evaluation procedures as "factorable." Their recommendation to use this structure in matrix factorizations and other tasks in optimization calculations has apparently not been adopted and implemented by other researchers. Setting $\bar{y} = 1$, we obtain from the lower part of Table 10.4 the adjoint values $\bar{v}_4 = 1$, $\bar{v}_3 = v_2$, $\bar{v}_2 = v_3$, and

$\bar{v}_0 = v_2 \cos(v_1)$. Substituting these values in the dyadic representation above, we obtain the correct Hessian

$$\begin{aligned} \nabla^2 f(x) &= \begin{bmatrix} v_3 v_2 - v_2 v_3 + 2 v_2 \cos(v_1), & -v_2 v_3 + v_2 \cos(v_1) \\ -v_2 v_3 + v_2 \cos(v_1), & -v_2 v_3 \end{bmatrix} \\ &= \exp(x_1) \begin{bmatrix} 2\cos(x_1 + x_2), & \cos(x_1 + x_2) - \sin(x_1 + x_2) \\ \cos(x_1 + x_2) - \sin(x_1 + x_2), & -\sin(x_1 + x_2) \end{bmatrix} . \end{aligned}$$

The product form (10.10) may be interpreted as one particular way of performing edge-elimination on the Hessian graph. To obtain $\dot{Z}$ all first-order edges c_{ij} with $j > 0$ are eliminated in the linearized graph for f. The same modification applies to its mirror image on the right. The results of this first accumulation are the arcs $(j-n, i)$ connecting independents directly to intermediates and their adjoint mirror images $(\bar{\imath}, \overline{\jmath - n})$ connecting intermediates directly to dependents with the identical values $c_{i,j-n} = c_{\overline{\jmath-n},\bar{\imath}}$. These first-order derivatives form the nonzero elements of the matrix $\dot{Z}$. All other remaining arcs are crossing from left to right with the second-order values $c_{\bar{k},i} = \bar{v}_j$ or $c_{\bar{k},k} = \bar{v}_j \cdot \varphi''_j$ for some j. Applying vertex-elimination to both endpoints of each crossing arc we obtain one dyadic contribution to the Hessian $\nabla^2 f(x)$. In general this will not be the most efficient way to accumulate Jacobians and, as one can see in the example above, there may be some cancellations.

In general one would probably want to maintain the symmetry of the Hessian graph by always eliminating vertices or edges simultaneously with their mirror images. Then only half of the graph needs to be kept and updated in storage. Recently Hovland et al. [BB+08] have explored and verified the benefits of exploiting symmetry in Hessian accumulation. Hence there is now empirical evidence for the following rule.

Rule 16

THE CALCULATION OF GRADIENTS BY NONINCREMENTAL REVERSE MAKES THE CORRESPONDING COMPUTATIONAL GRAPH SYMMETRIC, A PROPERTY THAT SHOULD BE EXPLOITED AND MAINTAINED IN ACCUMULATING HESSIANS.

While we observed in Chapter 3 that code for the nonincremental reverse is difficult to generate at precompile-time, all the necessary dependence information is available when we have the computational graph for f in store.

Summary and Conclusion

There is plenty of room for investigation and experimentation with regards to heuristics that are efficient and cheap to implement. Since greedy Markowitz is shown to be rather inefficient on a discretized evolution model problem in several space dimensions, alternative pivoting strategies must be developed. In section 10.1 we considered a few, but no conclusive winner has emerged so far. The discussed discretized evolution models are the only class of problems where a lower bound on the number of multiplications needed in the accumulation process appears to be known. It seems clear that certain preeliminations should be performed a priori and the local preelimination strategies discussed in section 10.2 should be incorporated into the next generation of AD tools. Even these local variations on the forward and reverse modes can generate significant savings in runtime and storage. We are convinced that many more improvements are possible.

In many applications Jacobians and Hessians are not ends in themselves but only means for computing Newton steps or solving other linear systems. As shown in several papers [DB89, Gri90, Ver99, Hos97], this goal can often be achieved more efficiently and directly without explicitly accumulating these derivatives matrices. Also, the accumulation process may destroy structural information that is still present in the linearized computational graph. For example, one may deduce the structural rank and even the size of Jordan blocks in dynamical systems directly from the computational graphs [RR99]. One may then ask which graph simplifications leave these structural properties intact and search for a corresponding minimal representation. One structural property that we have examined in section 10.3 is scarcity as represented by the set of Jacobians obtainable from the linearized computational graph. Here we allowed for perturbations of the elemental partials, as we did similarly in our backward error analysis in section 3.4. It remains to seen whether our collection of scarcity-preserving eliminations is large enough to always lead to a minimal representation.

10.5 Examples and Exercises

Exercise 10.1 (*Proof of Proposition* 10.1)
Let c_{ik} denote the number of multiplications needed to multiply the factors T_{l-j} for $j = i \ldots k$ in some chosen order. Then we must have for some particular index $j \in [i+1, k-1]$

$$c_{ik} = c_{ij} + c_{jk} + n\,(1 \,+\, 2(j \,-\, i))(1 \,+\, 2(k \,-\, j))\,,$$

since the partial products have $(j-i)$ and $(k-j)$ tridiagonal factors and thus a bandwidth of $1+2(j-i)$ and $1+2(k-j)$, respectively. Derive the corresponding recurrence relation for the reduced cost

$$r_{ik} \equiv (1 + c_{ik}/n)/2 - (k-i+1)\,(k-i)\,,$$

and show that $0 \le r_{ik} \le (k-i+1)\,(k-i)/2$ by induction on the product length $k-i$. Rewrite this result in terms of c_{ik} to complete the proof of the lemma.

Exercise 10.2 (*Rosenbrock Function*)
For the classical optimization test function

$$f(x_1, x_2) = 100(x_2 - x_1^2)^2 + (1 - x_1)^2$$

compute its gradient $F(x_1, x_2) \equiv \nabla f(x_1, x_2)$ symbolically.

a. Write a corresponding evaluation procedure and draw its computational graph.

b. Compare the number of arithmetic operations needed to accumulate the Jacobian of $F = \nabla f$, which is also the Hessian of f by node-elimination using the forward, reverse, or Markowitz ordering.

c. Verify that the three orderings considered in part **b** yield the same result by accumulating the Jacobian symbolically.

Exercise 10.3 (*Long-Term Evolution*)
Reconsider the case of a one-dimensional evolution discussed before and in Proposition 10.1 assuming that $l \ll n$. Show that the complexity for all accumulation methods is still the same up to a constant and grows linearly with l. Compare the slopes for forward, reverse, and greedy Markowitz elimination.

Exercise 10.4 (*Multidimensional Evolutions*)
Adopt the development after Proposition 10.1 to the 27-point stencil in three dimensions and show that the orders of magnitude remain the same if one uses the 5- and 7-point stencils, in two and three dimensions, respectively.

Exercise 10.5 (*Comparison of Tiebreakers*)
Given integers $l_i \geq 1$ for $i = 0 \ldots n-1$, consider the "absorption function"

$$f(x) = \prod_{i=0}^{n-1} \left[\psi_i(x_i) \prod_{j=1}^{l_i} x_i \right]$$

with ψ_i some univariate elementals.

```
y = 1
for i = 0...n-1
   y *= ψ_i(x_i)
   for j = 1...l_i
      y *= x_i
```

For the evaluation procedure displayed in the box, draw the computational graph in the case $n = 5$ and $l_i = 2$ for $i = 0 \ldots 4$.

a. Compare the costs of the following four vertex-elimination strategies: forward, reverse, absolute Markowitz, and relative Markowitz. The last two are defined by the requirement that $mark(v_j)$ or $relmark(v_j)$ (see (9.12) and (10.1)) is minimal for the node j being eliminated. Use forward as a tiebreaker when several vertices achieve the minimal absolute or relative Markowitz degree.

b. With n arbitrary and $l_i = 2(i+1)$ develop a general expression for the multiplications required by forward, reverse, and relative Markowitz.

c. On the problems considered in part **b** examine the behavior of absolute Markowitz with forward and reverse as tiebreakers, respectively.

Exercise 10.6 (*Proof of Lemma* 10.1)
To establish the lemma, prove first that i is independent if and only if $\bar{\imath}$ is dependent (maximal). Then show that for i, j independent, any path from i to $\bar{\jmath}$ corresponds to a path from j to $\bar{\imath}$, and both have the same value by Proposition 9.2.

Exercise 10.7 (*Consider Proposition* 10.3)

a. Prove (ii) by a counting argument assuming for simplicity that all edges are free.

b. Prove (iii) assuming that all edges are free.

Exercise 10.8 (*Eliminate 5 Intermediates*)
For the computational graph on the right find a vertex-elimination sequence that accumulates the Jacobian using only 34 multiplications (assuming all arc values are general real numbers).

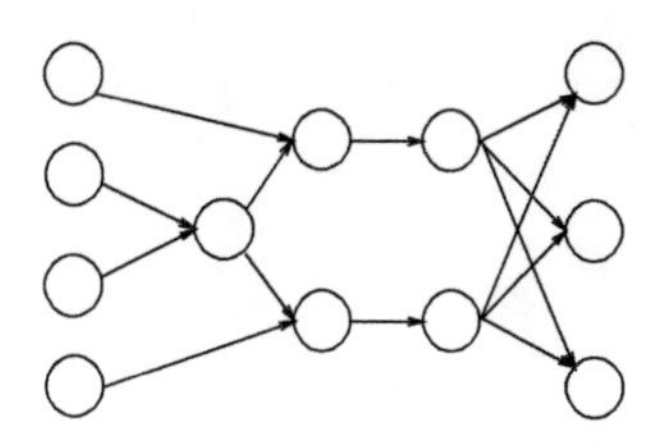

Exercise 10.9 (*Appolonius' Identity*)
Check that for any scaling $\alpha \neq 0$

$$u * w = \left[(u + \alpha w)^2 - (u - \alpha w)^2\right]/(4\alpha)$$

and show that this identity can be used to eliminate the first case in (10.9). Rewrite (10.6) accordingly with $\alpha = 1$ and draw the resulting modification of the symmetric graph displayed in Fig. 10.17.

Chapter 11

Observations on Efficiency

In Chapters 7 and 8 we examined the exploitation of sparsity and other structural properties in evaluating first and second derivatives by variations and combinations of the forward and reverse modes. Chapter 9 discussed some complexity results, whereas in Chapter 10, by allowing the elimination of intermediate vertices in any order, we widened the scope of possible derivative accumulation procedures even further. In this chapter we interpret the results obtained so far and derive recommendations for the users on how to best set up problems for the purposes of AD. In section 11.1 we show that the complexities for Jacobian calculation obtained in Chapters 7–10 are in certain cases up to a small constant optimal. A ramification of this rather elementary observation is that the cheap gradient principle of Rule 8 cannot be generalized to a cheap Jacobian principle. In section 11.2 we examine the exploitation of partial separability. One kind, named value separability, allows us to reduce the maximal domain size $\hat{n}$ toward the average domain size $\bar{n}$. Similarly, the other kind, named argument separability, can be used to reduce the maximal range size $\hat{m}$ toward the average value $\bar{m}$. Either way, the efficiency of the practically advantageous matrix compression approach is increased toward the theoretical efficiency of the dynamically sparse approach, which is hard to realize in practice. In section 11.3 we collect some advice for users of AD methods and software.

11.1 Ramification of the Rank-One Example

Consider the vector function

$$F(x) \equiv b\,\psi(a^\top x) \quad \text{for} \quad a, b, x \in \mathbb{R}^n \quad \text{and} \quad \psi \in C^1(\mathbb{R})\ .$$

This function might be evaluated by the procedure listed in Table 11.1. The index sets $\mathcal{X}_i$ and $\mathcal{Y}_i$ are easily determined as

$$\mathcal{X}_i = \{i\}\ , \quad \mathcal{X}_{n+i} = \{1, 2, \ldots, i\}\ , \quad \mathcal{X}_{2n+i} = \{1, 2, \ldots, n\} \quad \text{for} \quad i = 1 \ldots n$$

Table 11.1: Evaluation Procedure for Rank-One Example

$v_{i-n} = x_i$		$i = 1 \ldots n$
$v_i = a_i * v_{i-n}$		$i = 1 \ldots n$
$v_{n+i} = v_{n+i-1} + v_i$		$i = 1 \ldots n-1$
$v_{2n} = \psi(v_{2n-1})$		
$v_{2n+i} = b_i * v_{2n}$		$i = 1 \ldots n$
$y_i = v_{2n+i}$		$i = 1 \ldots n$

and

$$\mathcal{Y}_{n+i} = \mathcal{Y}_i = \{1, 2, \ldots, n\}\,, \quad \mathcal{Y}_{2n+i} = \{i\} \quad \text{for} \quad i = 1 \ldots n\,.$$

Let us assume that additions and multiplications require the same runtime so that $\pi = 1$ in (4.11).

The total number of additions and multiplications is $3n-1$. Weighting them with the sizes of the $\mathcal{X}_i$, we obtain $n\,1 + (2+3+\cdots+n) + n\,n = 3n^2/2 + O(n)$ runtime units. Weighting them with the sizes of the $\mathcal{Y}_i$, we obtain similarly $n\,n + n\,(n-1) + n\,1 = 2n^2$. With $|\psi| \equiv TIME\{eval(\psi)\}$ expressed in the same units, the separability measures $\bar{n}$ and $\bar{m}$ take the form

$$\bar{n} \equiv \frac{3n^2/2 + n|\psi| + O(n)}{3n - 1 + |\psi|} = n\left[1 - \frac{3n/2 + O(n^0)}{3n - 1 + |\psi|}\right] \tag{11.1}$$

$$\bar{m} \equiv \frac{2n^2 + n|\psi| + O(n)}{3n - 1 + |\psi|} = n\left[1 - \frac{n + O(n^0)}{3n - 1 + |\psi|}\right]. \tag{11.2}$$

Here we have used the fact that $\mathcal{X}_{2n} = \mathcal{Y}_{2n} = \{1, 2, \ldots, n\}$ contains the maximal number of elements n. Obviously we can push both $\bar{n}$ and $\bar{m}$ arbitrarily close to $n = m$ by selecting a univariate elemental ψ whose evaluation takes a lot longer than the total of $2n$ multiplications and n additions needed to compute y. In that situation, we would have

$$\begin{aligned} WORK\{eval_{+j}(F)\} &\approx WORK\{eval(F_i)\} \\ \approx WORK\{eval(F)\} &\approx WORK\{eval(\psi)\}\,. \end{aligned}$$

In other words, the task $eval_{+j}(F)$ of reevaluating $F(x)$ after changing just the component x_j, and the task $eval(F_i)$ of evaluating only the ith component $y_i = F_i$ are nearly as expensive as evaluating the whole of F from scratch. For all three tasks, exactly one evaluation of ψ is required, which may completely dominate the computational cost.

Since $\bar{n} \approx n = m \approx \bar{m}$, the estimation of $F'(x)$ by differencing carries, according to (7.12), the full price tag of an n-fold increase in complexity and runtime. On the other hand, one might expect that evaluating the scalar derivative ψ' costs no more than evaluating ψ. Then we can compile the Jacobian

$$F'(x) = b\,\psi'(a^\top x)a^\top$$

"by hand," using exactly $n + n^2$ multiplications. If ψ and ψ' together are much more expensive than n^2 multiplications, we find that the whole Jacobian can be obtained for no more than twice the cost of evaluating F itself.

Generally, we can expect that analytical methods benefit from the presence of many transcendentals and other expensive elementals φ. Usually, their derivatives $\dot{\varphi}, \bar{\varphi}$, and even $\dot{\bar{\varphi}}$ come quite cheaply once φ itself has been evaluated at the common argument. In contrast, difference quotients require reevaluation of φ at many nearby points without obtaining any savings from the proximity. Multiplications, on the other hand, push up the runtime factors for analytical methods, whereas additions and subtractions are more or less neutral. These considerations apply primarily to the forward mode, which is rather closely related to difference quotients. The reverse mode can easily beat difference quotients even on functions that consist exclusively of multiplications, as is the case for Speelpenning's product (see Exercise 3.6).

The Worst Case

On the other hand, we see from the explicit formula for the Jacobian in the last example that for general vectors a and b, its n^2 elements are all distinct, so that their calculation by any conceivable scheme requires an effort of n^2 arithmetic operations. Now let us switch our attention to a situation where ψ has a runtime of just a few arithmetic units. Then the cost of evaluating F is no less than $3n$ time units, and we observe by inspection of (11.1) and (11.2) that

$$\bar{n} \approx n/2 \quad \text{and} \quad \bar{m} \approx 2n/3 \, .$$

Then the estimates (7.14) and (7.15) for the forward or reverse modes imply

$$\frac{n}{3} = \frac{n^2}{3n} \leq \frac{TIME\{eval(F')\}}{TIME\{eval(F)\}} \leq \frac{\gamma\, n}{2} \, .$$

The left inequality says that for this example, calculating the Jacobian is at least $n/3$ as expensive as evaluating the underlying function. This observation pertains to any conceivable method for computing Jacobians. The right inequality says that our favorite, the forward or reverse mode, exceeds this lower complexity bound by at most the small factor $3\gamma/2$. Therefore, we face the following observation.

RULE 17

THE CHEAP GRADIENT RESULT DOES NOT YIELD CHEAP JACOBIANS AS WELL.

Of course, in other examples, the situation is completely different. One might object that we are making too much of an example that looks rather silly from a practical point of view. The peculiarity that F' always has rank-one can easily be fixed by adding x to $F(x)$ and correspondingly the identity matrix I to $F'(x)$. This modification barely affects our complexity counts and makes solving the system $F(x) + x = w$ for some right-hand side $w \in \mathbb{R}^n$ at least formally a meaningful computational task. To do this by Newton's method we would have to solve a sequence of linear systems whose matrix is a rank-one perturbation of the identity. Rather than forming and factoring the Jacobian, it is advisable to use the Sherman–Morrison–Woodbury formula. In this particular example it yields Newton steps at an expense of $O(n)$ arithmetic operations. This approach has been generalized to methods for directly computing Newton steps without explicitly forming (and even less factoring) the Jacobian [Utk96a].

Summarizing Conclusions

We make the following observations.

- The minimal complexity of Jacobians can be up to $\min\{n, m\}$ times that of the underlying vector function, irrespective of the differentiation methodology employed.

- Common subexpressions reduce the computational complexity of Jacobians less than that of the underlying vector functions. Hence, they tend to increase the runtime ratio between Jacobian and function, rather than being beneficial as one might expect.

- Transcendental intrinsics and other expensive elementals are beneficial for AD compared to differencing. Multiplications are costly to differentiate analytically and additions or subtractions are neutral.

- The degree to which the independents x_i are intertwined during the evaluation of F can be quantified in terms of the average domain size $\bar{n} \leq \hat{n} \leq n$ defined in (7.11). It determines the complexity of Jacobian evaluation in the sparse forward mode, as introduced in section 7.2.

- The degree to which the evaluation of the m dependents y_i relies on shared subexpressions can be quantified in terms of the average range size $\bar{m} \leq \hat{m} \leq m$ defined in (7.13). It determines the complexity of Jacobian evaluation in the sparse reverse mode introduced in section 7.2.

11.2 Remarks on Partial Separability

As we have seen in the preceding sections, the maximal row and column lengths $\hat{n}$ and $\hat{m}$ of the Jacobian determine the cost of its compressed evaluation. Hence, one may wonder whether they can be reduced by rewriting the function $F(x)$ or its evaluation procedure. As it turns out, on many problems such a reduction can be achieved through the exploitation of *partial separability*. There are two kinds of partial separability related to the average domain and range sizes $\bar{n} \leq \hat{n}$ and $\bar{m} \leq \hat{m}$ introduced in section 7.1. They determine the complexity estimate for the dynamically sparse forward and reverse methods, which is, unfortunately, hard to realize because of runtime overheads. By suitably decomposing $F(x)$

we can reduce $\hat{n}$ and $\hat{m}$ toward $\bar{n}$ and $\bar{m}$, so that the costs of the compressed Jacobian evaluation procedures go down correspondingly.

First, we consider the more established concept of partial *value separability,* which was introduced by Griewank and Toint in [GT82] (it was called partial function separability in [Gri93]). This structural property is always implied by sparsity of the Hessian, a fact often not appreciated. It is easy to derive a decomposition that is too fine because too many subfunctions are involved. Our practical aim is to obtain a structure that is likely to be reflected in the evaluation procedure and can be exploited for the cheaper calculation of Jacobian and Hessian alike. First we prove the following basic result.

Lemma 11.1 (Hessian Sparsity Implies Partial Separability)
Suppose $f : D \subset \mathbb{R}^n \longmapsto \mathbb{R}$ is twice continuously differentiable on an open convex domain $\mathcal{D}$ containing the origin. Then we have the following equivalences.

(i) $$\frac{\partial^2 f}{\partial x_i \partial x_j} \equiv 0 \quad \textit{for some pair} \quad i < j \quad \textit{and all} \quad x \in \mathcal{D}$$

$$\Longleftrightarrow$$

$$\begin{aligned} f(x_1, \ldots, x_n) &= f(x_1, \ldots, x_{i-1}, 0, x_{i+1}, \ldots, x_{j-1}, x_j, x_{j+1}, \ldots, x_n) \\ &+ f(x_1, \ldots, x_{i-1}, x_i, x_{i+1}, \ldots, x_{j-1}, 0, x_{j+1}, \ldots, x_n) \\ &- f(x_1, \ldots, x_{i-1}, 0, x_{i+1}, \ldots, x_{j-1}, 0, x_{j+1}, \ldots, x_n) \,. \end{aligned}$$

(ii) $$\frac{\partial^2 f}{\partial x_i^2} \equiv 0 \quad \textit{for some index } i \textit{ and all} \quad x \in \mathcal{D}$$

$$\Longleftrightarrow$$

$$\begin{aligned} f(x_1, \ldots, x_n) &= f(x_1, \ldots, x_{i-1}, 0, x_{i+1}, \ldots x_n) \\ &+ x_i \frac{\partial f}{\partial x_i}(x_1, \ldots, x_{i-1}, 0, x_{i+1}, \ldots x_n) \,. \end{aligned}$$

Proof. Without loss of generality we may assume $i = 1 < 2 = j$ and suppress the dependence on the extra variables $x_3, \ldots, x_n$. Then we have for any point $\bar{x}_1, \bar{x}_2 \in \mathcal{D}$

$$0 = \int_0^{\bar{x}_1} \int_0^{\bar{x}_2} \frac{\partial^2 f(x_1, x_2)}{\partial x_1 \partial x_2} dx_2 \, dx_1 = \int_0^{\bar{x}_1} \frac{\partial f(x_1, x_2)}{\partial x_1}\bigg|_0^{\bar{x}_2} dx_1$$

$$= \int_0^{\bar{x}_1} \frac{\partial f(x_1, \bar{x}_2)}{\partial x_1} dx_1 - \int_0^{\bar{x}_1} \frac{\partial f(x_1, 0)}{\partial x_1} dx_1 = f(x_1, \bar{x}_2)\bigg|_0^{\bar{x}_1} - f(x_1, 0)\bigg|_0^{\bar{x}_1}$$

$$= f(\bar{x}_1, \bar{x}_2) - f(0, \bar{x}_2) - f(\bar{x}_1, 0) + f(0, 0) \,.$$

After replacing $(\bar{x}_1, \bar{x}_2)$ by (x_1, x_2) and bringing $f(x_1, x_2)$ on one side, we have established the first implication "$\Longrightarrow$". The converse follows immediately by differentiation. The second equivalence follows directly from the Taylor expansion of

$$f(x_1, \ldots, x_n) \quad \text{at} \quad (x_1, \ldots, x_{i-1}, 0, x_{i+1}, \ldots, x_n)$$

with respect to x_i. ∎

The first implication in Lemma 11.1 shows that the (i, j) entry in the Hessian can vanish only when $f(x_1, \ldots, x_n)$ is a sum of subfunctions none of which depends jointly on both x_i and x_j. We believe that normally this is the only situation in which a modeler can even recognize that the (i, j) entry of the Hessian vanishes. Note that the negative term can be merged with either one of the other two. The case of a vanishing diagonal element is more complicated because there the decomposition is not strictly additive. Neither splitting destroys any other sparsity, so one may continue to split the subfunctions until their Hessians are essentially dense. The finest possible result of this decomposition is described in the following proposition.

Proposition 11.1 (Decomposition into Dense Subfunctions)
Suppose $f : \mathcal{D} \subset \mathbb{R}^n \longmapsto \mathbb{R}$ is n-times continuously differentiable on an open convex domain $\mathcal{D} \ni 0$. Then there exist index sets

$$I_k,\ J_k \subseteq \{1, 2 \ldots n\} \quad \text{with} \quad I_k \cap J_k = \emptyset \quad \text{and} \quad |I_k| + |J_k| > 0$$

for $k = 1 \ldots s$ and functions $f_k : \mathbb{R}^{|J_k|} \longmapsto \mathbb{R}$ such that at all $x \in \mathcal{D}$

$$f(x) = f(0) + \sum_{k=1}^{s} \prod_{i \in I_k} x_i\, f_k\,(x_j)_{j \in J_k}\,, \tag{11.3}$$

where the Hessian of each f_k is dense.

Proof. The proof is based on repeated splittings of $f(x)$ according to (i) and (ii) of Lemma 11.1. Whenever there is an off-diagonal element that is *essential* in that the corresponding diagonal elements are nonzero, we may split $f(x)$ or its additive components according to (i). Whenever no off-diagonals or only nonessential off-diagonals are left and there is a zero diagonal element, we may split according to (ii). Whenever a variable x_i has been factored out according to (ii), the decomposition process can be continued on the nonlinear remainder $\partial f/\partial x_i$. Splitting according to (i), we may always merge the third, negative term with one of the others so that we obtain a binary tree of subfunctions. Since the number of variables x_j that occur as arguments in any one subfunction is reduced by (at least) one during each split, the tree cannot have a depth greater than n. This procedure also ensures that the differentiation implicit in (ii) needs to be carried out at most n times, which is why f must be assumed to belong to $\mathcal{C}^n$, in the worst case.

The functions f_k corresponding to the leaves of the tree have neither diagonal nor off-diagonal zeros in their Hessians, since nonessential off-diagonals can exist only if there are zero diagonal entries. This completes the proof. ∎

During the repeated splittings that lead (at least theoretically) to the decomposition (11.3), the same index set pairs I_k, J_k can occur many times. Just as in repeated partial differentiation, we may merge terms with identical *profiles* (I_k, J_k) so that each one of them occurs at most once in the final decomposition. Moreover, we may want to go a bit further in combining terms with identical J_k in order to reduce the number s of additive constituents.

Relation to Average Domain Size

Let us suppose, for the moment, that the function $f(x)$ was given in the form (11.3), in the sense that its evaluation procedure first computes the intermediate values f_k and only at the end combines them to form the value $y = f(x)$. Assuming that these final multiplications and additions have a negligible cost, we find for the average domain size of f that

$$\bar{n} \leq \frac{\sum_{k=1}^{s} |J_k| \, TIME\{eval(f_k)\}}{\sum_{k=1}^{s} TIME\{eval(f_k)\}} \leq \max_{1\leq k\leq s} |J_k| \, .$$

Here we have used the fact that the index domains $\mathcal{X}_i$ for all intermediate v_i used to calculate f_k must be contained in J_k. The maximal size of any index set J_k is the maximal row length for the Jacobian of the vector function

$$F(x_1 \ldots x_n) = \left(f_k \left(x_j \right)_{j \in J_k} \right)_{k=1 \ldots s} ,$$

which maps $\mathcal{D} \subseteq \mathbb{R}^n$ into $\mathbb{R}^s$. Hence the rows of $F'(x)$ representing the gradients of the f_k can be computed by column compression (see section 8.3) at roughly $\max_k |J_k|$ times the effort of evaluating F by itself. Since

$$\frac{\partial f(x)}{\partial x_j} = \sum_{k=1}^{s} \prod_{j \neq i \in I_k} x_i \, f_k \left(x_{\tilde{\imath}} \right)_{\tilde{\imath} \in J_k} + \sum_{k=1}^{s} \prod_{\tilde{\imath} \in I_k} x_{\tilde{\imath}} \, \frac{\partial}{\partial x_j} f_k \left(x_{\tilde{\imath}} \right)_{\tilde{\imath} \in J_k} , \tag{11.4}$$

the cost of computing the gradient ∇f from the Jacobian $F'(x)$ may also be considered negligible. In many cases there are no zero diagonal elements, so all I_k are empty, and the formula (11.4) simply reduces to

$$\nabla f(x)^\top = e^\top F'(x) \quad \text{with} \quad e = (1, 1 \ldots 1) \in \mathbb{R}^s \, . \tag{11.5}$$

This relation has been exploited in [BB⁺97] to compute the gradients of scalar objective functions with sparse Hessian in the forward mode. Even though these gradients themselves are dense, $\hat{n} = \hat{n}(F)$, the maximal number of nonzeros in any one of the gradients ∇F_k may be quite small. For functions that arise as discretizations of functionals on infinite-dimensional spaces, $\hat{n}$ is often determined by a local stencil and thus unaffected by grid refinement.

Partial Separability of Computational Graphs

When the original function $F(x)$ is already vector-valued, the splitting process described above can be performed for each dependent component $F_i(x)$

separately. However, rather than discussing this in terms of sparsity of the second derivative tensor $F''(x)$, it seems more appropriate to discuss both kinds of partial separability in terms of the evaluation procedure of $F(x)$. In fact, we introduced partial separability as a consequence of Hessian sparsity mainly because it seems to be widely accepted that large problems are sparse. While Hessian sparsity is a sufficient condition for partial separability, it is by no means necessary. Any quadratic form

$$f(x) = \frac{1}{2} x^\top A x = \frac{1}{2} \sum_i a_{ii}\, x_i^2 + \sum_{i<j} a_{ij}\, x_i x_j$$

is partially value separable, irrespective of whether its Hessian $A = A^\top$ is sparse or not. In terms of the index domains $\mathcal{X}_i$ and the index ranges $\mathcal{Y}_i$, we may define partial separability as follows.

Definition (PS): PARTIAL SEPARABILITY
The function $F : \mathbb{R}^n \longmapsto \mathbb{R}^m$ given by an evaluation procedure is called

(i) *partially value separable if for at least one $y_i = v_{\hat\imath}$ with $\hat\imath = l - m + i$*
- *$\varphi_{\hat\imath}$ is an addition or subtraction,*
- *$|\mathcal{X}_k| < n$ for all $k \prec \hat\imath$;*

(ii) *partially argument separable if for at least one $x_j = v_{j-n}$*
- *$j - n$ has at least two successors,*
- *$|\mathcal{Y}_k| < m$ for all $k \succ j - n$.*

These two kinds of partial separability allow the following function rewrites.

Row Splitting by Value Separability

In case (i), there must be two immediate predecessors v_j and v_k of $v_{\hat\imath}$ so that $y_i = v_{\hat\imath} = v_j \pm v_k$. Then we may split y_i into two new dependents $y_{i-1/2} = v_j$ and $y_{i+1/2} = v_k$ and set

$$\tilde F \equiv \left(y_1,\, y_2, \ldots, y_{i-1},\, y_{i-1/2},\, y_{i+1/2},\, y_{i+1}, \ldots, y_n\right)^\top : \mathbb{R}^n \longmapsto \mathbb{R}^{m+1} .$$

The Jacobian $\tilde F'$ is identical to F' except that its ith row has been split in two. Therefore, we may say that partial value separability of F allows row splitting of its Jacobian. Since $\nabla_x y_i = \nabla_x y_{i-1/2} + \nabla_x y_{i+1/2}$, the maximal row length (of the Jacobian $\tilde F'$) is less than or equal to that of $F'(x)$. On the other hand, the maximal column length is likely to go up by 1 unless $\mathcal{X}_j$ and $\mathcal{X}_k$ are disjoint. Hence we may conclude that row splitting is beneficial for row compression but harmful for column compression.

Obviously the row splitting process can be repeated until, for all dependents $y_i = v_{\hat\imath}$, either $\varphi_{\hat\imath}$ is nonlinear or there is a predecessor $k \prec \hat\imath$ with $|\mathcal{X}_k| = n$. Denoting the completely split vector function by $\tilde F : \mathbb{R}^n \longmapsto \mathbb{R}^{\tilde m}$ with $\tilde m \geq m$, we have the following algebraic relation. There must be a matrix $Q \in \mathbb{R}^{m \times \tilde m}$ such that

$$F(x) = Q\,\tilde F(x) \quad \text{and} \quad F'(x) = Q\,\tilde F'(x) \quad \text{for all} \quad x \in \mathcal{D} . \tag{11.6}$$

Because of the way we defined the splitting process, $Q \in \mathbb{R}^{m\times\tilde{m}}$ is a selection matrix; in other words each one of its $\tilde{m} \geq m$ columns is a Cartesian basis vector in $\mathbb{R}^m$. It merely assigns each component of $\tilde{F}$ to the component of F from which it originated.

The notion of partial value separability can be generalized even more by allowing arbitrary post factors $Q \in \mathbb{R}^{m\times\tilde{m}}$. In other words, one tries to pull out a general linear transformation from the end of the evaluation procedure. This can be done by identifying the earliest intermediates v_k that impact the y_i only linearly and then elevating these v_k to the status of dependent variables. These indices form the set of minimal elements

$$I \equiv \min\{\, i \leq l \,|\, \bar{\mathcal{X}}_i = \emptyset \,\}\,, \tag{11.7}$$

where the index domains $\bar{\mathcal{X}}_i$ are defined by Table 7.4 and satisfy (7.21). The new Jacobian $\tilde{F}'$ has the width

$$\check{n} \equiv \hat{n}(\tilde{F}') = \max_{i\in I}\{|\mathcal{X}_i|\} \;\leq\; \hat{n} = \hat{n}(F')\,. \tag{11.8}$$

Provided the cost for the final multiplication by Q is small, we can expect that $\check{n} \geq \bar{n}$ still.

Column Splitting by Argument Separability

Since row splitting is bad for column compression, it seems natural to attempt column splitting in the partially argument separable case. According to Definition (PS) (ii), all immediate successors x_k of some independent x_j have $|\mathcal{Y}_k| < m$. Then we may introduce a replica $x_j^{(k)}$ of x_j for each $k \succ j - n$ and rewrite $v_k = \varphi_k(\ldots, v_j, \ldots)$ as $v_k = \varphi_k(\ldots, v_j^{(k)}, \ldots)$. Elevating the $v_j^{(k)}$ to the status of independent variables, we obtain a new vector function $\tilde{F} : \mathbb{R}^{\tilde{n}} \longmapsto \mathbb{R}^m$, where $\tilde{n} - n = |\{k \succ j - n\}| - 1$. Since clearly

$$\frac{\partial F}{\partial x_j} = \sum_{k\succ j} \frac{\partial \tilde{F}}{\partial x_j^{(k)}}\,,$$

we see that this time a column of the Jacobian has been split. Consequently, the maximal column length can only go down, whereas the maximal row length is quite likely to go up. This process can be repeated until all x_j have either a single immediate successor or $|\mathcal{Y}_k| = m$ for some $k \succ j - n$. This time we can express the relation between the new function $\tilde{F} : \mathbb{R}^{\tilde{n}} \longmapsto \mathbb{R}^m$ and the original F by

$$F(x) = \tilde{F}(P\,x) \quad \text{and} \quad F'(x) = \tilde{F}'(P\,x)P \quad \text{for all} \quad x \in \mathcal{D}\,.$$

Here $P \in \mathbb{R}^{\tilde{n}\times n}$ is also a selection matrix associating each replica independent with its original. Again one can generalize the concept of partial argument separability to allow for more general matrices $P \in \mathbb{R}^{\tilde{n}\times n}$. Their rows correspond to the set

$$J \equiv \max\{\, i \geq 1 - n \,|\, \dot{\mathcal{X}}_i = \emptyset \,\}\,, \tag{11.9}$$

representing exactly the maximal ones amongst those variables v_i that are linear functions of the independent variables, including the independents themselves. The corresponding gradients $\dot{V}_i$ can be viewed as rows of the matrix P. The tensor of new Jacobian $\tilde{F}'(x)$ has the width

$$\check{m} \equiv \hat{n}(\tilde{F}'(x)^\top) = \max_{j \in J}\{|\mathcal{Y}_j|\} \leq \hat{m} \equiv \hat{n}(F'(x)^\top), \tag{11.10}$$

which will usually be greater than $\bar{m}$.

Expansion to the Nonlinear Core

One can also allow both row splitting and column splitting to arrive at a multiplicative decomposition

$$F(x) = Q\,\tilde{F}(Px) \quad \text{and} \quad F'(x) = Q\,\tilde{F}'(Px)P \quad \text{for} \quad x \in \mathcal{D}\,.$$

Here $\tilde{F} : \mathbb{R}^{\tilde{n}} \longmapsto \mathbb{R}^{\tilde{m}}$ with $\tilde{n} \geq n$ and $\tilde{m} \geq m$ might be called the *nonlinear core* of F, while P and Q can be viewed as linear pre- and postfactors of the Jacobian. The expanded Jacobian $\tilde{F}' \in \mathbb{R}^{\tilde{m} \times \tilde{n}}$ can be expected to be much sparser than $F'(x)$ even though it has at least as many nonzero entries.

A trivial example of the situation is the rank-one function $F(x) = b\,\psi(a^\top x)$ discussed in section 11.1. Here we may set $\tilde{F} = \psi$, $P = a^\top$, and $Q = b$. Then the Jacobian is obtained in the factored form $F'(x) = b\psi'(a^\top x)a^\top$, which is obviously a cheap and much more appropriate representation than the corresponding dense rank-one matrix of order n.

Practical Exploitation of Partial Separability

The combination of value and argument separability discussed here is a generalization of the concept of group partial separability used in the optimization package LANCELOT [CGT92]. There it may be used to more efficiently evaluate or approximate derivatives symbolically or by secant updating methods.

In AD, separability has so far mostly been used to evaluate gradients in the forward mode using (11.5). With the notable exception of Gay's implementation in AMPL [Gay96] the detection of partial separability has so far not been automated in other AD systems, so that the user is expected to provide the function in partially separable form. Of particular importance is the least squares case where

$$f(x) = \sum_{i=1}^{m} F_i(x) = \sum_{i=1}^{m} \tilde{F}_i(x)^2\,,$$

with $\tilde{F}_i$ for $i = 1 \dots m$ being a collection of residuals that need to be minimized. Instead of simply squaring the residuals one may also set $F_i = \psi(\tilde{F}_i)$ with ψ being some usually convex functional like the ones used in robust regression. Provided each $\tilde{F}_i(x)$, and thus each $F_i(x)$, depends only on a few components of x, the Jacobian $F_i'(x)$ is sparse and can be computed cheaply in the forward mode with compression. This rather elementary observation has a significant effect on the balance between the efficiency of the forward and reverse modes with matrix compression. We may refer to $\check{n}$ and $\check{m}$ as defined in (11.8) and (11.10), representing the maximal domain and range size of the nonlinear core as the nonlinear width and height of F, respectively. Then we arrive at the following conclusion.

RULE 18

REVERSE BEATS FORWARD ONLY WHEN THE NONLINEAR WIDTH IS MUCH LARGER THAN THE NONLINEAR HEIGHT.

This rule of thumb can also be applied for second derivatives when they are calculated by the methods discussed in section 8.5 without exploitation of symmetry. The question of what exactly $\check{n} \ll \check{m}$ means can only be answered for particular packages and problems. Unless runtime is critical the forward mode seems a safer bet as long as $\check{n}/\check{m}$ is in the single digits. It usually incurs no problems with storage requirements, whereas current implementations of the reverse mode may confront the user with various unpleasant surprises. Apart from the notorious storage issue and general overhead costs, there is also more uncertainty about the meaning of "adjoint" derivatives in the nonsmooth case.

11.3 Advice on Problem Preparation

When differentiation of several dependent variables is desired with respect to a single independent variable (say, time), the choice and application of the forward, or tangent, mode of AD is comparatively straightforward. On the other hand, when only one dependent variable (say, energy) needs to be differentiated with respect to several independent variables, then the choice and application of the reverse or adjoint mode is quite easy. Nevertheless, to improve efficiency one might consider even in these simple situations some of the suggestions on code preparation discussed in the final section of this chapter.

Our main focus here is the evaluation of derivative matrices, which has turned out to be a considerably more complicated issue than tangents and gradients. However, we would like to stress that this complication arises only through the desire or need to obtain derivatives with more or less optimal efficiency. On very many small- to medium-size problems, evaluating Jacobians or Hessians in the dense forward or reverse mode will work just fine without any hassle to the user.

Identifying Your Needs

The first question users and algorithm designers should examine is whether or not a complete, although possibly sparse, Jacobian or Hessian really needs to be evaluated. Especially on large-scale problems, such matrices may represent an enormous amount of data that might not be utilized very well by the numerical algorithm, at least relative to the cost of obtaining them. Unfortunately, we have not been able to provide a hard-and-fast estimate for the cost of evaluating derivative matrices. Yet it is clear that any restriction to subspaces of the domain or projection onto subspaces of the range of the vector function F must help, except when sparsity is lost in the process. By using seed matrices S or weight matrices W whose columns span these subspaces one can obtain derivatives selectively and thus, in general, more efficiently.

On the other hand, one should be aware that some numerical methods avoid explicit derivative evaluations, especially of higher order, at almost any cost. The price for this outdated design principle may, for example, come due in low solution accuracy, slow convergence, or shaky robustness. Now, not only second derivatives, which we already considered, but, as we will see in Chapter 13, higher derivatives as well can be obtained accurately and at a reasonable cost, provided they are sparse or can be restricted to subspaces of moderate dimension. Some numerical software tools have an option for the user to provide first or second derivatives. This option should be selected when available.

Whenever possible, one should provide fixed upper bounds on array sizes and iteration counters. This is especially important for the reverse mode, where temporary storage requirements might otherwise become excessive. For this reason it is also crucial that dynamically allocated objects be deallocated as soon as possible, preferably in a LIFO fashion.

Finally, when smoothness is a concern, one might want to consider the discussion at the end of Chapter 14.

Selecting the Mode

For the evaluation of (possibly) reduced Jacobians we recommend the following rule of thumb. Use the forward mode whenever $\hat{n}(F'(x)) \leq n$ and the maximal number of nonzeros per row is not much larger than $\hat{m}(F'(x)) \leq m$, the maximal number of nonzeros per column. In many applications one will have fairly good estimates or at least upper bounds regarding the size of $\hat{n}$ and $\hat{m}$. Moreover, one may be able to specify the sparsity pattern by hand, which is typically also required by sparse linear algebra routines. Failing this, one may employ ADIFORs SparsLinC, ADOL-C, or a similar tool to determine the sparsity pattern algorithmically.

Using one of the coloring algorithms discussed in [GMP05], one can subsequently determine coloring numbers p and q for the row and column incidence graphs, respectively. If p or q is reasonably close to the minimum of $\hat{n}$ and $\hat{m}$, one should apply CPR seeding and thus obtain the Jacobian $F'(x)$ more or less directly in a compressed data structure. If there is a significant gap, one may prefer NR seeding and reconstruction for which standard utilities are available for example in ADOL-C [GJU96].

If an increase of order $\min(\hat{n}, \hat{m})$ in the memory requirement is too large, one may perform strip-mining by evaluating some 5 to 10 Jacobian-vector or vector-Jacobian products at a time. This modification should only slightly affect the runtime of one-sided compression, whose basic growth (relative to the original code) is also governed by $\min(\hat{n}, \hat{m})$. Should this factor also be considered too large in that regard, one might consider the combined compression approach discussed in section 8.4 or restructure the function as discussed below.

Rewriting the Function

It is widely understood that sometimes a Jacobian or Hessian may be dense only because of an additive term with a low rank compared to the number of independents or dependents. Then one would try to handle this term separately and might use the Sherman–Morrison–Woodbury [DS96] formula for solving the resulting linear systems. In section 11.2 we have seen that this kind of partial separability can be detected and exploited in a more general and systematic fashion, though as yet no AD tool does it automatically. Hence we restrict ourselves to sketching the approach verbally.

The use of partial value or argument separability is often quite simple. For example, an overall objective function might be computed by summing many "individual" contributions representing "local" energies or squared residuals. Then one should treat the individual contributions as dependent variables, thus performing a row splitting on the Jacobian, as discussed in the previous section. Conversely, an evaluation procedure might begin by evaluating an average of some or all of the independent variables and later use it repeatedly as an argument of elemental functions. Then this average should be treated as an independent variable for differentiation purposes, as this column splitting on the Jacobian might significantly increase its sparsity. Unfortunately, it is currently still incumbent on the user to put the pieces together at the end, i.e., to add together rows and columns of the expanded Jacobian, once they have been evaluated.

Going Ahead

Finally, we would like to stress again that in most cases the user does not have to resort to the more laborious measures of employing combined compression or expanding the function. Moreover, it can be expected that in the near future more and more of these techniques will be automated and incorporated into AD software tools.

11.4 Examples and Exercises

Exercise 11.1 (*Linear Least Squares*)
Consider the linear function $F = \mathbb{R}^n \longmapsto \mathbb{R}^n$ and the quadratic function $f = \mathbb{R}^n \longmapsto \mathbb{R}$ defined by

$$F(x) = B^{-1}(Ax) - b \quad \text{and} \quad f(x) = \frac{1}{2}\left\|F(x)\right\|_2^2 ,$$

where $B \in \mathbb{R}^{n\times n}$ is dense upper triangular with $\det(B) \neq 0$ and $A \in \mathbb{R}^{n\times n}$ sparse but with dense last row. Write down explicit expressions for the Jacobian

$F'(x)$, the gradient $\nabla f(x)$, and the Hessian $\nabla^2 f(x)$. Assuming that multiplying a dense vector by B^{-1} or its transpose dominates the cost of computing Ax, show that

$$\begin{aligned} TIME\{F'\} &\approx n\, TIME\{F\}, \\ TIME\{\nabla f\} &\approx 2\, TIME\{f\}, \\ TIME\{\nabla^2 f\} &\approx n\, TIME\{f\}. \end{aligned}$$

These results confirm the claims made in section 11.1.

Exercise 11.2 (*Separable Energy Revisited*)
Recall the scalar function $f(x)$ with its original and division-free evaluation procedures examined in Exercise 7.3.
a. Show that both formulations are partially value separable, but not partially argument separable. Using the former property split the dependent y into component functions $Y = (Y_i)_{i=1\ldots n}$ so that $y = e^\top Y$ with $Y = F(x)$ a vector function and e a vector of ones as in (11.5). Using obvious modifications of the two evaluation procedures, verify that the average domain sizes $\bar{n}$ remain essentially the same and compute the average range sizes $\bar{m}$, which were trivially equal to 1 for the unsplit original function.
b. Derive the Jacobian F' "by hand" and note that since it is dense we have trivially $\hat{n} = n = \hat{m}$. Replace $b^\top x$ by an extra variable x_{n+1} to obtain a new vector function $\tilde{F}(\tilde{x}) : \mathbb{R}^{n+1} \longmapsto \mathbb{R}^n$ with the variable vector $\tilde{x} = (x, x_{n+1})$ such that $f(x) = Q\tilde{F}(Px)$, where $P = [I_n, b]^\top \in \mathbb{R}^{(n+1)\times n}$ and $Q = e^\top \in \mathbb{R}^n$. Derive the Jacobian $\tilde{F}'(\tilde{x})$ of the nonlinear core $\tilde{F}$ "by hand" and characterize its sparsity in terms of its maximal row length, maximal column length, and the corresponding coloring numbers. Conclude that static row compression of $\tilde{F}'(\tilde{x})$ by a suitable seed matrix S has essentially the same operations count as dynamically sparse differentiation on $F'(x)$. In contrast column compression is expensive and no gain can be made through combined compression either.
c. Demonstrate how the full second derivative tensor $\tilde{F}''(\tilde{x})$ can be computed cheaply by symmetric two-sided compression with the seed matrix S constructed in part **b**. Show that, on the other hand, simultaneous row and column compression is quite expensive. Consider the Hessian Lagrangian

$$H = \nabla^2_{\tilde{x}}\, \bar{y}\, \tilde{F}(\tilde{x}) \in \mathbb{R}^{(n+1)\times(n+1)}$$

and verify the last inequality of Proposition 7.1. Compute its reduced width by the algorithm given in Table 8.2.

Part III

Advances and Reversals

Chapter 12

Reversal Schedules and Checkpointing

The key difficulty in adjoining large evaluation programs is their reversal within a reasonable amount of memory. Performing the actual adjoint operations is a simple task compared to (re)producing all intermediate results in reverse order. In fact, as one can see in Table 4.8, just a couple of additions and multiplications need to be performed for each value that was computed and "taped" on the way forward. As we saw in section 4.3, some of these memory moves can be avoided by inverting certain incremental operations as listed in Table 4.10. However, for general nonlinear problems, we cannot expect that even the most selective storage strategy reduces the required memory traffic by more than a small constant. The problem of reversing a program execution has received some perfunctory attention in the computer science literature (see, e.g., [Ben73] and [vdS93]). The first authors to consider the problem from an AD point of view were apparently Volin and Ostrovskii [VO85] and later Horwedel [Hor92].

Checking the Memory

A good-size function evaluation might easily run for 10 minutes on a superscalar chip sustaining a rate of several hundred million arithmetic operations per second. If all intermediate results were saved, this would amount to a total memory requirement of roughly a terabyte ($= 10^{12}$ bytes). If we were a lot more selective about savings and the processor was not performing quite as many flops, we might get by with several gigabytes, which could be available as external memory.

However, that would barely make the whole process computationally attractive since storing and retrieving some gigabytes may take a while, even under favorable circumstances. Here we have already taken into account the strictly sequential nature of "tape" accesses in the basic reverse mode specified in Table 4.4. Hence nothing can be gained by sophisticated "external memory algorithms" that have been developed for computational geometry and other applications with very large and nonsequential data access requirements.

We simply cannot afford to save one double-precision number for every handful of elementary operations without slowing the adjoint calculation down significantly. Hence the modifications of the reverse mode that we will consider in this chapter serve two purposes. First, they make adjoints feasible even for extremely long simulations by storing only an a priori fixed number of checkpoints that can be accommodated in main or external memory. Second, they ensure a reasonable speed of execution by generating and using the actual taping data in pieces that are small enough to remain in internal storage. Of course, the optimal strategy will strongly depend on the structure of the problem at hand and the characteristics of the computing resources available.

12.1 Reversal with Recording or Recalculation

If the considered evaluation procedure has no specific structure, one may allow the placement of checkpoints anywhere during the evaluation process to reduce the overall memory requirement. This simple approach is provided, for example, by the AD-tool TAF [GK98]. As an alternative, one may exploit the call graph structure of the function evaluation to place local checkpoints at the entries of specific subroutines. The resulting call graph reversals are discussed in section 12.2. This subroutine-oriented checkpointing is employed, e.g., by the AD-tools Tapenade [HP04] and OpenAD [NU$^+$04].

As soon as one can exploit additional information about the structure of the function evaluation, an appropriate adapted checkpointing strategy can be used. In particular, this is the case if a time-stepping procedure is contained in the evaluation procedure allowing the usage of time-stepping-oriented checkpointing described in section 12.3. If the number of time steps l is known a priori and the computational costs of the time steps vary little, one rather popular checkpointing strategy is to distribute the checkpoints uniformly through the time interval, an approach that is also known as windowing for optimal control problems based on PDEs [Ber98]. Here, we will call this technique uniform checkpointing. However, it was shown in [WG04] that this approach is not optimal. A more advanced but still not optimal technique is the successive halving checkpointing implemented, for example, in [Kub98]. Using dynamic programming one can construct binomial schedules that are provably optimal, as explained in section 12.3. Since all these nonuniform checkpointing strategies are determined before the actual evaluation we refer to them as offline checkpointing.

If the evaluation procedure relies on some adaptive time stepping the total number of time steps is a priori not known. It becomes available only after the complete time integration making nonuniform offline checkpointing impossible. Instead, one may place a checkpoint each time a certain number of time steps has been executed. For this checkpointing approach used, for example, by CVODES [SH03] the maximal memory required cannot be predicted. However, when the amount of memory per checkpoint is very large, one certainly wants to limit the number of checkpoints required a priori. For that purpose, online checkpointing approaches have been proposed [SH05, HW06] and will be briefly discussed also in section 12.3.

Moreover, the usage of parallel processors offers additional opportunities for minimizing the actual execution time of checkpointing schedules. Corresponding parallel reversal approaches will be sketched in section 12.4. Here we quote some nice theoretical observations and results, even through as far as we know they have not yet been implemented in practice.

For the formal description of reversal strategies our point of departure is Table 12.1, which is a slightly revised version of Table 4.4.

Table 12.1: Forward and Return Motions with Total Taping

	Statement	Range
$\overrightarrow{F}$	$v_{i-n} = x_i$	$i = 1 \ldots n$
	$v_i \rightarrowtail \circ\circ$ $v_i = \varphi_i(u_i)$	$i = 1 \ldots l$
	$y_{m-i} = v_{l-i}$	$i = m-1 \ldots 0$
$\overleftarrow{F}$	$\bar{v}_{l-i} = \bar{y}_{m-i}$ $\bar{v}_i = 0$ $\bar{v}_{i-n} = \bar{x}_i$	$i = 0 \ldots m-1$ $i = l-m \ldots 1$ $i = n \ldots 1$
	$v_i \leftarrowtail \circ\circ$ $\bar{u}_i \mathrel{+}= \bar{\varphi}_i(u_i, \bar{v}_i)$ $\bar{v}_i = 0$	$i = l \ldots 1$
	$\bar{x}_i = \bar{v}_{i-n}$	$i = n \ldots 1$

In contrast to Table 4.4 the new version, Table 12.1, emphasizes the separation between the recording sweep labeled $\overrightarrow{F}$ and the return sweep labeled $\overleftarrow{F}$. As we will see in section 12.2, the reversal of multilevel programs may require the separation of $\overleftarrow{F}$ from $\overrightarrow{F}$, in which case only the information stored on the tape $\circ\circ$ by $\overrightarrow{F}$ is accessible to $\overleftarrow{F}$. This means in particular that there are no common local variables. For the time being we wish to go in the other direction, i.e., to reduce storage by intertwining the forward and return sweep more closely. To eliminate tape storage completely we could use the return sweep listed in Table 12.2. The difference between the two return sweeps is that the retrieval operation $v_i \leftarrowtail \circ\circ$ in Table 12.1 has been replaced by the recalculation of v_i from scratch by the j loops in Table 12.2.

Table 12.2: Return Motion with Total Recalculation

	Statement	Range
$\overleftarrow{F}$	$\bar{v}_{l-i} = \bar{y}_{m-i}$ $\bar{v}_i = 0$ $\bar{v}_{i-n} = \bar{x}_i$	$i = 0 \ldots m-1$ $i = l-m \ldots 1$ $i = n \ldots 1$
	$v_j = \varphi_j(u_j) \quad j = 1 \ldots i-1$ $\bar{u}_i \mathrel{+}= \bar{\varphi}_i(u_i, \bar{v}_i)$ $\bar{v}_i = 0$	$i = l \ldots 1$
	$\bar{y}_{m-i} = \bar{v}_{i-n}$	$i = n \ldots 1$

Note that for clarity we have made the minor assumption that the first n local variables v_{i-n} for $i = 1 \ldots n$ are not overwritten. Since the elemental function φ_i winds up being (re)evaluated $l - i$ times, completely eliminating tape storage in this way is quite expensive. If all φ_i were of similar complexity, the runtime of $\overleftarrow{F}$ as defined by Table 12.2 would be about $\frac{1}{2}l$ times that of the common forward sweep $\overrightarrow{F}$. This growth in temporal complexity is probably unacceptable unless l is quite a small number, which is unlikely when the φ_i are truly elementary operations. However, in multilayered programs we may think of l as the number of functions called by any evaluation (sub)procedure, which may be reasonably small.

To compromise between the single-minded minimization of temporal or spatial complexity represented by Tables 12.1 and 12.2, respectively, one may try to store some of the intermediates v_i and recompute others from the last fully known intermediate states. Whether and how this is possible will depend very much on the data dependencies between the elemental functions φ_i and the allocation scheme &. Deriving an optimal, i.e., minimal with respect to memory requirement, reversal is a combinatorial problem, which is provable NP-hard [Nau06]. Hence, we face again a very complex problem.

The Single-Assignment Scenario

If there is no overwriting in that the mapping & is injective, there is no need for either taping or recomputing since the intermediate values v_i will still be there on the way back. We consider this situation rather exceptional since it means that the undifferentiated evaluation procedure is already burdened with way too much memory traffic relative to the number of operations being performed. Nevertheless, the basic versions of some adjoint generators like Tapenade and TAF do in fact convert all functions to single-assignment codes by replicating all variables that receive more than one value into local arrays.

There are two main difficulties with this particular reversal approach, which was apparently originally proposed by Volin and Ostrovskii [VO85]. First, at least in Fortran 77, it needs to be decided at compile-time how many replicas of a particular variable are needed in order to dimension the arrays appropriately. This typically is quite difficult if there are iterations depending on dimension parameters or numerical comparisons. Second, local variables exist only as long as a function call is executed, which means that the return sweep has to follow immediately after the forward sweep within one extended version of the function. On multilayered evaluation programs this typically requires some extra forward evaluations compared to the basic approach that saves all intermediate values during the first (and thus last) forward evaluation motion. The resulting multilayered reversal schedules for structured programs will be discussed in section 12.2.

The Time Evolution Scenario

The opposite extreme to no overwriting is the situation where all v_i are mapped into the same location. We have already discussed this possibility briefly in section 5.1 with the v_i representing a sequence of state vectors that are mapped into each other by transformations ψ_i. For the forward simulation there is normally no need to keep the intermediate states, so each new v_{i+1} may immediately overwrite v_i irrespective of whether the transition functions ψ_i have anything to do with each other.

Typical examples of such discrete evolutions are simulations of fluid flows in the atmosphere or under the ground. Here every state consists of velocities, densities, concentrations, etc., on a spatial grid with thousands or even millions of nodes. Then, saving all intermediate states over hundreds or thousands of time steps is out of the question. Instead one saves only a few carefully selected states as checkpoints from which the forward simulation is repeated once or several times. In some sense the tape ○○ is generated and utilized in pieces. In section 12.3 we will derive reversal schedules that minimize the number of repeating forward steps given an upper limit on the number of checkpoints that can be kept in storage. As it turns out a small number of checkpoints suffices to reverse simulations over thousands of time steps at a moderate increase in operation count. Checkpointing may be interpreted as multilevel reversal in the sense of section 12.2 with simulations over subranges of time being interpreted as function calls.

12.2 Reversal of Call Trees

In this section we assume that the structure of a multilayered evaluation process is characterized by a call tree. Its nodes represent actual calls at runtime, several of which may originate from the same source call, for example, within a loop. It is demonstrated that the reversal schedule for such evaluations may be determined by specifying for each function call whether the link to its parent is *strong* or *weak*. As a consequence its reversal is either *split* or *joint*, respectively. Before we discuss the resulting traversal of call trees let us look at a small example to illustrate this basic distinction. The terms *recording*, *returning*, and *reversing* sweep or motion should be understood from the context and will be explained in more detail later.

Consider the computation of the Euclidean distance r between two points in the plane that are given by their polar coordinates (ρ_1, α_1) and (ρ_2, α_2). To convert first to Cartesian coordinates, let us use the subfunction

$$(x, y) = F(\rho, \alpha) \equiv (\rho \cos(\alpha), \rho \sin(\alpha)) \,.$$

The corresponding adjoint operation is given by

$$\begin{aligned}(\bar\rho, \bar\alpha)^\top \mathrel{+}= \bar F(\rho, \alpha, \bar x, \bar y) &\equiv (\bar x, \bar y)^\top F'(\rho, \alpha)\\ &\equiv (\bar x \cos(\alpha) + \bar y \sin(\alpha), -\bar x\, \rho \sin(\alpha) + \bar y\, \rho \cos(\alpha))^\top \,.\end{aligned}$$

Treating F as an elemental function we may evaluate

$$r = f(\rho_1, \alpha_1, \rho_2, \alpha_2) \equiv \|F(\rho_1, \alpha_1) - F(\rho_2, \alpha_2)\|$$

by the code listed in the upper half of Table 12.3. Assuming that there is no overwriting, the forward evaluation of f can be followed immediately by the return sweep for the adjoint $\bar{f}$ calculation listed in the lower half of Table 12.3. Because all values occur only once as arguments we may use the nonincremental form. The corresponding code for the adjoint $\bar{F}$ is listed in Table 12.4, with the upper half again representing the original procedure for F alone.

Table 12.3: Procedure for f and $\bar{f}$

(v_{-3}, v_{-2})	$=$	(ρ_1, α_1)
(v_{-1}, v_0)	$=$	(ρ_2, α_2)
(v_1, v_2)	$=$	$F(v_{-3}, v_{-2})$
(v_3, v_4)	$=$	$F(v_{-1}, v_0)$
(v_5, v_6)	$=$	$(v_3 - v_1, v_4 - v_2)$
(v_7, v_8)	$=$	$(v_5 * v_5, v_6 * v_6)$
v_9	$=$	$\sqrt{v_7 + v_8}$
r	$=$	v_9
$\bar{v}_9$	$=$	$\bar{r}$
$(\bar{v}_7, \bar{v}_8)$	$=$	$(\frac{1}{2}\bar{v}_9/v_9, \frac{1}{2}\bar{v}_9/v_9)$
$(\bar{v}_5, \bar{v}_6)$	$=$	$(2 * \bar{v}_7 * v_5, 2 * \bar{v}_8 * v_6)$
$(\bar{v}_3, \bar{v}_4)$	$=$	$(\bar{v}_5, \bar{v}_6)$
$(\bar{v}_1, \bar{v}_2)$	$=$	$(-\bar{v}_5, -\bar{v}_6)$
$(\bar{v}_{-1}, \bar{v}_0)$	$=$	$\bar{F}(v_{-1}, v_0, \bar{v}_3, \bar{v}_4)$
$(\bar{v}_{-3}, \bar{v}_{-2})$	$=$	$\bar{F}(v_{-3}, v_{-2}, \bar{v}_1, \bar{v}_2)$
$(\bar{\rho}_2, \bar{\alpha}_2)$	$=$	$(\bar{v}_{-1}, \bar{v}_0)$
$(\bar{\rho}_1, \bar{\alpha}_1)$	$=$	$(\bar{v}_{-3}, \bar{v}_{-2})$

Table 12.4: Procedure for F and $\bar{F}$

(v_{-1}, v_0)	$=$	(ρ, α)
v_1	$=$	$\cos(v_0)$
v_2	$=$	$\sin(v_0)$
v_3	$=$	$v_{-1} * v_1$
v_4	$=$	$v_{-1} * v_2$
(x, y)	$=$	(v_3, v_4)
$(\bar{v}_3, \bar{v}_4)$	$=$	$(\bar{x}, \bar{y})$
$\bar{v}_2$	$=$	$\bar{v}_4 * v_{-1}$
$\bar{v}_{-1}$	$=$	$\bar{v}_4 * v_2$
$\bar{v}_1$	$=$	$\bar{v}_3 * v_{-1}$
$\bar{v}_{-1}$	$+=$	$\bar{v}_3 * v_1$
$\bar{v}_0$	$=$	$\bar{v}_2 * v_1$
$\bar{v}_0$	$-=$	$\bar{v}_1 * v_2$
$(\bar{\rho}, \bar{\alpha})$	$=$	$(\bar{v}_{-1}, \bar{v}_0)$

Here we have combined the calculation of intermediate variable pairs into a single statement in order to save space.

Clearly, evaluating $\bar{F}$ from scratch during the return sweep of f requires the repetition of the code for F and in particular the reevaluation of the trigonometric pair $(\cos(\alpha), \sin(\alpha))$, which is comparatively costly. To avoid this inefficiency, one might modify the forward sweep and the return sweep to the procedures $\overrightarrow{F}$ and $\overleftarrow{F}$ listed in Tables 12.5 and 12.6.

Joint and Split Reversals

The *recording sweep* $\overrightarrow{F}$ saves some intermediate results onto the tape, which are then restored and reused by the *returning sweep* $\overleftarrow{F}$ on the way back. In terms of direct input and output we consider them as equivalent to F and $\bar{F}$. However, there is a side effect, namely, the recording onto the tape $\circ\circ$, which may be either local to F or global. As we will see, all tape accesses are always LIFO so that all intermediates may be saved on one global $\circ\circ$ acting as a stack.

Now we may replace the calls to F in the upper half of Table 12.3 by calls to $\overrightarrow{F}$, and the calls to $\bar{F}$ in the lower half, by calls to $\overleftarrow{F}$. Here care must be taken that the two calls to $\overleftarrow{F}$ occur in the opposite order of the original

Table 12.5: Recording Motion of F

	(v_{-1}, v_0)	$=$	(ρ, α)
	v_1	$=$	$\cos(v_0)$
	v_2	$=$	$\sin(v_0)$
$\boxed{F}\triangleright$	(v_1, v_2)	$\rightarrowtail$	○○
	v_3	$=$	$v_{-1} * v_1$
	v_4	$=$	$v_{-1} * v_2$
	(x, y)	$=$	(v_3, v_4)

Table 12.6: Returning Motion of F

	(v_{-1}, v_0)	$=$	(ρ, α)
	$(\bar{v}_3, \bar{v}_4)$	$=$	$(\bar{x}, \bar{y})$
	(v_1, v_2)	$\leftarrowtail$	○○
	$\bar{v}_2$	$=$	$\bar{v}_4 * v_{-1}$
	$\bar{v}_{-1}$	$=$	$\bar{v}_4 * v_2$
$\triangleleft\boxed{F}$	$\bar{v}_1$	$=$	$\bar{v}_3 * v_{-1}$
	$\bar{v}_{-1}$	$+=$	$\bar{v}_3 * v_1$
	$\bar{v}_0$	$=$	$\bar{v}_2 * v_1$
	$\bar{v}_0$	$-=$	$\bar{v}_1 * v_2$
	$(\bar{\rho}, \bar{\alpha})$	$=$	$(\bar{v}_{-1}, \bar{v}_0)$

calls $\boxed{F}\triangleright$. Otherwise the storage and retrieval on a common tape ○○ would not produce the correct results.

Compatible Pairs

In general we have a lot of freedom in designing the pair $\boxed{F}\triangleright, \triangleleft\boxed{F}$. Provided they are *compatible*, i.e., correctly communicate with each other via the tape ○○. As a variation we might decide, for example, to make $\boxed{F}\triangleright$ save only the sinus value $v_2 = \sin(v_0)$ and to have $\triangleleft\boxed{F}$ recompute the cosine $v_1 = \cos(v_0)$ as

$$v_1 = \text{sign}(\pi/2 - |v_0|)\sqrt{1 - v_2^2}\,.$$

If the sign factor was omitted, the two buddies $\boxed{F}\triangleright, \triangleleft\boxed{F}$ would be compatible only when the angle $v_0 = \alpha$ may be assumed to lie in $[-\pi/2, \pi/2]$.

Even the original combination $F, \bar{F}$ is a compatible pair. However, we can make one critical distinction. Viewing the trigonometric function couple $T(\alpha) \equiv (\cos(\alpha), \sin(\alpha))$ also as a separate function, we note that $\bar{F}$ must call T for a second time, whereas, in the variations of $\triangleleft\boxed{F}$ we discussed above, this is not necessary. Now we may characterize the two approaches by the corresponding call trees. The symbols $\square$, $\square\!\triangleright$, $\triangleleft\!\square$, and $\triangleleft\!\square\!\triangleright$ represent *motions* that will now be defined in the context of general call trees.

Walking the Call Tree

Consider a runtime call tree of the structure given in Fig. 12.1. Here and in the derived call trees that will be considered later, the execution order is depth and left first. More precisely, starting with the unique root, each node begins by allocating its local variables, then executes its internal code interspersed with calls to its children from left to right, and finally terminates by deallocating its memory and returning control to its parent. Hence the set of functions that are allocated and being executed at any one time always represents an *active bough* of nodes starting at the root. The lines between the boxes represent the exchange of argument and function values.

Now the question arises of how we can reverse the forward execution of the call tree with an acceptable compromise between spatial and temporal complexity. To this end each call will be executed once or several times in the following four modes, which we will call *sweeps* or *motions*, to distinguish them from the modes *forward*, *reverse*, and *cross-country* that have been discussed before.

Advancing Motion $\square$

Simply calling a function F in the usual forward evaluation mode will be denoted by $\boxed{F}$. Normally the call $\boxed{F}$ will perform some internal calculations as well as calling some child functions. We have to distinguish the following second forward motion from advancing.

Recording Motion $\square\!\rhd$

Even during a multilayered adjoint procedure, each function F is performed exactly once in recording motion, which we will represent by the symbol $\boxed{F}\!\rhd$. During recording, certain variable values are saved on some tape $\circ\!\circ$ just before they are overwritten, producing a record of size $LOG(F)$. An implementation-dependent question is whether there is at least one tape for each function call, or just one tape operated like a stack. In either case these tapes must exist until they are consumed by the following backward motion.

Returning Motion $\lhd\!\square$

Using the recording on $\circ\!\circ$ produced by $\boxed{F}\!\rhd$, the returning motion call $\lhd\!\boxed{F}$ propagates adjoint values backward through F. We need not quantify the computational complexity of $\boxed{F}\!\rhd$ or $\lhd\!\boxed{F}$ because both need to be executed exactly once, no matter how the rest of the tree reversal is organized. What does differ is the number of times the advancing sweeps $\boxed{F}$ are executed and the period of time that the record of size $LOG(F)$ needs to be kept in memory.

Options for $\square\!\rhd$ and $\lhd\!\square$

In section 12.4 we will allow for the concurrent execution of several recording motions from suitable checkpoints. The subsequent returning call requires adjoint values supplied by the returning motion of its successor in the original call tree. Therefore, and because there is never any gain in repeating backward motions, the returning motions cannot be parallelized and their execution time is critical for the overall process. Hence it may make sense to burden the recording motion with all kinds of preparatory calculations to make the subsequent returning motion as fast as possible. As an extreme possibility the recording motion may calculate the (local) Jacobian of its output values with respect to its input values and store it on the tape. Then the corresponding returning motion would reduce to a simple vector-matrix multiply, which might be very fast, especially when the local Jacobian is sparse.

Reversing Motion $\langle\ \rangle$

To lower the external memory requirement one may strive to have $\langle F]$ executed right after $[F\rangle$, a combined reverse motion that we will symbolize by $\langle F\rangle$. In most cases this can only be achieved by repeating a forward sweep just before the reverse sweep, thus increasing the temporal complexity. So basically one has to decide whether this reduction in memory justifies an extra evaluation of $[F]$. Moreover, the combination of $[F\rangle$ and $\langle F]$ into a single call $\langle F\rangle$ allows one to implement the logging memory locally, an important advantage that we will nevertheless ignore in the following analysis.

Back to the Example

To illustrate the use of the four motions, let us consider Figs. 12.1, 12.2, 12.3 and 12.4. Fig. 12.1 represents the original function evaluation, where we have distinguished the two calls to F, and correspondingly to T, as F_i and T_i for $i = 1, 2$. The connecting arcs represent the exchange of real, undifferentiated values in both directions. The corresponding call tree for the evaluation of $\bar{f}$ according to Tables 12.3 and 12.4 is displayed in Fig. 12.2. Here we have reinterpreted the last two computational statements in Table 12.4 as

$$\bar{v}_0 = \bar{T}(v_0, \bar{v}_1, \bar{v}_2) \quad \text{with} \quad \bar{T}(\alpha, \bar{c}, \bar{s}) \equiv \bar{s}\, \cos(\alpha) - \bar{c}\, \sin(\alpha)$$

and affected the evaluation of $\bar{T}_i$ by $\langle T_i\rangle$. This reinterpretation would make more practical sense if T involved some substantial internal calculations which would be repeated three times when the program reversal proceeds according to Fig. 12.2. The double lines represent the exchange of values and adjoint derivative information in both directions.

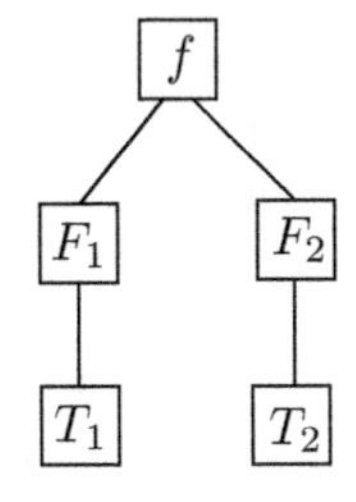

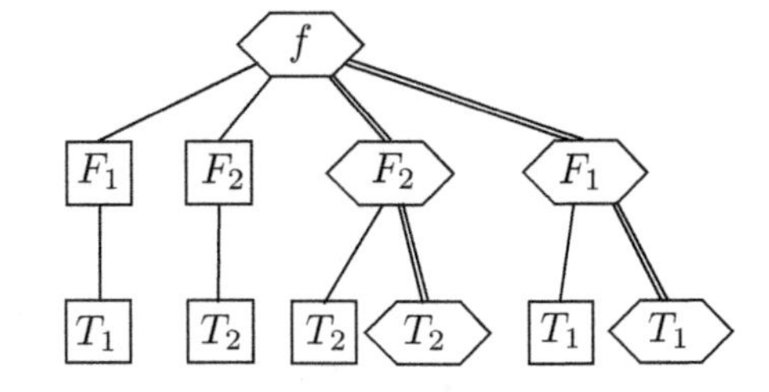

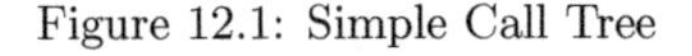
Figure 12.1: Simple Call Tree

Figure 12.2: Joint Reversal of F_i and T_i

Now suppose we split the reversals of both F and T by replacing the compatible pairs $[F]$, $\langle F\rangle$ and $[T]$, $\langle T\rangle$ by $[F\rangle$, $\langle F]$ and $[T\rangle$, $\langle T]$, respectively. The corresponding reversal tree is displayed in Fig. 12.3. As we can see, there is some saving in that T_1 and T_2 and the internal forward evaluations of F_1 and

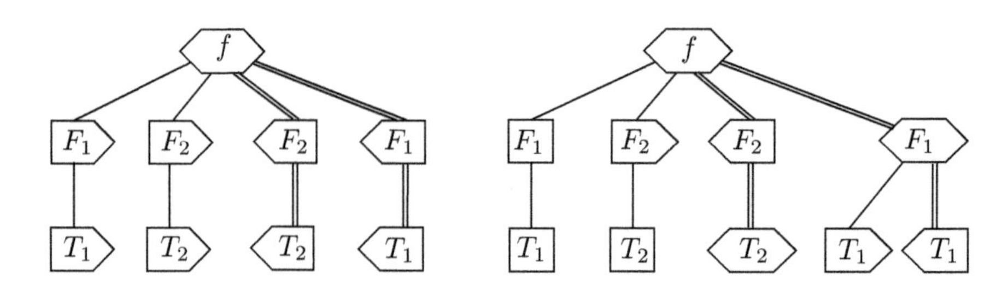

Figure 12.3: Split Reversal of F_i and T_i

Figure 12.4: T_1, F_2 Split; T_2, F_1 Joint

F_2 need only be performed once. On the other hand, we have an increased storage requirement as these values must all reside on the tape at the moment in the middle when we switch from the forward to the return sweep of the master function f. As a sweeping generalization we conclude Rule 19.

RULE 19

JOINT REVERSALS SAVE MEMORY, SPLIT REVERSALS SAVE OPERATIONS.

We should be careful to note that operations count is not synonymous with runtime. A reduction in storage to the size of the internal memory available may reduce the runtime even if it entails a moderate increase in the operations count as illustrated, for example, in [KW06].

Fortunately, there is no need to brush all calls the same way; rather, we may decide for each one of them whether to perform a joint or split reversal. The schedule resulting from a mixed strategy that splits F_2 and T_1, but not F_1 and T_2, is displayed in Fig. 12.4.

Below we will examine the consequences of the split/joint decisions for individual function calls on the overall reversal schedule and in particular its spatial and temporal complexity. It is found that unless the call tree is quite deep, reversing all calls in the joint mode is not a bad idea, a strategy that is, for example, employed in the preprocessor tools Tapenade [HP04] and TAF [GK98]. Overloading tools like ADOL-C [GJU96] on the other hand reverse as default everything in split mode by saving all intermediate values the first time through. This approach avoids any repeated forward sweeps but generates a large storage requirement, which must slow things down considerably once the internal memory has been exhausted.

Links and Inheritance

Rather than indicating for each function call whether it is to be reversed in *split* or *joint* mode, we find it more convenient to make this choice by specifying the link to its parent in the call tree as *strong* or *weak*, respectively. To explain this connection we observe first that a function call with split reversal must be recording exactly when it is called by a parent in recording motion, because this must be the last forward motion before the subsequent returning motion.

Similarly a child with split reversal will also follow the parent on earlier advancing motions and on the very last return motion. Therefore, we will characterize the relation between a parent and a child with split reversal as a *strong* link or connection.

A child with joint reversal will follow all its parent's forward (= advancing or recording) motions by merely advancing itself and only initiate its own recording when the parent is already returning on the way back. After the recording motion the child may immediately initiate its returning motion, which leads to Table 12.7.

Table 12.7: Inheritance Rules for Motions with ⬡ ≡ ⊳ + ⊲

Link	Parent	□	⊳	⊲
Strong	Child (split)	□	⊳	⊲
Weak	Child (joint)	□	□	⬡

By way of explanation, we note that it makes no sense to record a function that is called by an advancing parent since another forward motion will be needed before the parent can be run backward. Hence one may perform the recording of the child at that later stage, unless, of course, it is to be reversed in the joint mode. In that case the child function will be recorded as part of a joint reversal only when the parent is already in returning motion.

On the other hand, it may be quite useful to record at an early stage only on the upper levels of a call tree and to reverse low-level function calls in the joint mode. When the leaves are elementary functions, there is usually nothing to record at all, so there is no real difference between ⊲ and ⬡. Hence a parent in recording motion calls children with joint reversals merely in advancing motion, but the others must be called in recording motion, too.

For the time being we will assume that all child calls in the original call tree are followed by some internal calculations, so that they cannot be called in the reversing motion on the way forward even when their parents are "in reverse."

Active Boughs

To specify a tree reversal we have to add to the motion rules listed in Table 12.7 the natural requirement that parents call their children from left to right or from right to left depending on whether they themselves are in a forward or backward motion, respectively. At the end of each motion the function returns control to its parent, except in the middle of a joint reversal, where the recording motion is followed immediately by a returning motion. The whole process starts with the reversing motion of its root.

Then it follows immediately from the motion inheritance rules given above that the set of functions that are in motion at any time of the tree reversal form an *active bough* of the following structure.

Lemma 12.1 (Active Bough Structure)
The active bough always starts at the call tree root in recording or returning motion and consists of

- *a returning segment on the top,*
- *a recording segment in the middle,*
- *an advancing segment on the bottom.*

One or two of the segments, but not the recording one (in the middle) alone, may be empty and the links connecting the various segments are always weak.

We may visualize the active bough like an elephant's trunk that sweeps back and forth across the call tree, occasionally contracting towards the root before reaching towards the leaves again. Each transition from one active bough to its successor during the reversal corresponds to one of the following modifications.

— The *truncation* of the last element from the last segment (representing the return of control by the lowest function).

— The *extension* of the last segment by an extra element (representing a call of the "next" child by a parent).

— The *augmentation* of a different element onto the last segment (representing a new recording or an advancing child motion, initiated by a returning or recording parent, respectively).

— The *reversal* of the recording segment from $\square\!\!\rhd$ to $\lhd\!\!\square$ (provided it is a singleton and the advancing segment is empty).

To illustrate the evolution of the active bough, let us reconsider the call tree of Fig. 12.1 redrawn as Fig. 12.5 with the lower nodes renamed and a single strong link between f and G (formerly F_2). Hence, only G is reversed in split mode, whereas $F(= F_1)$, $T(= T_1)$, and $U(= T_2)$ are reversed in joint mode. The resulting sequence of active boughs is displayed in Fig. 12.6.

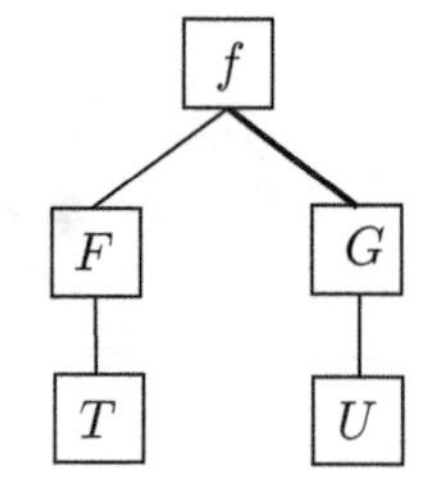

Figure 12.5: Example with Strong Link

As one can see, the active bough does consist of the three segments: returning, recording, and advancing as asserted. They are separated by the thick black lines. Whenever these separating lines coincide and/or run along the margin, one or even two of the segments are empty. In fact, only in the seventh last column all three segments are present. All the action is on the bottom, with the last segment being truncated (T), extended (E), augmented (A), or reversed (R) as specified in the top row.

It is apparent that the strongly linked node G always performs exactly the same motion as its parent f. In contrast F, U, and T follow the recording

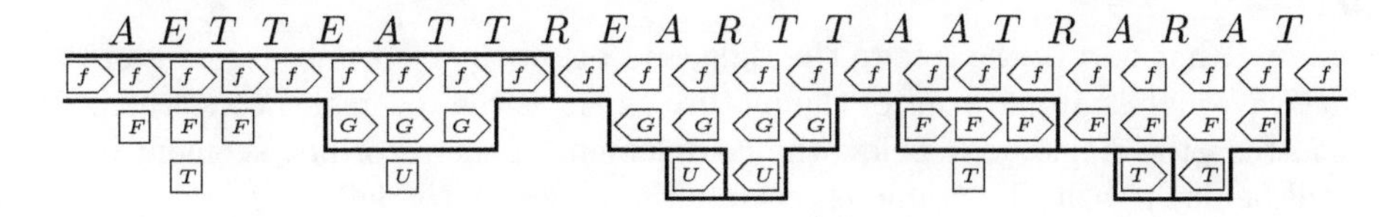

Figure 12.6: Active Boughs on Euclidean Distance Problem (Fig. 12.5). Above lines: "returning"; between lines: "recording"; below lines: "advancing"

motion of their parents with an additional advance and record in a joint fashion only at the last possible moment. This strategy limits the memory requirement but means that T and U advance two extra times, while F advances once more than in a fully split reversal characterized by all links being strong.

Clearly, Fig 12.6 is not a concise way to visualize the reversal schedule; we just wanted to highlight the properties and transitions of the active bough. Fig. 12.7 is much more concise and displays a snapshot during the reversal of a house-shaped tree with 23 vertices.

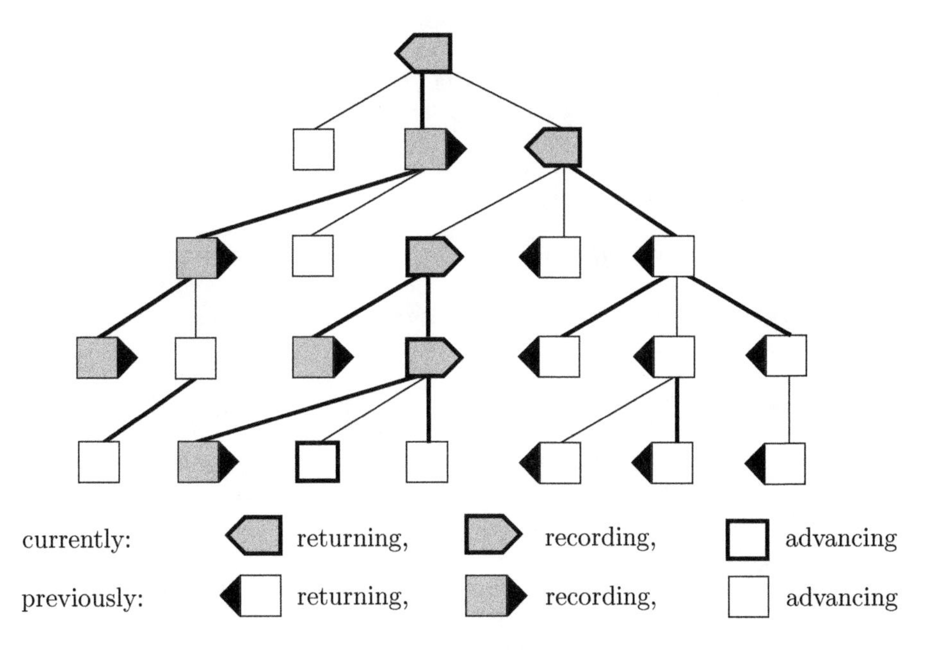

Figure 12.7: Partially Completed Tree Reversal with Weak and Strong Links

The black-bordered symbols represent the active bough, consisting of two returning, two recording, and one advancing function. In other words, we are globally on the way back as the root and its oldest (= rightmost) child are already in returning motion. Moreover, the youngest (= leftmost) grandchild belonging to that branch is already recording, even though the link to its parent is weak. Consequently, its two older siblings and all their six descendents must already have been reversed, which is indicated by the return boxes with solid noses. That part of the tree to the right and below the returning segment of the active bough is already out of the picture as far as the reversal is concerned. In contrast, the single node below the last element of the recording segment must still be dealt with. The same is true for the nodes on the left.

In Fig. 12.7 we have distinguished the functions that have so far performed only advancing motions and are represented by $\square$ from the ones that have already been recorded and are represented by the boxes with a solid nose to the right. The latter form a subtree $\mathcal{R}$ consisting of the returning and recording segments plus all their strongly linked descendents to the left. More precisely, a function is such a descendent exactly when its ancestral chain reaches the returning or recording segment via strong links at an element that has no active children to the left of the active bough. The ones that are only weakly linked to an element of the active bough remain so far unrecorded. These observations are true in general and lead to the following complexity results.

Proposition 12.1 (Tree Reversal Schedule)
The graph traversal described above always terminates and ensures the following properties:

(i) *Each function node is called exactly once in recording motion and also at some later time, once in returning motion.*

(ii) *The number of times a function is advanced equals the number of weak links in its ancestral chain.*

(iii) *The set $\mathcal{R}$ of functions that are being or have been recorded at any stage during the reversal is given by the actively returning and recording segments, plus all their strong left descendents as defined above.*

Proof. Like most other assertions about call trees we can prove the result by induction on the generations, a number which we may formally define for any particular vertex as the number of direct ancestors plus 1. All assertions are clearly true for the first generation, which consists only of the root call. Now suppose they are true for all calls up to a certain generation, which we will refer to as the parents in the following arguments. The children of these parents form the next generation, for which the validity of the three assertions comprising the proposition must be established. A child with split reversal always follows its parent, so it is also called exactly once in recording motion and also at some time later once in returning motion. A child with joint reversal on the other hand is called exactly once in reversing motion when its parent is in return motion. By definition, reversing consists of recording followed by returning, so that we have proven (i). To prove (ii) by induction, we note that strongly linked children are advanced exactly as often as their parents. Weakly linked once, on the other hand, are advanced once more when their parents are already in recording motion. The last assertion (iii) implies that the recordings can all be placed on a stack in a LIFO fashion. ■

As an immediate consequence of Proposition 12.1 (ii), we may formulate the following result.

Corollary 12.1 (Temporal Complexity of Reversal)
The extra temporal complexity caused by joint reversals is bounded by the original function evaluation cost multiplied by the maximal number of weak links in any chain of the call tree.

Here the fully split reversal of the same call tree, but without any weak links, serves as a baseline complexity reference.

Weak Depth and the Quotient Graph

We may define the maximal number of weak links in any chain as the *weak depth* of the tree. It represents the normal depth of the *quotient tree* that is obtained if one in-lines all strong calls, i.e., merges all children with split reversals with their parents. In other words, all strongly coupled subtrees are considered as supernodes of the quotient graph. The reversal of a general tree always corresponds to a completely joint reversal of its quotient graph, which has by definition no strong links.

The completely joint reversal displayed in Fig. 12.2 is generated by the call tree of Fig. 12.1, which does not have any strong links and is therefore identical with its quotient tree. The (weak) depth is thus equal to 2, a factor that bounds the number of extra function evaluations performed in Fig. 12.2 compared to the calculations required by the fully split reversal displayed in Fig. 12.3. In fact, only T_1 and T_2 need to be evaluated in the advancing mode two more times, because there are two weak links in their ancestral chain. In contrast, F_1 and F_2 are only evaluated one extra time.

While extra split reversals reduce temporal complexity they do increase the spatial complexity as more recordings need to be kept in memory. Before examining these requirements, let us first make the following observation.

Rule 20

> During call tree reversals all taping occurs in a Last-In-First-Out fashion.

In practical terms this means that one may use a single global tape ⚬⚬ to perform all taping for the recording and returning motions of the various functions. In contrast to the local tapes, the global one will be a true stack in that reading and writing may alternate, except when all functions are reversed in split mode.

At any stage during the reversal the set $\mathcal{R}$ of functions that have been completely recorded consists of the strong left predecessors of the active bough without its active segment. To achieve some duality with the temporal complexity bound given in Corollary 12.1 we would like to gauge the size of $\mathcal{R}$ by the distribution of strong links. Let us call two links independent if they do not belong to the same chain. In other words they are not connected by a path in the line-graph of the call tree. Thus we obtain the following corollary of Proposition 12.1 (iii).

Corollary 12.2 (Spatial Complexity of General Reversal)
The extra spatial complexity of a general reversal is bounded by the memory requirement of the fully joint reversal multiplied by the maximal number of mutually independent strong links in the call tree.

Proof. Any maximal chain can occur as an actively returning segment in the fully joint reversal serving as reference. Let L denote the maximum over the logging space needed for any such chain. During the reversal records must also be kept for all strong left predecessors of any such active bough, which is certainly contained in the set of all strong predecessors of the active bough. This predecessor set consists of strong chains emanating from any element of the active bough of which there can be no more than the maximal number of independent strong links, say w. Hence the extra spatial complexity is bounded by $w\,L$, as asserted. ∎

The bound on the spatial complexity given by Corollary 12.2 is not very tight. One of the effects that has been ignored is that at any stage during the recording and returning, elements in the active bough are usually only partially recorded. Also the detailed structure of the pattern formed by the weak and strong links has been ignored. One would hope that on any particular call tree this pattern could be chosen such that an acceptable compromise between temporal and spatial complexity is arrived at.

The fully joint reversal procedure that treats all function calls as weak links is really quite acceptable as a general approach and was adopted for the transformation tools Tapenade and TAF. We follow this preference by drawing links by default as thin and thus weak. Reasons to deviate from this policy by treating some calls as strong links might be the availability of extra memory and/or a call tree of great depth. On the other hand, when the call tree is rather wide we may have a problem anyway because the recording of parents with a large number of children may require too much memory, irrespective of whether we consider the links as weak or strong. Then one may have to reorganize the tree, which is one way to interpret the next section.

Total Recalculation Revisited

We end this section by interpreting the total recalculation option considered at the beginning of the chapter in terms of the tree reversal mechanism developed earlier. Suppose we consider all elemental functions φ_i in Table 12.1 as independent function calls without children of their own. Then the call tree for F has only two levels, namely, the root $\boxed{F}$ and the second generation $\boxed{\varphi_i}$ for $i = 1 \dots l$ as depicted in Fig. 12.8. Irrespective of whether we choose weak or strong links, the memory requirement corresponds to the one reflected in Table 12.1 since $\boxed{F}\!\triangleright$ must save the prevalues for the outputs of each φ_i.

To reduce this memory requirement we may reorganize the tree by bracketing the φ_i into the functions

$$F_i \equiv [F_{i-1}] + (\varphi_i) \quad \text{for} \quad i = 2 \ldots l \quad \text{with} \quad F_1 \equiv (\varphi_1) \, ,$$

where the + indicates successive execution. Through the introduction of those l artificial functions F_i we have turned the original completely flat call tree for F into the comb-like tree for the mathematically equivalent F_l depicted in Fig. 12.9.

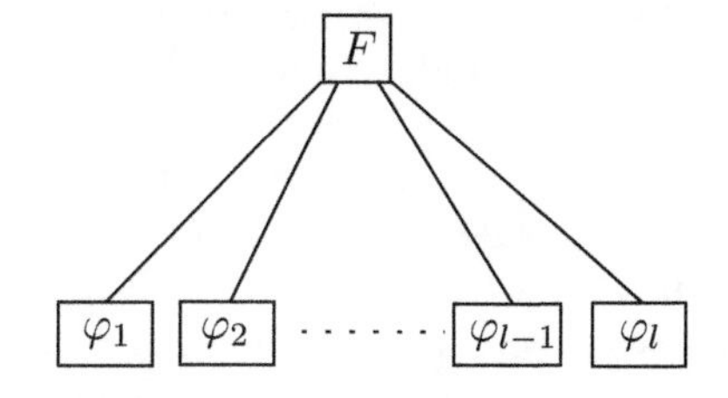

Figure 12.8: Flat Call Tree

We consider the links between the F_i and "their" elemental function φ_i as strong, which is reflected by the thick slanted lines in Fig. 12.9. By defining the links between the F_i and their children F_{i-1} as weak, we obtain a reversal schedule that corresponds almost exactly to the total recalculation option shown in Table 12.2.

According to Proposition 12.1 (ii) the function φ_i is called in advancing motion $l-i$ times since there are that many weak links between the node $\boxed{\varphi_i}$ and the root $\boxed{F_l}$. Obviously, Corollary 12.1 is satisfied with the weak depth given by $l-1$. A little more difficult is the question of spatial complexity. The whole idea of the total recalculation version listed in Table 12.2 was to avoid any taping at all. This requires that none of the artificial functions F_i may do any taping, which can in fact be achieved in the following way: Since F_i performs no internal calculation before calling F_{i-1}, the output variables of the latter are either uninitialized or have been initialized to input variables of F_i. In either case they need not be saved according to Rule 6. Subsequently, since F_i does not perform any internal calculation after the call to φ_i, it may call $\langle\varphi_i\rangle$ immediately without saving the prevalues of v_i. Hence any taping can indeed be avoided, and we may consider Table 12.2 as equivalent to reversing Fig. 12.9.

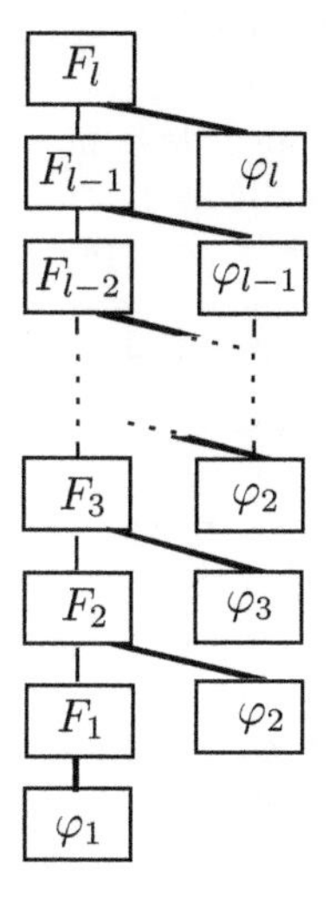

Figure 12.9: Vertical Call Tree

In general one may wish to find a compromise between the flat tree of Fig. 12.8 with minimal operations count and the vertical tree Fig. 12.9 with its minimal space requirement. In the following section we define a whole range of trees with varying depth and height for the same evolutionary problem.

12.3 Reversals of Evolutions by Checkpointing

A well-understood method for reducing memory at the expense of some increase in the operations count is checkpointing. It is usually motivated and applied for explicitly "evolutionary problems," where a large number of transition steps is taken between successive states of a physical system. Prime examples are atmospheric and oceanographic simulations, where checkpointing has been successfully used to avoid storing hundreds or even thousands of time steps on huge grids. An interesting application is the fast solution of Toeplitz systems [BB+95], where checkpointing can be used to reduce the memory requirement from $O(n^2)$ to $O(n \log n)$, with n being the number of variables. The optimal checkpointing strategy presented in this section has been used successfully, for example, in seismic research [Sym07].

Evolutionary systems correspond to a wide call tree on the highest level as displayed in Fig. 12.8. Since the top routine F is always reverted in a joint mode the memory requirement would be very large whether all intermediate states are stored in local arrays or saved on the tape ○○. Then it also does not matter very much whether the individual calls are reverted in the joint or split mode, corresponding to weak or strong links with their parents. Checkpointing strategies suggest storing only a few intermediate states and then repeating forward calls a number of times in order to reach all states in reverse order. In our context we may interpret this as recursively splitting the evolutionary simulation into subranges and considering them as functions. Then the flat tree displayed in Fig. 12.8 with $l = 6$ and φ_i rewritten as F_i might turn into the deeper tree displayed in Fig. 12.10. Applying the joint reversal strategy to all vertices in the new tree, one obtains the schedule displayed in Fig. 12.11. It is known to be optimal for serially reverting six time steps, while never storing more than two checkpoints.

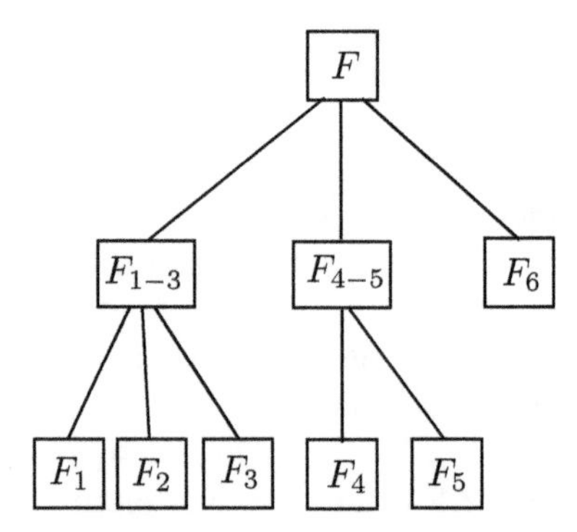

Figure 12.10: Bracketed Tree of Evolutionary System

In drawing Fig. 12.11 we have made a slight simplification; namely, we have left out the advancing motions F_6, F_5, and F_3 that should normally precede their reversals in a fully joint schedule. The reason for this omission is that $F \equiv F_{1-6}$, F_{4-5}, and F_{1-3} are by definition container functions that do nothing but call their children. In particular, they perform no internal calculation after calling their last (= rightmost) child F_6, F_5, and F_3, which can therefore be reversed immediately. More generally, one may recommend the split reversal of youngest children whenever the parents do not do much more after raising them, which seems a realistic scenario for larger families nowadays.

To show that the reversal schedule of Fig. 12.11 is in some sense optimal, we will adopt a more abstract point of view.

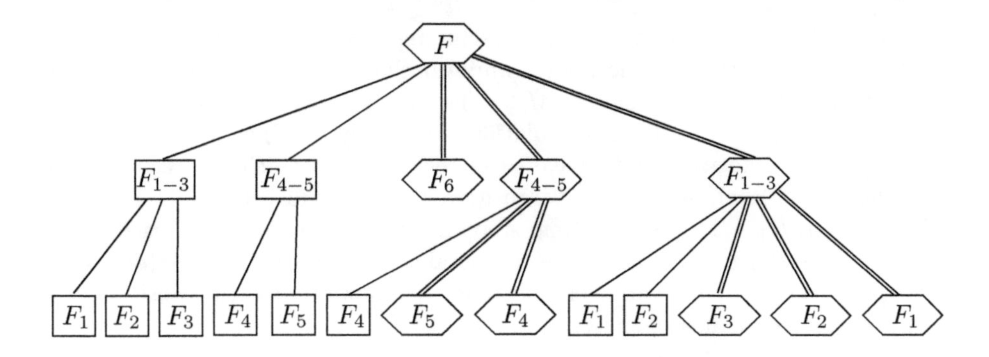

Figure 12.11: Fully Joint Reversal of Call Tree of Fig. 12.10

Chain Reversal Problem

Suppose we have a sequence of nodes v_i that are connected by pointers φ_i as displayed below.

$$v_0 \underset{\varphi_1}{\longrightarrow} v_1 \underset{\varphi_2}{\longrightarrow} v_2 \underset{\varphi_3}{\longrightarrow} \cdots \underset{\varphi_{l-1}}{\longrightarrow} v_{l-1} \underset{\varphi_l}{\longrightarrow} v_l$$

As we have suggested by the notation, we think of the v_i as large-dimensional state vectors of an evolutionary system and the pointers φ_i really represent mathematical mappings between v_{i-1} and v_i. The key property of these transition functions φ_i is that they cannot be inverted at a reasonable cost. Hence it is impossible to run backward from v_l to reach v_0 by simply applying the inverses φ_i^{-1}. We also assume that due to their size, only a certain number, say c, of states v_i can be kept in memory. In other words we can save up to c nodes as checkpoints, from which we may later restart the forward calculation.

To distribute the checkpoints uniformly over the given number of transition steps forms one obvious solution to the storage requirement problem. Subsequently the adjoints are computed for each of the resulting groups of transition steps separately. This checkpointing strategy has been called *windowing* in the PDE-related literature (see, e.g., [Ber98, BGL96]) or two-level checkpointing in the AD-oriented publications (see, e.g., [Kub98]). Naturally, one can apply two-level checkpointing recursively for the groups of transition steps that are separated by equidistant checkpoints. This approach is called *multilevel checkpointing* [GK98]. A multilevel checkpointing schedule is determined by the number of levels r and the number of checkpoints c_i that are uniformly distributed at level i, $i = 1, \ldots, r-1$. Therefore, we will call this approach uniform checkpointing with r levels. For a given r-level checkpointing, one easily derives the identities

$$l_u = \prod_{i=1}^{r}(c_i+1), \quad M_u = \sum_{i=1}^{r} c_i, \quad t_u = \sum_{i=1}^{r} \frac{c_i\, l_r}{c_i+1} = r\, l - \sum_{i=1}^{r} \prod_{\substack{j=1 \\ j \neq i}}^{r} (c_j+1)\,,$$

where l_u denotes the number of transition steps for which the adjoint can be calculated using the specific r-level uniform checkpointing. The corresponding memory requirement equals M_u. The number of transition step evaluations required for the adjoint calculation is given by t_u, since at the first level $c_1 l_u/(c_1+1)$ transition steps have to be evaluated to reach the second level. At the second level, one group of transition steps is divided into c_2+1 groups. Hence, $c_2(l_u/c_1+1)/(c_2+1)$ time steps have to be evaluated in each group to reach the third level. Therefore, we obtain $(c_1+1)c_2(l_u/c_1+1)/(c_2+1) = c_2 l_u/(c_2+1)$ at the second level and so on. It follows that each transition step φ_i is evaluated at most r times.

The two-level as well as the multilevel checkpointing techniques have the drawback that each checkpoint is assigned to a fixed level and cannot be reused at other levels. A method that optimally reuses the checkpoints as soon as possible is developed in the remainder of this section. Since we assume throughout that no backward pointers are available, one can derive the following characterization of corresponding reversal strategies.

Definition (RS): REVERSAL SCHEDULES
Given $c \geq 1$ and starting with $l \geq 1$, $i = 0$,
perform a sequence of basic actions

$$\begin{aligned}
A_k &\equiv \textit{increment } i \textit{ by } k \in [1, l-i-1],\\
C_j &\equiv \textit{copy } i \textit{ to checkpoint } j \in [1, c],\\
R_j &\equiv \textit{reset } i \textit{ to checkpoint } j \in [1, c],\\
D &\equiv \textit{decrement } l \textit{ by } 1 \textit{ if } i = l-1,
\end{aligned}$$

until l has been reduced to zero.

This task is always feasible, as one may use the sequence of actions

$$\begin{aligned}
C_1 + \mathcal{W}(l,1) \;&\equiv C_1 + A_{l-1} + D + R_1 + A_{l-2} + D + \\
&\quad \ldots + R_1 + A_k + D + \ldots + R_1 + A_1 + D + R_1 + D\,,
\end{aligned} \qquad (12.1)$$

where the $+$ indicates successive performance. This schedule corresponds to the total recalculation option discussed in section 12.2, as it makes do with a single checkpoint $c = 1$. The downside is that it involves many advances A_k over long distances $k > 0$.

In general we want to minimize the following two objectives:

primary: $t \equiv$ sum over k for all A_k performed,

secondary: $q \equiv$ number of times any C_j is performed.

In other words, t counts the total number of forward steps and q counts the number of checkpoint settings. For an evolutionary system, t represents the total number of times any one of the transition functions φ_i needs to be evaluated. For the trivial schedule above, we have $t = (l-1)+(l-2)+\cdots+1 = (l-1)l/2$ and $q = 1$. By defining their unweighted sum as the primary objective, we assume implicitly that the computational effort of these evaluations φ_i is similar for all $i = 1 \ldots l$.

Since it will turn out that for most combinations (l, c) minimizing t does not determine optimal offline reversal schedules uniquely, we introduce q, the total number of checkpoint settings, as a secondary objective. Hence we may define

$$t(l, c) \equiv \text{minimal } t \text{ over all schedules for given } (l, c),$$

$$q(l, c) \equiv \text{minimal } q \text{ over all schedules with } t = t(l, c).$$

For general combinations (l, c) the minimal value $q(l, c)$ is still attained by a variety of schedules, so that we could introduce a third objective as a tiebreaker. However, the number of checkpoint reads, or restarts R_j, is not suitable for that purpose, as it equals $l - 1$ for all schedules that minimize t.

In the context of adjoint calculation we are not merely interested in program reversal. We will interpret the decrement action D as reversing = recording + returning of the currently last computational step φ_{i+1} from state v_i to v_{i+1} with $i = l - 1$. Now we may reinterpret the tree reversal displayed in Fig. 12.11 as the sequence of actions on a linked list with six pointers and seven nodes sketched in Fig. 12.12.

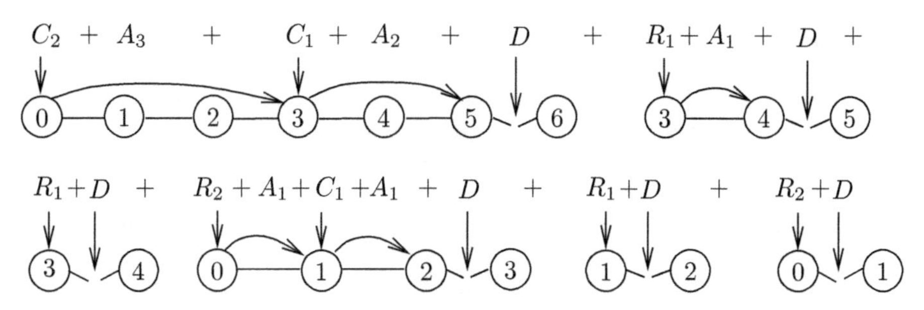

Figure 12.12: Reversal Schedule of Fig. 12.11 for the Problem of Fig. 12.10

By inspection of Fig. 12.12 we note that $c = 2$ and that the total number of forward steps equals $t = 3 + 2 + 1 + 1 + 1 = 8$. The number of checkpoint settings equals $q = 3$ and the number of restarts is $5 = 6 - 1$. We will often relate t to l, the number of advances in the underlying forward simulation run, and obtain the bound

$$t \leq r\, l\,,$$

where r is the maximal number any particular step is advanced; here its value is 2. This number r has the interesting property that it always bounds the ratio between the temporal complexity of the offline reversal schedule and the original simulation, even when the various time steps have widely varying temporal complexity. The reason is that, as one can see by inspection of Fig. 12.12, each of the individual steps is advanced at most $r = 2$ times. More specifically, only the first, second, and fourth are advanced twice, the last never, and the remaining ones only once. As we will see shortly, all schedules that minimize t have this desirable property, so we face only a very slight deterioration in the complexity ratio t/l when the steps have widely varying operation counts.

Dynamic Programming for Minimizing Time

Fortunately the minimization of t and q is not a hard problem in the combinatorial sense. In fact, under our assumption one can obtain explicit formulas for the optimal values t and q and the offline reversal schedules by which they are achieved. Even if we gave up the assumption of uniform step complexity and specified a cost for each link, one could minimize the accordingly weighted sum t in $O(cl^2)$ operations.

To obtain this quadratic dependence on l one may exploit a certain monotonicity property [Wal99] in a way that is familiar from the computation of optimal binary search trees [Knu98]. This desirable state of affairs hinges on our other assumption: all checkpoints have the same size. Using it, we can decompose the reversal schedules into a reasonable number of optimal substructures, a concept that lies at the heart of dynamic programming. Otherwise, the following crucial lemma need not hold.

Lemma 12.2 (Checkpoint Persistence)
Any offline reversal schedule that minimizes t for given (c, l) can be modified such that

> *After some checkpoint setting C_j at a state i, the next action C_j occurs only when l has already been reduced to i or below. Moreover, until that time no actions involving the states between 0 and i are taken.*

This assertion remains true when the cost of traversing individual links varies.

Proof. Suppose the first assertion is false in that the action C_j was taken at some state i and then later again at some other state with $l > i$ all the time. Since it is obviously always wasteful to advance through an existing checkpoint, all that can have happened concerning state i in between the two successive actions C_j are one or more resets R_j followed by advances over various lengths. Then these advances would all be reduced in length had we performed C_j at least one step further at state $i + 1$ in the first place. This violates the assumption that t was already minimized and thus leads to a contradiction irrespective of whether the steps have uniform costs. ∎

The simple observation of Lemma 12.2 has the powerful implication that optimal reversal schedules for some chain $(0, l)$ can without loss of generality be decomposed into subschedules dealing with the parts $(0, i)$ and (i, l). After taking the initial checkpoint, the state is advanced to some "midpoint" i, then the subchain (i, l) is reversed using the available $(c - 1)$ checkpoints, and finally $(0, i)$ is reversed using, again, the full set of c checkpoints. Since the objective function is in any case linear, the subschedules must also be optimal in the same sense as the overall schedule. Thus the optimal substructuring principle of dynamic programming can be applied, and under the assumption of uniform step costs, one even obtains an explicit solution.

Proposition 12.2 (Decomposition of Sequential Reversals)
As a consequence of Lemma 12.2 *we have the dynamic programming relation*

$$t(l,c) \equiv \min_{1\leq\hat{l}<l} \left\{\hat{l} + t(l-\hat{l}, c-1) + t(\hat{l}, c)\right\} ,$$

which yields the explicit binomial form

$$t(l,c) = r(l,c)\, l - \beta(c+1, r-1) , \tag{12.2}$$

where $\beta(c,r) \equiv (c+r)!/(c!r!)$ *and* $r = r(l,c)$, *the unique integer satisfying* $\beta(c,r-1) < l \leq \beta(c,r)$.

Proof. The three terms on the right-hand side of $t(l,c)$ represent $\hat{l}$, the cost for advancing to the interior checkpoint at $\hat{l}$; $t(l-\hat{l}, c-1)$, the cost for reversing the right subchain $(l-\hat{l}, l)$ with $c-1$ checkpoints; and finally $t(\hat{l}, c)$, the cost for reversing the left subchain $(0, \hat{l})$. The proof of the explicit formula for $t(l,c)$ proceeds by induction but is a little technical. It can be found in [GP+96] and, with notation close to that used here, in [GW00]. ■

The expression for $t(l,c)$ given in the proposition is certainly easy to compute, but due to the implicit dependence on r it is a little hard to "see" directly how t depends on c and more importantly on l. To obtain an intuitive understanding let us first look at the ratio r, which satisfies as a consequence of the proposition the relation

$$(r-1)c/(c+1) \leq t/l \leq r .$$

Provided c is of significant size, we may therefore consider r as the integer approximation to the cost of reversing an evolutionary system over l time steps using c checkpoints relative to the cost of its forward simulation.

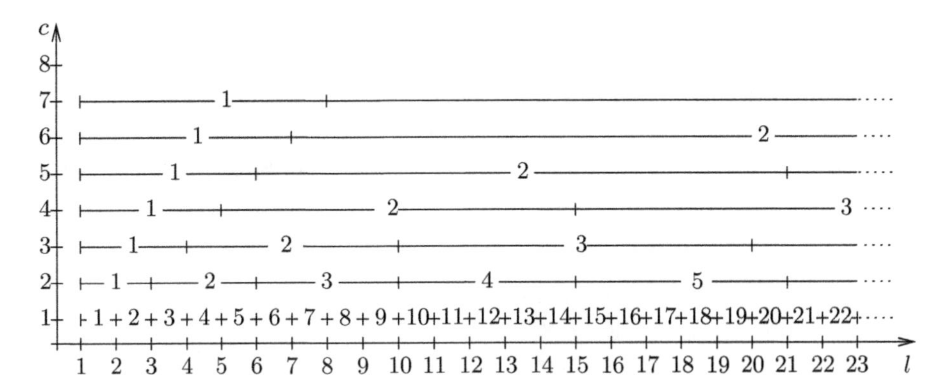

Figure 12.13: Relative Cost r for Chain Length l and Checkpoint Number c

The values of r for given l and c are displayed in Fig. 12.13. As one can see, the growth of the relative cost r as a function of l is quite slow once c has

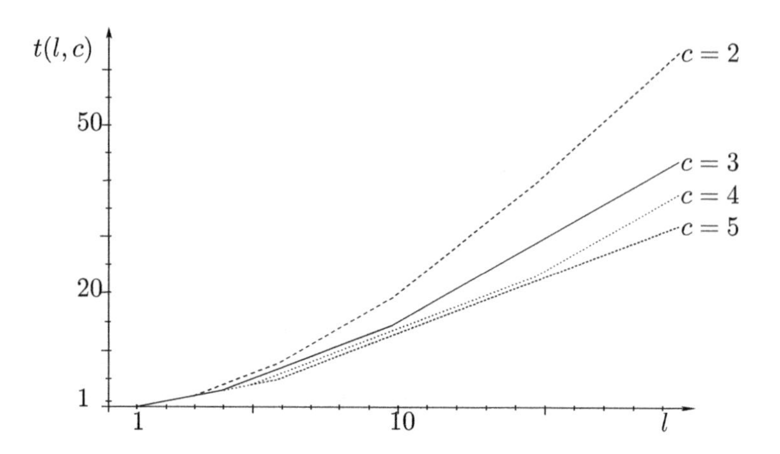

Figure 12.14: Absolute Cost t for Chain Length l and Checkpoint Number c

reached a nontrivial size. The resulting values of $t = t(l,c)$ as a function of l are displayed in Fig. 12.14 for $c = 2, 3, 4$, and 5.

As one can see in Fig. 12.14, the absolute cost $t(l,c)$ is a piecewise linear function of l whose slope is given by the monotonically growing relative cost $r(l,c)$. Due to the rapid growth of the intervals in which $t(l,c)$ is linear, the average slope $t(l,c)/c$ relative to the origin is still greater than

$$(r(l,c) - 1)c/(c+1) ,$$

so that the integer $r(l,c)$ is indeed a good measure of the cost penalty incurred in going from simulation to reversal.

Dual Approach for Maximizing Length

To gain a better understanding of the binomial coefficients occurring in Proposition 12.2 and the structure of the optimal reversal schedules, let us consider a *dual* optimization problem, which is a little simpler and was historically solved first.

Rather than minimizing t (and subsequently q) given l and c, one may ask for the maximal list length l that can be reversed using c checkpoints under the constraint $t \leq rl$ for some second integer parameter r. In fact, we will impose the even more stringent side constraint that none of the links in the chain can be traversed more often than r times. It can be easily checked (see Exercise 12.2) that Lemma 12.2 also applies to schedules that are optimal solutions of this dual problem. Hence we again have decomposition into optimal subschedules and obtain the recursive relationship

$$l(c,r) = l(c,r-1) + l(c-1,r) . \tag{12.3}$$

This equation may be visualized as follows.

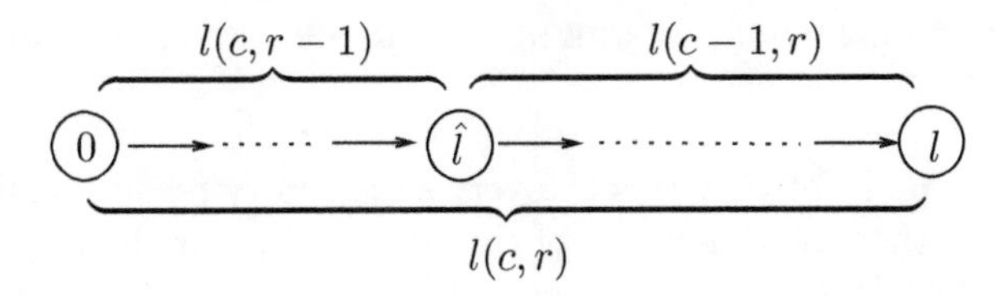

For computing a general expression of $l(c,r)$ we need optimal values of $l(c,r)$ for $c = 1$ and $r = 1$. When $c = 1$ one can only apply the total recalculation strategy yielding $l = r+1$, since the first link will then be traversed the maximum number of r times. Similarly, if $r = 1$, one can only advance once through each link yielding $l = c + 1$, as all but the last state must be placed in checkpoints. Hence we have the special values

$$l(1,r) = r + 1 \quad \text{and} \quad l(c,1) = c + 1 \quad \text{for} \quad c, r > 1 \; .$$

Starting with these boundary conditions we see that the recursion (12.3) generates the Pascal triangle and we have the following result.

Proposition 12.3 (Binomial Reversal Schedules)
The maximal solution $l = l(c,r)$ of the dual optimization problem is given by

$$\begin{aligned} l(c,r) = \beta(c,r) &\equiv \binom{c+r}{c} \equiv \frac{(c+r)!}{c!r!} \\ &\approx \frac{1}{\sqrt{2\pi}} \left[1 + \frac{r}{s}\right]^c \left[1 + \frac{c}{r}\right]^r \sqrt{\frac{1}{c} + \frac{1}{r}} \\ &\approx \frac{1}{\sqrt{2\pi \min(c,r)}} \left[\frac{\max(c,r)}{\min(c,r)}\, e\right]^{\min(c,r)} , \end{aligned} \tag{12.4}$$

where the first approximation is based on Stirling's formula for the factorial.

An appealing feature of Proposition 12.3 is the complete symmetry of l as a function of c and r, which we may view as growth factors with regard to spatial and temporal complexity, respectively. Moreover, if we accept a similar penalty in both, then l as a function of $c \sim r$ is exponential. If on the other hand, we keep one of these parameters, normally c, constant, then l grows like a power of the other. More specifically, we have

$$l \approx \begin{cases} \exp(r)/\sqrt{r} & \text{if} \quad c \sim r \; , \\ c^r & \text{if} \quad c \gg r \equiv \text{const} \; , \\ r^c & \text{if} \quad r \gg c \equiv \text{const} \; . \end{cases}$$

The last relation is probably the most important, and taking the cth root we obtain the following rule.

RULE 21

THE RELATIVE REVERSAL COST (r) GROWS LIKE THE CHAIN LENGTH (l) RAISED TO THE RECIPROCAL OF THE NUMBER OF CHECKPOINTS AVAILABLE (c).

From a practical point of view this is a very satisfactory result. Again it should be noted that neither the dimension of the state space nor the number of independent variables plays any role in this fundamental relation. Similar asymptotic estimates apply when one uses the simple bisection strategy $\hat{l} = l/2$ rather than the binomial partitioning employed here. However, our provably optimal strategy reduces the number of checkpoints needed by about a half and utilizes them more evenly.

Optimal Offline Reversal Schedules

So far we have emphasized the possible relations between the chain length l and the reversal costs t (absolute) or r (relative) for given checkpoint number c without worrying very much about the underlying schedules. Another consequence of Lemma 12.1 is that we may define the j occurring in the Definition (RS) of reversal schedules as the number of presently still-available checkpoints. It can be updated according to the following rules.

Update of Checkpoint Counter

- Initialize $j = c$ at the very beginning.
- Decrement j by 1 after each C_j.
- Terminate irregularly when $j = 0$.
- Increment j by 1 after R_j when $i = l - 1$.

In other words, the set of checkpoints can be managed in a LIFO fashion with only the last element being accessed at any stage. However, in contrast to a conventional stack, repeated nondestructive reads of the last element must be possible. Let $\mathcal{W}(c, l)$ denote an optimal schedule for reversing a chain of length $l > 0$ using $c > 0$ checkpoints, so that in particular $\mathcal{W}(c, 1) = D$ and $\mathcal{W}(1, l)$ is as given in (12.1). Then we may paraphrase Lemma 12.2 as saying that without loss of generality

$$\mathcal{W}(c, l) = A_{\hat{l}} + C_j + \mathcal{W}(c-1, l-\hat{l}) + R_j + \mathcal{W}(c, \hat{l}).$$

Here $\hat{l} < l$ is an optimal placing of the next checkpoint. When l is exactly equal to $\beta(c, r)$ for some integer r, then (12.3) requires that the next checkpoint is set at the "midpoint" $\hat{l} = \beta(c, r-1)$ so that the original range $(0, l)$ is partitioned into the subchains $(0, \hat{l})$ and $(\hat{l}, l)$ with the relative lengths given by

$$1 = \frac{r}{c+r} + \frac{c}{c+r} = \frac{\hat{l}}{l} + \frac{l-\hat{l}}{l}\,.$$

Hence we see that if we have many available checkpoints c compared to the size of the relative cost factor r, then the next checkpoint is set relatively early at a "midpoint" in the left part of the full chain $(0, l)$. If, on the other hand, checkpoints are scarce and we have to accept a relatively (to c) large number of repeats r, then the "midpoint" moves to the right. In the spirit of dynamic programming, the subranges $(0, \hat{l})$ and $(l - \hat{l}, l)$ are then subdivided in the same proportions but with (c, r) replaced by $(c, r-1)$ and $(c-1, r)$, respectively.

When l does not have one of the special binomial values $\beta(c, r)$, then the same partitioning rule applies in principle, but the "midpoint" that minimizes the absolute cost t is no longer unique. A certain arbitrariness remains even if we impose the secondary objective of minimizing q, the number of checkpoint settings C_j. The range of optimal values for $\hat{l}$ given arbitrary integers c and l is represented by the gray-shaded area in Fig. 12.15.

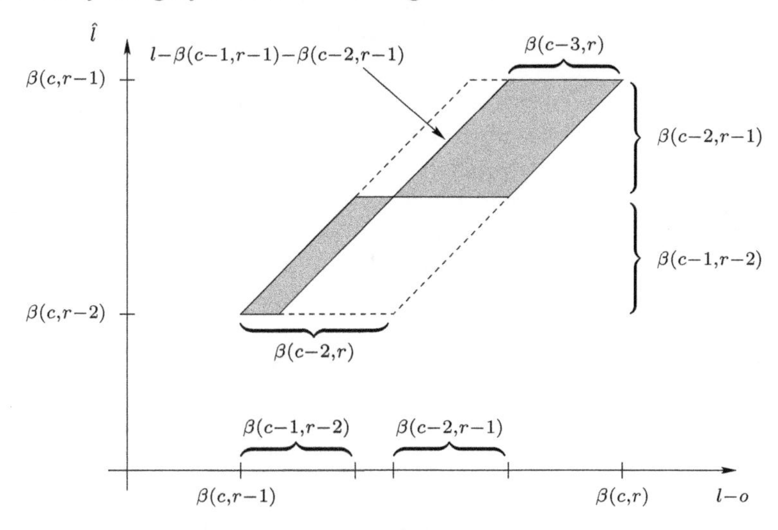

Figure 12.15: The Optimal Domain of $\hat{l}$ for $l > 2$

The outer parallelogram with the dashed boundary contains all combinations $(l, \hat{l})$ minimizing t, the total number of forward steps. The two smaller parallelograms inside contain the combinations that also minimize q, the number of checkpoint settings. The following result gives one particular value of $\hat{l}$ that was selected for algebraic simplicity. Also, both l and $\hat{l}$ have been shifted by a value o, representing the original state of a subchain (o, l) that need no longer be zero. Naturally this relabeling of states does not change any of the mathematical relations derived earlier.

Proposition 12.4 (Optimal Offline Chain Reversal Schedule)
To minimize first t and then q one may partition any subrange (o, l) at the state

$$\hat{l} = \begin{cases} o + \beta(c, r-2) & \text{if } (l-o) \leq \beta(c, r-1) + \beta(c-2, r-1)\,, \\ o + \beta(c, r-1) & \text{if } (l-o) \geq \beta(c, r) - \beta(c-3, r)\,, \\ l - \beta(c-1, r-1) - \beta(c-2, r-1) & \text{otherwise,} \end{cases}$$

where again $\beta(c, r-1) < l - o \leq \beta(c, r)$. *Provided this choice is applied recursively starting with* $o = 0$, *the resulting* r *is given by Proposition* 12.1 *and the optimized* q *has the value*

$$q(l, c) = \begin{cases} \beta(c-1, r-1) & \text{if} \quad l \leq \beta(c, r-1) + \beta(c-1, r-1)\,, \\ l - \beta(c, r-1) & \text{if} \quad l \geq \beta(c, r-1) + \beta(c-1, r-1)\,. \end{cases}$$

The number of checkpoint reads is always equal to $l - 1$.

Proof. See [GW00] and the exercises at the end of this chapter. ∎

The slope of $q(l, c)$ given in Proposition 12.4 alternates between 0 and 1, and we have the approximate relation

$$\frac{q}{l} \approx \frac{c}{c+r}\,,$$

which holds exactly when $l = \beta(c, r)$. Hence we see that in the desirable situation where c, the number of stored checkpoints, greatly exceeds the number r of repeated forward sweeps, almost all states serve once as checkpoints. Since we always have $(l-1)$ restarts, the total amount of data that needs to be transferred for the purpose of checkpointing varies in any case between l and $2l$ times the size of a single state. The upper bound is attained exactly for the basic reversal schedule, where every state is saved on the way forward and then read on the way back.

While selecting $\hat{l}$ to minimize q as secondary objective makes sense when l is exactly known a priori, other considerations may apply when this is not the case. It follows by inspection of the larger parallelogram in Fig. 12.15 that $\hat{l} = \beta(c, r-1)$ is a time-optimal choice for all l between the bounds $\beta(c, r) - \beta(c-2, r)$ and $\beta(c, r) + \beta(c-2, r+1)$. These intervals are quite large especially when $c > r$, so that an adaptive procedure for selecting $\hat{l}$ allows for a fair bit of uncertainly regarding l on the first forward sweep.

To visualize reversal schedules we may use the representation employed in Fig. 12.16. It displays once more the optimal schedule considered before in Figs. 12.11 and 12.12. However, now the vertical axis represents the time step counter of the original evolutionary system, and the horizontal axis represents execution time in units of a single forward simulation step.

Horizontal lines represent checkpoints set by some action C_j, and solid slanted lines represent advances A_k. The differentiation action D consists of a dotted line representing the recording motion and a dashed line representing the subsequent returning motion. Here we have assumed for simplicity that both parts of a step reversal take just as much time as a single advance. As a result, the total execution time is given by $t + 2\,l = 8 + 12 = 20$. If the reversal requires more or less than two units of time, one merely has to extend or reduce the hooks vertically. This adjustment changes the total execution time but does not affect the reversal schedule, which is optimal as long as all advances take the same time per step.

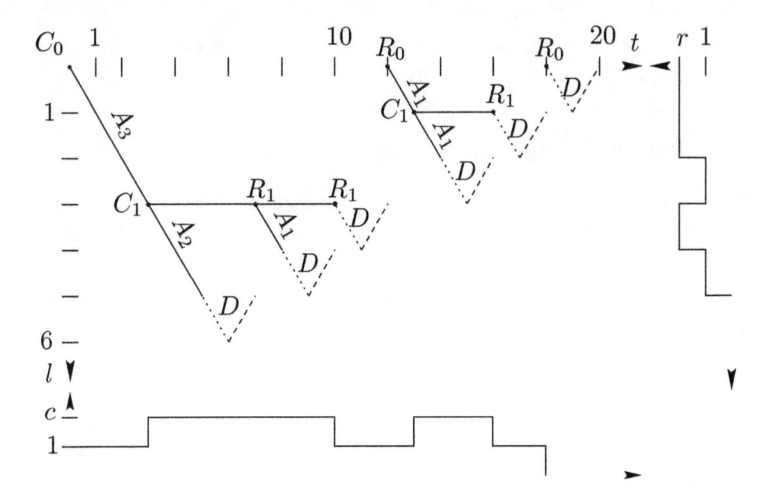

Figure 12.16: Optimal Schedule for $c = 2$ and $l = 6$ Yielding $t = 8$ and $q = 3$

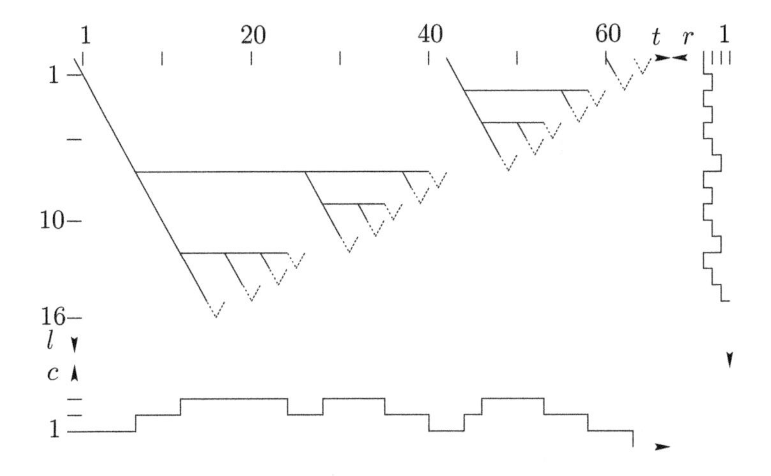

Figure 12.17: Optimal Sequential Schedule for $c = 3$ and $l = 16$

The profile on the lower margin of Fig. 12.16 depicts the number of checkpoints in use at any stage of the calculation, which varies between 1 and 2. The profile on the right margin depicts the number of times any particular time step has been advanced, which also varies between 1 and the upper bound $r = 2$. Here, as throughout, we have not counted the final recording motion among the advances. The uniform bound $r = 2$ on the number of repeats still limits the runtime ratio for the advances when the temporal complexity varies from step to step. Figure 12.16 represents a solution of the dual problem as it optimizes l for given $c = 2 = r$.

A more realistic scenario is depicted in Fig. 12.17, where a given $l = 16$ and $c = 3$ yield $t = 33$ and $r = 3$, so that the total execution time is $33 + 2 * 16 = 65$, again assuming that a reversal takes two units. The solution of the dual problem for $c = 3 = r$ would have been $l = 20$.

Comparison of Different Checkpointing Strategies

For uniform checkpointing and binomial checkpointing, the integer r has the same meaning, namely, the maximal number of times any particular transition step φ_i is evaluated during the reversal. Hence for comparing both approaches, assume at the beginning that r has the same value and that the same amount of memory is used, i.e., $M_u = M_b = c$.

Now, we examine the maximal number of time steps l_u and l_b for which a reversal can be performed using the two different strategies. To maximize l_u we need to choose the number of checkpoints c_i on the maximal level r such that

$$\text{Max } l_u = \prod_{i=1}^{r}(1 + c_i) \qquad \text{with} \qquad \sum_{i=1}^{r} c_i = c \, .$$

Assuming that r is a factor of c one can easily see that the maximum is attained when constantly $c_i = c/r$ so that then

$$l_u^* = \left(\frac{c}{r} + 1\right)^r = \left(\frac{c + r}{r}\right)^r .$$

Proposition 12.3 yields for binomial checkpointing

$$l_b^* = \binom{c + r}{r} = \prod_{i=0}^{r-1}\left(\frac{c}{r - i} + 1\right) .$$

Obviously, one has

$$\frac{c}{r} + 1 < \frac{c}{r - i} + 1 \qquad \text{for } 0 < i \leq r - 1 \, .$$

These inequalities yield $l_u^* < l_b^*$ if $r \geq 2$. Hence for all $r \geq 2$ and a given c, binomial checkpointing allows the reversal of a larger number of transition steps compared to multilevel checkpointing. In more detail, using Stirling's formula we obtain

$$\frac{l_b}{l_u} \approx \binom{c + r}{r}\left(\frac{c}{r} + 1\right)^{-r} = \frac{1}{\sqrt{2\pi r}}\left(\frac{c}{r} + 1\right)^c \approx \frac{1}{\sqrt{2\pi r}} \exp(r) \, .$$

Hence, the ratio of l_b^* and l_u^* grows exponentially in r without any dependence on the number of available checkpoints. Fig. 12.18 shows l_u^* and l_b^* for the most important values $2 \leq r \leq 5$.

Since r denotes the maximal number of times each transition step is evaluated, we have the following upper bounds for the number of transition steps evaluated during the reversal using r-level uniform checkpointing and binomial checkpointing, respectively:

$$t_u = c\left(\frac{c}{r} + 1\right)^{r-1} < r\, l_u^* \qquad \text{and} \qquad t_b = r l_b^* - \binom{c + r}{r - 1} < r\, l_b^* \, .$$

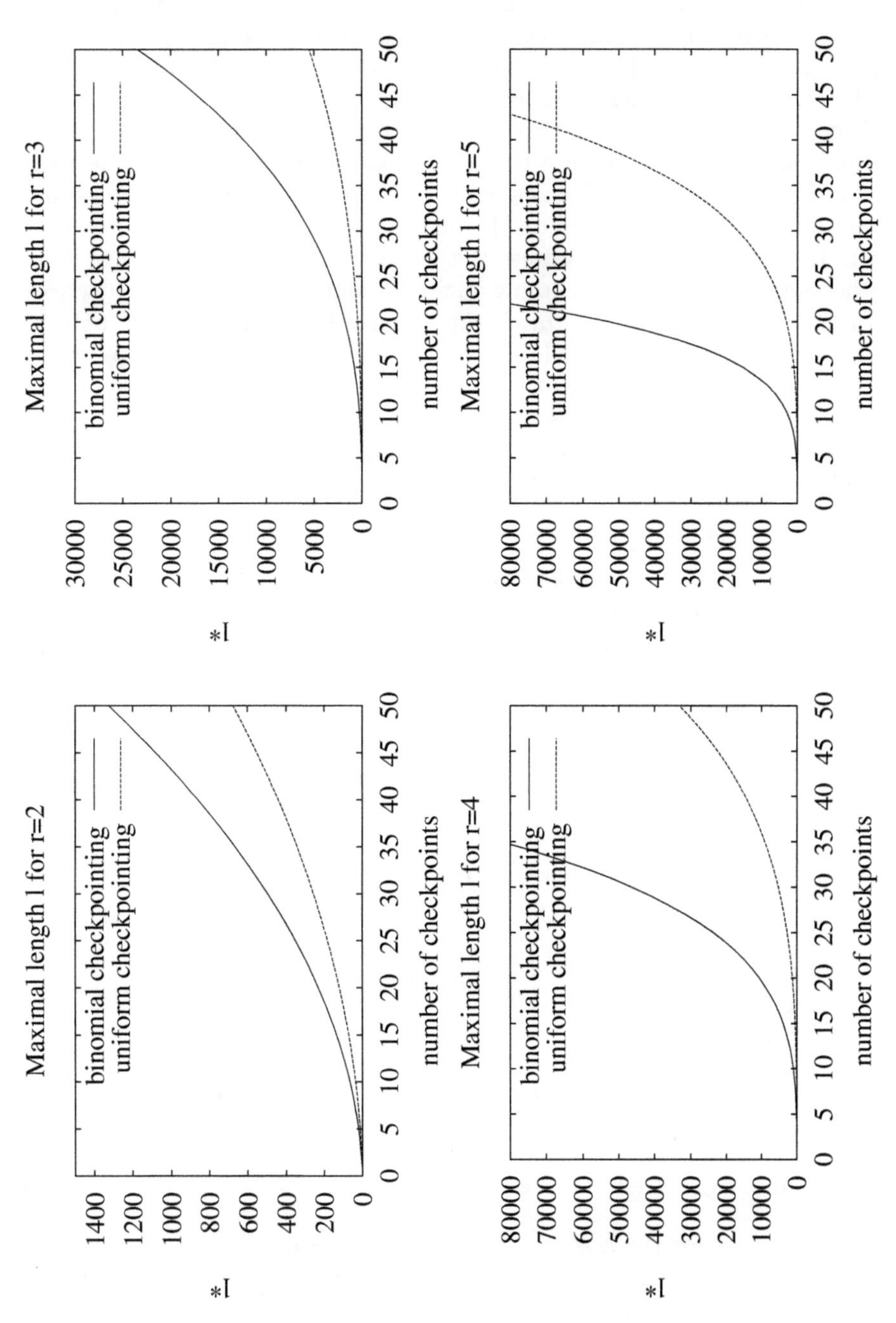

Figure 12.18: l_u^* and l_b^* for $u = 2, 3, 4, 5$

For example, it is possible to compute the reversal for $l = 23000$ transition steps with only 50 checkpoints, less than $3l$ time step evaluations, and l adjoint steps using binomial checkpointing instead of three-level checkpointing, where $l_u^* \leq 5515$. If we allow $4l$ transition step evaluations then 35 checkpoints suffice to compute the reversal of 80000 transition steps using binomial checkpointing, where $l_u^* \leq 9040$. These numbers are only two possible combinations taken from Fig. 12.18 to illustrate the really drastic decrease in memory requirement that can be achieved if binomial checkpointing is applied.

However, usually the situation is the other way round, i.e., one knows l and/or c and wants to compute the reversal as cheap as possible in terms of computing time. Here, the first observation is that r-level uniform checkpointing introduces an upper bound on the number of transition steps the reversal of which can be computed, because the inequality $l \leq (c/r+1)^r$ must hold. Furthermore, binomial checkpointing allows for numerous cases also a decrease in runtime compared to uniform checkpointing. For a given r-level uniform checkpointing and $M_u = c$, one has to compare t_u and t_b. Let r_b be the unique integer satisfying (12.2). Since at least one checkpoint has to be stored at each level, one obtains the bound $r \leq c$. That is, one must have $c \geq \log_2(l)$ to apply uniform checkpointing. Therefore, the following combinations of r and r_b are possible for the most important, moderate values of u:

$$r = 3 \;\Rightarrow\; r_b \in \{2,3\}, \qquad r = 4 \Rightarrow r_b \in \{3,4\}, \qquad r = 5 \Rightarrow r_b \in \{3,4,5\}\,.$$

For $3 \leq r \leq 5$, one easily checks that $t_u > t_b$ holds if $r_b < r$. For $r = r_b$, one can prove the following, more general result.

Proposition 12.5 (Comparison of Uniform and Binomial Checkpointing) *Suppose for a given l and an r-level uniform checkpointing with $M_u = c$ that the corresponding r_b satisfying (12.2) coincide with r. Then, one has*

$$\begin{array}{lllll} t_2 &=& 2l - c - 2 &=& t_b \quad \text{if } r = r_b = 2\,, \\ t_u &>& t_b & & \quad \text{if } r = r_b > 2. \end{array}$$

Proof. For $r_b = r = 2$ the identity $t_2 = t_b$ is clear. For $r = r_b > 2$, the inequality

$$\sum_{i=1}^{u}\prod_{\substack{j=1\\ j\neq i}}^{u}(c_j+1) = \frac{(r-1)!}{(r-1)!}\left(\prod_{j=1}^{r-1}(c_j+1) + (c_r+1)\sum_{i=1}^{r-1}\prod_{\substack{j=1\\ j\neq i}}^{r-1}(c_j+1)\right)$$
$$< \frac{1}{(r-1)!}\prod_{i=2}^{r}\left(\sum_{j=1}^{r} c_j + i\right) = \binom{c+r}{r-1}$$

holds. Using the definitions of t_u and t_b, this relation yields $t_u > t_b$. ■

Hence, except for the case $r = r_b = 2$, where t_u and t_b coincide, the runtime caused by binomial checkpointing is less than the one caused by uniform multilevel checkpointing if $r = r_b$.

Additional Runtime Effects

The enormous memory reduction that becomes possible when using checkpointing strategies may even lead to an overall runtime reduction despite the fact that significant recomputations are required. This runtime reduction is due to a well-known effect, namely, that computing from a level of the memory hierarchy that offers faster access cost may result in a significant smaller runtime.

The numerical example that serves to illustrate these runtime effects of the checkpointing procedure is discussed in detail in [KW06]. The underlying problem is a fast turnaround maneuver of an industrial robot, where one wants to minimize an energy-related objective. The complete equations of motion for this model can be found in [KB03]. The robot's task to perform a turnaround maneuver is expressed in terms of initial and terminal conditions in combination with control constraints. However, for illustrating the runtime effects of the binomial checkpointing integrated into ADOL-C, only the gradient computation of the objective with respect to the control was considered.

To compute an approximation of the trajectory, the standard Runge–Kutta method of order 4 was applied for time integration, resulting in about 800 lines of code. The integration and derivative computations were computed using an AMD Athlon64 3200+ (512 kB L2-Cache) and 1GB main memory. The resulting averaged runtimes in seconds for one gradient computation are shown in Fig. 12.19, where the runtime required for the derivative computation without checkpointing, i.e., the total taping (TT), is illustrated by a dotted line.

The runtime needed by the checkpointing approach (CP) using $c = 2, 4, 8, 16, 32, 64\ (128, 256)$ checkpoints is given by the solid line. To illustrate the corresponding savings in memory requirement, Table 12.8 shows the tape sizes for the basic approach as well as the tape and checkpoint sizes required by the checkpointing version. The tape size for the latter varies since the number of independents is a multiple of l due to the distributed control.

Table 12.8: Memory Requirements for $l = 100, 500, 1000, 5000$

# time steps l	100	500	1000	5000
	without checkpointing			
tape size (KByte)	4 388.7	32 741.9	92 484.7	1 542 488.1
	with checkpointing			
tape size (KByte)	79.3	237.3	434.8	2 014.9
checkpoint size (KByte)	11.4	56.2	112.2	560.2

One fundamental checkpointing characteristic, i.e., the more checkpoints that are used, the less runtime the execution needs, is clearly depicted by case $l = 1000$ in Fig. 12.19. The smaller runtime for the total taping completes the setting. However, the more interesting cases for this example are $l = 100$ and $l = 5000$, respectively. In these situations a smaller runtime was achieved even though checkpointing was used. For the robot example the computation could

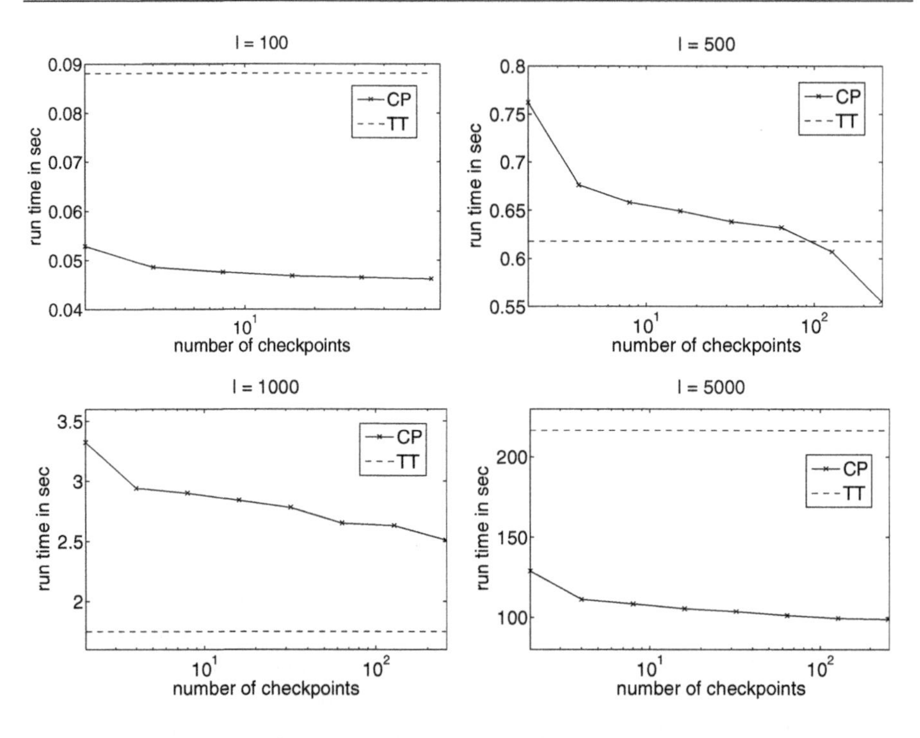

Figure 12.19: Comparison of Runtimes for $l = 100, 500, 1000, 5000$

be redirected from main memory mainly into the L2-cache of the processor ($l = 100$) and from at least partially hard disk access completely into the main memory ($l = 5000$). The remaining case from Fig. 12.19 ($l = 500$) depicts a situation where only the tape and a small number of the most recent checkpoints can be kept within the L2-cache. Hence, a well-chosen ratio between l and c causes in this case a significantly smaller recomputation rate and results in a decreased overall runtime, making the checkpointing attractive.

Online Checkpointing

As stated before, optimal offline checkpointing can be applied only if the number of transition steps l to be reversed is known in advance. This is usually not the case if an adaptive time-stepping procedure is applied. For example, in the context of flow control, the PDEs to be solved are usually stiff, and the efficiency of the solution process relies therefore on the appropriate adaptation of the time step size. Then one has to decide on the fly during the first forward integration where to place a checkpoint. Hence, without knowing how many transition steps are left to perform, one has to analyze the current distribution of the checkpoints. This fact makes an offline checkpoint optimization impossible. Instead, one may apply a straightforward checkpointing by placing a checkpoint each time a certain number of time steps has been executed. This transforms the uncertainty in the number of time steps to a uncertainty in the number of checkpoints needed as done, for example, by CVODES [SH03]. However, when the amount of memory per checkpoint is very high, one certainly wants to determine the number of checkpoints required a priori.

Alternatively, one may discard the contents of one checkpoint to store the current available state depending on the transition steps performed so far. Obviously, one may think that this procedure could not be optimal since it may happen that one reaches the final time just after replacing a checkpoint, in which case another checkpoint distribution may be advantageous. A surprising efficient heuristic strategy to rearrange the checkpoints is implemented by the online procedure arevolve [SH05]. Here, the efficiency of a checkpoint distribution is judged by computing an approximation of the overall recomputation cost caused by the current distribution. This number is compared with an approximation of the recomputation cost if one resets a checkpoint to the currently available state. Despite the fact that significant simplifications are made for approximating the required recomputations, the resulting checkpointing schemes are surprisingly efficient. Naturally, the minimal cost can be computed only afterwards when the number of time steps is known.

12.4 Parallel Reversal Schedules

In sections 12.2 and 12.3 we assumed that the original function evaluation, as well as its reversal for the purpose of adjoint calculations, are performed on a single-processor machine. In this final section of Chapter 12 we briefly examine ways of maintaining and enhancing parallelism in reverse differentiation. We have already noted in Chapter 9 that the associativity of the chain rule opens up another dimension of potential concurrency that could be exploited by cross-country methods for Jacobian calculations. These generalizations of the forward and reverse modes accumulate derivative quantities in a different fashion, possibly taking into account the seeds $\dot{x}$ or $\dot{X}$ and $\bar{y}$ or $\bar{Y}$ only at the very end. In contrast, we will continue here to consider only reorganizations of the function evaluation process itself, with the actual formulas for the backward propagation of adjoint quantities remaining completely unchanged.

Parallel Chain Reversal

The optimal reversal schedules derived in section 12.3 were defined as a sequence of actions that need to be performed exactly in the given order. Most of these actions, namely, the repeated forward advances, have nothing to do directly with the backward propagation of adjoints that we have incorporated into the action D. The latter must be performed for the links i in descending order $i = l \dots 1$ and cannot be parallelized. However, some of the various advances A_k can easily be performed concurrently. As an extreme case we may again consider the total recalculation schedule given in (12.1). Suppose we had only $c = 1$ checkpoints but $l - 1$ processors on a shared memory system. Then we could start the processors up with $i = 0$ one after the other to perform the actions $A_{l-1}, A_{l-2}, \ldots$, and A_1, respectively, and deliver the states $i = l - 1 \ldots 0$ in time for the final action D. For this delivery to be just in time the dispatch times would have to be staggered by the exact amount of time it takes one extra processor to perform the differentiation action D.

In general, we will say that a parallel chain reversal schedule is *uninterrupted* if one processor can run backwards, performing the returning part of the action D without any interruption. To this end the other processors must always deliver the previous states $i = l-1 \ldots 0$ at regular time intervals of d units relative to a simple forward step A_1. As it turns out (see Exercise 12.4), simply parallelizing the binomial schedules derived in section 12.3 may exceed the number of checkpoints that are available. Corresponding to the dual formulation discussed in section 12.3, one may now consider the following optimization problem:

What is the maximal chain length l that can be reversed uninterrupted using c checkpoints and p processors given the ratio d?

This combinatorial optimization problem appears to be much harder than those originating from the sequential formulation considered in section 12.3. Also, the ratio d, which was almost irrelevant in the sequential case, has a very strong influence, and it determines the minimal elapsed time. So far, explicit formulas for provably optimal parallel schedules have been obtained only when $p \geq c$ and for special values of d. They confirm the natural expectation that l again grows exponentially in the parameters $c \sim p$ for any given d. Moreover, the optimal parallel schedules involve Fibonacci numbers rather than binomials [Wal99], and the partitioning into subchains roughly follows a fixed ratio.

For the case $d = 2$, $c = 4$, and $p = 3$, an optimal schedule has been computed by exhaustive search, yielding the chain length $l = 17$. The corresponding schedule is displayed in Fig. 12.20. As in Fig. 12.17 the vertical axis counts the number of links along the chain and may therefore be thought of as physical time in a simulation. The horizontal axis represents elapsed time on a multiprocessor measured in units of one forward computational step. All slanted lines represent running processes, whereas all horizontal lines represent checkpoints. Processors may degenerate to checkpoints by simply standing still. Hence a vertical line drawn at any wall clock time may intersect at most p slanted lines and maximally $p + c$ vertical or slanted lines altogether. The two slanted borders underneath represent the actions of the special processor, namely, running forward to the end of the chain and then running backward uninterruptedly performing the returning motion D. The other $(p - 1)$ processors perform repeated advances from suitable chosen checkpoints, with one of them doing the recording step represented by the dotted line.

The profiles on the lower margin represent the number of processors employed and the number of checkpoints used at any stage of the calculation. As one can see, both computational resources are fully exhausted at the apex, where the returning motion begins. At the beginning and towards the end, some checkpoints and processors are not needed, as is typical for the warm-up

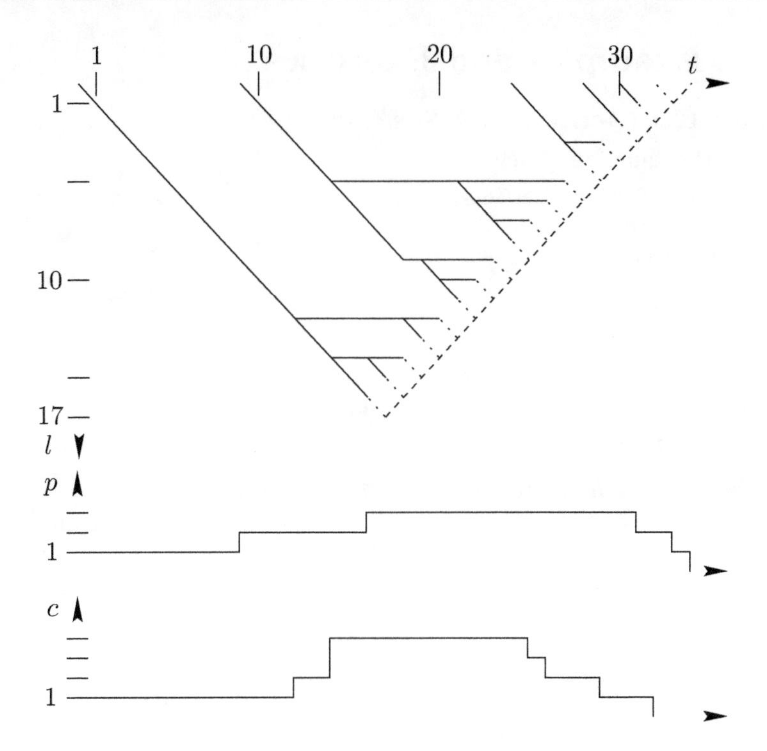

Figure 12.20: Optimal Parallel Schedule for $c = 4$, $p = 3$, and $l = 17$

and cool-down phases of parallel computations. Similar effects occur for the subschedules represented by triangles along the counterdiagonal in Fig. 12.20. They gradually obtain resources from their predecessors before their apex and then release them one by one to their successors on the way back.

The schedule plotted in Fig 12.20 is optimal under the assumption that storing and retrieving checkpoints does not cause any significant time delays. Our other assumption, namely, that the runtime ratio between performing a returning step and advancing equals 1, is not quite as unrealistic as it may appear at first. The reason is that the reversing step can be decomposed into a recording motion and a returning motion, as discussed in section 12.2. The former can be executed in parallel by some extra processor(s) and prepare the ground such that the subsequent returning motion, which actually involves adjoint quantities, may be performed by the backward runner in about the same time that it takes to advance one step.

12.5 Examples and Exercises

Exercise 12.1 (***Active Bough Structure***)

Suppose the function f has a call tree with weak and strong links such as is shown in the figure to the right. Write down the development of the active bough as in Fig. 12.6.

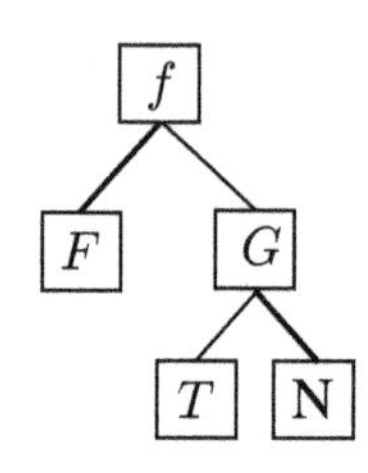

Exercise 12.2 (***Checkpoint Persistence for Dual***)
Consider a schedule that is an optimal solution of the dual optimization problem. Show that the assertion of Lemma 12.2 is also true for this schedule.

Exercise 12.3 (***Chain Reversal for $l = 20$ and $c = 3$***)
For $c = 3$, $o = 0$, and $l = 20$, construct an optimal checkpoint schedule according to Proposition 12.4. Arrange the intervals (o, l) as a binary tree starting at the root $(0, 20)$. Write down the corresponding reversal schedule using the actions according to Definition (RS).

Exercise 12.4 (***Parallelizing a Binomial Schedule***)
Figure 12.17 displays an optimal checkpoint schedule for the sequential case. One may parallelize this schedule by starting the unrecorded advances in such a way that the return motion can be performed without any delay. To this end translate all slanted lines leftward such that the endpoints of their solid parts form a straight line with slope 1 intersecting the horizontal axis at 30. The recording and returning motions then form a chain of diamonds to the right of and below this line. How many processes are needed? Are three checkpoints enough for this new parallel version of Fig. 12.17?

Chapter 13

Taylor and Tensor Coefficients

So far we have "only" considered the evaluation of first and second derivatives because they are beneficial, if not indispensable, for most nonlinear problems in scientific computing. This applies in particular to optimization and the numerical integration of stiff differential equations, possibly with algebraic side constraints. Third and higher derivatives may become important at degenerate minima and other singular points where one has to "dig deeper" to determine solution properties that may not be tied down by first and second derivatives. This leads into the domain of numerical bifurcation theory [SS+91], where singularities called cusps, swallow tails, and butterflies are determined by up to third, fourth, and fifth derivatives, respectively. By unfreezing problem parameters one may seek out these higher-order singularities and interpret them as "organizing centers" of the nonlinear problem in some larger region. With the help of AD this approach can now be applied to realistic computer models of complex systems. For an application of AD to computing Hopf bifurcation points see [GM96] and [GMS97].

A Motivating Bifurcation Example

As a motivating example consider the problem of computing a degenerate minimum of a scalar function $f(x, y, \lambda, \mu)$. The minimization is carried out only with respect to the first two variables x and y, with the last two λ and μ playing the role of bifurcation parameters. By that we mean that the following system of third-order optimality conditions becomes well defined in the full variable vector $(x, y, \lambda, \mu, \alpha)$ with $\dot{x} \equiv \cos(\alpha)$ and $\dot{y} \equiv \sin(\alpha)$.

$$
\begin{aligned}
0 &= f_x = f_y && \text{first-order necessary,} \\
0 &= f_{xx}\dot{x} + f_{yx}\dot{y} = f_{xy}\dot{x} + f_{yy}\dot{y} && \text{second-order degeneracy,} \\
0 &= f_{xxx}\dot{x}^3 + 3f_{xxy}\dot{x}^2\dot{y} + 3f_{xyy}\dot{x}\dot{y}^2 + f_{yyy}\dot{y}^3 && \text{third-order necessary.}
\end{aligned}
$$

The first condition simply requires that the gradient of f with respect to x and y vanish. The second degeneracy condition requires that the Hessian of f with respect to x and y has the null vector $(\dot{x}, \dot{y}) \equiv (\cos(\alpha), \sin(\alpha)) \in \mathbb{R}^2$. The third condition requires that the third directional derivative of f along $(\dot{x}, \dot{y})$ vanish, too. These three conditions are necessary for f to attain a degenerate minimum at the point (x, y) and can be turned into a sufficient condition by requiring that the fourth directional derivative along $(\dot{x}, \dot{y})$ be sufficiently large relative to some nonnegative cross term (see [Gri80]). Degenerate minima are important in various applications and represent, for example, critical points in phase bifurcation diagrams of mixed fluids [Mic82] and [GR87].

Now suppose we have been given f as an evaluation procedure and try to solve the "defining system of equations" [Bey84] above to locate a degenerate minimum (x, y) with suitably adjusted parameters (λ, μ). For this we obviously have to compute not only gradient and Hessian but also the third directional derivatives $\nabla_s^3 f$ with $s = (\dot{x}, \dot{y})$ representing a null vector of the Hessian. We view the fact that higher derivatives need only be evaluated on smaller subspaces as typical. As in the case of first and second derivatives this restriction can be built into the differentiation process and thus reduce the cost significantly.

Moreover, in order to compute the degenerate minimum by Newton's method we also need their gradients with respect to the full variable vector (x, y, λ, μ) for all first, second, and third derivatives involved in the defining system. For example, if $\dot{y}$ happened to be zero the gradient of the third-order necessary condition would be

$$\nabla f_{xxx} = (f_{xxxx}, f_{xxxy}, f_{xxx\lambda}, f_{xxx\mu}) \, .$$

By now it will be no surprise to the reader that we recommend the use of the reverse mode for evaluating such higher-order gradients. They are characterized by one level of differentiation with respect to all variables on top of several levels of differentiation in some special direction(s) (here the x-axis). The latter, directional differentiation will be carried out in the forward mode, and in agreement with the claims we made in Rule 10 in Chapter 5, everything can be arranged so that there is only one forward and one backward motion.

13.1 Higher-Order Derivative Vectors

Although we have not elaborated on the details, it should be clear that the degenerate minimization problem described above can be solved if we have the ability to calculate directional derivatives of the form

$$F_k(x, s) \equiv \frac{1}{k!} \nabla^k F(x) s^k = \frac{1}{k!} \frac{\partial^k}{\partial t^k} F(x + t\, s) \Big|_{t=0} \in \mathbb{R}^m \quad \text{for} \quad k \le d \tag{13.1}$$

and their adjoints

$$\begin{aligned} \bar{F}_k(\bar{y}, x, s) \equiv \bar{y}^\top \nabla_x F_k(x, s) &= \frac{1}{k!} \bar{y}^\top \nabla^k F(x)\, s^k \in \mathbb{R}^n \\ &= \frac{1}{k!} \frac{\partial^k}{\partial t^k} \bar{y}^\top \nabla_x F(x + t\, s) \Big|_{t=0} \end{aligned} \quad \text{for} \quad k < d \tag{13.2}$$

Here we have considered a general $k+1$ times continuously differentiable vector-valued function $F : \mathbb{R}^n \mapsto \mathbb{R}^m$ and used an arbitrary weight vector $\bar{y} \in \mathbb{R}^m$ for the adjoint. In the minimization example we have $m = 1$ and may thus choose simply $\bar{y} = 1$. The second equation in (13.2) relies on the commutativity of differentiation with respect to x and t. As we will see in section 13.4 the alternative interpretations of $\bar{F}_k$ as adjoint of a Taylor coefficient or Taylor coefficient of a gradient actually lead to different ways of calculating this vector.

Loosely speaking, a vector of the form $F_k(x, s)$ represents a derivative tensor contracted a number of times equaling its degree by the vector $s \in \mathbb{R}^n$, thus yielding an m-vector in the range space. In contrast, $\bar{F}_k(\bar{y}, x, s)$ represents a derivative tensor contracted one fewer times than its degree in the domain direction s but also once by the vector $\bar{y}$, thus yielding a vector of dimension n. These vectors $\bar{y}$ and $\bar{F}_k$ may be interpreted as elements in the duals of the range and domain spaces, respectively. As we will see, both kinds of vectors F_k and $\bar{F}_k$ can be computed at a cost that is proportional to the cost of evaluating $F(x)$ multiplied by the degree d squared. Theoretically, implementations with complexity order $d \log d$ based on fast convolution algorithms are possible, but it seems highly doubtful whether the crossover point is reached for any applications class of interest.

In the optical applications considered by Berz [Ber90a], the given F is the mapping from a geometrical object to its image, and the coefficient functions F_k with $k > 1$ represent "aberrations" from the desired linearity

$$F(x + ts) = F(x) + \gamma\, t\, s \quad \text{for} \quad s \in \mathbb{R}^p \times \{0\}^{n-p} \subset \mathbb{R}^n \,.$$

Here s represents the first p components of the variable vector x which range over the coordinates of the inverse image that are to be mapped with a magnification factor γ into the coordinates of the image $F(x + ts)$. The other $n - p$ components of x may be thought of as design parameters specifying, for example, the geometrical layout of a system of optical lenses. These parameters must then be tuned to minimize the sum of squares of the aberrations $F_k(x, s)$, typically up to a degree of about $d = 10$. For this design optimization process, one needs the gradient of the residual $\|F_k(x,s)\|^2/2$ with respect to x, which is exactly the adjoint vector $\bar{F}_k(\bar{y}, x, s)$ for $\bar{y} \equiv F_k(x, s)$. We consider this situation as representative of the optimization of designs whose evaluation involves higher derivatives, for example, in the form of geometrical curvatures.

Mixed Partials by Exact Interpolation

In the example above we also need the mixed derivative $f_{xy}(x, y, \lambda, \mu)$ and its gradient $\nabla f_{xy} = (f_{xyx}, f_{xyy}, f_{xy\lambda}, f_{xy\mu})$. Rather than evaluating the 2×2 Hessian of f with respect to (x, y) directly, one may calculate f_{xy} on the basis of the identity

$$f_{xy} = \frac{1}{2}\left[f_{ss} - f_{xx} - f_{yy}\right] \,. \tag{13.3}$$

Here $f_{ss} = 2f_2(x, s)$ with f_2, as defined in (13.1), denotes the second derivative of f in the "diagonal" direction $s = (1, 1, 0, 0)$. Obviously, one may also go the other way and compute f_{ss} given f_{xx}, f_{yy}, and f_{xy} without any complication. Hence we may view the triplets (f_{xx}, f_{xy}, f_{yy}) and (f_{xx}, f_{ss}, f_{yy}) as alternative representations of the same Hessian and convert back and forth to the format that is most suitable for a certain computational purpose.

For the propagation of derivatives through a function evaluation procedure the representation as a collection of univariate derivatives seems preferable, even though some redundant information may be carried along. The relation between the two formats is linear and can therefore be affected by problem-independent transformations as described in section 13.3. Moreover, exactly the same linear relations apply to the corresponding gradients, so that in our example

$$\nabla f_{xy} = \frac{1}{2}\left[\nabla f_{ss} - \nabla f_{xx} - \nabla f_{yy}\right] ,$$

where ∇ represents differentiation with respect to x, y, λ, μ, and possibly any other problem parameter whose impact on f_{xy} is of interest.

Taylor versus Tensor Coefficients

Higher derivatives have long been used in the form of univariate Taylor series for the solution and analysis of differential equations and other dynamical systems. The rules for propagating truncated univariate Taylor series forward through an evaluation procedure have been known for a long time [Moo66, Wan69] and can be implemented quite efficiently with very regular data access patterns. These formulas will be reviewed in section 13.2. Univariate derivatives of degree d can then be obtained by simply rescaling the corresponding kth Taylor coefficient with the factor $k!$. In this way the binomial weights occurring in Leibniz's formula for the derivatives of a product are avoided, and overflow is a little less likely.

Higher derivatives in several variables form tensors, starting with the gradient and Hessian. Throughout this chapter we will use the term *tensor coefficient* to denote derivatives and *Taylor coefficient* to describe the scaled coefficient in the corresponding Taylor expansion. Tensors will be denoted by the nabla operator ∇ as prefix and Taylor coefficients by subscripts as in (13.1) and (13.2). As we will see the propagation of Taylor coefficient vectors through an evaluation procedure can be interpreted as evaluating F in "Taylor series arithmetic."

Full higher derivative tensors in several variables can be evaluated directly by multivariate versions of the chain rule. This approach has been implemented in [Nei92] and [Ber91b], and by several other authors. They have given particular thought to the problem of addressing the roughly $p^d/d!$ distinct elements of a symmetric tensor of order $p \leq n$ and degree d efficiently. Since this problem cannot be solved entirely satisfactorily, we advocate propagating a family of roughly $p^d/d!$ univariate Taylor series instead. From their coefficients the elements of all tensors up to the degree d can be obtained by the conversion process described in section 13.3. Propagating the required family of univariate Taylor series usually involves about the same number of operations as the direct approach but requires roughly $dp/(d+p)$ times as much storage. However, the data access pattern is much more regular. One would expect that p is normally not the total number n of independent variables, but the dimension of a subspace in the domain on which the higher derivatives are really required. Otherwise the number of tensor elements $p^d/d!$ grows very rapidly with the degree d, despite the factorial $d!$ in the denominator.

In section 13.4 we explore methods of computing adjoint vectors of the form $\bar{F}_k(\bar{y}, x, s)$ defined in (13.2). As it turns out, performing a conventional reverse sweep on the vector function $F_k : \mathbb{R}^{n+n} \longmapsto \mathbb{R}^n$ defined in (13.1) requires an unnecessary amount of coding and computation. Instead the reverse mode can be applied in Taylor arithmetic, yielding the whole family of higher-order adjoint vectors $\bar{F}_k$ for $k \leq d$ in one sweep.

13.2 Taylor Polynomial Propagation

In this section we consider the task of deriving from a given polynomial

$$x(t) = x_0 + tx_1 + t^2 x_2 + \cdots + t^d x_d \in \mathbb{R}^n \tag{13.4}$$

the resulting expansion

$$y(t) \equiv y_0 + ty_1 + t^2 y_2 + \cdots + t^d y_d = F(x(t)) + O(t^{d+1}) \in \mathbb{R}^m \,. \tag{13.5}$$

Here and throughout this chapter subscripts represent the degree of Taylor coefficients rather than the numbering of intermediate variables or vector components. Also as a consequence of the chain rule, one can observe that each y_j is uniquely and smoothly determined by the coefficient vectors x_i with $i \leq j$. In particular we have

$$\begin{aligned}
y_0 &= F(x_0) \\
y_1 &= F'(x_0)x_1 \\
y_2 &= F'(x_0)x_2 + \frac{1}{2}F''(x_0)x_1x_1 \\
y_3 &= F'(x_0)x_3 + F''(x_0)x_1x_2 + \frac{1}{6}F'''(x_0)x_1x_1x_1 \\
&\dots
\end{aligned}$$

In writing down the last equations we have already departed from the usual matrix-vector notation. It is well known that the number of terms that occur in these "symbolic" expressions for the y_j in terms of the first j derivative tensors of F and the "input" coefficients x_i with $i \leq j$ grows very rapidly with j. Fortunately, this exponential growth does not occur in automatic differentiation, where the many terms are somehow implicitly combined so that storage and operations count grow only quadratically in the bound d on j.

Provided F is analytic, this property is inherited by the functions

$$y_j = y_j(x_0, x_1, \dots, x_j) \in \mathbb{R}^m,$$

and their derivatives satisfy the identities

$$\frac{\partial y_j}{\partial x_i} = \frac{\partial y_{j-i}}{\partial x_0} = A_{j-i}(x_0, x_1, \ldots, x_{j-i})$$

as will be established in Proposition 13.3. This yields in particular

$$\begin{aligned}
\frac{\partial y_0}{\partial x_0} = \frac{\partial y_1}{\partial x_1} = \frac{\partial y_2}{\partial x_2} = \frac{\partial y_3}{\partial x_3} &= A_0 = F'(x_0) \\
\frac{\partial y_1}{\partial x_0} = \frac{\partial y_2}{\partial x_1} = \frac{\partial y_3}{\partial x_2} &= A_1 = F''(x_0)x_1 \\
\frac{\partial y_2}{\partial x_0} = \frac{\partial y_3}{\partial x_1} &= A_2 = F''(x_0)x_2 + \frac{1}{2}F'''(x_0)x_1x_1 \\
\frac{\partial y_3}{\partial x_0} &= A_3 = F''(x_0)x_3 + F'''(x_0)x_1x_2 + \frac{1}{6}F^{(4)}(x_0)x_1x_1x_1 \\
&\cdots
\end{aligned}$$

The $m \times n$ matrices $A_k, k = 0 \ldots d$, are actually the Taylor coefficients of the Jacobian path $F'(x(t))$, a fact that is of interest primarily in the context of ODEs and DAEs.

Much of this chapter is concerned with the analysis of the derived functions F_k, defined as follows.

Definition (TC): TAYLOR COEFFICIENT FUNCTIONS
Under Assumption (ED) *let*

$$y_k = F_k(x_0, x_1, \ldots, x_k) \quad \textit{with} \quad F_k : \mathbb{R}^{n\times(k+1)} \longmapsto \mathbb{R}^m$$

denote for $k \le d$ *the coefficient function defined by the relations* (13.4) *and* (13.5). *When more than* $k+1$ *vector arguments are supplied the extra ones are ignored, and when fewer are supplied the missing ones default to zero.*

This definition is a generalization of $F_k(x, s)$ given in (13.1). In order to evaluate the functions F_k we need to propagate corresponding Taylor coefficients through all intermediate variables. They will be denoted in this chapter by u, v, and w since indices are used to number Taylor coefficients. More specifically, we use the following notation.

Definition (AI): APPROXIMANTS OF INTERMEDIATES
For given d-times continuously differentiable input path $x(t) : (-\varepsilon, \varepsilon) \longmapsto \mathbb{R}^n$ *with* $\varepsilon > 0$, *denote the resulting values of an intermediate variable* v *by* $v(x(t)) :$ $(-\varepsilon, \varepsilon) \longmapsto \mathbb{R}$ *and define its Taylor polynomial by*

$$v(t) = v_0 + tv_1 + t^2v_2 + \cdots + t^dv_d = v(x(t)) + o(t^d)\,.$$

In other words, we have overloaded the symbol v to represent three related mathematical objects: the intermediate variable itself if without argument, the exact value path in the form $v(x(t))$, and the Taylor polynomial of degree d approximating it when written as $v(t)$. The Taylor coefficients at $t = 0$ are unambiguously represented by v_k for $0 \le k \le d$. Since the coefficients v_k are specified for the independents by the path $x(t)$ they can be computed recursively according to the propagation rules developed below.

The efficient and stable propagation of formal power series has received considerable attention in the scientific literature (see, e.g., [BK78, Fat74, Oba93]). Much of this analysis is concerned with algorithms that are fast for rather large degrees d. It seems unlikely that the crossover points are reached even when 10 or 20 Taylor coefficients $y_k = F_k$ are computed, for example, for the numerical solution of initial value problems in ODEs [Mun90]. Even for those values of d, the number of memory moves, which is linear in d, seems to represent a substantial part of the execution time so that reducing the number of arithmetic operations may not pay off if it entails a less regular variable-access pattern. Hence we will consider here only the classical, straightforward algorithms for propagating Taylor series. Nevertheless, a good implementation of the convolution product displayed in the second row of Table 13.1 below seems of primary importance to AD, especially with regards to the reverse mode, as we will see in section 13.4.

Table 13.1: Taylor Coefficient Propagation through Arithmetic Operations

$v =$	Recurrence for $k = 1 \ldots d$	OPS	MOVES
$u + cw$	$v_k = u_k + c\, w_k$	$2d$	$3d$
$u * w$	$v_k = \sum_{j=0}^{k} u_j w_{k-j}$	d^2	$3d$
u/w	$v_k = \frac{1}{w_0}\left[u_k - \sum_{j=0}^{k-1} v_j w_{k-j}\right]$	d^2	$3d$
u^2	$v_k = \sum_{j=0}^{k} u_j u_{k-j}$	$\frac{1}{2}d^2$	$2d$
$\sqrt{u}$	$v_k = \frac{1}{2v_0}\left[u_k - \sum_{j=1}^{k-1} v_j v_{k-j}\right]$	$\frac{1}{2}d^2$	$2d$

Arithmetic Operations

Suppose $v(t) = \varphi(u(t), w(t))$ is obtained either as $v(t) = u(t) + c\,w(t)$ with some real constant c or as $v(t) = u(t) * w(t)$ or as $v(t) = u(t)/w(t)$ so that $w(t) * v(t) = u(t)$. Then one simply has to apply truncated polynomial arithmetic and obtain the recurrences listed in Table 13.1. The third column gives the number of multiplications followed by additions up to $O(d)$.

It should be noted that the division of Taylor polynomials is well defined exactly when it is okay for their leading coefficients, i.e., the evaluation without derivatives. Hence we see that differentiation does not exacerbate the situation, at least theoretically. Complexitywise, we note that the division operator involves exactly as many arithmetic operations as the multiplication operator. However, in the division case as well as the final square root elemental, we have proper recursions, whereas multiplication of Taylor polynomials is a simple convolution, which could be executed with some degree of parallelism. The reduction of the complexity from d^2 to $\frac{1}{2}d^2$ for $v = u^2$ and $v = \sqrt{u}$ takes advantage of the symmetry in the convolution sums on the right-hand side.

Nonlinear Univariate Elementals

Simply differentiating "symbolically," we obtain for general smooth $v = \varphi(u)$ the derivative expressions

$$\begin{aligned}
v_0 &= \varphi(u_0) \\
v_1 &= \varphi_1(u_0)\, u_1 \\
v_2 &= \varphi_2(u_0)\, u_1\, u_1 + \varphi_1(u_0)\, u_2 \\
v_3 &= \varphi_3(u_0)\, u_1\, u_1\, u_1 + 2\, \varphi_2(u_0)\, u_1\, u_2 + \varphi_1(u_0)\, u_3 \\
v_4 &= \varphi_4(u_0)\, u_1\, u_1\, u_1\, u_1 + 3\varphi_3(u_0)\, u_1\, u_1\, u_2 \\
&\quad + \varphi_2(u_0)\, (u_2\, u_2 + 2\, u_1\, u_3) + \varphi_1(u_0)\, u_4 \\
v_5 &= \quad \ldots ,
\end{aligned} \tag{13.6}$$

where $\varphi_i = \varphi^{(i)}/i!$ is the ith Taylor coefficient of φ. As one can see the number of terms grows quite rapidly with the degree of the derivative. This effect is even more pronounced in the multivariate case, where u_2u_2, u_1u_3, and u_3u_1 cannot simply be added to each other. The general formula was originally given by Faa di Bruno [dB56] and has since been extended to the multivariate case [Fra78]. The task of computing the coefficients v_i for $i = 0 \ldots d$ from the pairs (u_i, φ_i) for $i = 0 \ldots d$ is known in the literature as the composition problem. It was shown in [BK78] to have the same arithmetic complexity as the problem of inverting a power series $u(t)$, i.e., finding d derivatives of φ such that $\varphi(u(t)) = 1 + o(t^d)$. The classical methods for both tasks require order-d^3 arithmetic operations without any restriction on the coefficients of φ or u.

Fortunately the general case with completely unrelated Taylor coefficients $\varphi_0, \varphi_1, \ldots, \varphi_d$ never occurs in practice because all elemental functions $\varphi(u)$ of interest are solutions of linear ODEs. More specifically, they all satisfy an identity of the form

$$b(u)\, \varphi'(u) - a(u)\, \varphi(u) = c(u) \; . \tag{13.7}$$

Here, the coefficient functions $b(u)$, $a(u)$, and $c(u)$ are supposed to be "known" i.e., expressible in terms of arithmetic operations and univariate functions that we can already differentiate. In other words, we can determine the Taylor coefficients b_k, a_k, and c_k from the coefficients u_k of u.

Differentiating the identity $v(x(t)) = \varphi(u(x(t)))$ and multiplying by the function $b(u(x(t)))$ we obtain the relation

$$b(u(x(t)))\,\frac{d}{dt}v(x(t)) = \Big[c(u(x(t))) + a(u(x(t)))v(x(t))\Big]\,\frac{d}{dt}u(x(t))\,.$$

Since $u(x(t))$ and thus $v(x(t))$ are as a consequence of Assumption (ED) at least d-times continuously differentiable, their derivatives have the expansions

$$\frac{d}{dt}u(x(t)) = \tilde{u}_1 + t\tilde{u}_2 + \cdots + t^{d-1}\tilde{u}_d + O(t^d) \quad \text{with} \quad \tilde{u}_j \equiv j u_j \quad \text{for} \quad j > 0$$

and similarly

$$\frac{d}{dt}v(x(t)) = \tilde{v}_1 + t\tilde{v}_2 + \cdots + t^{d-1}\tilde{v}_d + O(t^d) \quad \text{with} \quad \tilde{v}_j \equiv j v_j \quad \text{for} \quad j > 0\,.$$

Substituting these expansions into the above identity and identifying coefficients, we obtain the following specialization of Theorem 5.1 in [BK78].

Proposition 13.1 (Taylor Polynomials of ODE Solutions)
Provided $b_0 \equiv b(u_0) \neq 0$ *we have*

$$\tilde{v}_k \;=\; \frac{1}{b_0}\left[\sum_{j=1}^{k}(c_{k-j} + e_{k-j})\tilde{u}_j - \sum_{j=1}^{k-1} b_{k-j}\tilde{v}_j\right] \quad \textit{for} \quad k = 1\ldots d\,,$$

where

$$e_k \;\equiv\; \sum_{j=0}^{k} a_j\, v_{k-j} \quad \textit{for} \quad k = 0\ldots d-1\,.$$

The scaling of the coefficients u_j and v_j to $\tilde{u}_j = j u_j$ and $\tilde{v}_j = j v_j$, respectively, reduces the total number of multiplications by half. It is easily undone afterwards. When $a(u) \equiv 0$ in (13.7) the elemental function is simply a rational quadrature of the form

$$\varphi(u) = \int \frac{c(u)}{b(u)}du\,.$$

Then we may set $a_j = 0$ and thus $e_j = 0$ in the formula above, which reduces the number of arithmetic operations by $d^2 + O(d)$. The same reduction occurs when $a(u) = a_0$ or $b(u) = b_0$ or both are constant, so that the total number of arithmetic operations is either $3d^2 + O(d)$, $2d^2 + O(d)$, or $d^2 + O(d)$, respectively. In all cases, the growth is quadratic in d, which compares favorably with the exponential growth we observed for the "symbolic" differentiation of a general elemental $\varphi(u)$.

For the standard elementals we have listed the coefficient functions $a(u)$, $b(u)$, $c(u)$, and the resulting recurrences for the coefficients in Table 13.2. The operations count in the penultimate column is again up to $O(d)$.

Table 13.2: Taylor Coefficient Propagation through Univariate Elementals

v	a	b	c	Recurrence for $k = 1 \ldots d$	OPS	MOVES
$\ln(u)$	0	u	1	$\tilde{v}_k = \frac{1}{u_0}\left[\tilde{u}_k - \sum_{j=1}^{k-1} u_{k-j}\,\tilde{v}_j\right]$	d^2	$2d$
$\exp(u)$	1	1	0	$\tilde{v}_k = \left[\sum_{j=1}^{k} v_{k-j}\,\tilde{u}_j\right]$	d^2	$2d$
u^r	r	u	0	$\tilde{v}_k = \frac{1}{u_0}\left[r\sum_{j=1}^{k} v_{k-j}\tilde{u}_j - \sum_{j=1}^{k-1} u_{k-j}\,\tilde{v}_j\right]$	$2d^2$	$2d$
$\sin(u)$	0	1	$\cos(u)$	$\tilde{s}_k = \sum_{j=1}^{k} \tilde{u}_j c_{k-j}$	$2d^2$	$3d$
$\cos(u)$	0	-1	$\sin(u)$	$\tilde{c}_k = \sum_{j=1}^{k} -\tilde{u}_j s_{k-j}$		

Note that the recurrences for the coefficients s_k and c_k of $\sin(u)$ and $\cos(u)$ must be applied in an alternating fashion. Because the recurrences are very simple and regular, an optimizing compiler can implement them quite efficiently. As a result the quadratic growth in the operations count is for small d masked by the constant overhead and the linearly growing number of memory moves.

Fig. 13.1 lists the runtime ratios for propagating $d + 1$ Taylor coefficients through the vector function

$$y = F(x) \equiv \begin{bmatrix} -\sin(x_3) + 10^8 * x_3 * (1 - 1/x_1) + \arcsin(x_2) \\ -10 * x_2 * \exp(x_3) + \ln(x_1) + 3 * 10^7 * x_3 * (1 - x_2) * \sqrt{x_1} \end{bmatrix}.$$

As one can see the runtime ratio grows initially with a slope a little less than a half, which then increases to a little more than half as d approaches 20. The quadratic complexity term only becomes dominant when d is even larger. These results were obtained using ADOL-C [GJU96] with the runtime for $d = 0$ normalized to 1. Due to the smallness of the code this runtime was about 20 times larger than that of a compilable code for the function itself. For an application of higher-order Taylor polynomial approximations based on overloading, see [GM97].

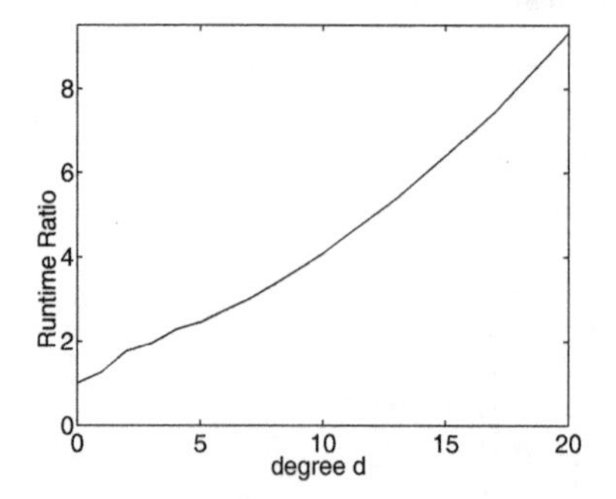

Figure 13.1: Runtime Ratio for Taylor Coefficients

Taylor Ring Arithmetic

We conclude this section with an abstract interpretation of the differentiation formulas developed above. This way of looking at things is not needed for the evaluation of multivariate tensors as discussed in section 13.3, but it is very helpful for understanding the material on higher-order gradients in section 13.4. With respect to the addition, subtraction, and multiplication formulas listed in Table 13.1, the polynomials

$$\mathcal{P}_d \equiv \left\{ v(t) = \sum_{j=0}^{d-1} v_j\, t^j \;\middle|\; v_j \in \mathbb{R} \right\}$$

form for every order $d > 0$ a commutative ring. Obviously we have the inclusions

$$\mathbb{R} \;\equiv\; \mathcal{P}_1 \subset \mathcal{P}_2 \subset \cdots \subset \mathcal{P}_d \subset \mathcal{P}_{d+1} \subset \cdots ,$$

with $\mathbb{R}$ already containing the neutral elements 0 and 1 with regards to addition and multiplication, respectively. Since $t \cdot t^{d-1} = t^d$ is truncated to zero in $\mathcal{P}_d$, all rings other than $\mathbb{R}$ itself contain zero divisors and can therefore not be extended to fields. Defining the modulus

$$\big|v(t)\big| \;\equiv\; \sum_{j=0}^{d-1} \big|v_j\big| \;\in \mathbb{R}\,, \tag{13.8}$$

we may turn $\mathcal{P}_d$ into a complete algebra over the real numbers. Then we have not only the triangle inequality $|u(t)+w(t)| \leq |u(t)|+|w(t)|$ but also the consistency relation $|u(t) * w(t)| \leq |u(t)|\,|w(t)|$, as shown in Exercise 13.3.

Hence we may think of truncated Taylor series as scalars on which one can perform arithmetic just like on real or complex numbers. In modern computer languages like C++, one may simply overload real variables with Taylor variables of a certain order d. This parameter d may be fixed at compile-time for efficiency or be selected at runtime for flexibility. The contingency "division by zero" now becomes "division by zero divisor," i.e., division by a power series with vanishing constant coefficient. This calamity is not necessarily more likely to occur because the subset of troublesome denominators has in either case a codimension of 1 in the real space of scalars viewed as a Euclidean space of dimension d.

Vectors $(v^{(i)}(t))_{i=1\ldots n}$ of Taylor polynomials $v^{(i)}(t) \in \mathcal{P}_d$ form the free module $\mathcal{P}_d^n$, which can be endowed with the Euclidean norm

$$\big\|\big(v^{(1)}(t), \ldots, v^{(n)}(t)\big)\big\| \;\equiv\; \big[|v^{(1)}(t)|^2 + \cdots + |v^{(n)}(t)|^2\big]^{\frac{1}{2}} \tag{13.9}$$

or any other equivalent norm. Having fixed the topology we may then define $\mathcal{C}(\mathcal{P}_d^n, \mathcal{P}_d^m)$ as the set of functions between the free modules $\mathcal{P}_d^n$ and $\mathcal{P}_d^m$ that are continuous on some open domain in $\mathcal{P}_d^n$.

By the rules of multivariate calculus any d-times continuously differentiable vector function $F : \mathbb{R}^n \mapsto \mathbb{R}^m$ has a unique extension to a derived mapping

$E_d(F) : \mathcal{P}_d^n \mapsto \mathcal{P}_d^m$. In other words, there exists for each $d > 0$ a linear extension operator

$$E_d \; : \; C^d(\mathbb{R}^n, \mathbb{R}^m) \; \mapsto \; C(\mathcal{P}_d^n, \mathcal{P}_d^m) \, .$$

Here $C^d(\mathbb{R}^n, \mathbb{R}^m)$ represents the space of functions $F : \mathbb{R}^n \mapsto \mathbb{R}^m$ that are d-times continuously differentiable on some open domain in $\mathbb{R}^n$. As we will note in section 13.4, one can perform not only arithmetic but also multivariate analysis on such free modules. For the time being we note the following rule.

RULE 22

FORWARD DIFFERENTIATION CAN BE ENCAPSULATED IN TAYLOR ARITHMETIC.

Not only univariate but also multivariate Taylor polynomials with specified limits on the exponents, e.g., the total degree, form complete algebras. Hence propagating them forward may also be viewed and implemented as executing the original evaluation procedure on an extended ring of scalars. For reasons that are explained in the next section we prefer the propagation of families of univariate Taylor polynomials, which allow us to stick with univariate expansions as far as the ring interpretation is concerned.

Complex Approximation to Taylor Arithmetic

First-degree Taylor arithmetic can be approximated by complex arithmetic, as suggested in [NAW98]. They noted that with i, the imaginary unit

$$F(x + i\,t\dot{x}) = F(x) - \frac{1}{2}t^2 F''(x)\dot{x}\dot{x} + i\,t\left[F'(x)\dot{x} - \frac{1}{6}t^2 F'''(x)\dot{x}\dot{x}\dot{x}\right] + O(t^4).$$

Consequently, the imaginary part

$$\mathrm{Im}[F(x + i\,t\dot{x})] = t\big[F'(x)\dot{x} + O(t^2)\big]$$

yields a second-order (in t) approximation

$$\dot{y} = F'(x)\dot{x} = \frac{1}{t}\mathrm{Im}[F(x + i\,t\,\dot{x})] + O(t^2). \tag{13.10}$$

It was found in [NAW98] that the resulting derivative values were much more accurate than the customary central differences evaluated in real arithmetic. For example, when $F(x) = \sqrt{x}$ one obtains at $x = 1$ the Taylor expansion

$$F(1 + i\,t\,\dot{x}) = \big(1 + i\,t\,\dot{x}\big)^{\frac{1}{2}} = 1 + \frac{i}{2}\,t\,\dot{x} + \frac{1}{8}t^2\dot{x}^2 - \frac{i}{16}t^3\dot{x}^3 + O(t^4)$$

so that

$$\frac{1}{t}\,\mathrm{Im}(F(1 + i\,t\,\dot{x})) = \frac{1}{2}\dot{x} - \dot{x}^3\,t^2\frac{1}{16} + O(t^3) = F'(1)\dot{x} + O(t^2)\,.$$

However, the computational effort is larger than in the corresponding Taylor arithmetic of order 2, as we will see in Exercise 13.8.

13.3 Multivariate Tensor Evaluation

Since we imagine that the number n of independent variables of a given function F may be rather large, we may assume that the differentiation can be restricted to a subspace spanned by the columns of some seed matrix $S \in \mathbb{R}^{n \times p}$. This term was originally introduced in Chapter 8, where sparsity of the Jacobian $F'(x)$ allowed the reconstruction of all its nonzero elements from the product $B = F'(x)S$. In section 8.5 we found that the same approach can be used to reconstruct Hessians from their restriction to a subspace. Here we will not place any restrictions on the seed matrix S and consider the linear parameterization

$$x = x(z) = x(0) + Sz \quad \text{with} \quad z \in \mathbb{R}^p .$$

Then all intermediate variables $v_i = v_i(z)$ may be viewed as functions of the "reduced" variables z. We will annotate their derivatives with respect to z at $z = 0$ with subscript S as, for example, in

$$\nabla_S \, v \equiv \left. \frac{\partial v(z)}{\partial z} \right|_{z=0} \in \mathbb{R}^p$$

for some intermediate v. Now let us briefly review the temporal and spatial complexity of multivariate Taylor propagation as a motivation for our univariate approach.

Complexity of Multivariate Tensor Propagation

As we have mentioned in the introduction to this chapter, one can quite easily generalize the propagation rules derived in the previous section for Taylor polynomials in a single parameter t to the case of several variable parameters z. For example, we have for the multiplication operation $v = u * w$ the well-known rules

$$\nabla_S v = u \, \nabla_S w + w \, \nabla_S u$$

and

$$\nabla_S^2 v = u \, \nabla_S^2 w + \nabla_S u (\nabla_S w)^\top + \nabla_S w (\nabla_S u)^\top + w \, \nabla_S^2 u \, .$$

For third and higher derivative tensors, the matrix-vector notation is no longer sufficient, but the following observations are still valid. To calculate $\nabla_S^k v$, each element of $\nabla_S^j u$ with $0 \le j \le k$ has to be multiplied with all elements of $\nabla_S^{k-j} w$. The result is then multiplied by a binomial coefficient given by Leibniz's theorem and finally incremented to the appropriate element of $\nabla_S^k v$. It is well known (see, for example, [Ber91b] and Exercise 13.4) that the number of operations needed to perform this multivariate convolution is given by

$$\binom{2p+d}{d} \approx \frac{(2p)^d}{d!} \quad \text{if} \quad d \ll p \, . \tag{13.11}$$

One may consider this number as a rough approximation to the factor by which the cost to evaluate F grows when d derivatives in p variables are propagated forward. Just as in the univariate case, the propagation of multivariate Taylor polynomials through a division and most nonlinear intrinsics involves about the same number of operations as propagating them through a multiplication. Only linear elementals like $v = u + cw$ are significantly cheaper.

Symmetric Tensor Storage

If the dth derivative tensor $\nabla_S^d v$ was stored and manipulated as a full p^d array, the corresponding loops could be quite simple to code, but the memory requirement and operations count would be unnecessarily high. In the case $d = 2$, ignoring the symmetry of Hessians would almost double the storage per intermediate (from $\frac{1}{2}(p+1)p$ to p^2) and increase the operations count for a multiplication by about 100%. This price may be worth paying in return for the resulting sequential or at least constant stride data access. However, by standard combinatorial arguments $\nabla_S^d v \in \mathbb{R}^{p^d}$ has exactly

$$\binom{p+d-1}{d} \approx \frac{p^d}{d!} \quad \text{if} \quad d \ll p \tag{13.12}$$

distinct elements. Hence, the symmetry reduces the number of distinct elements in $\nabla_S^d v$ almost by the factor $d!$. Therefore, in the case $d = 3$ the number of distinct elements is reduced almost by 6, and in the case $d = 4$, the storage ratio becomes almost 24. Since higher-order tensors have very many entries, one therefore has to use symmetric storage modes, such as the classical tetrahedral scheme [Knu73].

The drawback of symmetric storage modes is that the access of individual elements is somewhat complicated, requiring, for example, three integer operations for address computations in the implementation of Berz [Ber91b]. Moreover, the resulting memory locations may be further apart with irregular spacing so that significant paging overhead can be incurred. None of these difficulties arises when $p = 1$, so S has a single column vector s. Then for any intermediate value v the directional derivatives $v, \nabla_s v, \nabla_s^2 v, \ldots, \nabla_s^d v$ or the corresponding Taylor coefficients can be stored and manipulated as a contiguous array of $(d+1)$ reals.

Multivariate Tensors via Univariate Taylors

Consequently, one may ask whether it is possible to propagate univariate Taylor polynomials of degree d along a family of directions $s \in \text{range}(S) \subset \mathbb{R}^n$ and then to utilize the coefficients obtained to reconstruct the multivariate Taylor polynomial represented by the tensors $\nabla_S^k F(x)$ for $k = 1 \ldots d$. In other words, we wish to determine a polynomial of degree d in p variables from its restriction to a hopefully minimal set of rays through the origin. This approach goes back at least to Rall and was developed further in [BCG93].

When $p = 2 = d$ the task is easy, as we have for any $S = [s_1, s_2] \in \mathbb{R}^{n\times 2}$ the following generalization of (13.3):

$$\left.\frac{\partial^2 F(x + s_1 z_1 + s_2 z_2)}{\partial z_1 \partial z_2}\right|_{z=0} = s_1^\top \nabla^2 F(x) s_2$$

$$= \frac{1}{2}\left[\left.\frac{\partial^2 F(x + (s_1 + s_2)t)}{\partial t^2}\right|_{t=0} - \left.\frac{\partial^2 F(x + s_1 t)}{\partial t^2}\right|_{t=0} - \left.\frac{\partial^2 F(x + s_2 t)}{\partial t^2}\right|_{t=0}\right].$$

Thus we see that the only mixed second partial derivative of the bivariate function $F(x + s_1 z_1 + s_2 z_2)$ can be expressed in terms of univariate derivatives along the three directions s_1, s_2, and $(s_1 + s_2)$. The other two entries of the Hessian $\nabla_S^2 F$, and of course the gradient $\nabla_S F$, are obtained directly as univariate derivatives. It should be noted that there is a certain redundancy as the first directional derivative $\partial F(x + (s_1 + s_2)t)/\partial t$ along the "diagonal" direction $(s_1 + s_2)$ is simply the sum of the first derivatives along s_1 and s_2. Whether or not the function value $F(x)$ itself is evaluated repeatedly depends on the software implementation.

Since only the highest-order coefficient in each univariate Taylor polynomial of degree d provides information about the derivatives tensor $\nabla_S^d F$, it is clear that we need at least as many directions s as this symmetric tensor has distinct entries. In fact this number, given in (13.12) for general p, is also sufficient, and we may select the directions in the special form

$$s = \sum_{j=1}^{p} s_j \, \mathbf{i}_j \quad \text{with} \quad 0 \le \mathbf{i}_j \in \mathbb{Z} \quad \text{and} \quad \sum_{j=1}^{p} \mathbf{i}_j = d\,.$$

These vectors s correspond exactly to the multi-indices $(\mathbf{i}_1, \mathbf{i}_2 \ldots \mathbf{i}_p)$ representing the entries of the highest derivative tensor. The case $p = 2$ and $d = 3$ is depicted in Fig. 13.2. The numbers l_i, q_i, and c_i represent the linear, quadratic, and cubic coefficients of the third-order Taylor expansions along the four directions $3s_1$, $2s_1 + s_2$, $s_1 + 2s_2$, and $3s_2$. For simplicity we have assumed that the common constant term $F(x)$ vanishes.

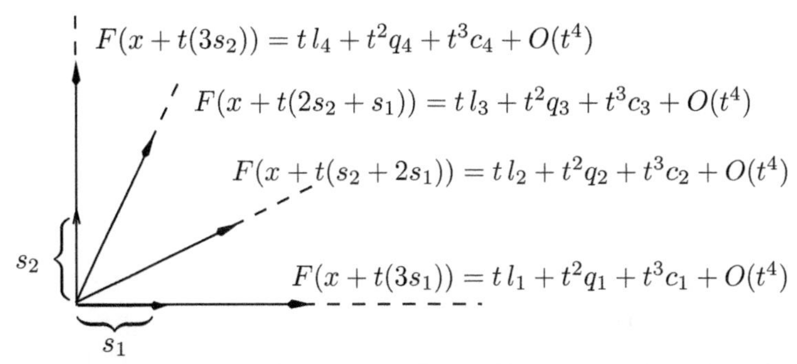

$F(x + t(3s_2)) = t\,l_4 + t^2 q_4 + t^3 c_4 + O(t^4)$
$F(x + t(2s_2 + s_1)) = t\,l_3 + t^2 q_3 + t^3 c_3 + O(t^4)$
$F(x + t(s_2 + 2s_1)) = t\,l_2 + t^2 q_2 + t^3 c_2 + O(t^4)$
s_2
$F(x + t(3s_1)) = t\,l_1 + t^2 q_1 + t^3 c_1 + O(t^4)$
s_1

Figure 13.2: Family of Univariate Taylor Expansions along Rays from Origin

Let us label the elements of the derivative tensors individually as

$$\nabla_S F(x) = (g_1, g_2), \nabla_S^2 F(x) = (h_{11}, h_{12}, h_{22}), \nabla_S^3 F(x) = (t_{111}, t_{112}, t_{122}, t_{222}) .$$

In order to relate these coefficients to the coefficients l_i, q_i, and c_i defined above, we consider the Taylor expansion

$$F(x + ts) = F(x) + t\nabla F(x)\, s + \frac{t^2}{2}\,\nabla^2 F(x)\, s\, s + \frac{t^3}{6}\nabla^3 F(x)\, s\, s\, s + O(t^4) .$$

Here we have stretched matrix-vector notation a little bit by assuming that symmetric tensors can be multiplied by a sequence of vectors from the right.

Identifying terms along the four special directions, shown in Fig. 13.2, we obtain the linear relations depicted in Table 13.3 in spreadsheet style, e.g., $q_2 = 2h_{11} + 2h_{12} + h_{22}/2$.

Table 13.3: Relations between Tensor and Univariate Taylor Coefficients

	g_1	g_2
l_1	3	0
l_2	2	1
l_3	1	2
l_4	0	3

	h_{11}	h_{12}	h_{22}
q_1	9/2	0	0
q_2	2	2	1/2
q_3	1/2	2	2
q_4	0	0	9/2

	t_{111}	t_{112}	t_{122}	t_{222}
c_1	9/2	0	0	0
c_2	4/3	2	1	1/6
c_3	1/6	1	2	4/3
c_4	0	0	0	9/2

As one can see, only the third linear system in Table 13.3 is square, whereas the first two have more rows than columns and are thus overdetermined. The computed coefficients l_i and q_i for $i = 1, 2, 3, 4$ obtained from univariate Taylor expansions must nevertheless be consistent up to rounding errors. Also, as one can easily check, all three systems have full column rank so that the tensor coefficients are uniquely determined. One possible representation of gradient and Hessian in terms of the Taylor coefficients is $g_1 = l_1/3$, $g_2 = l_4/3$, and

$$h_{11} = 2q_1/9, \quad h_{12} = [q_2 + q_3 - 5(q_1 + q_4)/9]/4, \quad h_{22} = 2q_4/9 .$$

The third system in Table 13.3 has the nonzero determinant 243/4 and always yields the unique solution

$$t_{111} = 2\,c_1/9, \quad t_{112} = (-5c_1/9 + 2c_2 - c_3 + 2c_4/9)/3,$$
$$t_{122} = (2\,c_1/9 - c_2 + 2\,c_3 - 5\,c_4/9)/3, \quad t_{222} = 2\,c_4/9 .$$

Hence we see that in the special case $p = 2$ and $d = 3$, the tensor coefficients making up $\nabla_S F(x), \nabla_S^2 F(x)$, and $\nabla_S^3 F(x)$ can be computed as linear combinations of univariate Taylor coefficients along four special directions in the range of S. The nonzero elements in the transformation matrix are small rational numbers and there are quite a few structural zeros. Provided the evaluation of F itself is of any complexity, the final conversion from univariate Taylor coefficients to multivariate tensor coefficients is very cheap by comparison.

Fortunately, the observations in the last paragraph are valid not only for our toy scenario $(p,d) = (2,3)$ but for all parameter pairs (p,d) that seem of practical interest in the foreseeable future. This result has been derived in the paper [GUW00], from which we quote the following proposition without proof. It uses boldface variables $\mathbf{i}$, $\mathbf{j}$, and $\mathbf{k}$ to represent multi-indices with p nonnegative components, whose sum is denoted by the modulus sign $|\cdot|$. For any vector $z \in \mathbb{R}^p$ and a multi-index $\mathbf{k}$ the standard multinomial notation is defined by

$$\binom{z}{\mathbf{k}} \equiv \binom{z_1}{\mathbf{k}_1}\binom{z_2}{\mathbf{k}_2}\cdots\binom{z_p}{\mathbf{k}_p}.$$

Proposition 13.2 (Taylor-to-Tensor Conversion)
Let $F : \mathbb{R}^n \mapsto \mathbb{R}^m$ be at least d-times continuously differentiable at some point $x \in \mathbb{R}^n$ and denote by $F_r(x,s)$ the rth Taylor coefficient of the curve $F(x+ts)$ at $t = 0$ for some direction $s \in \mathbb{R}^n$. Then we have for any seed matrix $S = [s_j]_{j=1}^p \in \mathbb{R}^{n\times p}$ and any multi-index $\mathbf{i} \in \mathbb{N}^p$ with $|\mathbf{i}| \le d$ the identity

$$\left.\frac{\partial^{|\mathbf{i}|} F(x + z_1 s_1 + z_2 s_2 + \cdots + z_p s_p)}{\partial z_1^{\mathbf{i}_1} \partial z_2^{\mathbf{i}_2} \ldots \partial z_p^{\mathbf{i}_p}}\right|_{z=0} = \sum_{|\mathbf{j}|=d} \gamma_{\mathbf{ij}}\, F_{|\mathbf{i}|}(x, S\,\mathbf{j}), \qquad (13.13)$$

where the constant coefficients $\gamma_{\mathbf{ij}}$ are given by the finite sums

$$\gamma_{\mathbf{ij}} \equiv \sum_{0<\mathbf{k}\le\mathbf{i}} (-1)^{|\mathbf{i}-\mathbf{k}|} \binom{\mathbf{i}}{\mathbf{k}} \binom{d\mathbf{k}/|\mathbf{k}|}{\mathbf{j}} \left(\frac{|\mathbf{k}|}{d}\right)^{|\mathbf{i}|}.$$

The family of coefficients $\gamma_{\mathbf{ij}}$ occurring in the proposition depends only on the degree d and the number of directions $p \le n$. Hence it can be precomputed for all problems with these same integer parameters. Table 13.4 lists the number of coefficients in all tensors $\nabla_S^e v$ with $0 \le e \le d$, the cost of a multivariate product (13.11), the number of nonzero conversion coefficients $\gamma_{\mathbf{ij}}$, and their L_1 norm, i.e., maximal sum of the moduli $|\gamma_{\mathbf{ij}}|$ over $\mathbf{j}$ for $p \equiv d = 1\ldots 7$. When $p < d$ the fully mixed partial in d directions does not occur. When $p > d$ the same coefficients $\gamma_{\mathbf{ij}}$ occur repeatedly on subspaces of dimension d.

Table 13.4: Number and Size of Conversion Coefficients

Degree $d = p$	1	2	3	4	5	6	7
Tensor size	2	6	20	70	252	924	3 432
Cost of product	3	15	84	495	3 003	18 564	116 280
$\lvert\{\gamma_{\mathbf{ij}} \ne 0\}\rvert$	1	7	55	459	4 251	38 953	380 731
$\max_{\mathbf{i}} \sum_{\mathbf{j}} \lvert\gamma_{\mathbf{ij}}\rvert$	1.00	1.50	2.22	3.28	4.84	7.13	10.50

In general the number of nonzero coefficients $\gamma_{\mathbf{ij}}$ for given p and d can be bounded by a small multiple of (13.11), the cost of propagating a multivariate Taylor polynomial through a single multiplication operator. As shown in Exercise 13.5 the cost of propagating $\binom{p+d-1}{d}$ univariate Taylor polynomials through one multiplication operation is for most pairs (p, d) amazingly close to the multivariate reference cost (13.11). Hence the cost of one final conversion from Taylor family to tensor format is indeed negligible. The only drawback is that the total memory grows by a factor close to the harmonic mean $p\,d/(p+d)$ of p and d. However, should storage be a problem one can easily strip-mine, i.e., perform several forward sweeps propagating only a manageable portion of the $\binom{p+d-1}{d}$ univariate Taylor polynomials each required time. Strip-mining is also possible in multivariate Taylor propagation, but may double the operations count, as shown in [BCG93].

In many applications, one needs the derivatives of variables $y \in \mathbb{R}^m$ that are implicitly defined as functions of some variables $x \in \mathbb{R}^{n-m}$ by an algebraic system of equations

$$G(z) \;=\; 0 \in \mathbb{R}^m \quad \text{with} \quad z = (y, x) \in \mathbb{R}^n .$$

Naturally, the n arguments of G need not be partitioned in this regular fashion and we wish to provide flexibility for a convenient selection of the $n - m$ *truly* independent variables. Let $P \in \mathbb{R}^{(n-m)\times n}$ be a $0-1$ matrix that picks out these variables so that it is a column permutation of the matrix $[0, I_{n-m}] \in \mathbb{R}^{(n-m)\times n}$. Then the nonlinear system

$$G(z) \;=\; 0, \quad Pz = x,$$

has a regular Jacobian, wherever the implicit function theorem yields y as a function of x. Hence, we may also write

$$F(z) \equiv \begin{pmatrix} G(z) \\ Pz \end{pmatrix} = \begin{pmatrix} 0 \\ Pz \end{pmatrix} \equiv S\,x,$$

where $S = [0, I_p]^\top \in \mathbb{R}^{n\times p}$ with $p = n - m$. Now, we have rewritten the original implicit functional relation between x and y as an inverse relation $F(z) = Sx$. Given any $F : \mathbb{R}^n \mapsto \mathbb{R}^n$ that is locally invertible and an arbitrary seed matrix $S \in \mathbb{R}^{n\times p}$ we may evaluate all derivatives of $z \in \mathbb{R}^n$ with respect to $x \in \mathbb{R}^p$ using the computation of higher-order derivatives as described in this section. Similar to the situation that arises in the differential equation context discussed in section 13.5 below one has to perform several sweeps using Taylor polynom arithmetic where only one extra coefficient is calculated each time. Solving a linear system of equations these higher derivatives are then transformed into the desired derivatives of implicitly defined variables. This iterative approach can be organized so that the correpsoning complexity grows only quadratically in the highest derivative degree. Corresponding drivers are provided for example by the tool ADOL-C.

13.4 Higher-Order Gradients and Jacobians

In the previous sections, we considered first the calculation of directional derivatives in the form of univariate Taylor coefficients and then their combination and conversion to compute tensors that represent all derivatives up to a certain degree on some subspace of the domain. As we have seen in the introduction to this chapter, there are certain situations in which one may wish for something in between, namely, the result of several directional differentiations subsequently differentiated once more with respect to a large number of independent variables.

In principle, there is no difficulty in cheaply computing the gradient vector $\bar{F}_k(\bar{y}, x, s)$ defined in (13.2) as adjoint of the directional derivative $F_k(x, s)$ defined in (13.1). We merely have to apply the reverse mode to the whole forward propagation process described in section 13.2. This involves propagating adjoint values backward through each one of the formulas in Tables 13.1 and 13.2. Appropriate adjoint formulas were in fact derived and implemented in the early versions of ADOL-C [GJU96], a package that has always provided derivative vectors of the form (13.1) and (13.2) with arbitrary degree d. While this effort yielded correct results at reasonable runtimes, the hand-coding of adjoints was a waste of time. The reason is not the usual one, namely, the availability of adjoint generating software, but a simple mathematical property, which we will first derive on a particular elementary function. Consequently, we will find that it is more elegant and more economical to evaluate $\bar{F}_k$ in the last form given in (13.2), namely, as Taylor coefficient of a gradient map.

Adjoint of the Taylor Exponential

Consider, for example, the Taylor coefficient recurrence for the exponential relation $v = \varphi(u) = \exp(u)$ from Table 13.2. It may be rewritten in the incremental form displayed on the left side of Table 13.5, which yields the adjoint version on the right by the rules described in section 5.2.

Table 13.5: Taylor Exponential with Componentwise Adjoint

$v_0 = \exp(u_0)$	for $k = d \ldots 1$
for $k = 1 \ldots d$	for $j = k \ldots 1$
$v_k = 0$	$\bar{v}_{k-j} \mathrel{+}= \bar{v}_k u_j\, j/k$
for $j = 1 \ldots k$	$\bar{u}_j \mathrel{+}= \bar{v}_k v_{k-j}\, j/k$
$v_k \mathrel{+}= u_j v_{k-j}\, j/k$	$\bar{u}_0 \mathrel{+}= \bar{v}_0 v_0$

As we have come to expect, the adjoint involves roughly twice as many arithmetic operations, namely, $2d^2 + O(d)$, as the underlying matrix exponential. Here we have neglected the $d^2 + O(d)$ multiplications by the ratio j/k, which can again be avoided by suitable scaling of the u_j and v_k, costing only $O(d)$ operations. Collecting the incremental sums and reordering the indices we obtain the recurrences

$$\bar{v}_j \mathrel{+}= \sum_{k=j}^{d} \bar{v}_k \, u_{k-j}(1 - j/k) \quad \text{for} \quad j = d-1 \ldots 0 \tag{13.14}$$

and

$$\bar{u}_j \mathrel{+}= \sum_{k=j}^{d} \bar{v}_k \, v_{k-j} \, j/k \quad \text{for} \quad j = 0 \ldots d \, . \tag{13.15}$$

Now we see more clearly that the update calculation for the $\bar{v}_j$ is actually recursive, whereas the formulas for the $\bar{u}_j$ look almost like a convolution. Fortunately, the former calculation can be gotten rid off and the second turned into a proper convolution. To arrive at this desirable situation we use the following result.

Lemma 13.1 (Jacobian of Taylor Exponential)
The recurrence relations $v_0 = \exp(u_0)$ *and*

$$v_k = \sum_{j=0}^{k-1} v_j \, u_{k-j} \, (1 - j/k) \quad \textit{for} \quad k = 1 \ldots d$$

yield the total derivative values

$$\frac{dv_k}{du_j} = \begin{cases} v_{k-j} & \textit{if} \quad 0 \le j \le k \le d \, , \\ 0 & \textit{if} \quad 0 \le k < j \le d \, . \end{cases}$$

Proof. The proof is by induction on k, as performed in Exercise 13.7. ■

The hypothesis of Lemma 13.1 is just a minor rewrite of the formula for the exponential listed in Table 13.2. The corresponding dependencies between the local input variables u_j and the local output variables v_k are displayed on the left-hand side of Fig. 13.3. As we can see, there are about d^2 arcs, of which roughly half connect local "dependents" v_k among each other.

The adjoint evaluation listed in the right-hand side of Table 13.5 amounts to starting with the adjoint values $\bar{v}_k$ and then propagating them backward through the subgraph on the left of Fig. 13.3. This involves first updating the $\bar{v}_k$ themselves before they may finally be reset to zero by the adjoint $\bar{v}_k = 0$ of the statement $v_k = 0$, which we have omitted for simplicity. These intermediate values of the $\bar{v}_k$ are really of no interest, and one may strive to express the final values $\bar{u}_j$ directly in terms of the original values $\bar{v}_k$. This amounts to preaccumulating the local Jacobian $\left(\partial v_k / \partial u_j\right)_{0 \le j \le k \le d}$ in the sense of section 10.2. According to Lemma 13.1 this local Jacobian is lower triangular Toeplitz and its entries are given directly by the v_k. Hence we may simplify the pair of relations (13.14) and (13.15) to the formula

$$\bar{u}_j \mathrel{+}= \sum_{k=j}^{d} \bar{v}_k \, v_{k-j} \, . \tag{13.16}$$

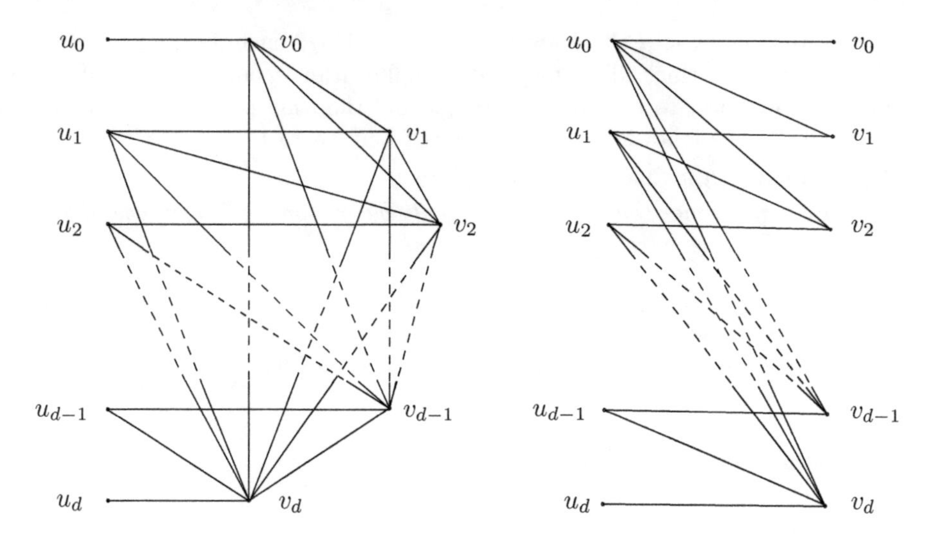

Figure 13.3: Computational Graph of Taylor Recurrence and Preaccumulate

In order to turn the right-hand side into the exact form of a convolution we set

$$\hat{u}_j \equiv \bar{u}_{d-j} \quad \text{and} \quad \hat{v}_k \equiv \bar{v}_{d-k} \, .$$

Then we obtain after some elementary manipulations the equivalent formula

$$\hat{u}_j \mathrel{+}= \sum_{k=0}^{j} \hat{v}_k \, v_{j-k} \, .$$

The savings and simplification that we have achieved by preaccumulating the Jacobian of the Taylor exponential can be achieved for all other intrinsics as well. Moreover, there is actually no need to preaccumulate because the factor with which the reversed vector of adjoint values needs to be convoluted is simply the derivative of the intrinsic function in the sense of the Taylor calculus sketched below.

Taylor Calculus

Since their evaluation involves elemental derivatives of up to order k, the vector of the coefficient functions $y_k = F_k(x_0, x_1, \ldots, x_k)$ are $(d-k)$-times continuously differentiable at all points x_0 where Assumption (ED) applies. To quantify the dependence of the y_k on the x_j with $j \leq k$ we may examine the partial derivative matrices $\partial y_k / \partial x_j \in \mathbb{R}^{m \times n}$ for $0 \leq j \leq k < d$. Since $\partial y_0 / \partial x_0 = F_0'(x_0)$ is the standard Jacobian of $F \equiv F_0$ we will refer to these $m \times n$ matrices as higher-order Jacobians. As we will see in section 13.5, these Jacobians may be useful for the numerical solution of ODEs and are almost indispensable for the numerical analysis of the more general problem class of DAEs.

For $0 \leq j \leq k < d$ there are formally $(d+1)d/2$ such Jacobians, but only d of them are distinct. This fortunate simplification occurs because differentiation with respect to the *space* vectors x and the *time* parameter t commutes according to the following generalization of Lemma 13.1.

Proposition 13.3 (Jacobians of Taylor Functions)
Suppose F is d-times continuously differentiable on some neighborhood of a point $x_0 \in \mathbb{R}^n$. Then we have

$$\frac{\partial y_j}{\partial x_i} = \frac{\partial F_j}{\partial x_i} \equiv A_{j-i}(x_0, \ldots, x_{j-i}) \quad \textit{for all} \quad 0 \leq i \leq j < d \tag{13.17}$$

with $A_i \equiv F_i'(x_0, \ldots, x_i)$ the ith Taylor coefficient of the Jacobian of F at $x(t)$, i.e.,

$$F'(x(t)) = \sum_{i=0}^{d-1} A_i t^i + o(t^{d-1}) \, .$$

Proof. Under our assumptions, the value $y(x(t)) = F(x(t))$ is d-times continuously differentiable with respect to t and x_i, provided t is sufficiently small. Hence differentiation with respect to t and x_i commutes, so that for $j < d$

$$\begin{aligned} \frac{\partial y_j}{\partial x_i} &= \frac{1}{j!}\frac{\partial}{\partial x_i}\left[\frac{\partial^j}{\partial t^j}\, y(x(t))\right]_{t=0} = \frac{1}{j!}\left[\frac{\partial^j}{\partial t^j}\left(\frac{\partial}{\partial x_i}\, y(x(t))\right)\right]_{t=0} \\ &= \frac{1}{j!}\left[\frac{\partial^j}{\partial t^j}\left(F'(x(t))t^i\right)\right]_{t=0} = \frac{i!}{j!}\binom{j}{i}\left[\frac{\partial^{j-i}}{\partial t^{j-i}}\, F'(x(t))\right]_{t=0} \\ &= \frac{1}{(j-i)!}\left[\frac{\partial^{j-i}}{\partial t^{j-i}}\, F'(x(t))\right]_{t=0} = A_{j-i}. \end{aligned}$$

The last two equations follow from Leibniz's rule and the fact that only the ith derivative of t^i does not vanish at $t = 0$. ∎

Considering the d vectors $y_i \in \mathbb{R}^m$ for $i = 0 \ldots d-1$ as dependents of the independent vectors $x_i \in \mathbb{R}^n$ for $i = 0 \ldots d-1$ one obtains according to the proposition for their adjoints the relation

$$\bar{x}_j = \sum_{i=j}^{d-1} \bar{y}_i A_{i-j} \quad \text{for} \quad 0 \leq j < d \, .$$

This is a generalization of the special case (13.16) even though we have written it nonincrementally. Again the convolution-like structure motivates us to look for a more intuitive interpretation.

Another consequence of the proposition is that full sensitivity information about the relation between the truncated Taylor polynomials

$$x(t) = \sum_{j=0}^{d-1} t^j x_j \in \mathcal{P}_d^n \quad \text{and} \quad y(t) = \sum_{j=0}^{d-1} t^j y_j \in \mathcal{P}_d^m$$

is surprisingly cheap. More specifically we have the following corollary.

Corollary 13.1 (Derivative of Taylor Calculus)
Under the assumptions of Proposition 13.3, *let* $\Delta x(t) = \sum_{j=0}^{d-1} t^j \Delta x_j \in \mathcal{P}_d^n$ *be any perturbation that is sufficiently small with respect to the norm* $\|\Delta x(t)\|$ *as defined by* (13.9). *Then the resulting polynomial*

$$\Delta y(t) = \sum_{j=0}^{d-1} t^j \Delta y_j = F(x(t) + \Delta x(t)) - y(t) + o(t^{d-1}) \in \mathcal{P}_d^m$$

satisfies

$$\Delta y_k - \sum_{j=0}^{k} A_{k-j} \Delta x_j = o(\|\Delta x(t)\|) \quad \textit{for all} \quad 0 \le k < d\,.$$

Proof. For each k the vector $y_k + \Delta y_k$ is a continuously differentiable function of the Δx_j with x_j fixed for $j = 0 \ldots k$. According to the preceding proposition, the Jacobians with respect to these real variable vectors are the A_{k-j}, so the assertion holds by standard multivariate calculus on reals. ■

In terms of the power series themselves we may rewrite the corollary as

$$\Delta y(t) = A(t)\, \Delta x(t) + o\|\Delta x(t)\|$$

where

$$A(t) \equiv \sum_{j=0}^{d-1} t^j A_j = F'(x(t)) - o(t^{d-1}) \in \mathcal{P}_d^{m\times n}\,.$$

Recall the definition of $E_d(F) : \mathcal{P}_d^n \longmapsto \mathcal{P}_d^m$ as the extension of the real vector function $F \in \mathcal{C}^{d-1}(\mathbb{R}^n, \mathbb{R}^m)$ to the free module domain $\mathcal{P}_d^n$ and range $\mathcal{P}_d^m$. Now we find that if F is at least d-times differentiable, then $E_d(F)$ is at least once differentiable as a function between the modules $\mathcal{P}_d^n$ and $\mathcal{P}_d^m$. Denoting differentiation by the operator D we have the commutative relation $DE_d = E_d D$, as shown in Fig. 13.4.

$$\begin{array}{ccc}
\mathcal{C}^d(\mathbb{R}^n, \mathbb{R}^m) & \xrightarrow{\quad D \quad} & \mathcal{C}^d(\mathbb{R}^n, \mathbb{R}^{m\times n}) \\
\big\downarrow E_d & & \big\downarrow E_d \\
\mathcal{C}^1(\mathcal{P}_d^n, \mathcal{P}_d^m) & \xrightarrow{\quad D \quad} & \mathcal{C}^0(\mathcal{P}_d^n, \mathcal{P}_d^{m\times n})
\end{array}$$

Figure 13.4: Commuting Extension and Differentiation Operators E_d and D

The path through the top right corner corresponds to first developing a scheme for evaluating the matrix-valued function $F'(x) = D\,F(x)$ and then replacing the real point $x \in \mathbb{R}^n$ with the polynomial argument $x(t) \in \mathcal{P}_d^n$. Hence we may utilize any one of the accumulation schemes discussed in Part II for the evaluation of Jacobians and simply execute it in Taylor series arithmetic. Naturally, the elementary partials c_{ij} must also be evaluated in Taylor arithmetic with $\cos(u(t)) \in \mathcal{P}_d$ serving as the derivative of $\sin(u(t))$ at $u(t) \in \mathcal{P}_d$, for example. The complexity will normally be bounded by the number of arithmetic operations in the real Jacobian accumulation multiplied by $(d+1)^2$.

What do we get for our troubles? The collection of real matrices $A_j = F_j'$ for $j = 0 \dots d-1$, which form the Jacobian in the Taylor calculus sense depicted by the lower path in the commuting diagram. These higher-order Jacobians are quite useful in the context of dynamical systems, as analyzed in the final section 13.5. Before that, let us revisit our original goal of computing the derivative vectors $\bar{F}_k(\bar{y}, x, s)$ defined in (13.2).

Higher-Order Adjoints

Given any weight vector $\bar{y} \in \mathbb{R}^m$ the weighted average $f(x) \equiv \bar{y}^\top F(x)$ has an extension $E_d(f)$ to $\mathcal{P}_d^n$. Its gradient $\nabla[E_d(f)] \in \mathcal{P}_d^n$ can be computed by the reverse mode in Taylor arithmetic at any polynomial argument $x(t) \in \mathcal{P}_d^n$. For this the adjoint value of the scalar result $f(x(t))$ must be initialized to the Taylor polynomial $1 = 1 + t \cdot 0 + \cdots + t^d \cdot 0$. Then multiplying it with the components of $\bar{y}$ representing the constant partials of f with respect to the components of F yields corresponding adjoint polynomials that are still constant, i.e., have no nontrivial powers of t. These start occurring only when one reaches a nonlinear elemental in the computational graph of F yielding adjoint values that are nonconstant polynomials themselves.

At a cost proportional to $(1+d)^2\, TIME(F)$ we obtain

$$\nabla[E_d(f)](x(t)) = \sum_{j=0}^{d-1} t^j g_j = \nabla f(x(t)) + o(t^{d-1}) ,$$

where $g_j \in \mathbb{R}^n$ for $j = 0 \dots d-1$.

For linear arguments $x(t) = x + ts$ we find that

$$\begin{aligned} g_j &= \frac{\partial^j}{j!\partial t^j} \nabla f(x+t\,s)\bigg|_{t=0} = \frac{\partial^j}{j!\partial t^j} \nabla \bar{y}^\top F'(x+t\,s)\bigg|_{t=0} \\ &= \frac{1}{j!} \bar{y}^\top \nabla^{j+1} F(x) s^j = \bar{F}_j(\bar{y}, x, s) \, . \end{aligned} \tag{13.18}$$

Thus we see that all desired adjoint vectors $\bar{F}_j(\bar{y}, x, s)$ for $j < d$ are obtained in one reverse sweep in Taylor arithmetic of degree d. The lowest derivatives are found in the first and the highest derivatives in the last coefficients of the gradient polynomial $\nabla f(x(t))$. This ordering certainly seems natural but it should be noted that it is the opposite of what one obtains by the naive approach of initializing the adjoint of the highest derivative $f_{d-1} = \bar{y}^\top F_{d-1}$ to 1 and then propagating adjoints backward in real arithmetic. Then the adjoint of an intermediate coefficient v_j corresponds to the $(d-j)$th coefficient of the adjoint polynomial $\bar{v}(t)$ obtained by the reverse Taylor approach suggested earlier. An instance of this coefficient reversal was observed for the exponential in Lemma 13.1.

Gradient of Acceleration on the Lighthouse Example

In order to illustrate the correctness and utility of the reverse mode executed in Taylor arithmetic let us again consider the well-worn lighthouse example from Chapter 2. Suppose we specialize $\nu = 2$ and $\omega = 1$ to simplify matters a little bit. Then we may ask what the acceleration of the light point at time $t = t_0 = 0$ is when $\gamma = \gamma_0 = 1$ and how it depends on the values of these remaining two variables. Hence we first wish to calculate the second Taylor coefficient of the distance $r = y_1(t)\sqrt{1+\gamma^2}$ along the straight line

$$x(t) \;=\; (1, t) \;= (\gamma_0, t_0 + t)$$

and subsequently compute its gradient with respect to γ_0 and t_0. Here $y_1(t)$ denoted the first coordinate of the light point. The naive way of achieving this goal is to apply the conventional reverse mode to the chain of real operations leading from γ_0 and t_0 to the second Taylor coefficient r_2. As we shall demonstrate the same result can be obtained much more succinctly by applying the reverse mode in Taylor arithmetic with the number of elemental operations not being inflated compared to the undifferentiated evaluation procedure.

Since differentiating the intrinsic function tan directly is a little tedious we will replace it with the ratio sin/cos. Another deviation from the original formulation is that we give each intermediate variable a single character name, leaving subscripts free to number the corresponding Taylor coefficients. In the upper part of Table 13.6 we have listed the corresponding componentwise calculations for propagating the first three Taylor coefficients in the first three columns and the numerical results at the arguments $\gamma_0 = 1$ and $t_0 = 0$ in the last three columns. The formulas for the higher coefficients are in agreement with Tables 13.1 and 13.2. The two lines for sine and cosine must be read together as c_0 enters into the calculation of s_1, s_0 into c_1, c_1 into s_2, and so on. This interconnection is also important for the corresponding adjoint calculations.

Strictly speaking the Taylor coefficient r_2 equals only half of the acceleration. Also, to make the adjoint values come out as simple fractions we set $\bar{r}_2 = 1/\sqrt{2}$. Hence we will calculate the gradient of $r_2/\sqrt{2}$, which would have to be multiplied by $\sqrt{2}$ if one wanted to obtain the gradient of the acceleration itself.

Adjoining the componentwise calculation in the upper part of Table 13.6 yields the nonincremental reverse form listed in the lower part of the table. The final values $(\bar{t}_0, \bar{\gamma}_0) = (4, -1.5)$ multiplied by $2\sqrt{2}$ represent the gradient of the acceleration $2r_2$ with respect to (t_0, γ_0) at $(0, 1)$. The values $(\bar{t}_1, \bar{\gamma}_1) = (2, -0.5)$ multiplied by $\sqrt{2}$ represent the gradient of the velocity r_1 with respect to (t_0, γ_0) at $(0, 1)$. The values $(\bar{t}_2, \bar{\gamma}_2) = (1, 0)$ multiplied by $\sqrt{2}$ represent the gradient of the position r_0 with respect to (t_0, γ_0) at $(1, 0)$.

Table 13.6: Nonincremental Reverse Sweep on Simplified ($\gamma = 2, w = 1$) Lighthouse with $d = 2$, $\bar{r}_2 = 1/\sqrt{2}$, and $\bar{r}_1 = \bar{r}_0 = 0$

$\gamma_0 = 1$	$\gamma_1 = 0$	$\gamma_2 = 0$	1	0	0
$t_0 = 0$	$t_1 = 1$	$t_2 = 0$	0	1	0
$s_0 = \sin(t_0)$	$s_1 = t_1 * c_0$	$s_2 = t_1 * c_1/2 + t_2 * c_0$	0	1	0
$c_0 = \cos(t_0)$	$c_1 = -t_1 * s_0$	$c_2 = -t_1 * s_1/2 - t_2 * s_0$	1	0	−0.5
$v_0 = s_0/c_0$	$v_1 = (s_1 - v_0 * c_1)/c_0$	$v_2 = (s_2 - v_0 * c_2 - v_1 * c_1)/c_0$	0	1	0
$u_0 = \gamma_0 - v_0$	$u_1 = \gamma_1 - v_1$	$u_2 = \gamma_2 - v_2$	1	−1	0
$w_0 = v_0/u_0$	$w_1 = (v_1 - w_0 * u_1)/u_0$	$w_2 = (v_2 - w_0 * u_2 - w_1 * u_1)/u_0$	0	1	1
$a_0 = \gamma_0 * \gamma_0$	$a_1 = 2 * \gamma_0 * \gamma_1$	$a_2 = \gamma_1 * \gamma_1 + 2 * \gamma_0 * \gamma_2$	1	0	0
$b_0 = 1 + a_0$	$b_1 = a_1$	$b_2 = a_2$	2	0	0
$e_0 = \sqrt{b_0}$	$e_1 = 0.5 * b_2/e_0$	$e_2 = 0.5 * (b_2 - e_1 * e_1)/e_0$	$\sqrt{2}$	0	0
$r_0 = e_0 * w_0$	$r_1 = e_0 * w_2 + e_1 * w_0$	$r_2 = e_0 * w_0 + e_1 * w_1 + e_2 * w_0$	0	$\sqrt{2}$	$\sqrt{2}$
$\bar{r}_2 = 1/\sqrt{2}$	$\bar{r}_1 = 0$	$\bar{r}_0 = 0$	$1/\sqrt{2}$	0	0
$\bar{e}_2 = \bar{r}_2 * w_0$	$\bar{e}_1 = \bar{r}_2 * w_1 + \bar{r}_1 * w_0 - \bar{e}_2 * e_1/e_0$	$\bar{e}_0 = \bar{r}_2 * w_2 + \bar{r}_1 * w_1 + \bar{r}_0 * w_0 - \bar{e}_2 * e_2/e_0 - \bar{e}_1 * e_1/e_0$	0	$1/\sqrt{2}$	$1/\sqrt{2}$
$\bar{a}_2 = \bar{b}_2 = 0.5 * \bar{e}_2/e_0$	$\bar{a}_1 = \bar{b}_1 = 0.5 * \bar{e}_1/e_0$	$\bar{a}_0 = \bar{b}_0 = 0.5 * \bar{e}_0/e_0$	0	0.25	0.25
$\bar{w}_2 = \bar{r}_2 * e_0$	$\bar{w}_1 = \bar{r}_2 * e_1 + \bar{r}_1 * e_0 - \bar{w}_2 * u_1/u_0$	$\bar{w}_0 = \bar{r}_2 * e_2 + \bar{r}_1 * e_1 + \bar{r}_0 * e_0 - \bar{w}_2 * u_2/u_0 - \bar{w}_1 * u_1/u_0$	1	1	1
$\bar{u}_2 = -\bar{w}_2 * w_0/u_0$	$\bar{u}_1 = -\bar{w}_2 * w_1/u_0 - \bar{w}_1 * w_0/u_0$	$\bar{u}_0 = -\bar{w}_2 * w_2/u_0 - \bar{w}_1 * w_1/u_0 - \bar{w}_0 * w_0/u_0$	0	−1	−2
$\bar{v}_2 = \bar{w}_2/u_0 - \bar{u}_2$	$\bar{v}_1 = \bar{w}_1/u_0 - \bar{u}_1 - \bar{v}_2 * c_1/c_0$	$\bar{v}_0 = \bar{w}_0/u_0 - \bar{u}_0 - \bar{v}_2 * c_2/c_0 - \bar{v}_1 * c_1/c_0$	1	2	3.5
$\bar{c}_2 = -\bar{v}_2 * v_0/c_0$	$\bar{c}_1 = -\bar{v}_2 * v_1/c_0 - \bar{v}_1 * v_0/c_0 + \bar{s}_2 * t_1/2$	$\bar{c}_0 = -\bar{v}_2 * v_2/c_0 - \bar{v}_1 * v_1/c_0 - \bar{v}_0 * v_0/c_0 + \bar{s}_2 * t_2 + \bar{s}_1 * t_1$	0	−0.5	0
$\bar{s}_2 = \bar{v}_2/c_0$	$\bar{s}_1 = \bar{v}_1/c_0 - \bar{c}_2 * t_1/2$	$\bar{s}_0 = \bar{v}_0/c_0 - \bar{c}_2 * t_2 - \bar{c}_1 * t_1$	1	2	4
$\bar{t}_2 = -\bar{c}_2 * s_0 + \bar{s}_2 * c_0$	$\bar{t}_1 = -\bar{c}_2 * s_1/2 + \bar{s}_2 * c_1/2 - \bar{c}_1 * s_0 + \bar{s}_1 * c_0$	$\bar{t}_0 = -\bar{c}_0 * s_0 + \bar{s}_0 * c_0$	1	2	4
$\bar{\gamma}_2 = 2 * \bar{a}_2 * \gamma_0 + \bar{u}_2$	$\bar{\gamma}_1 = 2 * \bar{a}_2 * \gamma_1 + 2 * \bar{a}_1 * \gamma_0 + \bar{u}_1$	$\bar{\gamma}_0 = 2 * \bar{a}_2 * \gamma_2 + 2 * \bar{a}_1 * \gamma_1 + 2 * \bar{a}_0 * \gamma_0 + \bar{u}_0$	0	−0.5	−1.5

The relation

$$0 = \sqrt{2}\,\bar{\gamma}_2 = \partial r_2/\partial \gamma_2 = \partial r_0/\partial \gamma_0$$

reflects that the position r_0 stays constant at 0 when the slope γ_0 is altered. In contrast the velocity

$$\sqrt{2} = \sqrt{2}\,\bar{t}_2 = \partial r_2/\partial t_2 = \partial r_0/\partial t_0$$

is nonzero and agrees with the value $r_1 = \sqrt{2}$ obtained by propagating Taylor coefficients starting from $(\gamma, t) = (1, t)$. The result represented by the last two rows of Table 13.6 can be obtained in a much simpler way by executing the formal adjoint of the first column in the upper part of Table 13.6 in Taylor arithmetic. By applying the nonincremental inverse we obtain

$$\begin{aligned}
\bar{r} &= (1/\sqrt{2}, 0, 0)\,, \\
\bar{e} &= \bar{r}{*}w = (1/\sqrt{2}, 0, 0){*}(0, 1, 1) = (0, 1/\sqrt{2}, 1/\sqrt{2})\,, \\
\bar{b} &= 0.5{*}\bar{e}/e = 0.5{*}(0, 1/\sqrt{2}, 1/\sqrt{2})/(\sqrt{2}, 0, 0) = (0, 0.25, 0.25)\,, \\
\bar{a} &= \bar{b} = (0, 0.25, 0.25)\,, \\
\bar{w} &= \bar{r}{*}e = (1, 0, 0)\,, \\
\bar{u} &= -\bar{w}{*}w/u = -(1, 0, 0){*}(0, 1, 1)/(1, -1, 0) = -(0, 1, 2)\,, \\
\bar{v} &= \bar{w}/u - \bar{u} = (1, 0, 0)/(1, -1, 0) + (0, 1, 2) = (1, 2, 3)\,, \\
\bar{c} &= -\bar{v}{*}v/c = -(1, 2, 3){*}(0, 1, 0)/(1, 0, -0.5) = -(0, 1, 2)\,, \\
\bar{s} &= \bar{v}/c = (1, 2, 3)/(1, 0, -0.5) = (1, 2, 3){*}(1, 0, 0.5) = (1, 2, 3.5)\,, \\
\bar{t} &= \bar{s}{*}c - \bar{c}{*}s = (1, 2, 3.5){*}(1, 0, -0.5) + (0, 1, 2){*}(0, 1, 0) = (1, 2, 4)\,, \\
\bar{\gamma} &= 2{*}\bar{a}{*}\gamma + \bar{u} = (0, 0.5, 0.5){*}(1, 0, 0) - (0, 1, 2) = (0, -0.5, -1.5)\,.
\end{aligned}$$

As we can see, the results are consistent if, in the "naive" adjoint, components are ordered from the highest to the lowest index, as we have done in the lower part of Table 13.6. We already observed this reversal on the Taylor exponential before Lemma 13.1.

Here there is not much difference in the operations count, but the second approach is certainly preferable in terms of conceptual economy. It is also interesting to note that the adjoint values for the intermediates w, v, and s are not the same in the two calculations. The reason is that, for example, the value $\bar{w}_1 = 0$ on page 325 does not reflect the role that w_1 plays in the calculation of w_2. These internal dependences are preeliminated in Taylor arithmetic as we observed in Lemma 13.1 for the exponential.

Section Summary and Conclusion

In this section we have shown that the various procedures for calculating gradients and Jacobians discussed throughout this book also can be performed in Taylor arithmetic. Moreover, the results make sense in that they can be interpreted as higher-order Jacobians and gradients or adjoint vectors.

This extension of the scalar type from the field of reals $\mathbb{R}$ to a ring of truncated polynomials $\mathcal{P}_d$ has little bearing on most issues considered before, especially those regarding data dependencies and joint allocations. While data transfers and memory requirements scale with d, operations counts scale theoretically with d^2, unless the evaluation procedure in question involves almost exclusively linear operations.

During the accumulation of Jacobians and gradients, all Taylor operations are additions or subtractions and multiplications, so optimized convolution routines for each value of d up to some reasonable upper bound, say 9, promise a significant reduction in the runtime. It should also be noted that there is never any need to append the topological information encapsulated by the precedence relation $\prec$ and the corresponding computational graphs with extra arcs and nodes for higher derivatives. Hence we conclude with the following rule.

RULE 23

WHAT WORKS FOR FIRST DERIVATIVES IS FIT TO YIELD HIGHER DERIVATIVES BY EXTENSION TO TAYLOR ARITHMETIC.

Conceptually, Taylor arithmetic is always implemented as operator overloading. This remains true even when extra function calls are avoided and compiler optimization is enhanced through in-lining or direct source generation. In any case the transition from real to Taylor arithmetic is a very local affair. Normally, little can be gained by optimizing across several statements or even blocks, except when such improvements were already possible on the real code (see Rule 6). Of course, the same cannot be said about the reverse mode, which is best implemented by generating an adjoint code with optimized save and recompute decisions. As we have indicated in Rule 23, we do not think that one should ever generate more than one adjoint source code, e.g., for second-order adjoints. Although there are some shifts in the temporal and spatial complexities, one adjoint source compiled and linked with libraries for various degrees d should provide great flexibility and good efficiency while avoiding a proliferation of source codes for various differentiation modes.

13.5 Special Relations in Dynamical Systems

So far we have not imposed any extra condition on the relation between the domain path $x(t)$ and its image $F(x(t)) = y(t) + o(t^d)$. In this section we will consider first the ODE situation where

$$\dot{x}(t) \equiv dx(t)/dt = F(x(t)) \quad \text{with} \quad F : \mathbb{R}^n \mapsto \mathbb{R}^n \, .$$

Later we will examine the more general case of DAEs that are defined by an identity

$$0 = y(t) = F(z(t), \dot{z}(t)) \quad \text{with} \quad F : \mathbb{R}^{n+n} \mapsto \mathbb{R}^n \, .$$

In both cases we assume *autonomy* in the sense that the "time" t may not enter explicitly into the equations, an assumption that is innocuous from an analytical point of view. Moreover, we assume that all our calculations are performed at time $t = 0$. The latter normalization entails no loss of generality since our considerations are local to a current solution point. In contrast to the previous section the paths $x(t)$ and $y(t)$ are in general no longer polynomials.

Ordinary Differential Equations

Given a sufficiently smooth autonomous right-hand side function $F : \mathbb{R}^n \longmapsto \mathbb{R}^n$, we consider the initial value problem

$$\dot{x}(t) = F(x(t)), \quad x(0) = x_0 \in \mathbb{R}^n . \tag{13.19}$$

It has been known since Newton that the exact solution can be approximated by the Taylor series

$$x(t) = \sum_{i=0}^{d} x_i t^i + o(t^d) \tag{13.20}$$

and the coefficient vectors $x_i \in \mathbb{R}^n$ can be computed recursively by the relation

$$x_{i+1} = \frac{1}{1+i}\, y_i = \frac{1}{1+i}\, F_i\left(x_0, \ldots, x_i\right) . \tag{13.21}$$

In this way all coefficients $x_i = x_i(x_0)$ are obtained as smooth functions of the base point x_0. The usual procedure is to calculate $x_1 = y_0$ from x_0, then to plug in $x_0 + tx_1$ on the right in order to obtain $x_2 = \frac{1}{2}y_1$, and so on. In this way one has to perform d sweeps through the evaluation algorithm for F with the degree of the Taylor arithmetic growing by 1 each time. The cost for computing the solution approximation $x(t)$ of degree d in t can thus be bounded by $d^3/3$ times the cost of evaluating F in real floating point arithmetic provided there are only a few sinusoidal elementals. If fast convolution methods were used the cost ratio might come down to $O(d^2 \log d)$. However, this prospect is not very enticing because a reduction to $O(d^2)$ can be achieved in at least two elementary ways, as will be explained below.

The explicit Taylor series method performs a step of size $\Delta t = h$ by progressing from the current point x_0 to the next iterate $\sum_{0 \le j \le d} h^j x_j$. This classical method is presently not very often applied, partly because it is still believed that second and higher coefficients x_j with $j > 1$ are hard to come by. A better reason for not using it is that, like all explicit numerical integrators, Taylor expansion methods are forced to make the stepsize excessively small, wherever the system is stiff [HW96]. There are A-stable implicit variants that overcome this limitation (see, for example, [GC$^+$97, Gri95, Ned99]). They are based on solving a nonlinear equation of the form

$$\sum_{j=0}^{d} \gamma_j\, x_j(x_0) \left(\frac{h}{2}\right)^j = r$$

with respect to the next point x_0. Here the coefficients γ_j for $j = 0 \ldots d$ may depend on d. The right-hand side $r \in \mathbb{R}^n$ depends on the method and the previous point along the solution trajectory.

The solution of the above system by a Newton-based method requires the Jacobian of the left-hand side, which is obviously a linear combination of the total derivatives

$$X_i = \frac{dx_i}{dx_0} \in \mathbb{R}^{n \times n} \quad \text{for} \quad i = 0 \ldots d \, . \tag{13.22}$$

With $A_i \equiv \partial x_{i+k}/\partial x_k$ for $k \geq 0$ the partial derivatives occurring in Proposition 13.3, we derive by the chain rule (see Exercise 13.6) the recursion

$$X_{i+1} = \frac{1}{i+1}\left[\sum_{j=0}^{i} A_{i-j}X_j\right] \quad \text{for} \quad i = 0 \ldots d-1 \, , \tag{13.23}$$

which ordinarily starts with $X_0 = dx_0/dx_0 = I$. In terms of the differential equation one can easily check that the X_i are simply the Taylor coefficients of the fundamental system $X(t)$ that solves for $X_0 = I \in \mathbb{R}^{n \times n}$ the matrix differential equation

$$\dot{X}(t) \equiv \frac{d}{dt}X(t) = F'(x(t))\,X(t) \quad \text{with} \quad X(0) = X_0 \, . \tag{13.24}$$

Here $x(t)$ is the exact analytical flow from a given $x(0) = x_0$. Note that in (13.24) the initial fundamental matrix X_0 need not be the identity or even square. Then the X_i can be interpreted as the total derivative of the x_i with respect to some problem parameters $z \in \mathbb{R}^p$ with $dx_0/dz = X_0$. If these parameters also enter directly into the right-hand side, the matrix differential equation (13.24) must be augmented and becomes inhomogeneous. It is then often called the *sensitivity equation* associated with the ODE.

Storing or Doubling via Newton

The numerical integration of an ODE based on Taylor series may well involve thousands or even millions of individual Taylor coefficients. Hence they should be obtained as efficiently as possible. A rather natural way to reduce the operations count for their recursive calculation is to store for each intermediate variable $v(t)$ the coefficients v_j that are already known from earlier sweeps. Then they can be restored on subsequent sweeps and only one extra coefficient needs to be calculated each time. In this way the total operations count for calculating all Taylor coefficients x_i for $i = 1 \ldots d$ from x_0 is exactly the same as that for propagating a given Taylor polynomial $x(t)$ through the right-hand side once. This approach was first implemented in the Fortran package ATOMFT [CC94]. In very many applications the functions describing right-hand sides are of manageable complexity so that storing all intermediate results may not be much of a problem. However, on the repeated sweeps the ratio between the number of operations and value fetches is almost exactly 1, just like in the reverse mode. This ratio may be of concern on modern workstations.

Especially when the Jacobians A_i need to be computed anyway, we may prefer another approach based on the following corollary of Proposition 13.3.

Corollary 13.2 (Linearity in Higher Coefficients)
Under the assumption of Proposition 13.3 *we have for any* $k < d$ *and* $k/2 < j \leq k+1$

$$y_k = F_k(x_0, x_1, \ldots, x_{j-1}, 0, \ldots, 0) + \sum_{i=j}^{k} A_{k-i}(x_0, \ldots, x_{k-i})\, x_i\,. \qquad (13.25)$$

Proof. The assertion is trivially satisfied for $j = k+1$ by the definition of F_k. Hence we make the induction hypothesis that it holds for some $j+1 > k/2+1$ and consider the remaining nonlinear part $y_k(x_0, x_1, \ldots, x_{j-1}, x_j, 0, \ldots, 0)$. By Proposition 13.3 the function y_k has with respect to x_j the Jacobian $A_{k-j} = A_{k-j}(x_0, x_1, \ldots, x_{k-j})$. Since $k - j < j$ by assumption on j, the Jacobian A_{k-j} is in fact constant with respect to x_j, and we must have by the mean value theorem

$$\begin{aligned} &F_k(x_0, \ldots, x_{j-1}, x_j, 0, \ldots, 0) \\ &\qquad = F_k(x_0, \ldots, x_{j-1}, 0, \ldots, 0) + A_{k-j}(x_0, \ldots, x_{k-j})x_j\,. \end{aligned}$$

Thus we can keep reducing j and splitting off one linear term at a time until the condition $j > k/2$ is no longer satisfied. This completes the proof by backward induction. ∎

According to the corollary each coefficient $y_k = F_k(x_0, \ldots, x_k)$ is linear with respect to the upper half of its arguments, namely, the x_j with $j > k/2$. Moreover, its nonlinear arguments x_j with $j \leq k/2$ uniquely determine the matrices A_i, with $i \leq k/2$ specifying this linear dependence. This effect can be directly observed in the expansions (13.6), where v_1 is linear in u_1, v_2 in u_2, v_3 in (u_2, u_3), and v_4 in (u_3, u_4). Note that so far we have not used any differential relation between $x(t)$ and $y(t)$.

Now, suppose the coefficients $x_0, x_1, \ldots, x_{j-1}$ have been calculated such that (13.21) is satisfied for $i \leq j-1$. Setting the next j coefficients $x_j, x_{j+1}, \ldots, x_{2j-1}$ tentatively to zero, we may now compute by Taylor arithmetic of order $(2j-1)$ the following preliminary coefficient vector values:

$$\hat{y}_k = F_k(x_0, \ldots, x_{j-1}, \underbrace{0, \ldots, 0}_{j \text{ zeros}}) \quad \text{for} \quad k = j-1 \ldots 2j-1\,.$$

Thus we can immediately compute the next coefficient $x_j = \hat{y}_{j-1}/j$ since $\hat{y}_i = y_i$ for $i < j$. Moreover, a subsequent reverse sweep yields the Jacobians

$$F_i' \equiv A_i = \frac{\partial y_i}{\partial x_0} = A_i(x_0, \ldots, x_i) \quad \text{for} \quad i < j\,.$$

Using the differential relation and the linearity property described above, we can now obtain the correct values of the next j coefficients by the updating loop

$$x_{k+1} = \frac{1}{k+1}\left[\hat{y}_k + \sum_{i=j}^{k} A_{k-i}\, x_i\right] \quad \text{for} \quad k = j-1 \ldots 2j-1\,.$$

Hence we can more than double the number of correct solution coefficients x_i by a single forward and return sweep in Taylor arithmetic. Starting from x_0 with $j = 1$ we have the sequence

$$1, 3, 7, 15, 31, 63, \ldots = 2^{s+1} - 1 \quad \text{for} \quad s = 0, 1, \ldots\ ,$$

where s on the right denotes the number of reverse sweeps. Thus we may perform the variant of Newton's method displayed in Table 13.7 to obtain d Taylor coefficients of the ODE solution in roughly $\log_2(d{+}2)$ extended evaluation sweeps.

Table 13.7: Coefficient Doubling by Newton's Method

initialize x_0
for $s = 0, 1, \ldots$
  $j = 2^{s+1} - 1$
  $(\hat{y}_0, \hat{y}_1, \ldots, \hat{y}_{2j-1}) = [E_{2j-1}F]\,(x_0, \ldots, x_{j-1})$
  $(A_0, A_1, \ldots, A_{j-1}) = [E_{j-1}F]'(x_0, \ldots, x_{j-1})$
  for $k = j-1 \ldots 2j-1$
    $x_{k+1} = \dfrac{1}{k+1}\left[\hat{y}_k + \sum_{i=j}^{k} A_{k-i} x_i\right]$

The extensions $(E_{2j-1}F)$ and $(E_{j-1}F)'$ require the evaluation and differentiation of F in Taylor series arithmetic of degree $2j-1$ and $j-1$, respectively. When all corresponding Jacobians A_k up to $k = d = 2^{s+1} - 1$ are also needed, the total operations count is only about 33% more than a single forward and return sweep in Taylor arithmetic of degree d. The partial linearity property described can and has been exploited not only for ODEs but also on the more general problems described below [PR97].

Differential Algebraic Equations

In contrast to the well-researched area of ODEs, the task of numerically solving a system of the form $y = F(z(t), \dot{z}(t)) = 0$ has only recently attracted a lot of attention [BCP96]. In the interesting cases where the trailing square Jacobian $C_0 \equiv \partial y/\partial \dot{z}$ is singular the existence and characteristic of solutions may depend on arbitrary high derivatives. Consequently this area should become a prime application of AD in the future.

To reduce the difficulty of the problem as measured by its index many authors recommend that certain components of the vector function $F : \mathbb{R}^{n+n} \mapsto \mathbb{R}^n$

be differentiated in a preprocessing stage. The general understanding tends to be that this needs to be done "by hand" or "symbolically." Normally this can be achieved only when the equation $F(z, \dot{z}) = 0$ separates into a differential part and an algebraic part, so that for some partitioning $z = (\hat{z}, \check{z})$,

$$\frac{d}{dt}\hat{z} = G(\hat{z}, \check{z}) \quad \text{and} \quad H(\hat{z}, \check{z}) = 0 .$$

Then H may be differentiated once or several times until the resulting extended system also allows the computation of $d\check{z}/dt$, possibly under some consistency condition on z.

Using the functions $F_k : \mathbb{R}^{2n(k+1)} \mapsto \mathbb{R}^n$ defined in Definition (TC), one obtains in the general unpartitioned case for $k = 0 \ldots d-1$

$$0 \;=\; F_k(z_0, z_1; z_1, 2z_2; \ldots; z_{j-1}, jz_j; \ldots; z_k, (k+1)z_{k+1}) \,. \tag{13.26}$$

Here the z_j and jz_j are the Taylor coefficients of the solution curve $z(t)$ and its derivative $\dot{z}(t)$, so the full argument $x(t) = (z(t), \dot{z}(t))$ of F has the ith Taylor coefficient $x_i = [z_i, (i+1)z_{i+1}]$. Partitioning the higher-order Jacobians as

$$F_k'(z_0, z_1; \ldots; z_k, (k+1)z_{k+1}) = [B_k, C_k] \quad \text{with} \quad B_k, C_k \in \mathbb{R}^{n\times n} \,,$$

one obtains the total derivative of $y_k = F_k$ with respect to z_i as

$$\begin{aligned} \frac{dy_k}{dz_i} &= \frac{\partial y_k}{\partial x_i}\frac{\partial x_i}{\partial z_i} + \frac{\partial y_k}{\partial x_{i-1}}\frac{\partial x_{i-1}}{\partial z_i} \\ &= \left[B_{k-i}, C_{k-i}\right]\begin{bmatrix} I \\ 0 \end{bmatrix} + \left[B_{k-i+1}, C_{k-i+1}\right]\begin{bmatrix} 0 \\ iI \end{bmatrix} \\ &= B_{k-i} + i\,C_{k-i+1} \quad \text{for} \quad 0 \le i \le k+1 \,, \end{aligned} \tag{13.27}$$

where $B_{-1} \equiv 0$ so that the superdiagonal blocks with $i = k+1$ reduce to $(k+1)C_0$ for $k = 0 \ldots d-1$. All blocks with $i > k+1$ vanish identically so that the Jacobian of the system (13.26) has the following lower block Hessenberg structure:

$$\begin{bmatrix} B_0 & C_0 & 0 & \cdots & 0 & 0 \\ B_1 & B_0{+}C_1 & 2C_0 & \cdots & 0 & 0 \\ \vdots & \vdots & \vdots & \ddots & \vdots & \vdots \\ B_{d-2} & B_{d-3}+C_{d-2} & B_{d-4}{+}2C_{d-3} & \ddots & (d{-}1)C_0 & 0 \\ B_{d-1} & B_{d-2}+C_{d-1} & B_{d-3}{+}2C_{d-2} & \cdots & B_0{+}(d{-}1)C_1 & dC_0 \end{bmatrix}$$

Since the Jacobian has d block rows and $(d+1)$ block columns, the system formally always has n degrees of freedom, which may be taken up by initial conditions. However, in the interesting cases, the square matrix C_0 and hence the block-triangular system obtained by eliminating the first block column, which corresponds to considering z_0 as constant, is always singular. Hence d must be gradually increased until the base point z_0 and the tangent z_1 are uniquely determined. In practice, this is a rather tricky process because one has to determine the rank of certain submatrices numerically.

Several authors (see [DPS89] and [Ned99]) have used AD within their methods for solving DAEs. Our purpose is not to advocate any one of them but to emphasize that all derivatives necessary for analyzing and solving such difficult equations can be provided by the techniques advocated here at a cost that grows "merely" quadratically with the degree d.

The provision of qualitative and quantitative derivative information for the structural method of Pryce [Pry01] was provided in [GW04].

13.6 Summary and Outlook

In this chapter we considered the evaluation of third-order and higher derivatives with respect to one or more variables. In the univariate case discussed in section 13.2 one propagates vectors of Taylor coefficients using recurrences that have been known and used for a long time, mostly for the solution of initial value problems in ODEs. Since all elementary intrinsics are solutions of linear ODEs, the effort for propagating d Taylor coefficients grows "only" like $O(d^2)$. This bound could be reduced to an asymptotic estimate like $O(d \log(1+d))$ if fast convolution algorithms rather than straightforward polynomial arithmetic were implemented. In either case, the corresponding derivatives can be obtained by simply rescaling the Taylor coefficients.

For differentiation with respect to several variables or directions in the domain we have recommended in section 13.3 the propagation of families of univariate Taylor polynomials. The coefficients of the desired derivative tensors can then be obtained by a stable conversion or interpolation process, whose cost is negligible compared to the propagation of multivariate (or the corresponding family of univariate) Taylor polynomials through any nontrivial evaluation procedure. The main advantages of our approach are regular memory access and conceptual simplicity, as no multivariate versions of the chain rule for higher derivatives need to be considered at all. The operations count is essentially the same but the memory requirement is slightly elevated compared to the direct computation of multivariate Taylor polynomials for each intermediate variable.

We have not considered at all efforts to compute internal enclosures for Taylor coefficients or remainder terms for whole Taylor approximations. A question that warrants further investigation is the exploitation of interface contraction as discussed in section 10.2 for first derivatives, which promises even bigger savings in the propagation of higher derivatives in several global directions.

In contrast to the first sections of this chapter, section 13.4 contains material that has not previously been published, but may possibly be viewed as folklore by some AD experts. It is found that applying one extra level of differentiation (with respect to many variables) to a higher-order directional derivative can be effected by performing the reverse mode in truncated Taylor series arithmetic. This approach reduces the operations count by a factor of between 1 and 2, apart from allowing the reuse of standard adjoint code. It also means that univariate convolutions become the workhorse of the reverse mode for higher gradients and Jacobians and should therefore be implemented with the utmost efficiency. It seems clear that further elaboration and clarification of this calculus on Taylor rings is needed before it reaches its full computational potential.

The final section 13.5 collects some relations that are useful for the theoretical analysis and numerical solution of ODEs and DAEs. In particular, it gives formulas for the matrix coefficients of the fundamental solution matrix of derived ODEs, which may also be used for the implementation of implicit Taylor series methods. Numerical experiments by Guckenheimer and his coworkers [CG97] suggest that these methods are particularly suitable when high-accuracy solutions are desired. In the context of DAEs, some level of implicit differentiation is always unavoidable, and we expect AD to become an essential ingredient of future, general-purpose analysis and integration schemes.

13.7 Examples and Exercises

Exercise 13.1 (*Inversion of Taylor Recurrences*)
Consider the recurrences in Table 13.1.
a. Rederive the formulas for u/w and $\sqrt{u}$ from those for $u * v$ and u^2.
b. Develop a recurrence for u^3 using the decomposition $u^3 = u^2 * u$. Find the corresponding inverse recurrence for $u^{1/3}$.
c. Compare the complexity of the recurrences for u^2, u^3, $u^{1/2}$, and $u^{1/3}$ obtained above with that for a general power u^r given in Table 13.2.

Exercise 13.2 (*Taylor Recurrence for Inverse Trigonometrics*)
Use Proposition 13.1 to develop a recurrence for $v = \arctan(u)$ and also $v = \arcsin(u)$.

Exercise 13.3 (*Consistency of Modulus*)
Verify that the modulus $|v(t)|$ defined by (13.8) for Taylor polynomials $v(t) \in \mathcal{P}_d$ satisfies the consistency relation $|u(t) * w(t)| \leq |u(t)||w(t)|$. Give examples where the inequality holds strictly and where it holds as equality.

Exercise 13.4 (*Complexity of Multivariate Convolution*)
For scalar-valued F the tensor $\nabla_S^j F$ has, according to (13.12), $\binom{p+j-1}{j}$ distinct elements. Prove by hand or check using a computer algebra package that
a. the collection $(\nabla_S^j F)_{j \leq d}$ contains $\binom{p+d}{d}$ distinct elements;
b. multiplying each element of $\nabla_S^j F$ with each element of $\nabla_S^{k-j} F$ for $j = 0, 1 \ldots k$ involves $\binom{2p+k-1}{k}$ arithmetic operations;
c. performing part **b** for $k = 0, 1 \ldots d$ involves a total of $\binom{2p+d}{d}$ arithmetic operations.

Exercise 13.5 (***Cost of Univariate Taylor Family Propagation***)
Based on (13.11), consider the ratio $q(d,p)$ between the cost of propagating $\binom{p+d-1}{d}$ univariate Taylor polynomials and that of propagating a single polynomial in p variables of the same degree d through a single multiplication operation.
a. Try to minimize $q(d,p)$ over positive integers p and d using your favorite computer algebra system.
b. Show that for fixed d the integer function $q(d,p)$ satisfies $q(d,p+1) \geq q(d,p)$ if and only if $(d-3)p \leq 2$, with equality occurring simultaneously if at all. Conclude that for $d \leq 3$ we have monotonic growth and for $d \geq 5$ the function is nondecreasing. Compute location and value of the maximum in the only remaining case, $d = 4$.
c. Show that $1/4 < q(p,d) < 3/2$ for all $d \leq 7$ and arbitrary p.

Exercise 13.6 (***Using Proposition* 13.3**)
Show that the fundamental solution $X(t)$ determined by (13.24) satisfies

$$X(t) \equiv \sum_{i=0}^{d} t^i X_i + O(t^{d+1}) \in \mathbb{R}^{n\times n} ,$$

with the X_i defined by the recurrence (13.23) starting from the initial value $X_0 \in \mathbb{R}^{n\times n}$. Check that this assertion remains true if $X_0 \in \mathbb{R}^{n\times p}$ is rectangular with any $p > 0$.

Exercise 13.7 (***Proof of Lemma* 13.1**)
Prove the lemma by induction on $k = 0, 1, \ldots$

Exercise 13.8 (***First-Order Taylor versus Complex Arithmetic***)
For $y = F(x) = \exp(x)$, compare the accuracy and computational effort in computing $\dot{y} = F'(x)\dot{x}$ by Taylor arithmetic and complex arithmetic according to (13.10), respectively. Determine the value of t for which the complex arithmetic approach yields the best approximation to $\dot{y}$ at $x = 2 = \dot{x}$, both theoretically and experimentally.

Chapter 14

Differentiation without Differentiability

Hitherto we have assumed that the elemental differentiability Assumption (ED) holds at all arguments $x \in \mathbb{R}^n$ of interest, so that we have according to Proposition 2.1 proper differentiation on some neighborhood. In this chapter we will examine what can be done if this assumption is violated at or very near a point of interest x. Rather than modeling F on a neighborhood we will then concentrate our attention on a ray $x + t\dot{x}$ or other smooth arc $x(t)$ for $t \geq 0$ nominated by the user. We will consider the direction of the input tangent $\dot{x} \in \mathbb{R}^n$ as an indication of where the user expects the desired derivatives to reflect the properties of the underlying function F. As we will see the provision of such a *preferred direction* and its scale may also be useful for the reverse mode; it is not needed at all in the smooth case.

By contrasting our situation with the "smooth" case, we have already implied that we are considering here cases that are somehow "nonsmooth." Unfortunately, on this topic the communication between AD users and AD tool providers, and even among AD researchers themselves, is often quite confused and sometimes clouded by wishful thinking. To reassure present users or woo new ones, AD tool providers are prone to assert that "you always get the derivatives of the smooth piece that the current argument lies in." (Potential) users, on the other hand, are likely to presume that the functions evaluated by their programs are not only piecewise but also globally differentiable.

Unfortunately, both parties are probably indulging in some wishful thinking, which need not reflect reality on any particular problem. Even leaving aside the thorny issue of numerical rounding, we must still face the fact that most sizable codes have iterative or otherwise adaptive aspects that cause at least small jumps in the output variables as functions of the input variables.

Even if users, or rather applications programmers, have carefully glued the pieces together in a differentiable fashion, AD tools may not always yield the correct results (see section 14.2). Depending on what she or he was led to believe about the power of the AD tool in use, such a careful applications programmer may well have good reasons for complaints. On the other hand AD providers are sometimes exasperated by forceful demands to generate derivatives of codes that are admittedly discontinuous.

Necessity, Sufficiency, and Locality

Of course, the chance of ever hitting a crack between two smooth pieces exactly appears to be zero at a generic argument x, whose components have been perturbed away from special values such as 0 and 1. However, the resulting domain of differentiability may be very small, possibly not exceeding the level of roundoff errors. In the same way a square matrix whose entries are computed as functions of generic variable values is almost certain to have a nonzero determinant, however it is evaluated. Obviously, it would be very foolish to conclude from this observation that the concept of matrix singularity is of no importance in numerical computations.

Just as numerical linear algebra routines alert the user to near-singularity and ill conditioning, AD tools should provide some indication of the validity of the derivative values produced. The analogy carries a bit further in that warnings that the computational task at hand may be difficult to resolve reflect not only the "conditioning" of the problem itself but also the limitations of the methodology employed by the tool. For example, a Hilbert matrix of size 12 may represent an insurmountable difficulty for a matrix factorization in floating point arithmetic, but a linear system with a suitable right-hand side might be solvable by a "symbolic" system employing rational arithmetic.

The principal method of AD is the application of the chain rule to the given elemental decomposition. If the chain rule does not apply in a suitable form, the procedures examined in this book thus far cannot work, no matter how smooth the resulting composite function $y = F(x)$ is. Here we have tacitly imposed the crucial restriction that AD tools may not perform any analysis regarding the interaction between various elementals, other than examining the propagation of activity and the to-be-recorded status discussed in Chapter 6.

In section 2.3 we considered the example $f(x) = |x|^3 = \sqrt{x^6}$. Fortunately, things are not quite as hopeless in this case as we made it sound in that introductory chapter. Suppose the user specifies not only the base point x but also some preferred direction $\dot{x}$. Then $|x + t\dot{x}|$ is differentiable for almost all values of t, and we have the unique one-sided derivative $|\dot{x}|$ at the origin $x = 0$. Taking the third power only improves matters and makes the directional derivatives identical to zero for all $\dot{x} \in \mathbb{R}$ when $x = 0$. As we will show in section 14.2, directional derivatives exist along all arcs that are regular in a certain sense. This regularity can be verified and turns out to be sufficient for the existence of generalized gradients.

For the most part, existence and uniqueness of derivatives are lost when the original coding for the function evaluation leaves something to be desired. The power $v = u^6$, for example, pushes roughly five-sixths of all floating point numbers into overflow or underflow. A subsequent root $w = \sqrt{v}$ reduces the range to one-half of all positive numbers. Hence a lot of information is unnecessarily lost, since the pair of assignments $v = |u|$ and $w = v^3$ loses only one-third of the possible arguments to overflow or underflow. Loosely speaking, the gap between necessary conditions and constructive, i.e., verifiable sufficient, conditions is widened whenever trouble spots are compounded so that their negative effects may or may not cancel each other out.

Of course there are many classes of mathematical models where nonsmoothness is genuine and not just a result of numerical approximations or sloppy coding. For example, in stochastic optimization one considers chance constraints of the form

$$P\{G(x) \geq \zeta\} \;\geq\; 1 - \varepsilon \in (0,1) \tag{14.1}$$

with $G : \mathbb{R}^n \longmapsto \mathbb{R}^m$ a constraint function, $\zeta \in \mathbb{R}^m$ a random variable, and $\varepsilon \in (0,1)$ a fixed threshold. When ζ has the distribution function $f : \mathbb{R}^m \longmapsto \mathbb{R}$, one may rewrite the chance constraint as

$$f(G(x)) \;\geq\; 1 - \varepsilon\,.$$

This scalar inequality could be viewed as a conventional algebraic constraint, were it not for the fact that accumulated probability densities are often continuous but not differentiable. A simple example of this effect is given in Exercise 14.4. See also the papers [Mar96] and [Yos87].

This chapter is organized as follows. In section 14.1 we use a modification of the lighthouse example to illustrate the problems and techniques to be examined in more detail later. In section 14.2 we show that the set of troublesome arguments has measure zero.

One-sided Laurent expansions with a leading fractional factor are developed in section 14.3 under certain assumptions on the direction taken. As it turns out, in most directions these lead also to generalized gradients.

14.1 Another Look at the Lighthouse Example

To illustrate the various difficulties that may arise and preview some of our results, let us revisit the lighthouse example introduced in Chapter 2. To make things a little more interesting, we set $\nu = 2$ and replace the linear function $y_2 = \gamma\, y_1$, describing the quay, by the cubic relation $y_2 = \psi(y_1)$ with $\psi(z) \equiv (z-2)^3 + 0.4$. In other words, we have replaced the straight quay line with a cubic curve and moved the origin of the coordinate system to the lighthouse position. This *cubic lighthouse problem* is sketched in Fig. 14.1, with ω set to 1 for simplicity from now on. To compute the point (y_1, y_2) where the light beam hits the quay at time t, we now have to solve the cubic equation

$$0 = f(z,t) \equiv \psi(z) - z\tan(t)$$

for the first coordinate $z \equiv y_1$.

As one can see from Fig. 14.2 there may be one, two, or three distinct real roots, of which only the smallest one is of interest. The nonphysical solutions are represented by dotted or dashed lines.

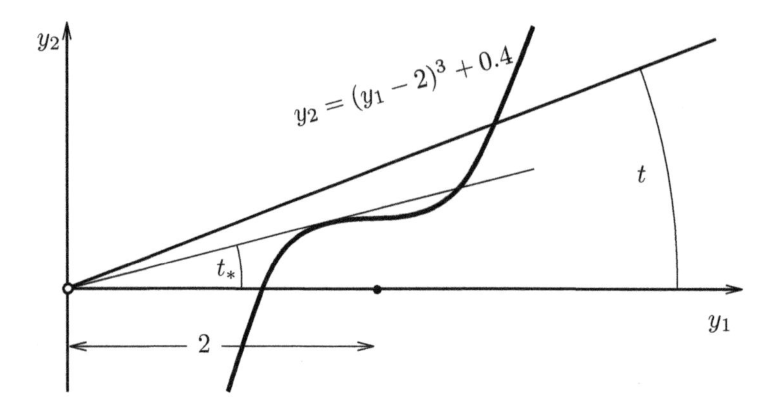

Figure 14.1: Lighthouse Geometry with Cubic Quay Wall

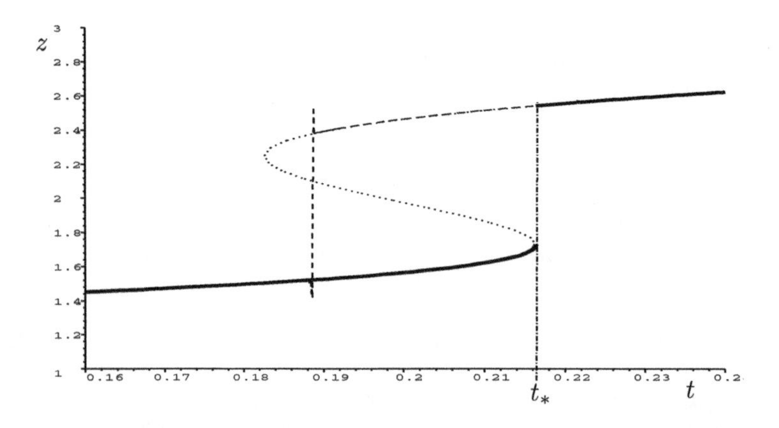

Figure 14.2: Light Point Jump at Critical Time from Lower to Upper Branch

Using Cardan's Formula

Suppose we code up Cardan's formula to express this smallest root in terms of standard elementals including $\sqrt{u}$ and $\sqrt[3]{u}$. Then we might obtain the evaluation procedure listed in Table 14.1. Deviating from our usual practice we have given the variables single-character names that are used repeatedly to improve readability. The variable d represents the discriminant, whose sign determines the number of real roots: three if $d < 0$, two if $d = 0$, and one if $d > 0$.

In order to stay within the conceptual framework used throughout this book we will not allow explicit jumps in evaluation procedures. Instead we assume that all possible branches are executed sequentially and that the correct results are selected at the end using the Heaviside function. Since some of these branches may "blow up" in one way or another, we will have to provide rules for propagating infinities and NaNs [Sta85] and also make sure that these special values can be annihilated in the end by constant zeros. In particular, we will consider integer return values of step functions as constants. All this will be rigorously defined in section 14.2 of this chapter.

According to Table 14.1, after computing d, we treat first the case $d > 0$, then the case $d \leq 0$, and finally select the relevant value for z in the last segment. These four segments are separated by dashed lines. When $d < 0$ the square root in the second segment will return a NaN that percolates through the segment to the root candidate c. It is subsequently annihilated by a constant zero in the next-to-last statement of our little procedure. Conversely, when $d > 0$, the argument of arccos in the third segment will have a modulus greater than 1 resulting in a NaN that infects the root candidate z at the end of the third segment. It is then annihilated in the fourth statement from the end.

Table 14.1: Cubic Lighthouse by Cardan

p	$=$	$\tan(t)$
q	$=$	$p - 0.2$
p	$/=$	3
d	$=$	q^2
d	$-=$	p^3
- - -	- - -	- - -
r	$=$	$\sqrt{d}$
u	$=$	$q + r$
u	$=$	$\sqrt[3]{u}$
v	$=$	$q - r$
v	$=$	$\sqrt[3]{v}$
c	$=$	$u + v$
- - -	- - -	- - -
p	$=$	$\mathrm{abs}(p)$
p	$=$	$\sqrt{p}$
q	$/=$	p^3
a	$=$	$\arccos(q)$
a	$/=$	3
z	$=$	$\cos(a)$
b	$=$	$a + \pi/3$
b	$=$	$-\cos(b)$
z	$=$	$\min(z, b)$
b	$=$	$a - \pi/3$
b	$=$	$-\cos(b)$
z	$=$	$\min(z, b)$
z	$=$	$2 * z * p$
- - -	- - -	- - -
d	$=$	$\mathrm{heav}(-d)$
z	$=$	$d * z$
d	$=$	$1 - d$
z	$+=$	$d * c$
z	$+=$	2

The limiting case $d = 0$ occurs at time $t_* \approx 0.2164$, as one can find out by solving the two equations $f(z,t) = 0 = f'(z,t)$ simultaneously with respect to z and t. At this special point, both branches run through without NaNs, but the root proposed by the second one, which is the smaller of two real solutions, is taken. The shape of the values $z = z(t)$ obtained from our version of Cardan's formula is displayed in Fig. 14.2 by a thick line. As one can see, there is a jump at the critical value t_*, where the light beam first touches the quay wall tangentially at $z_- \approx 1.729$ and then hits it properly at about $z_+ \approx 2.578$. The physics does not single out one of the values as being the proper value of the implicit function being calculated. Within the interval $(0.1827, t_*)$ the cubic equation has two other real roots, which are nonphysical in that the light beam could only reach them if the wall were transparent. Nevertheless, these additional roots have a bearing on the smoothness near the discontinuity and strongly influence the behavior of Newton's method, as we will see later.

To the right of the discontinuity the function is smooth and should have a well-defined, one-sided derivative. On the left the situation is different, as the slope tends to plus infinity as one approaches the critical value.

Now one may ask whether and how AD applied to the Cardan code can bring out these important features automatically, i.e., without anybody looking at a plot, which is unrealistic when there are several independent variables anyway.

There is certainly no difficulty in detecting that the Heaviside function, or more generally a numerical comparison, is critical in that arbitrarily small perturbations can cause a switch from one value or program branch to another. The AD tool may then raise the flag "potential jump." Similarly, it might warn "potential kink" when the absolute value of p at the beginning of the third section or the min further down are critical, i.e., evaluated at the origin or a tie, respectively. However, these necessary conditions for nonsmoothness are by no means sufficient, in that the resulting function may well be smooth even where the alarm goes off. For example, in our Cardan code, $p = 0$ can occur only when $d > 0$, so that the kink occurring in the second segment is of no consequence for the behavior of the final function value.

A similar but much more subtle effect occurs at root singularities that arise at the critical value t_*. Here the argument of the square root in the second segment is exactly 0 and the argument of the arccos in the third segment is exactly 1. Hence the AD tool may raise in both cases the flag "potential vertical tangent." However, as we have seen in Fig. 14.2, only the second singularity is reflected in the solution curve, whereas the first is not. The reason is that, in adding the intermediate values u and v, the root term $r = \sqrt{d}$ is canceled out, so that $c = u + v$, which represents the root on the right branch of the cubic, is quite nicely differentiable. How would or could an AD tool know this? Only if it propagated power series with fractional powers of t, and even then it would not be certain that the offending coefficients would cancel each other exactly in floating point arithmetic. For simplicity we will shy away from such fully blown Levi–Civita numbers [Ber96]. On the code above our model will merely yield the continuity information

$$z(t_* - t) = z_- + o(t^0) \quad \text{and} \quad z(t_* + t) = z_+ + o(t^0) \quad \text{for} \quad t > 0\,.$$

Note that the leading constant term has become dependent on the direction in which we "differentiate" in the sense of computing an approximating expansion of some sort.

So which tentative conclusions can we draw from the lighthouse example with a cubic quay wall? First, the physically relevant solutions of nonlinear systems may have jumps and vertical tangents. Some of them we might be able to detect automatically, by propagating derivative information of some sort through the evaluation program and keeping track of elementals that are being evaluated at critical arguments where they are not smooth. However, there may be cancellation of significant information that can be avoided only by propagating infinite series with fractional powers. We will return to the cubic lighthouse example at the beginning of the final Chapter 15, where we apply Newton's method rather than Cardan's formula.

14.2 Putting the Pieces Together

Sometimes program branches are introduced for the purpose of computational efficiency or numerical stability, without affecting global differentiability of the function being evaluated. Consider, for example, the linear system

$$\begin{bmatrix} a_{11}(x) & a_{12}(x) \\ a_{21}(x) & a_{22}(x) \end{bmatrix} \begin{bmatrix} y_1 \\ y_2 \end{bmatrix} = \begin{bmatrix} b_1(x) \\ b_2(x) \end{bmatrix},$$

where $x \in \mathbb{R}^n$ is some vector of independent variables on which the coefficients a_{ij} and b_i are assumed to depend smoothly.

Considering the diagonal element a_{11} as a natural first pivot, one might solve for y_1 and y_2 using the code listed in Table 14.2. Here we have left out tests on the other denominators for the sake of simplicity. Wherever the determinant

$$a_{11}(x)\, a_{22}(x) - a_{12}(x)\, a_{21}(x)$$

does not vanish, the solution components y_1 and y_2 are smooth functions of x and can be evaluated as suggested in Table 14.2. However, the corresponding derivatives do not come out correctly if one simply differentiates the two branches according to the rules we have developed so far in this book. To see this, let us assume that $a_{11}(x) \equiv x_1$ is the only coefficient depending on the variable x_1. Then at $x_1 = 0$ the "else" clause applies and its derivative yields $\partial y_1/\partial x_1 = 0 = \partial y_2/\partial x_1$ since a_{11} does not occur at all in the second branch. However, by implicit differentiation of the first equation $a_{11}y_1 + a_{12}y_2 = b_2$, this would imply $y_1 \partial a_{11}/\partial x_1 = y_1 = 0$, which is clearly not true in general. Assuming that there is only one additional independent variable x_2 on which all other coefficients depend, we observe that the domain $\mathbb{R}^2$ of our vector function $y = (y_1, y_2)$ has been split into two subsets: the x_2-axis $\{x_1 = 0\}$ where the "else" branch applies, and its complement $\{x_1 \neq 0\}$ where the "then" branch applies. Since the "else" branch formula only applies on the x_2-axis, it is not surprising that differentiating transverse to it, namely, in the x_1 direction, does not give the right result.

Table 14.2: Pivoting Branch on 2×2 System

```
if (a11 ≠ 0)  then
    a22 = a22 − a12 ∗ a21/a11
    y2 = (b2 − b1 ∗ a21/a11)/a22
    y1 = (b1 − a22 ∗ y2)/a11
else
    y2 = b1/a12
    y1 = (b2 − a22 ∗ y2)/a21
```

To correct the situation, we may expand the axis to a tube by replacing the test "if ($a_{11} \neq 0$)" with "if ($|a_{11}| > \varepsilon$)" for some positive tolerance $\varepsilon > 0$. Naturally, the first statement in the "else" clause must be replaced by the pair

$$a_{12} = a_{12} - a_{22} * a_{11}/a_{21}\,,$$
$$y_2 = (b_1 - b_2 * a_{11}/a_{21})/a_{12}\,.$$

The subsequent original assignment to y_1 is still correct. In contrast to the x_2-axis by itself, the ε-tube enclosing it has a nonempty interior, which turns out to be sufficient for the correctness of derivatives, as shown below.

Finite Selections

To put our observations concerning program branches into a general framework, we consider a function $F : \mathcal{D} \longmapsto \mathbb{R}^m$ defined on the open domain $\mathcal{D} \subseteq \mathbb{R}^n$ that is a *selection* of finitely many functions $F^{(k)} : \mathcal{D}_k \longmapsto \mathbb{R}^m$ that are $(d > 0)$-times continuously differentiable on their open domains $\mathcal{D}_k \subseteq \mathbb{R}^n$. Here selection means that for all $x \in \mathcal{D}$, there exists at least one selection function $F^{(k)}$ with $F(x) = F^{(k)}(x)$. Let $\mathcal{A}_k \subseteq \mathcal{D} \cap \mathcal{D}_k$ denote the *joint domain* of F and $F^{(k)}$, that is, the domain where both functions coincide. We call $\{(F^{(k)}, \mathcal{D}_k, \mathcal{A}_k)\}$ a representation of F on $\mathcal{D}$. Of course, such representations are not unique.

Definition (PD): PIECEWISE DIFFERENTIABILITY
The function F is said to be piecewise differentiable *on the open domain $\mathcal{D}$ if there exists a representation $\{(F^{(k)}, \mathcal{D}_k, \mathcal{A}_k)\}$ of F as a selection of finitely many $(d > 0)$-times continuously differentiable vector-valued functions $F^{(k)}$. A selection function $F^{(k)}$ is called* essential *if the joint domain $\mathcal{A}_k$ has a nonempty interior, and* essentially active *at $x \in \mathcal{D}$ if x belongs to the boundary of the interior of $\mathcal{A}_k$.*

Obviously each particular computational trace resulting from a sequence of branch decisions can be considered as a selection function. For instance, the "then" branch of the determinant example in Table 14.2 determines an essential selection function, whereas the selection function corresponding to the "else" branch is not essential in the original formulation without $\varepsilon > 0$. However, note that in general the selection functions defined by computational traces may not have open domains. For example, $F(x) = |x|^{\frac{1}{2}}$ should be represented as

$$F(x) = \begin{cases} \sqrt{x} & \text{if} \quad x > 0 \\ 0 & \text{if} \quad x = 0 \\ \sqrt{-x} & \text{if} \quad x < 0 \end{cases}$$

where all three functions $F^{(1)}(x) = \sqrt{x}$, $F^{(2)}(x) = 0$, and $F^{(3)}(x) = \sqrt{-x}$ are real analytic on $\mathcal{D}_1 = (0, \infty)$, $\mathcal{D}_2 = \mathbb{R}$, and $\mathcal{D}_3 = (-\infty, 0)$, respectively. If the origin had been included with the domain of either root branch, it would not have been open, as required by our definition of piecewise differentiability. By defining the constant 0 as a separate selection function, we avoid this problem, but since the joint domain is the singleton $\mathcal{A}_2 = \{0\}$ the new selection function is not essential.

Our definition of piecewise differentiability is considerably weaker than similar concepts in the literature [Sch94, Cha90] in that we do not require the continuity of F on $\mathcal{D}$; that is, we allow jumps and poles at the boundaries between the smooth pieces. Nevertheless, we can state the following proposition regarding the derivatives of F.

Proposition 14.1 (Fischer's Result [Fis91])
Suppose F is piecewise differentiable and $\{(F^{(k)}, \mathcal{D}_k, \mathcal{A}_k)\}$ is a representation of F on $\mathcal{D}$. Then the first d derivatives of F are well defined and identical to those of the $F^{(k)}$ on the interior of the joint domain $\mathcal{A}_k$. Moreover, we get the following statements.

(i) *If F is d-times continuously differentiable on $\mathcal{D}$, then the first d derivatives of $F^{(k)}$ have unique limits at all boundary points of the interior of $\mathcal{A}_k$ that also belong to $\mathcal{D}$, which again coincide with the derivatives of F.*

(ii) *If $F^{(k)}$ is essentially active at $x \in \mathcal{D}$ and d-times continuously differentiable on some neighborhood of $x \in \mathcal{D}_k$, then the first d derivatives of $F^{(k)}$ have unique limits for all sequences $\{x_j\}$, $x_j \in \mathrm{int}(\mathcal{A}_k)$, $x_j \longmapsto x$, which must coincide with the derivatives of $F^{(k)}$ at x.*

Proof. Both assertions hold by standard real analysis. ∎

In case of a differentiable function F, assertion (i) of the proposition ensures the computability of the derivatives when using a representation of F just containing essential selection functions. In (ii) we do not assume even continuity of F but obtain limits of derivatives in the neighborhood of the point x, which might be thought of as generalized derivatives at x. Hence we conclude with the following rule.

Rule 24

When program branches apply on open subdomains, algorithmic differentiation yields useful derivatives.

It is sometimes believed that the goal of subdomains with nonempty interiors can be achieved by only using inequality rather than equality tests. The latter are a little dubious on a finite precision floating point system anyhow. Unfortunately, life is not that easy, because any equality test can be rephrased as two nested inequality tests.

Consider, for example, the following conditional assignment:

$$\text{if} \quad (x \neq 0) \quad \text{then} \quad y = \sin(x)/x \quad \text{else} \quad y = 1\,.$$

The resulting value y is a globally analytical function $f(x)$ with $f'(0) = 0$ and $f''(0) = -1/3$. Fixing the control flow at $x = 0$ means differentiating the constant function $\tilde{f}(x) = 1$. While $f'(0) = 0$ happens to coincide with $\tilde{f}'(0) = 0$, the second derivative $\tilde{f}''(0) = 0$ differs from $f''(0)$. Now one might be tempted to blame the user for poor coding and suggest that $f(x)$ should be approximated by its Taylor series for all $|x| < \varepsilon$. However, few people will go through this trouble, and the higher derivatives are still likely to be off. Moreover, as we will see in Exercise 14.2, there is no pretext for blaming the user, as the conditional statement above works very well in floating point arithmetic.

A Lipschitz Continuous Example

To get some idea of how piecewise differentiable functions can be defined and what they may look like, let us consider the example $y = f(x) = |x_1^2 - \sin(|x_2|)|$. This scalar-valued function might be evaluated by the procedure listed in Table 14.3. It is globally defined but contains two absolute values as nonsmooth elementals. The first modulus is critical when x_2 vanishes, and the second when $v_1 \equiv x_1^2 = \sin(v_2) \equiv \sin(|x_2|)$. Hence the set of critical arguments consists near the origin of the x_1-axis and two smooth curves where $x_2 \approx \pm x_1^2$.

These six kinks can be seen in the plot in Fig. 14.3. There are four globally analytic selection functions, which correspond to the signs for each absolute value. The function is everywhere Lipschitz continuous, so that F is in fact almost everywhere differentiable. While the example has no proper tangent plane at the origin and along the critical curves, it is clear that along any other oriented direction one may compute a unique tangent. In other words, the function is Gâteaux and even Bouligand differentiable but not Fréchet differentiable. Nevertheless, we can compute directional derivatives of any order.

Table 14.3: A Lipschitz Continuous Example

v_{-1}	$=$	x_1
v_0	$=$	x_2
v_1	$=$	v_{-1}^2
v_2	$=$	$\|v_0\|$
v_3	$=$	$\sin(v_2)$
v_4	$=$	$v_1 - v_3$
v_5	$=$	$\|v_4\|$
y	$=$	v_5

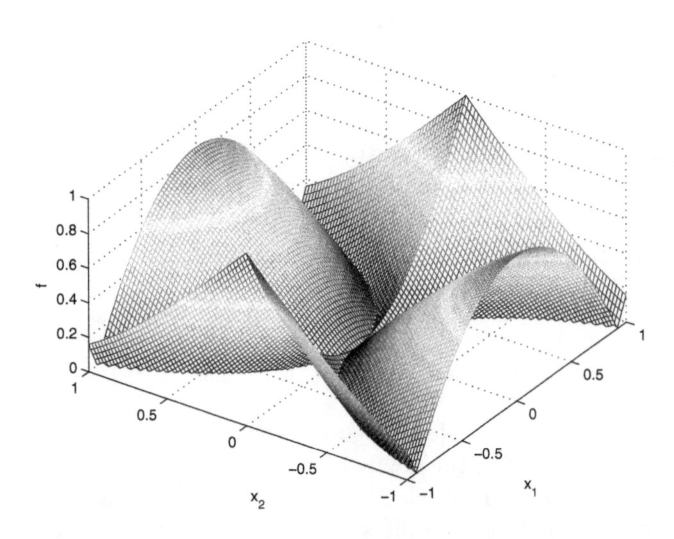

Figure 14.3: Nonsmooth Example $f = |x_1^2 - \sin(|x_2|)|$

In fact, we want to be even more ambitious and compute directional derivatives and generalized gradients of functions that are not even continuous. In this context assertion (ii) of Proposition 14.1 is of particular interest for points x not belonging to the joint domain $\mathcal{A}_k$, but where $F^{(k)}$ is essentially active.

Generalized Differentiation

The reader may wonder by now why we do not use the well-established concept of generalized gradients and Jacobians [Cla83]. First let us consider the good news. As long as the chain rule never involves, on the elemental level, more than one generalized gradient, the accumulation of the overall generalized derivative can proceed without ambiguity. Moreover, the structural complexity of these derivative objects remains simple, so that they can be represented by arrays of real numbers. A typical example is robust regression, where some norm-like functional

$$f(x) \equiv \psi(F(x)) \ : \ \mathbb{R}^n \longmapsto \mathbb{R}$$

of a smooth residual $F : \mathbb{R}^n \longmapsto \mathbb{R}^m$ is to be minimized. Here ψ may represent the L_1 norm $\sum_i |F_i|$, or the L_∞ norm $\max_i |F_i|$, or some other Lipschitz continuous function. By Rademacher's theorem, any function $f : \mathbb{R}^n \longmapsto \mathbb{R}$ that is locally Lipschitzian is almost everywhere differentiable. Moreover, even where it is not, it has a generalized gradient

$$\partial f = \mathrm{conv}\,\{g \in \mathbb{R}^n |\; g = \lim \nabla f(x_k),\; x_k \longmapsto x\}\,.$$

Here the sequences $x_k \longmapsto x$ consist of points at which f is differentiable and conv denotes the convex hull.

There are several other concepts of generalized differentiation (see, e.g., [Mor06]), which all have the following common characteristics.

- The derivative objects can become quite complex.
- The chain rule holds only as a set inclusion.

For example, the weighted Euclidean norm $v = \sqrt{u^2 + 2w^2}$ has at the origin $u = 0 = w$ the ellipse $\partial v = \{(v_u, v_w) \,|\, v_u^2 + \frac{1}{2}v_w^2 \leq 1\}$ as generalized gradient. This set has infinitely many extreme points and can therefore not be represented as a convex hull span of finitely many vertices. The second difficulty applies, for example, when $w = |u|$ and $v = w - w$. Here, Clarke's generalization of the chain rule yields the set inclusion

$$\partial w = [-1,\, 1] \quad \text{and} \quad \partial v \subseteq [-1,\, 1] - [-1,\, 1] = [-2,\, 2]\,.$$

This result is correct but not very useful since the singleton $\partial v = \{0\}$ is in fact a proper derivative.

Both characteristics mentioned above are quite troublesome from our point of view. In particular, we would like to limit the complexity of derivative objects a priori to allocate appropriate data structures. Nonsmooth analysis is usually rather sophisticated, often involving nested limit superiors and inferiors, which are not that easy to interpret geometrically. Fortunately, much of this sophistication becomes unnecessary if one restricts the analysis to functions defined by evaluation procedures over a simple set of elemental functions. Then most theoretical assertions can be shown by induction over the intermediate variables v_i and lead to corresponding derivative calculations in the forward mode. There are really only the following four kinds of calamities that can occur through nonsmooth elemental functions.

Kinks, Roots, Steps, and Poles

As customary in numerical software, assertions of mathematical properties and relations are based on the fiction of exact arithmetic, albeit with an extension to $\pm\infty$ in the style of IEEE arithmetic. In fact, the dreaded result NaN will turn out to be comparatively benign and stable.

Recall that in Table 2.3 the elemental functions were listed in three categories: *smooth, Lipschitz,* and *general.* This categorization was valid only under the assumption that all elemental functions are restricted to the interior of their maximal domains, which excluded trouble with roots and reciprocals at the origin. By eliminating this somewhat artificial restriction we wind up with two extra categories of contingencies yielding kinks, roots, poles, and steps, as discussed in the following subsections.

Kink Functions

The classical kink is the absolute value function $v = |u|$. Because of the identities

$$\min(u, v) \equiv \frac{1}{2}\,(u + v - |v - u|)$$

and

$$\max(u, v) \equiv \frac{1}{2}\,(u + v + |v - u|)\ ,$$

there is no need to consider min and max separately in discussing mathematical properties. Naturally, AD tools should deal with these important elementals directly for the sake of efficiency. The rules for treating them are easily derived from the ones we will establish for $\text{abs}(u) \equiv |u|$. This simple inheritance of properties applies also to the L_1 and L_∞ norms in several variables, which are directly defined in terms of abs and max.

Root Functions

In this category we include all univariate functions that are defined only on a closed subinterval of the real line. The most important members of this class are the square root $\sqrt{u}$ and the inverse trigonometric functions $\arcsin(u), \arccos(u)$. As for the general fractional power u^r with fractional exponent $0 < r \notin \mathbb{Z}$, the values at the boundaries of the domain are bounded, but the one-sided derivatives may be infinite.

From now on we will assume that at arguments outside their maximal real domain, root functions and other elementals return the special value NaN and are thus globally defined in some generalized sense after all. Normally NaNs can never be eliminated, in that a NaN occurring among the arguments of any elementary function forces the result to have the same special value. The only exception to this rule will be made for multiplication by zero.

Step Functions

Because of the identities

$$\mathrm{sign}(u) = \mathrm{heav}(u) - \mathrm{heav}(-u)$$

and

$$\mathrm{heav}(u) = 1 + (1 - \mathrm{sign}(u)) * \mathrm{sign}(u)/2 \ ,$$

we may use the sign and Heaviside functions interchangeably when it comes to discussing mathematical properties. Let us consider a program branch of the form

$$\text{if} \quad (w \geq 0) \quad \text{then} \quad v = \varphi(u) \quad \text{else} \quad v = \psi(u) \ . \tag{14.2}$$

It can be rewritten in C shorthand as

$$v = (w \geq 0)? \ \varphi(u) : \psi(u) \ .$$

This single assignment is mathematically equivalent to

$$v = \mathrm{heav}(w) * \varphi(u) + (1 - \mathrm{heav}(w)) * \psi(u) \ , \tag{14.3}$$

provided the values $\varphi(u)$ and $\psi(u)$ are both well defined whether $w \geq 0$ or not. However, this may well not be the case, for example, in the conditional assignment

$$v = (-u \geq 0)? \ 0 : u * \ln(u) \ .$$

Rewritten in terms of the Heaviside function as above and evaluated at $u = 0$, we would obtain in IEEE arithmetic

$$\begin{aligned} v &= \mathrm{heav}(0) * 0 + (1 - \mathrm{heav}(0)) * (0 * \ln(0)) \\ &= 1 * 0 + 0 * (0 * (-\infty)) = 0 * \mathrm{NaN} = \mathrm{NaN} \ . \end{aligned}$$

The problem here is that $u * \ln(u)$ is not well defined at $u = 0$. To avoid this problem we use the convention that for (extended) reals

$$\pm\infty * 0 = 0 = 0 * \mathrm{NaN} \ ,$$

which means that 0 annihilates anything.

Pole Functions

Another kind of contingency occurs when functions are unbounded near the boundary of their proper domains of definition. Prime examples are the reciprocal $\text{rec}(u) = 1/u$, the logarithm $\ln(u)$, and the inverse tangent $\arctan(u)$. Within their open domains they are analytic like the root functions. With the exception of logarithmic singularities we exclude the critical arguments of pole functions from their domains. In contrast to $\text{rec}(u)$ and $\ln(u)$, the inverse tangent function has infinitely many poles. Of course, in the actual evaluation of $\arctan(u)$, there will be tests on the size of the argument and corresponding shifts to a normalized argument in the semiopen interval $(-\frac{\pi}{2}, \frac{\pi}{2}]$. Then the only remaining critical argument is zero, a characteristic that we will assume for all univariate functions.

Assumption (IC): ISOLATED CRITICALITIES
Except possibly at the origin all univariate elementals are everywhere either real analytic or undefined. All multivariate elementals are globally real analytic.

By the composition of step functions and constant shifts one may represent any univariate function with finitely many critical arguments in this way, i.e., so that Assumption (IC) is satisfied. The key idea of the following analysis is that assuming (IC) the status of evaluation procedures is almost everywhere stable in that small perturbations in the argument do not alter the result dramatically, even if the result is a NaN. Hence we define the following concept.

Definition (SD): STABLE DOMAIN
The stable domain *of a vector function consists of all arguments in whose neighborhood each component of the function is either real analytic or constantly NaN.*

Obviously, stable domains are always open by definition and we find that the remaining trouble spots have measure zero according to the following result.

Proposition 14.2 (FULL MEASURE OF STABLE DOMAIN)
Suppose the function $F : \mathbb{R}^n \longmapsto \mathbb{R}^m$ is defined by an evaluation procedure of the form of Table 2.3*, satisfying Assumption* (IC)*. Then the closed complement of the stable domain of F has zero measure.*

Proof. Recall the numbering of the variables v_i for $i = 1 - n \ldots l$ introduced in section 2.2 with the first n representing independents and the last m dependents. Let $\mathcal{S}_i$ denote the open stable domain of the intermediate function $v_i = v_i(x)$, and $\tilde{\mathcal{C}}_i \equiv \mathbb{R}^n \setminus \mathcal{S}_i$, the corresponding closed complement. Since the stable domain of F and its complement are given by

$$\mathcal{S} = \bigcap_{j=0}^{m-1} \mathcal{S}_{l-j} \quad \text{and} \quad \tilde{\mathcal{C}} \equiv \mathbb{R}^n \setminus \mathcal{S} = \bigcup_{j=0}^{m-1} \tilde{\mathcal{C}}_{l-j} \,, \tag{14.4}$$

respectively, it suffices to show the zero measure of all $\tilde{\mathcal{C}}_i$ to ensure the zero measure of $\tilde{\mathcal{C}}$. We will use induction on i to show that all sets $\tilde{\mathcal{C}}_i$, $1 - n \le i \le l$, have zero measure. Obviously $\tilde{\mathcal{C}}_i = \emptyset$ for the independent nodes $1 - n \le i \le 0$. Let us suppose the assertion is true for all $\tilde{\mathcal{C}}_j$, $1-n \le j \le i-1$, and consider the elemental operation $v_i = \varphi_i\,(v_j)_{j \prec i}$. Here $\prec$ denotes the dependence relation introduced also in section 2.2. There are two possible cases:

1. The elemental function φ_i is globally real analytic and possibly multivariate. Then we have

$$\mathcal{S}_i \supseteq \bigcap_{j \prec i} \mathcal{S}_j \quad \text{and} \quad \tilde{\mathcal{C}}_i = \mathbb{R}^n \setminus \mathcal{S}_i \subseteq \mathbb{R}^n \setminus \bigcap_{j \prec i} \mathcal{S}_j = \bigcup_{j \prec i} (\mathbb{R}^n \setminus \mathcal{S}_j) = \bigcup_{j \prec i} \tilde{\mathcal{C}}_j \,.$$

   As a subset of a finite union of zero measure sets, $\tilde{\mathcal{C}}_i$ also has zero measure.

2. The univariate elemental function $v_i = \varphi_i(v_j)$ is critical at the origin with $j < i$ some fixed index. Let us consider the inverse image

$$v_j^{-1}(0) \equiv \{x \in \mathcal{S}_j \,:\, v_j(x) = 0\} \,.$$

   On each connected component of $\mathcal{S}_j$ the intermediate function v_j is either real analytic or constantly NaN, and hence either it is constant (possibly NaN) or its level sets all have zero measure and are closed in $\mathcal{S}_j$.

   Let $\mathcal{A}_j^{crit}$ denote the set of points $x \in \mathcal{S}_j$ where v_j attains the critical value zero but does not vanish identically on an open neighborhood of x. This set has zero measure and is relatively closed in $\mathcal{S}_j$. Then we get

$$\mathcal{S}_i \supseteq \mathcal{S}_j \setminus \mathcal{A}_j^{crit}$$

   and

$$\tilde{\mathcal{C}}_i = \mathbb{R}^n \setminus \mathcal{S}_i \subseteq \mathbb{R}^n \setminus \left(\mathcal{S}_j \setminus \mathcal{A}_j^{crit}\right) = \mathbb{R}^n \setminus \mathcal{S}_j \cup \mathcal{A}_j^{crit} = \tilde{\mathcal{C}}_j \cup \mathcal{A}_j^{crit} \,.$$

   Since $\mathcal{A}_j^{crit}$ has zero measure, $\tilde{\mathcal{C}}_i$ must have zero measure, too.

Finally, the equalities (14.4) ensure that $\tilde{\mathcal{C}}$ has zero measure, which completes the proof. ∎

We may interpret the result as follows.

RULE 25

FUNCTIONS GIVEN BY EVALUATION PROCEDURES ARE ALMOST EVERYWHERE REAL ANALYTIC OR STABLY UNDEFINED.

At points where functions are neither analytic nor stably undefined, one may obtain a warning sign by trapping all elementals in a critical state. This is practically quite easy to do whether AD is implemented by overloading or by source transformation, as discussed in Chapter 6. There is a slight possibility that the alarms may go off at points where the function is in fact locally smooth, for example, when $y = |x - x|$. The sequence of statements

$$v_0 = x; \quad v_1 = v_0; \quad v_2 = v_1 - v_0; \quad v_3 = \text{abs}(v_2); \quad y = v_3$$

is always critical. Yet the constant value $y = 0$ is obviously everywhere differentiable with respect to x. The difficulty here is that criticalities cancel each other out—a rather exceptional situation—possibly due to poor coding.

In general, we would expect that the following sufficient condition for stability is also necessary and therefore applies almost everywhere.

Corollary 14.1 (Noncriticality Implies Stability)
Under the assumptions of Proposition 14.2 the complement of the stable domain is contained in the closed set $\mathcal{C}$ of "critical" points x, where at least one of the resulting intermediate values $v = v(x)$ is a critical argument to a nonsmooth univariate elemental function. On connected components of the open complement $\mathbb{R}^n \setminus \mathcal{C}$, the same program branches apply and represent essential selection functions of F.

Proof. See the proof of Proposition 14.2. ∎

At the end of the following section we will show that all points along certain "regular" arcs are noncritical and belong to the same essential selection function. Moreover, the fact that an arc is regular can be verified constructively, at least if we ignore the effect of roundoff.

14.3 One-Sided Laurent Expansions

Suppose that at a given point $x_0 \in \mathbb{R}^n$ we also have a preferred direction $x_1 \in \mathbb{R}^n$ and we wish to examine the univariate function $F(x(t))$ with $x(t) \equiv x_0 + tx_1$ for small values of $t \in \mathbb{R}$. For example, we may have $x_1 = F(x_0)$ when solving ODEs numerically by some method involving first derivatives of the right-hand side along the solution curve. Alternatively, x_1 may be a prospective step of an optimization algorithm and one wants to predict the behavior of F on the basis of first and possibly higher directional derivatives. For a fixed, smooth arc $x(t)$ we may consider all resulting intermediate values $v_i(x(t))$ as univariate functions of t whenever they are defined at all.

A Laurent Model with Variable Exponent Bounds

For each intermediate variable v taking the exact values $v(x(t))$, we try to compute a Laurent approximation $v(t)$ of the form displayed in Table 14.4. Hence we slightly generalize the convention introduced by Definition (AI) in Chapter 13.

Apart from the coefficients v_j, there are two crucial parameters $e = e_v$ and $d = d_v$. As indicated by the subscripting they may vary from intermediate to

Table 14.4: Laurent Model with Fractional Leading Term

$$v(t) = t^{e_v} \sum_{j=0}^{d_v} t^j v_j = v(x(t)) + o(t^{e_v+d_v}) \quad \text{for} \quad t > 0$$

with

leading exponent	$e_v \in \mathbb{R} \cup \{\pm\infty\}$
significant degree	$d_v \in \{-1, 0, 1, \ldots, \infty\}$
leading coefficient	$v_0 \in \mathbb{R} \setminus \{0\}$ if $d_v > 0$
inner coefficients	$v_j \in \mathbb{R}$ for $0 < j < d_v$
trailing coefficient	$v_{d_v} \in \mathbb{R} \cup \{\pm\infty\}$ if $d_v \geq 0$

intermediate and have to be carried along together with the coefficients v_j. All parameter values will also depend heavily on the arc $x(t)$ specified by the user. We will usually consider it as fixed, but sometimes one may wish to augment it with higher-order terms to avoid degeneracies.

The sum $e_v + d_v$ may be equal to ∞ or $-\infty$, in which case we interpret $o(t^{e_v+d_v})$ for small t as exactly zero or completely arbitrary, respectively. For finite $e_v + d_v$, the first line in Table 14.4 implies the limit

$$\lim_{t \to 0} \left(v(x(t)) - t^{e_v} \sum_{j=0}^{d_v - 1} t^j \, v_j \right) \Big/ t^{e_v + d_v} = v_{d_v} \, . \tag{14.5}$$

This relation makes sense, even when $|v_{d_v}| = \infty$, and may thus be considered as the defining property of the Laurent approximation when the trailing coefficient indicates a logarithmic singularity, as discussed below. After these classifications we may make the following formal definition.

Definition (LN): LAURENT NUMBER
Any collection of values $(e_v, d_v, v_0, v_1, \ldots, v_{d_v-1}, v_{d_v})$ satisfying the conditions in Table 14.4 will be referred to as a Laurent number. It will be called "integral" if e_v is an integer, "exact" if $d_v + e_v = \infty$, and "bounded" if $e_v > 0$ or $e_v = 0 = d_v$ and $|v_o| < \infty$. In the opposite cases, it is called "fractional," "truncated," and "unbounded," respectively.

Obviously an integral and bounded Laurent number is a Taylor series if it is exact, and a Taylor polynomial if it is truncated.

Let us briefly discuss the parameters and their special values listed in Table 14.5. The integer d_v yields the number of significant terms in the expansion as $d_v + 1$. We will normally restrict d_v by some constant upper bound $\hat{d}$ representing the number of nonzero coefficients that we are able and willing to store for each intermediate. Sometimes d_v may be larger than $\hat{d}$ when we know that the extra coefficients v_j for $\hat{d} < j \leq d_v$ are zero and thus need not be stored. This fortuitous situation arises in particular with $d_v = \infty$ when $v(x(t))$ with $t > 0$ is known to be constant, for example, because it is obtained as the value of a step function. In the realistic scenario $\hat{d} < \infty$, we have $d_v = \infty$ exactly if $v(x(t))$ is a rational function multiplied by t^{e_v}.

Table 14.5: Special Values of Laurent Model

$e_v = -\infty$	:	"Not a Laurent number"
$e_v = \infty$	:	zero
$\lvert v_{d_v}\rvert = \infty$	:	logarithmic singularity
$d_v = -1,$	:	term of order e_v
$-\infty < e_v < 0$	:	pole of degree e_v

Normally we expect to maintain $d_v = \hat{d}$, which typically holds in the theoretical scenario $\hat{d} = \infty$. However, the possibility $d_v < \hat{d}$ arises when there is a loss of significant terms through exact cancellation as discussed near Table 14.6. In contrast to floating point arithmetic where one always fills mantissas up to full length, we choose not to carry insignificant terms, whose coefficients may require a considerable effort to compute. In extreme cases all significant terms may be lost, in which case we will set $d_v = -1$, but our Laurent model may still convey some information via the leading exponent e_v. When $|e_v| = \infty$, as discussed below, we set $d_v = -1$ throughout. The (for small t) dominating term in $v(t)$ has the exponent e_v, a key characteristic, where e_v may be any extended real number including $+\infty$ or $-\infty$. Normally e_v is uniquely defined by the condition $v_0 \neq 0$. When $d_v = -1$ there is no such leading nonzero coefficient and e_v is merely required to satisfy $v(t) = O(t^{e_v}) = o(t^{e_v - 1})$; hence v is called a *term of order* e_v. Such intermediates have an order of magnitude but no sign, which causes difficulties for roots, quotients, and the step functions. In contrast, the infinite value $e_v = \infty$ indicates a *zero*, as already discussed in section 14.2. It represents the neutral element with respect to addition and thus an annihilator of all other elements in the field of proper Laurent series.

Although one might also use $e = -\infty$ to represent an essential singularity, we use it as an encoding of the NaN equivalent NaL (not a Laurent number). In selecting the values of arithmetic operations and special functions for Laurent arguments, we will bail out by setting $e_v = -\infty$ whenever no informative approximation of $v(x(t))$ can be formed with a reasonable amount of effort and complication. It is understood that NaLs, like NaNs in IEEE arithmetic, cannot be eliminated and percolate through subsequent calculations except when multiplied by a zero.

Finite negative values of e_v represent *poles of degree* e_v. They are hopefully transients that later get reciprocated or multiplied by intermediates w with $e_w > e_v$, yielding in either case (again) bounded values. This may happen in the lighthouse example if $\cot(u) = \cos(u)/\sin(u)$ is first incremented by some constant and then reciprocated, yielding a function that is globally analytic. A related question is the representation of logarithmic singularities.

Encoding of Logarithmic Singularities

Trying to balance simplicity with sophistication, we have decided to use the special values $\pm\infty$ of the last significant coefficient v_{d_v} for this purpose.

Suppose we have $u(t) = t + t^2/2 + o(t^2)$, so that $e_u = 1$, $d_u = 1$, $u_0 = 1$, and $u_1 = 1/2$. Then we may set for $v = \ln(u)$ the parameters $e_v = 0$, $d_v = 0$, and $v_0 = -\infty$, which encodes the approximation $v(t) = -\infty + o(t^0)$ in the sense of (14.5). All higher-order terms are suppressed, but we have still propagated some useful information about v. Subsequently multiplying with u and then adding a 1 would yield the approximation

$$w = 1 + u * v = 1 + u * \ln(u) = 1 + t * (-\infty) + o(t)$$

with $e_w = 0$, $d_w = 1$, $w_0 = 1$, and $w_1 = -\infty$. This may be interpreted as saying that the right-hand derivative of w along the arc $u(t) = t + t^2/2$ is $-\infty$. When two variables with trailing coefficients of the same order and the "logarithmic value" $+\infty$ are subtracted, we set the result to a NaL. Since all leading coefficients are by definition nonzero, their multiplication with a trailing coefficient cannot introduce any new NaNs or NaLs.

It should be noted that we have not restricted the leading exponent e_d to be integral, so that leading roots are admitted. We nevertheless refer to our approximating functions as Laurent series, since the exponents in the sum are restricted to being integral. Therefore, the sum is in fact a polynomial. As a consequence we have to reduce the number of significant terms whenever we add or subtract two variables u and w, for which the difference $e_w - e_u$ is not an integer.

Arithmetic Operations on Laurent Numbers

Because of Assumption (ED), in section 13.2 we had therefore for all intermediates v the relations $e_v \geq 0$ and d_v equal to its upper bound $\hat{d}$. Here the relation between d_v and $\hat{d}$ will be much more complicated, with d_v occasionally even exceeding $\hat{d}$ for variables with special values.

The recurrence for $v = u * w$ in Table 13.1 is still applicable, with

$$d_v = \min\{d_u, d_w, \hat{d}\} \quad \text{and} \quad e_v = e_u + e_w \,. \tag{14.6}$$

Here e_v must be set to ∞ if the expression above yields $\infty - \infty$. Thus multiplication by zero ($e_v = \infty$) again yields a zero, even if the other factor is a NaL. Otherwise we assume throughout the remainder of this section that none of the arguments u and w are NaLs since otherwise the result values have the same unsavory character too. This convention means, for example, that in the Helmholtz energy considered in Exercises 7.3 and 11.2, an entropy term $x_i \ln x_i$ simply drops out if the corresponding species i is not present at all in the feed mixture of current interest. Multiplication by a constant c and specifically a sign change are very straightforward, with d and e remaining unaltered. For simplicity we again interpret division as a reciprocal followed by a multiplication. For $v = 1/u$ we may set

$$e_v = (|u_0| < \infty)? - e_u : -\infty \,.$$

Table 14.6: Evaluation of $v = u + w$ in Laurent Arithmetic

$$
\begin{array}{l}
\text{if } (e_w < e_u) \text{ then } w \longleftrightarrow u \\
e_v = e_u \\
s \;= e_w - e_u \\
\text{if } (s \text{ not integral}) \\
\text{then} \\
\quad d_v = \min\{d_u, \lfloor s \rfloor\} \\
\quad v_i \;= u_i \text{ for } 0 \le i \le d_v \\
\text{else} \\
\quad d_v = \min\{d_u, d_w + s\} \\
\quad v_i \;= u_i \text{ for } 0 \le i \le \min\{s-1, \hat{d}\} \\
\quad v_i \;= u_i + w_{i-s} \text{ for } s \le i \le \min\{d_v, \hat{d}\} \\
\quad s \;= \inf\{i \ge 0 : v_i \ne 0 \vee i = d_v + 1\} \\
\quad d_v = (s < \infty)?\, d_v - s : -1 \\
\quad e_v = e_v + s \\
\quad v_i \;= v_{i+s} \text{ for } 0 \le i \le \min\{d_v, \hat{d}\} \\
\quad d_v = (v_{d_v} \ne v_{d_v})? d_v - 1 : d_v
\end{array}
$$

Here writing down u_0 implicitly implies $d_u \ge 0$, and we have $d_v = d_u$ unless $e_v = -\infty$, in which case $d_v = -1$ by convention.

Whereas so far no significant terms could be lost or gained, this may happen through linear combinations $v = u + c\,w$, just like in floating point arithmetic. Without loss of generality, we set $c = 1$ and arrive at the algorithm for evaluating sums in Laurent arithmetic displayed in Table 14.6. The condition $(v_{d_v} \ne v_{d_v})$ in the last line of Table 14.6 is satisfied only if v_{d_v} is a NaN, which may arise when both u and w have logarithmic singularities of opposite sign.

The possible reduction in d indicates loss of significance, which is familiar from floating point arithmetic. In some sense this troublesome case seems much less likely to occur here. After all, two real coefficients rather than just a few leading bits in a mantissa must be exactly equal to each other. On the other hand, this is precisely what happens if one divides $(1 - \cos(x))$ by x for $x(t) = t$. In extreme cases all significant terms might be lost, and the resulting approximation is an order term as defined in Table 14.5. Whether or not this might be considered a disaster depends on the subsequent usage. To avoid order terms one may restart the propagation of the Laurent series after an increase in the initial $\hat{d}$. However, this approach might theoretically lead to an infinite loop, as we will discuss further down.

Laurent Propagation through Univariates

Now let us look at the rules for propagating Laurent models through univariate functions $v = \psi(u)$. Whenever the leading exponent e_v or the significant degree d_v differs from the corresponding parameters e_u and d_u, we list them in Table 14.7.

Table 14.7: Elementaries $v = \psi(u)$ at Laurent Numbers (Defaults: $e_v = e_u$ and $d_v = d_u$; Convention: $d_v \leq \bar{d}$ means $d_v = \min\{\bar{d}, \hat{d}\}$)

u \ $\psi(u)$	Regular $d_u \geq 0$ $e_u = 0$ $\|u_0\| < \infty$	Root/power $d_u \geq 0$ $0 < e_u < \infty$	Pole $d_u \geq 0$ $-\infty < e_u < 0$	Logarithmic pole $d_u = 0$ $e_u = 0$ $\|u_0\| = \infty$	Order term $d_u = -1$ $-\infty < e_u < \infty$	Zero $e_u = \infty$ $d_u = -1$
$\ln(u)$	NaL if $u_0 < 0$ $e_v \geq 0$ $d_v = d_u - e_v$	NaL if $u_0 < 0$ $e_v = d_v = 0$ $v_0 = -\infty$	NaL if $u_0 < 0$ $e_v = d_v = 0$ $v_0 = \infty$	NaL if $u_0 = -\infty$ $v_0 = \infty$	NaL	NaL
$\exp(u)$		$e_v = 0$ $d_v \leq d_u + e_u$	NaL	NaL	NaL if $e_u < 0$ $e_v = 0$ $d_v \leq e_u - 1$	$e_v = 0$ $d_v = \infty$ $v_0 = 1$
u^r	NaL if $u_0 < 0$	NaL if $u_0 < 0$ $e_v = e_u/r$	NaL if $u_0 < 0$ $e_v = e_u/r$	NaL if $u_0 = -\infty$ $v_0 = \infty$	NaL	Zero
$\sin(u)$	$e_v \geq 0$ $d_v \leq d_u + e_u - e_v$		$e_v = 0$ $d_v = -1$	$e_v = 0$ $d_v = -1$	$e_v = 0$ if $e_u < 0$	Zero
$\cos(u)$	$e_v \geq 0$ $d_v \leq d_u + e_u - e_v$	$e_v = 0$ $d_v \leq d_u + e_u$	$e_v = 0$ $d_v = -1$	$e_v = 0$ $d_v = -1$	$e_v = 0$ $d_v \leq e_u - 1$	$e_v = 0$ $d_v = \infty$

Within the resulting coefficient range v_i for $0 \leq i \leq d_v$ the recurrences listed in Table 13.2 are then still applicable without any changes.

While we have seen that the arithmetic operations can usually be applied to Laurent series, we immediately run into trouble with most of the nonlinear univariate functions $v = \psi(u)$ when $e_u < 0$. Then the Laurent series for u has a nonvanishing principal part and neither $\ln(u)$ nor $\exp(u)$ have Laurent expansions even when $\hat{d} = \infty$. Hence one has to return a NaL in all of these cases. For the uniformly bounded functions $\sin(u)$ and $\cos(u)$ we obtain order terms with $e_v = 0$. The various special cases are listed in Table 14.7.

Laurent Propagation through Conditionals

As we discussed in section 14.2, all program jumps and discontinuous elementaries can be encapsulated in the Heaviside or, equivalently, the sign function, provided we ensure that their zero return value annihilates even infinite multipliers. We have introduced such a mechanism in the form of the exact zero $u = 0$ with $e_u = \infty$, which annihilates any other Laurent number w even if $e_w < 0$ or $w_0 = \pm\infty$, which represent proper poles and logarithmic singularities. Hence we only have to define step functions for Laurent numbers in order to make all evaluation procedures that are defined on reals also executable in Laurent arithmetic. Using the same conventions as in Table 14.7, we obtain the four different cases listed in Table 14.8.

Table 14.8: Step Functions on Laurent Number

u	Regular		Zero	Order term
	$u_0 > 0$	$u_0 < 0$	$e_u = \infty$	$d_u = -1,\ \lvert e_u \rvert < \infty$
$\text{sign}(u)$	$e_v = 0$ $v_0 = 1$ $d_v = \infty$	$e_v = 0$ $v_0 = -1$ $d_v = \infty$	Zero	$e_v = 0$ $d_v = -1$
$\text{heav}(u)$	$e_v = 0$ $v_0 = 1$ $d_v = \infty$	Zero	$e_v = 0$ $v_0 = 1$ $d_v = \infty$	$e_v = 0$ $d_v = -1$

In the first two columns of Table 14.7, the sign of $u(t)$ is unambiguously defined for small positive by t, the leading coefficient, which is by definition nonzero. We can only be sure that the sign of $u(t)$ is zero if it is an exact zero in that $e_u = \infty$. That leaves us with the case where u is an order term with $d_u = -1$ and $e_u < \infty$. Then we can only assign the value NaL encoded by $e_v = -\infty$.

The sign function defines an ordering

$$u > w \quad \Leftrightarrow \quad \text{sign}(u - w) = 1 \,,$$

so that all conditional assignments based on comparison operators are uniquely extended to the linear space of truncated Laurent series. The result $v = \text{heav}(u)$ of the comparison operator $(u \geq 0)$? will thus be defined either as a NaL when u is an order term, or as an exact 1 with $e_v = 0, d_v = \infty$, and $v_0 = 1$, or as an exact zero with $e_v = \infty$ and $d_v = -1$. Since the latter annihilates poles and other nonessential singularities, we may evaluate the conditional assignment (14.2) in Laurent arithmetic in the form (14.3) without incurring a contingency, unless u is a NaL or an order term.

Now that we have specified rules for propagating Laurent numbers through all arithmetic operations and the most important nonlinear univariates let us formally state without proof the consistency of our modeling effort.

Proposition 14.3 (Consistent Propagation of Laurent Model)
Suppose $v(x(t))$ is obtained from $u(x(t))$ by one of the univariate functions listed in Tables 14.7 *and* 14.8 *or from $u(x(t))$ and $w(x(t))$ by an arithmetic operation $u * w$, u/w, or $u + c\,w$ with c a constant. Then the validity of the Laurent model of Table* 14.4 *for $u(x(t))$ and possibly $w(x(t))$ implies the validity of the resulting Laurent model for $v(x(t))$ obtained by the rules specified in* (14.6) *and Tables* 14.6, 14.8, *and* 14.7.

As a first example, let us consider the problem of differentiating the Euclidean norm y along the spiral path $(x, z) = (t * \cos(t), t * \sin(t))$ in the plane. The result happens to be obtained exactly as displayed in Table 14.9.

Table 14.9: Laurent Evaluation of $y = \sqrt{x^2 + z^2}$ for $(x, z) = t(\cos(t), \sin(t))$

	e	d	0	1	2
$x = t\cos(t)$	1	2	1	0	$-\frac{1}{2}$
$z = t\sin(t)$	2	2	1	0	$-\frac{1}{6}$
$u = x * x$	2	2	1	0	-1
$w = z * z$	4	2	1	0	$\frac{1}{3}$
$v = u + w$	2	2	1	0	0
$y = \sqrt{v}$	1	2	1	0	0

Now let us look at the results of our Laurent arithmetic when applied to some of the problems we considered in the introduction along the parabolic arc $x(t) = t - t^2$ and with four significant terms, in that we set $\hat{d} = 3$. As one can see, the cancellation of the constant term in the subtraction $1 - \exp(x)$ in Table 14.10 (a) causes the loss of one significant term, and similarly the logarithmic term, $\ln(x)$ immediately reduces d to 0 in Table 14.10 (b). In contrast the division by t in example (c) of Table 14.10 causes no loss of significant terms and all Laurent approximants are in fact polynomials.

To examine the problems with order terms in the denominator and under the sign, let us consider the functions $y = x^2/(1 - \cos(x))$ for the values $\hat{d} = 1$ and $\hat{d} = 2$. As one can see in Table 14.11 the value $\hat{d} = 1$ is not sufficient to fix the sign of the denominator w, which results in a NaL.

Table 14.10: Laurent Evaluation along $x(t) = t - t^2$ with $\hat{d} = 3$

(a) $y = (1 - \exp(x))/x \Rightarrow$ Quadratic Polynomial

	e	d	0	1	2	3
$x = t - t^2$	1	∞	1	-1	0	0
$u = \exp(x)$	0	3	1	-1	$\frac{1}{2}$	$\frac{7}{6}$
$v = 1 - u$	1	2	-1	$\frac{1}{2}$	$\frac{7}{6}$	–
$y = v/x$	0	2	-1	$\frac{1}{2}$	$\frac{7}{6}$	–

(b) $y = x^2 \ln(x) \Rightarrow$ Infinite Coefficients

	e	d	0	1	2	3
$x = t - t^2$	1	∞	1	-1	0	0
$u = \ln(x)$	0	0	∞	–	–	–
$v = x^2$	2	∞	1	-2	4	0
$y = v * u$	2	0	∞	–	–	–

(c) $y = \sin(x)/x \Rightarrow$ Cubic Polynomial

	e	d	0	1	2	3
$x = t - t^2$	1	∞	1	-1	0	0
$u = \sin(x)$	1	3	1	-1	$-\frac{1}{6}$	$\frac{1}{2}$
$y = u/x$	0	3	1	0	$-\frac{1}{6}$	$\frac{1}{3}$

(d) $y = x \sin(1/x) \Rightarrow$ Order Term

	e	d	0	1	2	3
$x = t - t^2$	1	∞	1	-1	0	0
$u = 1/t$	-1	3	1	1	1	1
$v = \sin(u)$	0	-1	–	–	–	–
$y = x * v$	1	-1	–	–	–	–

Table 14.11: Laurent Evaluation of $y = x^2/(1 - \cos(x))$ along $x(t) = t$

(a) $\hat{d} = 1$

	e	d	0	1
$x = t$	1	∞	1	0
$v = x * x$	2	∞	1	0
$u = \cos(x)$	0	1	1	0
$w = 1 - u$	2	-1	–	–
$y = v/w$	$-\infty$	-1	–	–

(b) $\hat{d} = 2$

	e	d	0	1	2
$x = t$	1	∞	1	0	0
$v = x * x$	2	∞	1	0	0
$u = \cos(x)$	0	2	1	0	$-\frac{1}{2}$
$w = 1 - u$	2	0	$\frac{1}{2}$	–	–
$y = v/w$	0	0	2	–	–

Once $\hat{d} = 2$ or greater, this ambiguity is removed and we obtain a well-defined approximation of order $\hat{d} = 2$ for the result y.

As we have seen, most of the trouble arises through order terms that occur in the denominator of divisions, under fractional roots, or as arguments of step functions. Note, however, that they may be harmlessly absorbed in a subsequent addition, so there is no reason to sound the alarm as soon as an order term is computed. Hopefully most order terms will turn into regular Laurent numbers when $\hat{d}+1$, the maximal number of stored coefficients, has been chosen sufficiently large. To get this result we first state the following result, again using the convention that $d_v = -1$ when $|e_v| = \infty$.

Lemma 14.1 (Limits of Leading Exponent and Significant Degree)
For each variable v the parameter pair (e_v, d_v) is componentwise monotonic as a function of $\hat{d}$. Moreover, once d_v exceeds its minimal value -1 the leading exponent e_v stays constant and the same is true for all coefficients v_i with $0 \leq i \leq d_v$. Consequently there exists a unique limit

$$(e_v^*, d_v^*) \equiv \lim_{\hat{d} \longmapsto \infty} (e_v, d_v) , \tag{14.7}$$

which is bounded by the values (e_v^∞, d_v^∞) obtained in Laurent arithmetic of degree $\hat{d} = \infty$.

Proof. Obviously, the assertion must be proven by induction on the intermediates with respect to their partial ordering. For the multiplication (weak) monotonicity in the components of (e_u, d_u) and (e_w, d_w) with simultaneous increase in $\hat{d}$ clearly implies monotonicity of (e_v, d_v). If $d_v = \min(d_u, d_w, \hat{d})$ was already nonnegative for the previous value of $\hat{d}$, the same must be true for the previous values of d_u and d_w. Then it follows by the induction hypothesis that neither e_u nor e_w can move, and the same must be true for their sum, $e_v = e_u + e_w$. At first the reciprocal seems a little more tricky as e_v is proportional to $-e_u$. However, e_u can only have grown if the old value of d_u was -1, in which case the division by an order term previously returned the absolutely minimal value $e_v = -\infty$. Hence e_v cannot decrease whether e_u is constant or not. The monotonicity of d_v and the stationarity of e_v once d_v is nonnegative follow just as for the multiplication. Similar arguments can be applied for all other elementaries. ∎

Clearly one would like to choose $\hat{d}$ at least large enough such that no order terms occur in denominators or as critical arguments. Unfortunately, this cannot always be achieved, even if things go well in the theoretical scenario $\hat{d} = \infty$. Moreover, a variation in the path $x(t)$ may not help either. To see this, let us consider the trivial example function $y = \text{sign}(1 - \sin^2(x) - \cos^2(x))$, whose differentiability leaves nothing to be desired, as $y = 0$ for all x. However, no matter how we choose $x(t)$ and $\hat{d} < \infty$ the intermediate u representing the argument of the sign function will always be an order term with $e_u < \infty$, whose sign is undefined, so that $(e_y^*, d_y^*) = (-\infty, -1)$. Yet when $\hat{d} = \infty$ the argument

u will have $e_u = \infty$ and sign(u) will be zero with $(e_y^\infty, d_y^\infty) = (\infty, -1)$ according to the third column in Table 14.8. The example shows in particular that the limit (14.7) might be less than its upper bound obtained in Laurent arithmetic of degree $\hat{d} = \infty$ in that $e_y^* = -\infty < \infty = e_y^\infty$.

This way of defining sign is in line with Proposition 14.2, which showed that functions may well be smooth, even though some of their arguments are critical, provided they are constant. Unfortunately, this constancy can only be guaranteed in practice as long as the intermediates leading up to such a critical argument are all constant or are at least polynomials with $d = \infty$. For example, we would get the right result for $f(x) = \text{sign}(x - x)$, provided x is initialized as a polynomial of a degree no greater than $\hat{d}$. We first discussed the cancellation of trigonometric functions because both $\sin(x)$ and $\cos(x)$ have infinitely many nonzero coefficients, so that their significant degree must drop down immediately to the upper bound $\hat{d}$. This cancellation occurs for all arcs, so the attempt to make all critical order terms significant by successively varying $x(t)$ and increasing $\hat{d}$ would lead to an infinite loop. Hence one must preselect a reasonable upper bound on $\hat{d}$ and simply give up once it has been exceeded. We may handle the situation as follows.

Definition (RA): REGULAR ARC AND DETERMINACY DEGREE
Given an evaluation procedure for a function F, an analytic arc $x(t)$ will be called regular *if for all intermediates v*

$$e_v^* = e_v^\infty \quad \text{and} \quad d_v^* = \infty \quad \text{unless} \quad |e_v^\infty| = \infty\,.$$

The smallest value d_* of $\hat{d}$ for which $e_v = e_v^\infty$ for all intermediates v will be called the determinacy degree *of the arc $x(t)$.*

The fact that an arc is regular can be verified by one forward propagation of our Laurent numbers with a corresponding upper bound on d_*. The converse, i.e., the nonexistence of a suitable $\hat{d}$, cannot be constructively established and we might wind up in an infinite loop by trying larger and larger values of $\hat{d}$.

In the example $1 - \sin^2(x) - \cos^2(x)$ all arcs are irregular. In the cubic lighthouse example evaluated by Cardan's formula, $t_* + t$ would also be an irregular arc because the second condition in Definition (RA) is violated. The dependent value z is always a nonvanishing constant plus an order term, so that $d_z^* = 0$ due to our handling of roots. In that case the loss of significant terms occurs during the addition of two Laurent numbers $u(t)$ and $w(t)$ with nonintegral exponent difference $e_w - e_u$. Then the AD tool could register that at least in this elemental function, increasing $\hat{d}$ won't help, so an infinite loop might be avoided. In the regular case we obtain the following result.

Proposition 14.4 (REGULAR ARCS AND ESSENTIAL SELECTIONS)
Suppose $x(t)$ is a regular arc with determinacy degree $d_* < \infty$. Then for all small $t > 0$, the points $x(t)$ belong to the same connected component of the noncritical stable set $\mathbb{R}^n \setminus C$ defined in Corollary 14.1.

Proof. Because of our definition of a regular arc, all intermediates $v(x(t))$ have an infinite Laurent series representation, unless they attain the values NaN for all small t due to $e_v^\infty = -\infty$. This Laurent expansion is only one-sided and may involve a leading fractional exponent. Since $d_v^* = -1$ is excluded except when $|e_v^*| = |e_v^\infty| = \infty$, all zeros are obtained exactly, i.e., not as limit-of-order terms with e_v tending to infinity and d_v constantly equal to -1. Hence all $v(x(t))$ that are not NaLs have a well-defined constant sign for all small t. This applies in particular to the arguments of nonsmooth univariates, so that none of them can be critical, which completes the proof. ■

Generalized Jacobians

Regularity of arcs, even with $d_v^* = \infty$ and e_v integral for all intermediates v, is not sufficient for the existence of generalized Jacobians, the grand prize we wish to gain at the end of this section. Let us consider the Raleigh quotient

$$y = f(x) \equiv (x \neq 0)?\, \frac{x^\top A x}{x^\top x} : 0\,,$$

where we have again used the conditional assignment notation of C. Here A is an $n \times n$ matrix, which may be assumed to be symmetric without loss of generality. It is well known that the stationary points $x \in \mathbb{R}^n \setminus \{0\}$ of f, where the gradient

$$\nabla f(x) = \frac{2}{x^\top x}\left[A - \frac{x^\top A x}{x^\top x}\right] x$$

vanishes, are exactly the eigenvectors of A with $f(x)$ representing the corresponding eigenvalue [Wil65]. This function $f(x)$ is piecewise differentiable as defined in Definition (PD) with the essential selection $x^\top A\, x / x^\top x$ at all $x \neq 0$ and the nonessential selection $y = 0$ at the origin itself.

Similarly, functions of the form $(\tilde{f}(x) - \tilde{f}(x_0))/\|x - x_0\|^2$ have been used in global optimization to eliminate a local minimum x_0 of some smooth objective function $\tilde{f}$ from subsequent calculations [GLC85]. Unfortunately, the results of such deflations are usually violently nonsmooth at the root of the denominator, as one can see from the gradient of the Raleigh quotient. Except when A happens to be a multiple of the identity, the gradient norm $\|\nabla f(x)\|$ attains arbitrarily large values for $x \approx 0$. Hence, we will disallow the reciprocal and, more generally, divisions as well as root singularities, as we wish to calculate generalized gradients.

Definition (FS): FINITE SLOPES
An evaluation procedure of a function F is said to have finite slopes along a regular arc $x(t)$ if no roots or poles but possibly kinks and steps are critical at $t = 0$. Consequently, for all intermediate variables v, the leading exponent e_v is either a natural number or possibly $\pm\infty$.

Whether or not F has finite slopes along a regular arc is easily determined. One simply has to see whether, during a forward propagation of the Laurent numbers starting with $x(t)$, any one of the root or pole functions is critical, which usually means that the argument is an order term. If no such critical root or pole is encountered (or it is later annihilated by multiplication with a zero), the following constructive result applies.

Proposition 14.5 (Finite Slopes Yield Generalized Jacobians)
Suppose F has finite slopes along a regular arc $x(t)$ and $e_v \geq 0$ for all dependents $y_i = v_i$. Then the corresponding essential selection function $F^{(k)}$ is differentiable at x_0, and its Jacobian can be calculated as

$$\frac{\partial y_{d_*}}{\partial x_{d_*}} = (F^{(k)})'(x_0) .$$

Here we have without loss of generality denormalized the independents and dependents such that $e = 0$ for all of them.

Proof. We may a priori eliminate all intermediates v with $e_v^* = e_v^\infty = -\infty$ since they must be annihilated later by a constant zero. Then one can easily check by induction that for all remaining intermediates v, the leading exponents e_v are natural numbers satisfying $e_v + d_v \geq \hat{d}$. In other words, all Laurent numbers are in fact truncated Taylor series, and we may replace the v_j, d_v, and e_v by v_{j+e_v}, $\hat{d}$, and 0, respectively. Now our assertion follows from Proposition 13.3 since varying x_{d_*} does not change the branch being evaluated by the various nonsmooth univariates. Hence we indeed obtain the Jacobian of the selection function $F^{(k)}$. ∎

According to Fischer's result, Proposition 14.1, being the derivative of an essential selection function $F^{(k)}$, is sufficient for being identical to the derivative of F if the latter is in fact continuously differentiable. Otherwise we have a generalized Jacobian or gradient that can be fed into a "bundle method" for optimization [SZ92]. These methods (see also [Kiw86]) assume (Lipschitz) continuity of F, a key property, which we have not imposed and can verify only if all nonsmoothness is restricted to kink functions rather than general branches.

Proposition 14.5 facilitates the evaluation and differentiation of the energy function considered in Exercise 7.3 even when one of the x_i is zero because the ith species does not occur at all in the mixed fluid. Then the corresponding term $x_i \ln x_i$ would vanish exactly and also yield a zero component in the gradient of the energy.

14.4 Summary and Conclusion

Starting from the observation that most practical evaluation codes are not really as differentiable as their programmers and users claim, we searched for constructive ways of nevertheless computing some kind of unambiguously defined derivatives. After introducing a very weak concept of piecewise differentiability, we first stated Fischer's result, namely, that derivative values can be obtained correctly by AD if the program branch taken is essential, i.e., applies on a subdomain with nonempty interior. In section 14.3 we found that this condition can be constructively verified along arcs that are regular and have a finite determinacy degree. One can then also compute elements of the generalized gradient and Jacobian sets used in nonsmooth analysis.

As shown in Proposition 14.2, there is only a zero probability of chancing upon an argument where an evaluation procedure is in an uncertain state, i.e., neither analytical nor yielding NaN on some open neighborhood for all intermediate and dependent variables. Nevertheless, we developed a constructive concept of Laurent numbers that allows the one-sided expansion of such functions as $|x|$, $\sin(x)/x$, and $x^2 \log(x)$ at the origin. The key requirement on users is that they nominate not only the point x_0 at which the evaluation procedure is to be differentiated, but also a direction x_1, or more generally an arc $x(t)$. Then we can perform directional differentiation effectively, applying l'Hôpital's rule whenever numerators and denominators vanish. When both are constantly equal to zero, our Laurent modeling effort breaks down and we signal this contingency with the flag "NaL." This calamity arises whenever denominators or arguments of step functions and fractional roots have uncertain sign. The (possibly fractional) exponent of the leading power and the number of significant terms may vary from intermediate to intermediate. Nevertheless, the model is implementable and should incur only a minimal runtime overhead as long as we deal in effect with truncated Taylor polynomials.

The main objection to the Laurent model is that, like many other proposed mechanisms used to deal with nonsmoothness, it barely ever kicks in at "real" arguments. Hence one should look for implementations that estimate the distance to the next critical value, and should this distance be very small, model the function locally from there. In this way one might be able to automatically generate higher derivatives of functions like $x/(\exp(x)-1)$ at values x near the origin. Another possible way to avoid the cancellation in the denominator is, of course, to perform genuinely symbolic formula manipulations, which we have excluded from consideration throughout.

There is a close connection between the constructive recipes we had developed for the first edition and the theory of semi-algebraic and subanalytic functions that has been evolved more recently [Cos00] and [BDL08].

14.5 Examples and Exercises

Exercise 14.1 (*Forward Differentiation of general Euclidean Norm*)
a. Show that the general Euclidean norm

$$\|(u_1, u_2, \ldots, u_n)\| = \sqrt{\sum_{k=1}^{n} u_k^2}$$

can be computed recursively using the special function $|u, w| = \sqrt{u^2 + w^2}$.
b. Write down the formula for propagating a directional derivative $(\dot{u}, \dot{w})$

through $|u, w|$ even at $(u, w) = (0, 0)$.
c. Check that the directional derivative of the general Euclidean norm is obtained correctly by applying the procedure developed in **a**.

Exercise 14.2 (*Numerics of Singular Compositions*)
a. On your favorite computing platform write a program that evaluates $\sin(x)/x$ for $x = 2^{-j}\sqrt{3}$ with $j = 0, 1, 2, \ldots$ until underflow, i.e., $\mathrm{fl}(x) = 0$. Compare the computed ratios with the limiting value 1 for $x = 0$ and discuss the numerical stability of the whole process.
b. Repeat the experiment for $x \ln(x)$ and $(\exp(x) - 1)/x$. Notice that the last ratio soon deviates significantly from its theoretical limit 1 due to cancellation in the numerator.

Exercise 14.3 (*Stable and Unstable Directions*)
Define a direction x_1 as *stable at a point* x_0 if $x_0 + tx_1 \in \mathcal{S}$ for all sufficiently small $t > 0$, where $\mathcal{S}$ is the stable domain of the function considered.
a. Show that the point x_0 is stable if and only if all directions x_1 including the trivial one $x_1 = 0$ are stable at x_0.
b. Consider the function $y = z\,|\sin(1/\max(0, x))|$ in the variables (x, z). It is stably unreal at all points (x, z) with nonpositive x. The complement of $\mathcal{S}$ consists of all points with x equaling 0 or having a reciprocal of the form πk with natural $k > 0$. At points with $x = 1/(\pi k)$ the directions (x_1, z_1) are unstable or not depending on whether x_1 is zero or not. However, at the origin the directions (x_1, z_1) are unstable depending on whether x_1 is nonnegative or not. This shows in particular that the set of unstable directions may well be large in the sense that its measure on the sphere is positive.

Exercise 14.4 (*Example of Henrion and Römisch* [HR99])
Consider the chance constraint (14.1) with $n = 2 = m$, G the identity mapping $G(x_1, x_2) = (x_1, x_2)$, and $\zeta = (\zeta_1, \zeta_2)$ uniformly distributed on the region $[0, 2] \times [0, 2] \setminus [0, 1] \times [0, 1]$. Verify that ζ has the distribution function

$$f(\zeta_1, \zeta_2) = \frac{1}{3}\begin{cases} (\tilde{\zeta}_1 - 1)\tilde{\zeta}_2 & \text{if } \tilde{\zeta}_2 \le 1 \le \tilde{\zeta}_1 \\ (\tilde{\zeta}_2 - 1)\zeta_1 & \text{if } \tilde{\zeta}_1 \le 1 \le \tilde{\zeta}_2 \\ \tilde{\zeta}_1\,\tilde{\zeta}_2 - 1 & \text{if } \tilde{\zeta}_1 \ge 1 \le \tilde{\zeta}_2 \\ 0 & \text{otherwise}\,, \end{cases}$$

where $\tilde{\zeta}_i = \max\{0, \min\{2, \zeta_i\}\}$ for $i = 1, 2$. Show that f is Lipschitz continuous but not differentiable in the open square $(0, 2) \times (0, 2)$.

Exercise 14.5 (*Delay Differential Equation*)
For some $\nu > 1$ consider the scalar system

$$\dot{x}(t) = x(t/\nu) \quad \text{with} \quad x(t) = 1 \quad \text{for} \quad 0 \le t \le 1\,.$$

a. Develop explicit (polynomial) formulas for $x(t)$ within the intervals $(1, \nu)$, (ν, ν^2), and (ν^2, ν^3).

b. Show that $\dot{x}(t)$, $\ddot{x}(t)$, and $\dddot{x}(t)$ are the first discontinuous derivatives of $x(t)$ at $t = 1$, $t = \nu$, and $t = \nu^2$, respectively.
c. For some fixed $t_0 > 1$ examine the dependence of $x(t_0)$ on $\nu > 1$.

Exercise 14.6 (*Numerical Quadrature*)
Consider a family of numerical quadratures of the form

$$F(x) \equiv \int_0^x f(t)dt = \frac{x}{n}\sum_{i=0}^{n} w_i f(x\,\xi_i) + c_n\left(\frac{x}{n}\right)^p f^{p+1}(\zeta)$$

Here the $\xi_i \in [0,1]$ are abcissas and the w_i are weights independent of the upper bound x, which is considered as variable. The $\zeta \in (0,x)$ is a mean value and f^{p+1} denotes the $(p+1)$st derivative, which we will assume to exist. Examples of such formulas are the composite trapezoidal or Simpson's rules. The question is now to what extent differentiation and numerical integration commute. Note that we cannot simply differentiate the above equation because it is not clear whether the mean value ζ is a differentiable function of the upper bound x.
a. Differentiate the sum on the right by hand with respect to x and interpret the resulting sum as the same quadrature applied to an integral closely related to the one with the integrand. Determine the corresponding error term under the assumption that f is in fact $(p+2)$-times continuously differentiable. Show using integration by parts that the new integral has exactly the value $f(x) = F'(x)$.
b. Repeat part **a** for more general integrals of the form

$$F(x) \equiv \int_0^x f(x,t)\,dt$$

where $f(x,t)$ is assumed to be sufficiently often differentiable.
c. Verify your results for the Simpson rule on

$$f(x) = 1/(1+x),\; f(x) = x^3,\; f(x) = \cos(x)$$

and another test case of your choice.
d. Show by example that for Lipschitz continuous $f(t)$ any quadrature might yield a nondifferentiable result even though $F(x) = \int_0^x f(t)\,dt$ is of course differentiable.

Chapter 15

Implicit and Iterative Differentiation

In the previous chapter we considered the differentiation of programs that contain kinks, poles, step functions, branches, and other nonsmooth elements. As we saw, it is then still possible to automatically generate derivative information in the form of directional derivatives and sometimes also generalized gradients. Here "automatically" means that the AD tool has no indication of the mathematical "purpose" of the code that it is called upon to differentiate. Often nonsmooth program constructs are used to efficiently approximate the values of a smooth mathematical function that is only defined implicitly. As we noted in the prologue this is, for example, true for eigen- and singular values, which cannot by their very mathematical nature be evaluated in a finite number of steps.

Using Newton's Method

Although the roots of cubic and even quartic equations can be expressed in terms of arithmetic operations and fractional powers, it is usually more efficient to calculate them by an iterative process. In the cubic lighthouse example we may iterate

$$\begin{aligned} z_{k+1} &= g(z_k, t) \equiv z_k - f(z_k, t)/|f'(z_k, t)| \\ &= z_k - \frac{(z_k - 2)^3 + 0.4 - z_k \tan(t)}{|3(z_k - 2)^2 - \tan(t)|} \,. \end{aligned} \tag{15.1}$$

This recurrence represents Newton's method with the slight modification that we take the modulus of the derivative $f' \equiv \partial f/\partial z$ in the denominator to make the middle root a repulsive fixed point if it exists. In real implementations of Newton's method one would use line searches or trust regions to stabilize the iteration. Such safeguards are not necessary here, as we observed convergence with $z_0 = 2.1$ for all t in the interesting range between $[0.1, 0.3]$, except $t \approx 0.189$. If one lets the iteration run for the fixed number of 40 steps one obtains the dashed curve in Fig. 14.2. It is distinguishable from the curve obtained by

Cardan's formula only where there are several roots, and the one converged to is sometimes not the one we want from a physical point of view. Rather than discussing ways of steering Newton's method in the right direction we wish to concentrate instead on the quality of the Newton approximation where it does converge.

As we will discuss in section 15.2, each Newton step is quite expensive to calculate on large-scale problems, and one usually wants to get away with just taking a handful of them. Suppose we terminate our Newton iteration as soon as the residual size $|f|$ no longer exceeds a given tolerance $\varepsilon > 0$. For the three values $\varepsilon \in \{0.05, 0.02, 0.003\}$ we obtain the solution curves displayed in Fig. 15.1. Since $|f(0.1, 0)| \approx 0.4$, this means that we try to reduce the residual to a few percents of its original value, which is quite acceptable in many modeling efforts. In the calm regions outside the critical interval $[0.14, 0.23]$ the approximations generated by Newton's method look quite reasonable, even though they are certainly not continuous. Wherever the number of steps actually taken changes there is a jump discontinuity, but the function values themselves, as well as their derivatives, don't differ dramatically.

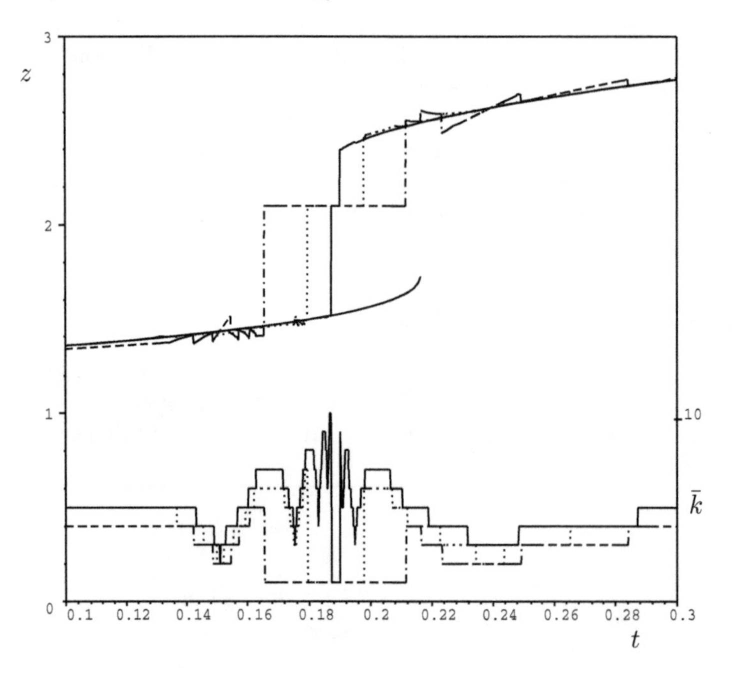

Figure 15.1: Newton Approximants and Number of Newton Steps for $\varepsilon = 0.05$ (dash-dot), $\varepsilon = 0.02$ (dot), and $\varepsilon = 0.003$ (solid)

The number of Newton steps $\bar{k}$—taken for the three tolerances—is displayed as a histogram at the bottom of Fig. 15.1. At first it may appear surprising that the fewest steps are taken in the critical interval. This effect is to some extent a consequence of our stopping criterion since, when the derivative $f'(z, t)$ is small, the residual $|f(z, t)|$ may be comparatively small even at points z far from the solution $z(t)$. Nevertheless, if the modeling only makes the values of f significant up to an absolute tolerance ε there is really no good reason to continue the iteration. Whereas in the calm regions the final curves for $\varepsilon = 0.02$ and $\varepsilon = 0.003$ are barely distinguishable from the exact solution, there are big differences in the critical interval. Here Newton's method may be at least temporarily sidetracked by the other two roots, and the final curves have a sawtooth structure rather than approximating the perfectly smooth derivative as well. Eventually, the derivatives will also converge but that may take many more iterations than is required for the convergence of the values themselves.

This delay effect can be observed in Fig. 15.2. Here we have plotted the derivatives $\dot{z}_k = \partial z_k / \partial t$ of the iterates z_k with respect to t after 4, 6, 8, and 16 steps. They can be obtained from $\dot{z}_0 = 0$ by the derivative recurrence

$$\dot{z}_{k+1} \; = \; g'(z_k, t)\dot{z}_k + \dot{g}(z_k, t) \; , \tag{15.2}$$

where we have denoted the two partial derivatives of the iteration function g by

$$g'(z,t) \equiv \frac{\partial}{\partial z} g(z,t) \quad \text{and} \quad \dot{g}(z,t) \equiv \frac{\partial}{\partial t} g(z,t) \; .$$

The properties of these recurrences will be examined in section 15.3. As one can see, the derivatives $\dot{z}_k$ converge like the z_k quite rapidly when $t > 0.24$, but they do take quite a few iterations to sort themselves out where there are several real roots and even when t is close to zero. The same is true for the Newton iterates themselves.

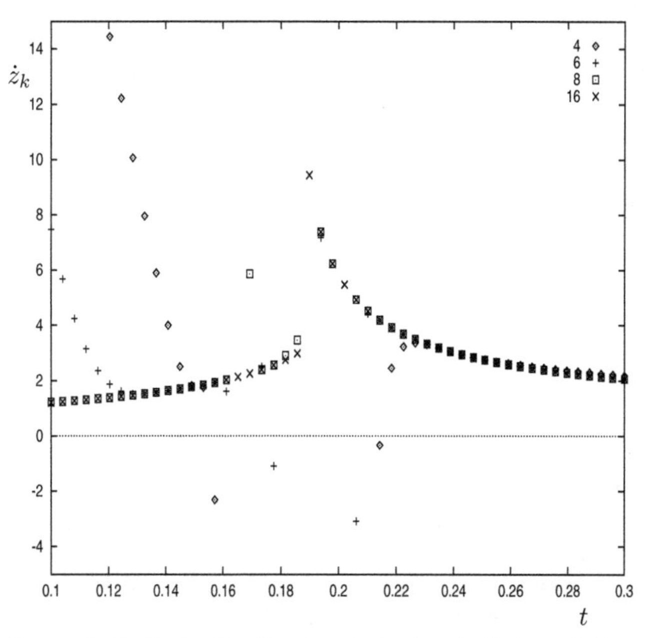

Figure 15.2: Derivative Iterates $\dot{z}_k$ for $k = 4,\ 6,\ 8,$ and 16

Rather than carrying derivative information along with the iterations, one may also try to compute the final value $\dot{z}$ from the so-called sensitivity equation

$$(g'(z,t)-1)\dot{z} \;=\; -\dot{g}(z,t) \quad \text{at} \quad z=z_k\approx z(t)\;,$$

which follows by implicit differentiation of the identity $g(z(t),t) = z(t)$. In the case of Newton's method, this is what (15.2) reduces to at an exact root because we have $g'(z,t)=0$ there also. In other words, if z_k were the exact solution, one extra Newton step with differentiation would give us the exact value $\dot{z}(t) = \dot{z}_{k+1}$ without $\dot{z}_k$ playing any role. Things are not as tidy when the iteration function g is only some approximation to Newton's method. However, as long as it is contractive in that $|g'| < 1$ near the solutions, then the approximating derivatives $\dot{z}_k$ still converge to the correct value, as we will see in section 15.2.

When implicit relations are solved iteratively with a number of steps controlled by a stopping criterion, then the resulting function approximation may be very rough, at least in certain critical regions. Even in calm regions the results are usually discontinuous, but the derivatives of the smooth pieces can be expected to approximate the derivatives of the implicit function reasonably well. As we will show in this section 15.1, calm regions can be characterized by the fact that the iterative method in question is nicely contractive, a desirable property that is then inherited by the corresponding derivative iteration. Carrying derivatives $\dot{z}_k$ along with the iterates z_k is one way of solving the sensitivity equation, which will be discussed as the piggyback approach in section 15.2. Subsequently, direct derivative recurrences are derived in section 15.3. Corresponding adjoint recurrences are presuited in section 15.4 for first-order iteratives and in section 15.5 for second-order derivatives.

15.1 Results of the Implicit Function Theorem

Let us consider a parameter-dependent system of nonlinear equations

$$w \;\equiv\; F(z,x)=0 \quad \text{with} \quad F:\mathbb{R}^l\times\mathbb{R}^n \mapsto \mathbb{R}^l\;, \tag{15.3}$$

where x represents, as before, the vector of independent variables or parameters with respect to which we wish to differentiate. When we discuss the reverse mode it will be important that frequently there is another "response" function

$$y=f(z,x) \quad \text{with} \quad f:\mathbb{R}^l\times\mathbb{R}^n \mapsto \mathbb{R}^m \tag{15.4}$$

that evaluates a few key quantities of the "state vector" z. Although this is not necessary for our theoretical statements, one may usually assume that l, the dimension of the state space, is orders of magnitudes larger than the number n of parameters x and the dimension m of y, which we may think of as a vector of objectives and soft constraints.

For example, consider the design optimization of an aircraft, one of the prime challenges in scientific computing. Here x parametrizes the shape of the wing and fuselage and its structural realization; z may represent the three-dimensional flow field and the elastic deformations of the aircraft in flight; and y could be a handful of performance indices like the lift/drag ratio, weight, cost of production, etc. With h the grid size in one spatial dimension, one may then assume for a full three-dimensional model that n, l, and m are of order h^{-1} (or h^{-2}), h^{-3}, and h^0, respectively. To optimize the response y by varying x, we need the reduced Jacobian dy/dx possibly post- or premultiplied by vectors $\dot{x}$ or $\bar{y}$, respectively. In order to effectively eliminate the variables z and w, at least theoretically we need the following assumption.

Assumption (JR): Jacobian Regularity

On some neighborhood of a solution (z_*, x) with $F(z_*, x) = 0$, the Jacobians of $F(z, x)$ and $f(z, x)$ with respect to $z \in \mathbb{R}^l$ and $x \in \mathbb{R}^n$ are once Lipschitz continuously differentiable and the square matrix

$$F'(z, x) \equiv \frac{\partial}{\partial z} F(z, x) \in \mathbb{R}^{l \times l}$$

is nonsingular at all points in that neighborhood.

By the implicit function theorem, (15.3) implies at regular roots $z_* \in \mathbb{R}^l$ was not idented at all.

$$\frac{dz}{dx} \equiv -\left(\frac{\partial w}{\partial z}\right)^{-1} \frac{\partial w}{\partial x} = -F'(z_*, x)^{-1} F_x(z_*, x) \, ,$$

where we abbreviate $F_x(z, x) \equiv \partial F(z, x)/\partial x \in \mathbb{R}^{l \times n}$. Then it follows by the chain rule for the total derivative of the objectives y with respect to the design parameters x that

$$\frac{dy}{dx} = \frac{\partial y}{\partial x} + \frac{\partial y}{\partial z}\frac{\partial z}{\partial x} = f_x(z_*, x) - f_z(z_*, x)\, F'(z_*, x)^{-1} F_x(z_*, x) \, . \qquad (15.5)$$

Here the subscripts of f_x and f_z denote differentiation with respect to x and z, respectively. Comparing this equation with the Jacobian representation (9.1) we note that we have a generalization of a Schur complement to the nonlinear case.

As in the linear situation the triple matrix product $f_z \left[F'\right]^{-1} F_x$ on the right-hand side can be bracketed in two different ways, and we again face the alternative of first computing

$$\frac{dz}{dx} = -F'(z_*, x)^{-1} F_x(z_*, x) \in \mathbb{R}^{l \times n} \qquad (15.6)$$

or

$$\frac{dy}{dw} = f_z(z_*, x) F'(z_*, x)^{-1} \in \mathbb{R}^{m \times l} \qquad (15.7)$$

and then multiplying these matrices by $f_z(z_*, x)$ or $F_x(z_*, x)$, respectively. The notation $dy/dw = (dy/dz)(dz/dw)$ is appropriate since $dz/dw = F'(z, x)^{-1}$ for fixed x by the inverse function theorem.

Direct and Adjoint Sensitivity Equation

Just as in the linear case, we find that (15.6) is best calculated in the forward mode and (15.7) by something akin to the reverse mode. In the forward mode we may restrict the differentiation to a single direction vector $\dot{x} \in \mathbb{R}^n$, and in the reverse mode to a single adjoint vector $\bar{y} \in \mathbb{R}^m$. Hence we compute or approximate the *implicit tangent*

$$\dot{z}_* \equiv -F'(z_*, x)^{-1} F_x(z_*, x)\,\dot{x} \in \mathbb{R}^l \tag{15.8}$$

and the *implicit adjoint*

$$\bar{w}_* \equiv F'(z_*, x)^{-\top} f_z(z_*, x)^\top \bar{y} \in \mathbb{R}^l\,. \tag{15.9}$$

In other words we wish to solve the *direct sensitivity equation*

$$0 = \dot{F}(z_*, x, \dot{z}_*, \dot{x}) \equiv F'(z_*, x)\,\dot{z}_* + F_x(z_*, x)\,\dot{x} \tag{15.10}$$

or the *adjoint sensitivity equation*

$$0 = \bar{F}(z_*, x, \bar{w}_*, \bar{y}) \equiv F'(z_*, x)^\top \bar{w}_* - f_z(z_*, x)^\top \bar{y}\,. \tag{15.11}$$

The definition of the symbols $\dot{F}$ and $\bar{F}$ is similar but not exactly the same as their usage in Chapter 3. The adjoint sensitivity equation has been analyzed and used occasionally in the engineering literature [CW$^+$80, NH$^+$92].

Hence we see that $\dot{z}_*$ and $\bar{w}_*$ can be computed by solving a linear system involving the Jacobian $F'(z_*, x)$ and its transpose, respectively. Correspondingly, the right-hand sides $\dot{F}(z_*, x, 0, \dot{x}) \in \mathbb{R}^l$ and $\bar{F}(z_*, x, 0, -\bar{y}) \in \mathbb{R}^l$ can be evaluated by a single forward or reverse sweep on the evaluation procedure for (F, f). From now on we will assume that a single vector $\dot{x}$ or $\bar{y}$ is given and the corresponding vectors $\dot{z}_*$ or $\bar{w}_*$ defined in (15.8) and (15.9) are desired. Then they immediately yield

$$\dot{y}_* = f_x\dot{x} + f_z\dot{z}_* \quad \text{and} \quad \bar{x}_*^\top = \bar{y}^\top f_x - \bar{w}_*^\top F_x \quad \text{with} \quad \bar{y}^\top \dot{y}_* = \bar{x}_*^\top \dot{x}\,, \tag{15.12}$$

where all derivatives are evaluated at (z_*, x).

Our adjoint sensitivity equation (15.11) might also be interpreted as the Karush–Kuhn–Tucker condition, [KT51, NS96], for the optimization problem

$$\min\{\bar{y}^\top f(z, x) \mid F(z, x) = 0,\ z \in \mathbb{R}^l\} \quad \text{for} \quad x \in \mathbb{R}^n\,.$$

Here x is considered fixed, so that the feasible set is locally the singleton $\{z = z_*(x)\}$ because of our regularity Assumption (JR). Then the Karush–Kuhn–Tucker condition reduces to the representation of the gradient of an (arbitrary) objective function as a linear combination of the constraint gradients, which form in fact a complete basis of $\mathbb{R}^l$. The coefficients in this representation are the Lagrange multipliers that form our vector $\bar{w}_*$. If one lets the x vary freely too, one obtains the extra condition that the reduced gradient

$\bar{x}_*$ given in (15.12) must also vanish. The resulting stationary solutions are heavily dependent on the objective $\bar{y}^\top f$, so that we have a more conventional Karush–Kuhn–Tucker point. This observation leads to the question of whether one really wishes to solve the equation pair $F(z,x) = 0$ and $\bar{F}(z,x,\bar{w},\bar{y}) = 0$ accurately for fixed x rather than to adjust these parameters simultaneously, thus hopefully making headway towards an optimal solution. From this point of view the "adjoint fixed point iteration" to be described in section 15.4 might be generalized to a procedure for updating Lagrange multipliers in a constrained optimization calculation.

Derivative Quality Criteria

No matter by what method an approximation $\dot{z}$ to $\dot{z}_*$ or an approximation $\bar{w}$ to $\bar{w}_*$ has been obtained, its quality can be assessed by evaluating the residuals

$$\|F(z,x)\| \quad \text{and} \quad \|\dot{F}(z,x,\dot{z},\dot{x})\| \quad \text{or} \quad \|\bar{F}(z,x,\bar{w},\bar{y})\| \,,$$

where $z \approx z_k$ is presumably the best approximation to z_* yet available. More precisely, we have the following lemma.

Lemma 15.1 (Forward and Reverse Consistency Check)
Under Assumption (JR) *and with $\dot{x} \in \mathbb{R}^n$ or $\bar{y} \in \mathbb{R}^m$ fixed there exist constants $\delta > 0$ and $\gamma < \infty$ such that with $\dot{F}$ defined in* (15.10),

$$\|\dot{z} - \dot{z}_*\| \;\le\; \gamma\,(\|F(z,x)\| + \|\dot{F}(z,x,\dot{z},\dot{x})\|)\,,$$

or with $\bar{F}$ defined in (15.11),

$$\|\bar{w} - \bar{w}_*\| \;\le\; \gamma\,(\|F(z,x)\| + \|\bar{F}(z,x,\bar{w},\bar{y})\|)$$

for all z with $\|z - z_*\| < \delta$ and $\dot{z} \in \mathbb{R}^l$ or $\bar{w} \in \mathbb{R}^l$ arbitrary.

Proof. The Jacobian of the combined system

$$F(z,x) = 0, \quad \dot{F}(z,x,\dot{z},\dot{x}) = 0$$

at a solution $z_* = z_*(x), \dot{z}_* = \dot{z}_*(x)$ takes the partitioned form

$$\frac{\partial(F,\dot{F})}{\partial(z,\dot{z})} = \begin{bmatrix} F_z & 0 \\ F_{zz}\,\dot{z} + F_{xz}\,\dot{x} & F_z \end{bmatrix} \in \mathbb{R}^{(2l)\times(2l)}\,.$$

This $(2l) \times (2l)$ matrix is block-triangular with the two diagonal blocks being identical and, by Assumption (JR), nonsingular as well as Lipschitz continuous. Thus the first inequality holds by Lemma 4.1.16 in [DS96]. The second inequality holds by analogous arguments applied to the combined system

$$F(z,x) = 0, \quad \bar{F}(z,x,\bar{w},\bar{y}) = 0\,. \qquad \blacksquare$$

The constant γ is a function of local Lipschitz constants and of the size of the inverse $F'(z,x)^{-1}$. As always in nonlinear equation solving, good estimates for these quantities are hard to come by.

As one can see in Lemma 15.1, both derivative vectors $\dot{z}$ and $\bar{w}$ are usually affected by error in the underlying z, which explains why there is often a time-lag in their convergence. The delay has been observed on most iterative schemes other than Newton's method.

Approximating Reduced Functions

As an immediate consequence of Lemma 15.1 we note that by replacing the exact vectors $\dot{z}_*$ and $\bar{w}_*$ in the formulas (15.8) and (15.9) by approximations $\dot{z}$ and $\bar{w}$ one obtains also first-order approximations to vectors of first-order reduced derivatives. Moreover, as originally suggested by Christianson in [Chr01], one can use the approximate derivatives obtained to compute corrected reduced function and Jacobian values whose error is essentially that of the first-order estimates squared.

Corollary 15.1 (CORRECTED FUNCTION ESTIMATE)

Under the assumptions of Lemma 15.1 *and with any vector* $\bar{w} \in \mathbb{R}^n$ *for given* $\bar{y} \in \mathbb{R}^l$*, the corrected value*

$$\sigma \equiv \sigma(z,x,\bar{w},\bar{y}) \equiv \bar{y}^\top f(z,x) - \bar{w}^\top F(z,x) \tag{15.13}$$

satisfies

$$\left|\bar{y}^\top f(z,x) - \sigma(z,x,\bar{w},\bar{y})\right| \le \Gamma \|F(z,x)\| \, \|\bar{F}(z,x,\bar{w},\bar{y})\| + \mathcal{O}(\|F(z,x)\|^2)$$

for a constant $\Gamma > 0$.

Proof. The assertion follows from the Taylor expansions

$$\begin{aligned} f(z_*,x) &= f(z,x) + f_z(z,x)(z_* - z) + \mathcal{O}(\|z - z_*\|^2) \,, \\ 0 = F(z_*,x) &= F(z,x) + F_z(z,x)(z_* - z) + \mathcal{O}(\|z - z_*\|^2) \end{aligned}$$

by the definition of $\bar{F}$ and with Γ as an upper bound on the inverse Jacobian norm. ∎

As we will see in section 15.5 one can expect that within an iterative procedure the corrected estimate $\bar{y}^\top f(z,x) - \bar{w}^\top F(z,x)$ converges roughly twice as fast as $\bar{y}^\top f(z,x)$ to the actual reduced function value $\bar{y}^\top f(z_*,x)$. Corresponding numerical tests were mute for the standard test case of the NACA0012 airfoil [GF02]. The expected convergence behavior of the reduced response derivative is shown in Fig. 15.3.

In our setting the discrepancies $\dot{z} - \dot{z}_*$ and $\bar{w} - \bar{w}_*$ come about through iterative equation solving. The same duality arguments apply if $\dot{z}_*$ and $\bar{w}_*$ are solutions of operator equations that are approximated by solutions $\dot{z}$ and $\bar{y}$ of corresponding discretizations. Under suitable conditions elaborated in [GP01,

VD00, GS02, BR01] the adjoint correction technique then doubles the order of convergence with respect to the mesh-width. In both scenarios solving the adjoint equation provides accurate sensitivities of the weighted response with respect to solution inaccuracies. For discretized PDEs this information may then be used to selectively refine the grid where solution inaccuracies have the largest effect on the weighted response [BR01].

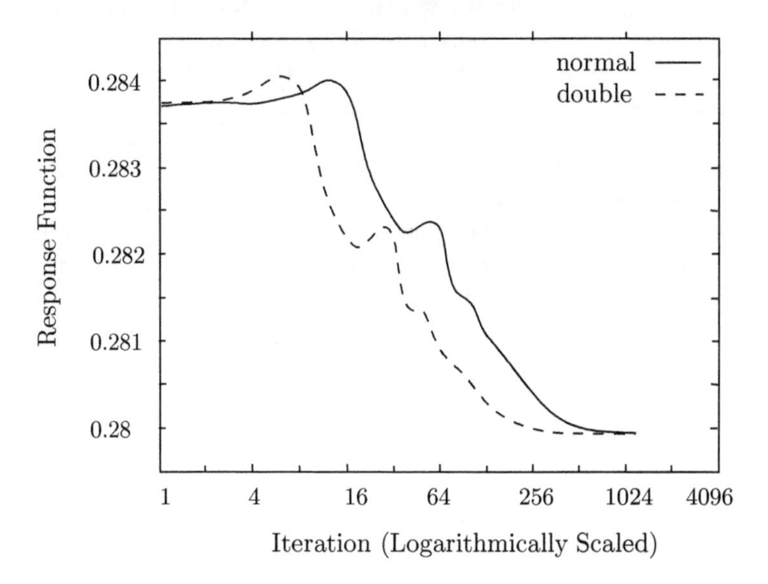

Figure 15.3: Normal and Corrected Values of Drag Coefficient

15.2 Iterations and the Newton Scenario

Many iterative solvers can be written in the form

$$z_{k+1} = G_k(z_k, x) \quad \text{with} \quad G_k : \mathbb{R}^{l \times n} \mapsto \mathbb{R}^l\ . \tag{15.14}$$

Being optimistic we will assume throughout that given a value of x we have an initial guess z_0 such that the resulting iterates z_k converge to some solution

$$z_* = z_*(x) \equiv \lim_{k \to \infty} z_k \quad \text{with} \quad F(z_*, x) = 0\ .$$

Newton's method is obtained if one defines

$$G_k(z_k, x) \equiv z_k - F'(z_k, x)^{-1} F(z_k, x) \tag{15.15}$$

so that equivalently

$$F'(z_k, x)(z_{k+1} - z_k) = -F(z_k, x)\ . \tag{15.16}$$

Hence computing a Newton step is exactly as difficult as solving either sensitivity equation (15.10) or (15.11).

Normally, one forms and factors the square Jacobian $F'(z_k, x)$ and then performs forward and backward substitutions for each new right-hand side. Whether or not this is possible with reasonable programming effort and runtime depends on the problem. However, it never seems to make sense to do one but not the other. In other words, if one can compute Newton steps exactly, one might as well solve either sensitivity equation directly, and vice versa.

In this highly desirable "Newton scenario" the iteration (15.14) can and should be executed with differentiation turned off, i.e., with x treated as a constant. After the corresponding solution $z \approx z_*$ has been obtained with satisfactory accuracy, one merely has to differentiate $F(z, x + t\dot{x})$ with respect to t, considering z as fixed, and then feed the resulting vector $\dot{F}(z, x, 0, -\dot{x})$ into one extra substitution process. For that one may use the factorization from the last Newton iterate, so that the cost for computing one $\dot{z}$ on top of z is probably negligible. Naturally this effort scales with n, the number of design parameters if one sticks with the forward mode and needs the derivatives with respect to all components of x. Similar observations apply to the reverse mode with m right-hand sides $\bar{F}(z, x, 0, -\bar{y})$.

Two-Phase or Piggyback

Compared to the Newton scenario things are not nearly as clear if one uses iterative solvers for either the Newton system (15.16) or the underlying nonlinear equation $F(z, x) = 0$ itself. In shape optimization and most other large-scale applications this is the only realistic possibility, as forming and factoring the Jacobian is simply out of the question. Then one may adopt one of the following two basic strategies.

Two-phase:	Solve the sensitivity system independently with the same solver after convergence on the nonlinear system has been achieved.
Piggyback:	Propagate derivatives along with the main iterates once the convergence is regular and possibly with some extra iterations.

In the Newton scenario just discussed, the first choice is clearly preferable, provided one is willing and able to make the effort to modify the code accordingly. The same is still true if the Jacobian F' is well enough understood to allow an efficient and accurate solution of linear systems $F'(z, x)a = b$ for any given right-hand side b. For example, the spectral properties might be known, possibly after a suitable preconditioning. Naturally, we would recommend the use of AD techniques discussed in section 10.3 for computing the Jacobian-vector and/or vector-Jacobian products required in Krylov space methods and other iterative solvers. Other than that, the only AD-specific recommendation concerning the solution of the sensitivity equation in a separate second phase is to utilize Lemma 15.1 in the stopping criterion.

There are several reasons for preferring the piggyback approach. A very important one is the user's inability or unwillingness to analyze and modify the various iterative processes that may be going on in several places in a larger evaluation program. Then one may try to get by with black box differentiation—simply processing the whole code with an AD tool like any other program with jumps and branches. One minor adjustment (that one would probably also make if one wanted to compute divided differences) is to tighten accuracy parameters and possibly increase upper bounds on iteration counters. This may be needed, because derivatives of iterative processes typically lag behind the iterates themselves, though they tend to have the same asymptotic rate, as we shall see.

Delayed Piggyback

On the other hand, it is usually not very beneficial to carry along derivatives as long as the iterative solver is still far away from any particular solution, quite likely using discontinuous adjustments like stepsize halving as part of its search strategy. All this tells us nothing about the implicit derivative $\dot{z}_*$, which is completely determined by the local properties of F near (z_*, x). Hence it makes sense to let the iteration initially run undifferentiated until it starts to exhibit a regular convergence pattern with a consistent reduction in the residual $F(z_k, x)$. Then one may halt, initialize the derivative $\dot{z}_k$ to zero, and recommence the iteration from the last iterate, this time computing derivative approximations for all subsequent iterates.

An example of this approach is displayed in Fig. 15.4. The results were obtained on a two-dimensional Navier–Stokes (NS) code with several thousand variables for simulating a transonic flow over an aircraft wing. The convergence of the iterative solver is quite slow throughout and particularly irregular during the first 900 iterations. This can be explained by the shock location not having settled down and crossing lines of the grid. Therefore, differentiation was only turned on after some 1 420 steps, when the iteration had already gained 6 digits of accuracy.

In this case black box differentiation was employed in that the total derivatives $\dot{z}_k = dz_k(x + t\dot{x})/dt \in \mathbb{R}^l$ were carried forward starting from $\dot{z}_k = 0$ for the initial $k = 1{,}420$. By differentiating $z_{k+1} = G_k(z_k, x)$ we find that the $\dot{z}_k$ must satisfy the recurrence

$$\dot{z}_{k+1} \;=\; G_k'(z_k, x)\dot{z}_k + \dot{G}_k(z_k, x, \dot{x}) \tag{15.17}$$

where

$$G_k'(z, x) \equiv \frac{\partial}{\partial z} G_k(z, x) \quad \text{and} \quad \dot{G}_k(z, x, \dot{x}) \equiv \left[\frac{\partial}{\partial x} G_k(z, x)\right] \dot{x}\,. \tag{15.18}$$

As one can see in Fig. 15.4, the norm of the derivative residuals as defined in (15.10), i.e., $\dot{F}_k \equiv \dot{F}(z_k, x, \dot{z}_k, \dot{x})$, jumped up on the first step but then started to decline immediately and exhibited asymptotically about the same linear rate of convergence as the residuals $F(z_k, x)$ themselves. This is no coincidence, as we can see from the following analysis.

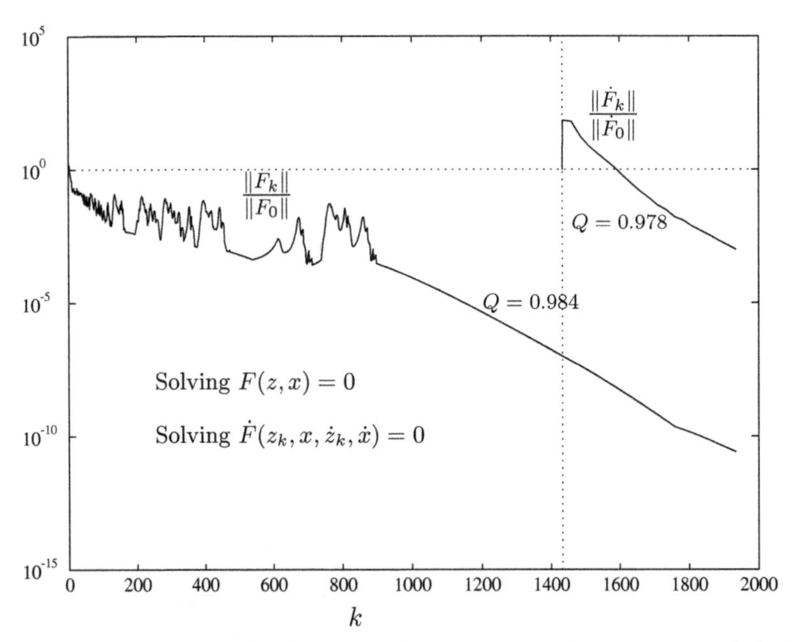

Figure 15.4: Convergence Behavior of Function and Derivative on NS Code

Contractions and Their Convergence Rates

The iterates $z_k \in \mathbb{R}^l$ generated by many practical methods for approximating a solution $z_* = z_*(x)$ with $F(z_*, x) = 0$ satisfy a quasi-Newton recurrence of the form

$$z_{k+1} = G_k(z_k, x) \equiv z_k - P_k F(z_k, x) . \qquad (15.19)$$

Here the preconditioner P_k is some $l \times l$ matrix that approximates the inverse of the Jacobian $F'(z_k, x)$. The closer that approximation, the more the iteration resembles Newton's method with its excellent local convergence properties. From a more abstract point of view P_k is a mapping from the range of F back into its domain. To ensure stable convergence from within a vicinity of z_* we make the following assumption.

Assumption (GC) : Global Contractivity

The Jacobians G'_k satisfies at all arguments (z, x) in some neighborhood of (z_*, x)

$$|||G'_k||| \leq \rho < 1 \quad \text{for all} \quad k \qquad (15.20)$$

with respect to some induced matrix norm $|||\cdot|||$ defined by (15.22).

When $G_k = G$ is the same for all k the assumption is essentially equivalent to the condition that the spectral radius of $G'(z, x) \in \mathbb{R}^{l \times l}$, i.e., the maximal modulus of any eigenvalue, is smaller than $\rho < 1$. Moreover, the assumption can be satisfied for $G_k(z,x) = z - P_k F(z,x)$ with some suitable preconditioners P_k if and only if the Jacobian $F'(z,x)$ is nonsingular and Lipschitz continuous on some neighborhood of the solution $z_* = z_*(x)$, as implied by Assumption (JR). Note in particular that then null vectors of $F'(z,x)$ are eigenvectors with eigenvalue 1 of $[I - P_k F'(z_*,x)]$. For a proof that the spectral radius of a matrix A is less than 1 if and only if there exists an ellipsoidal norm $\|\cdot\|$ with $\|A\| < 1$, see Proposition 2.2.8 in [OR70]. According to Ostrowski's theorem (see Propositions 10.1.3 and 10.1.4 in [OR70]), it follows from Assumption (GC) that all initial guesses z_0 whose distance to z_* is less than some bound lead to convergence with

$$Q\{z_k - z_*\} \equiv \limsup_k \frac{\|z_{k+1} - z_*\|}{\|z_k - z_*\|} \le \rho \,. \tag{15.21}$$

Here the vector norm $\|\cdot\|$ must be consistent with the matrix norm used in Assumption (GC) so that for any square matrix $A \in \mathbb{R}^{l\times l}$

$$\|A\| \;=\; \max_{0 \ne z \in \mathbb{R}^l} \|Az\|/\|z\| \,. \tag{15.22}$$

Quotient and Root Convergence Factors

The chosen norm strongly influences the so-called Q-factor $Q\{z_k - z_*\}$ defined by (15.21) for any iteration sequence $\{z_k - z_*\}$ with $z_k \ne z_*$ for all k. In contrast, it follows from the equivalence of all norms on finite-dimensional spaces that the R-factor

$$R\{z_k - z_*\} \;\equiv\; \limsup_k \sqrt[k]{\|z_k - z_*\|} \le Q\{z_k - z_*\} \le \rho \tag{15.23}$$

is norm independent. The last inequality holds by (15.21), and the other one is established as Proposition 9.3.1 in [OR70]. In both (15.21) and (15.23) we may replace the uniform bound $\rho < 1$ from Assumption (GC) by the corresponding limit superior

$$\rho_* \equiv \limsup_k \| G_k' \| \;\le \rho \,, \tag{15.24}$$

so that in conclusion

$$R\{z_k - z_*\} \;\le\; Q\{z_k - z_*\} \;\le\; \rho_* \;\le\; \rho < 1 \,.$$

When $R\{z_k - z_*\} = 0$, the convergence is said to be *R-superlinear*, and when even $Q\{z_k - z_*\} = 0$, it is called *Q-superlinear*. The latter, highly desirable property is again norm invariant and can be established for certain secant updating methods [DS96] without the even stronger condition $\rho_* = 0$ necessarily being satisfied.

Since the R-factor $R\{z_k - z_*\}$ is norm independent, it cannot exceed the infimum of the factor $Q\{z_k - z_*\}$ with $\|\cdot\|$ ranging over all vector norms. Moreover, one can easily construct examples (see Exercise 15.2) where there is still a gap between that infimum and the R-factor $R\{z_k - z_*\}$. In general we have

$$R\{z_k - z_*\} = Q\{z_k - z_*\} \quad \Leftrightarrow \quad Q\{z_k - z_*\} = \lim_k \frac{|||z_{k+1} - z_*|||}{|||z_k - z_*|||} ,$$

in which case the error $|||z_k - z_*|||$ is asymptotically reduced by the same nonzero factor $R\{z_k - z_*\} = Q\{z_k - z_*\}$ at each iteration. Otherwise the convergence is more irregular, with the reduction ratio $|||z_{k+1} - z_*|||/|||z_k - z_*|||$ jumping up and down, possibly even exceeding 1 over infinitely many iterations. Nevertheless, the average rate of convergence as measured by $R\{z_k - z_*\}$ may still be quite good, but one should be a little more circumspect in using the step norm $|||z_{k+1} - z_k|||$ as an estimate of the remaining error $|||z_k - z_*|||$.

Convergence Rates of Residuals

Unfortunately, one can rarely compute the solution error $|||z_k - z_*|||$ and must therefore be content to gauge the quality of the current approximation z_k in terms of the residual $F_k \equiv F(z_k, x)$. Under our Assumption (JR) one may view $\|F(z_k, x)\|$ as an equivalent norm to $|||z_k - z_*|||$ since there must be constants $\delta > 0$ and $1 \leq \gamma < \infty$ such that

$$\frac{1}{\gamma} \leq \frac{\|F(z, x)\|}{|||z - z_*|||} \leq \gamma \quad \text{for} \quad |||z - z_*||| < \delta \, .$$

This implies for any sequence z_k converging to, but never reaching, z_* that

$$R\{F_k\} \;=\; R\{z_k - z_*\} \;\leq\; Q\{F_k\} \;\leq\; \gamma^2 \, Q\{z_k - z_*\} \, .$$

In particular, we have the equivalent superlinear convergence conditions

$$Q\{z_k - z_*\} = 0 \quad \Leftrightarrow \quad Q\{F_k\} = 0$$

and

$$R\{z_k - z_*\} = 0 \quad \Leftrightarrow \quad R\{F_k\} = 0 \, .$$

In practice one may use the estimates

$$R\{F_k\} \approx \sqrt[k]{\frac{\|F_k\|}{\|F_0\|}} \quad \text{and} \quad Q\{F_k\} \approx \max\left\{\frac{\|F_k\|}{\|F_{k-1}\|}, \frac{\|F_{k+1}\|}{\|F_k\|}\right\}$$

to track the progress of the iteration. Here we have somewhat arbitrarily chosen to maximize over two successive residual reduction ratios for the Q-factor. The reason is that some kind of alternating approach seems to be a convergence pattern that occurs reasonably often, especially in the vicinity of singularities or severe ill conditioning [Gri80, NS96].

In the calculation reported in Fig. 15.4, both $R\{F_k\}$ and $Q\{F_k\}$ were asymptotically close to 0.984. However, it is clear that during the first 900 iterations there were quite a few stretches where the ratio $\|F_{k+1}\|/\|F_k\|$ exceeded 1 for several steps in a row. As depicted in the top right corner, the derivative residual $\dot{F}_k \equiv \dot{F}(z_k, x, \dot{z}_k, \dot{x})$ also converged linearly with the just slightly smaller factor $R\{\dot{F}_k\} \approx Q\{\dot{F}_k\} \approx 0.978$.

15.3 Direct Derivative Recurrences

Let us first consider the two-phase approach to calculating the directional derivative $\dot{z}_*$ for given $\dot{x}$ after $z_k \approx z_*$ has been computed with satisfactory accuracy. Then the recurrence (15.17) with $x_k = x_{\bar{k}}$ fixed is in fact a linear contraction and we must have

$$R\{\dot{z}_k - \dot{z}_*\} \leq Q\{\dot{z}_k - \dot{z}_*\} \leq \varrho < 1\,.$$

In other words we have Q-linear convergence with respect to the vector norm $\|\!| \;\; |\!\|$.

More interesting is the behavior of the piggyback iteration

$$\begin{aligned} z_{k+1} &= G_k(z_k, x) \\ \dot{z}_{k+1} &= G'_k(z_k, x)\, \dot{z}_k + \dot{G}_k(z_k, x, \dot{x}) \end{aligned} \tag{15.25}$$

with G' and $\dot{G}$ as defined in (15.18). The Jacobian of this coupled criterion is the block 2×2 matrix

$$\frac{\partial(z_{k+1}, \dot{z}_{k+1})}{\partial(z_k, \dot{z}_k)} = \begin{bmatrix} G'_k & 0 \\ B_k & G'_k \end{bmatrix} \in \mathbb{R}^{2l \times 2l}\,, \tag{15.26}$$

where we assume that the G_k are twice continuously differentiable with respect to (z, x) so that

$$B_k \equiv \frac{\partial^2}{\partial z^2} G_k(z, x)\, \dot{z}_k + \frac{\partial^2}{\partial z \partial x} G_k(z, x)\, \dot{x} \bigg|_{z = z_k}\,.$$

The off-diagonal block $B_k \in \mathbb{R}^{2l \times 2l}$ quantifies the sensitivity of the equation $G_k(z_k, x)\dot{x} + \dot{G}(z_k, x, \dot{x}) = 0$ with respect to changes in the primal variables z. Unless it vanishes the $\dot{z}_k$ are normally prevented from converging Q-linearly, one effect that occurs similarly for Lagrange multipliers in nonlinear programming. However, it is simple to establish a monotonic decline in the weighted norm

$$\|\!| z_k - z_* |\!\| + \omega\, \|\!| \dot{z}_k - \dot{z}_* |\!\| \quad \text{for} \quad \omega > 0\,. \tag{15.27}$$

More specifically we obtain the following generalization of a result by Gilbert [Gil92].

Proposition 15.1 (Convergence Factor of Derivatives)
Under Assumption (GC) *suppose that* $\lim z_k \to z_*$ *and the* G_k *have uniformly bounded second derivatives near* (z_*, x). *Then for all sufficiently small* ω *the weighted norms given in* (15.27) *converge* Q*-linearly to zero and we have*

$$R\{\dot{z}_k - \dot{z}_*\} \leq \rho_* \leq \rho < 1$$

when ρ_* *is defined in* (15.24).

Proof. First we note that $\dot{z}_0 = 0$ implies that $\sup \|\dot{z}_k\| \le c_0 \|\dot{x}\|$ for some $c_0 \in \mathbb{R}$. This follows by induction from the bound

$$\|\dot{z}_{k+1}\| \le \rho \, \|\dot{z}_k\| + \left\| \frac{\partial}{\partial x} G_k(z_k, x) \right\| \|\dot{x}\| \, .$$

As a consequence we obtain also a uniform upper bound

$$\|B_k\| \le c_1 \|\dot{x}\| \quad \text{for some} \quad c_1 \in \mathbb{R} \, .$$

Hence we may bound the induced norm of the partitioned Jacobian according to

$$\begin{aligned}
\sup_{(u,v) \in \mathbb{R}^{2l}} & \left(\|G'_k \, u\| + \omega \, \|B_k \, u + G'_k v\| \right) \Big/ \left(\|u\| + \omega \, \|v\| \right) \\
& \le \left(\rho \|u\| + \omega \, \rho \, \|v\| + \omega \, c_1 \, \|\dot{x}\| \, \|u\| \right) \Big/ \left(\|u\| + \omega \, \|v\| \right) \\
& \le \left(\rho + \omega \, c_1 \, \|\dot{x}\| \right) .
\end{aligned}$$

By making ω small we may push the induced norm below 1 and arbitrary close to ρ. For any such ω we get R-linear convergence of the component $\|\dot{z}_k - \dot{z}_*\|$ with factor $\rho + c_1 \, \omega \, \|\dot{x}\|$. ∎

The comparison with (15.21) suggests that the derivatives $\dot{z}_k$ converge at the same speed as the underlying iteratives z_k. Both vector sequences and similarly the adjoints $\bar{z}_k$ to be considered later have apparently the same R-factor. However, a closer analysis reveals that generally the derivatives do in fact lag a little bit behind as one would expect for nonlinear problems.

Without going into details, which are given in [GK05], we can explain this effect as follows. Suppose the iteration function $G_k = G$ does not depend on k and is twice continuously differentiable. Then the asymptotic convergence is controlled by the block Jacobian (15.26) that takes at (z_*, x) the form

$$\begin{bmatrix} A & 0 \\ B & A \end{bmatrix} \quad \text{with} \quad A, B \in \mathbb{R}^{l \times l} \, .$$

In the case of adjoints to be discussed later the bottom right block is transposed and B is symmetric. In either case the eigenvalues of the coupled Jacobian are exactly those of $A = G'(z_*, x)$ but each one of them has an algebraic multiplicity of at least two. For general B the geometric multiplicities are still simple so that the eigenvalues are in fact defective and give rise to nontrivial Jordan blocks. Consequently, the powers of the block Jacobian do not decline strictly geometrically but exhibit some polynomial growth. When A and B commute we obtain explicitly

$$\begin{bmatrix} A & 0 \\ B & A \end{bmatrix}^k = \begin{bmatrix} A^k & 0 \\ kBA^{k-1} & A^k \end{bmatrix}$$

so that approximately

$$z_k - z_* \approx A^k(z_0 - z_*) \quad \text{and} \quad \dot{z}_k - \dot{z}_* \approx A^k(\dot{z}_0 - \dot{z}_*) + k\,B\,A^{k-1}\,(z_0 - z_*)\,.$$

It should be noted that the second term in the expression for $\dot{z}_k - \dot{z}_*$ may grow until $k\|A^{k-1}\| \leq k\,\rho^{k-1}$ begins to decline. For $\rho \approx 1$ this happens only when $k \approx -1/\ln\rho \approx 1/(1-\rho)$. Hence, for slowly convergent fixed point solvers we must expect that derivatives may be severely perturbed by inaccuracies in the primal iterates. Asymptotically, we must expect $\|\dot{z}_k - \dot{z}_*\| \sim k\|z_k - z_*\|$, and the same error factor of k applies to the adjoints $\bar{z}_k$ derived later. Before we consider adjoint recurrences we will describe a simplification that is quite important in the quasi-Newton scenario.

Deactivation of Preconditioner

Suppose the P_k are at least locally smooth functions of (z, x) and thus t, provided $z_k = z_k(t)$ and $x = x(t) = x(0) + t\dot{x}$ are continuously differentiable functions of t. Then the matrices $\dot{P}_k \equiv d\,P_k(z_k(t), x(t))/dt$ are continuous, and the derivatives $\dot{z}_k = \dot{z}_k(t)$ exist and must satisfy the recurrence

$$\dot{z}_{k+1} \;=\; \dot{z}_k - P_k\,\dot{F}(z_k, x, \dot{z}_k, \dot{x}) - \dot{P}_k\,F(z_k, x)\,. \tag{15.28}$$

This is a specialization of (15.17) to an iteration function G_k of the form (15.19). The term $\dot{F}(z_k, x, \dot{z}_k, \dot{x})$ is defined by (15.10). The last term $\dot{P}_k\,F(z_k, x)$ is in some way the most interesting. If the preconditioner is fixed so that (15.19) reduces to a simple substitution method, the last term vanishes, since clearly $\dot{P}_k \equiv 0$. Even if the $\dot{P}_k$ are nonzero but their size is uniformly bounded, the term $\dot{P}_k\,F(z_k, x)$ disappears gradually as $F_k = F(z_k, x)$ converges to zero.

This happens, for example, in Newton's method, where $P_k = F'(z_k, x)^{-1}$ is continuously differentiable in (z, x), provided F itself is at least twice continuously differentiable. However, second derivatives should not really come into it at all, as the implicit derivative $\dot{z}_*$ is, according to the explicit representation (15.8), uniquely defined by the extended Jacobian of F. Hence we may prefer to simply drop the last term and use instead the *simplified recurrence*

$$\dot{z}_{k+1} = \dot{z}_k - P_k\,\dot{F}(z_k, x, \dot{z}_k, \dot{x})\,. \tag{15.29}$$

Its implementation requires the *deactivation* of P_k when this preconditioner depends on x, as it usually does. By this we mean that the dependence on x is suppressed so that it looks as through P_k consists of passive elements. Whether and how this can be done depends on the particular AD tool. It should not be made too easy, because the unintentional suppression of active dependencies can lead to wrong derivative values. For both derivative recurrences (15.28) and (15.29) Proposition 15.1 still applies as was shown in [GB$^+$93]. In both cases the root convergence factor is bounded by the limiting spectral radius ρ_*, which bounds the quotient convergence factor of the iterates z_k themselves. If $\rho_* = 0$ we have Q-superlinear convergence of the z_k and the slightly weaker property of R-superlinear convergence for the $\dot{z}_k$. This result applies in particular for Newton's method, where the z_k in fact converge quadratically. Abbreviating $\omega_k = F(z_k, x)$ and defining $\dot{\omega}_k$ correspondingly, we obtain the differentiated iteration listed in Table 15.1.

Table 15.1: Direct Fixed Point Iteration

$$
\begin{array}{l}
\text{fix } x, \dot{x} \in \mathbb{R}^n \\
\text{initialize } z_0, \dot{z}_0 \in \mathbb{R}^l \\
\text{for } k = 0, 1, 2, \ldots \\
\quad w_k = F(z_k, x) \\
\quad \dot{w}_k = \dot{F}(z_k, x, \dot{z}_k, \dot{x}) \\
\quad \text{stop if } \|w_k\| \text{ and } \|\dot{w}_k\| \text{ are small} \\
\quad z_{k+1} = z_k - P_k w_k \\
\quad \dot{z}_{k+1} = \dot{z}_k - P_k \dot{w}_k \\
y = f(z_k, x) \\
\dot{y} = f_z(z_k, x)\dot{z}_k + f_x(z_k, x)\dot{x}
\end{array}
$$

Multistep Contractivity

Naturally (15.29) can also be applied after convergence of the z_k, i.e., with z_k as the argument of F' and $\dot{F}$ kept constantly equal to the best approximation to z_* that has been obtained. In other words, we apply an iterative solver to the linear sensitivity equation (15.10). Storing and recalling a succession of the P_k rather than just the last one makes sense because the matrix product

$$D_k\, D_{k-1} \ldots D_0 \quad \text{with} \quad D_k \equiv I - P_k F'(z_*, x)$$

may be much more contractive than the upper bound ρ^{k+1} on its spectral norm suggests. This effect can be expected for methods that alternate between preconditioners that reduce various components of the solution error $z_k - z_*$ and thus also $\dot{z}_k - \dot{z}_*$. For example, one frequently thinks of high and low frequency modes in the error. In this scenario ρ may be barely below 1, but a cycle over several steps can nevertheless reduce the error $\dot{z}_k - \dot{z}_*$ by a significant fraction. For Q-superlinear convergent methods based on secant updating, one has in general $\rho_* > 0$, so that one cannot deduce R-superlinear convergence of the derivatives from Proposition 15.1. Nevertheless, it may sometimes be observed, as shown in Figs. 15.5 and 15.6.

These numerical experiments were conducted on the test function

$$F(z, t) \equiv \nabla_z f(z, t) \quad \text{with} \quad f(z, t) \equiv \frac{1}{2}\left(z^\top H z + t\|z\|^4\right) ,$$

where $H = [1/(i + j - 1)]$ is the Hilbert matrix of order 2, and $\|z\|$ denotes the Euclidean norm. There is only one parameter, $x = t$. Locally the minimizers of f are characterized as roots of the stationarity conditions $F = 0$, so that

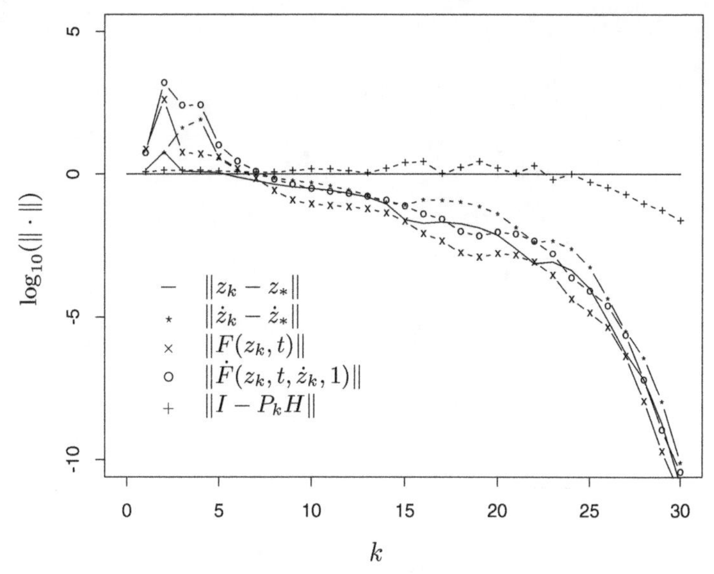

Figure 15.5: Results of Simplified Derivative Recurrence on Hilbert Example

minimization methods behave eventually like equation solvers. In the general nonlinear equation case, the progress toward the solution is usually gauged in terms of some norm of the residual vector F.

Since the unique solution $z_* = 0$ is independent of the parameter t, its derivatives $\dot{z}_*$ must also vanish, a situation that makes monitoring its error exceedingly simple. The approximate inverse Hessian was initialized as $P_0 = \text{diag}(i)_{i=1,2}$, which is somewhat "smaller" than the exact inverse H^{-1}. Consequently, the inverse form of the Davidon–Fletcher–Powell update takes a very long time before P_k and the resulting steps $z_{k+1} - z_k = -P_k F(z_k, t)$ become large enough to achieve superlinear convergence. The starting point was always the vector of ones $z_0 = (1, 1)$, and the parameter was set to $t = 1$.

The iteration depicted in Fig. 15.5 proceeds rather slowly until the Frobenius norm of the error matrix $D_k = I - P_k H$ drops below 1 at about the 25th step. Hence, our theoretical results apply at most for the last five iterations. Note that D_k is not exactly equal to $I - P_k F'(z_k, 1)$ since we have neglected the nonquadratic term. Over the whole range, the iterates z_k, their "derivatives" $\dot{z}_k$, and the corresponding residuals $F(z_k, t)$ and $\dot{F}(z_k, t, \dot{z}_k, 1)$ converge more or less monotonically at about the same rate. Since the iterates themselves converge so slowly, the derivatives do not lag noticeably behind. The situation is not radically different when the iteration is fully differentiated, as depicted in Fig. 15.6. However, as one can see from the top line, the derivative $\dot{P}_k$ of the preconditioner P_k grows to a Frobenius norm in the hundreds before it finally begins to decline. As a result, the first derivative of iterates and residuals seems to behave a little bit more erratically in the intermediate stage of the iteration. While we have not made a timing comparison to see how much overhead the

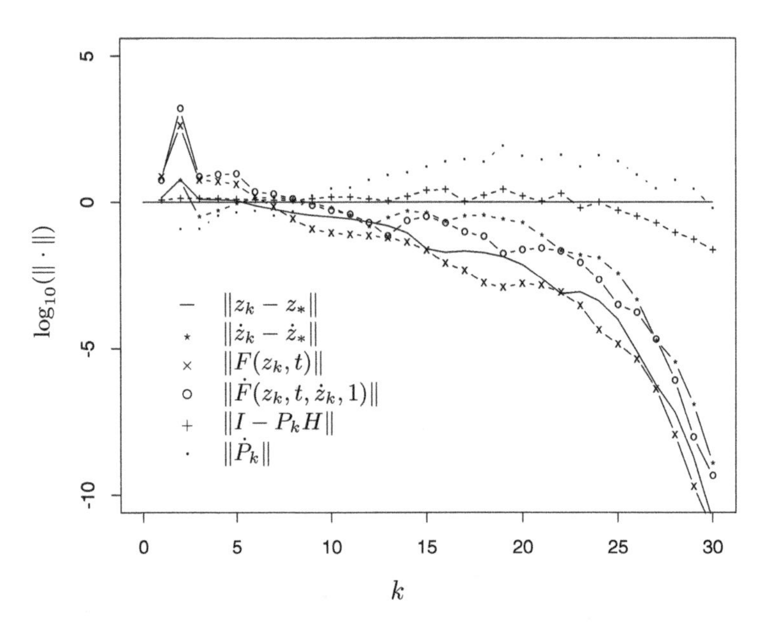

Figure 15.6: Results of Full Derivative Recurrence on Hilbert Example

differentiation of the preconditioner entails, it would seem so far that there is no reward for incurring that extra cost. On the other hand, if the identification and "deactivation" of P_k appear to require a significant recoding effort, one may also just differentiate the whole iteration. In both Figs. 15.5 and 15.6, the residual derivative $\|\dot{F}(z_k, t, \dot{z}_k, 1)\|$ is a fairly reliable indicator of the actual error $\|\dot{z}_k - \dot{z}_*\|$.

The growth of the norms $\|\dot{P}_k\|$ indicates that the assumptions of Proposition 15.1 were actually not satisfied even though the results confirm its assertion. It was shown in Proposition 2 of [GB+93] that the assertion is still valid when the P_k are generated by the standard secant updates, as the $\dot{P}_k$ may grow slowly towards infinity but are still asymptotically annihilated by the faster declining residuals F_k. Apparently, secant methods are the only class of nonlinear equation solvers other than memoryless contractive fixed point iterations with $P_k = P(z_k, x)$, for which convergence of the derivatives has been established. For a comparison of the full and simplified derivative recurrence on a CFD problem see the study by Carle and Fagan [CF96].

15.4 Adjoint Recurrences and Their Convergence

To compute $\bar{w}_*$ one may of course solve the adjoint sensitivity equation

$$F'(z_*, x)^\top \bar{w} \;=\; f_z(z_*, x)^\top \bar{y} \;=\; \bar{F}(z_*, x, 0, -\bar{y}) \tag{15.30}$$

obtained from (15.11). The transposed Jacobian $F'(z_*, x)^\top$ has the same size, spectrum, and sparsity characteristics as $F'(z_*, x)$ itself. Hence the task of solving the adjoint sensitivity equation (15.30) is almost exactly equivalent to the task of solving the direct sensitivity equation (15.10). Because of the similarity relation

$$P_k^\top F'(z_k,x)^\top \;=\; P_k^\top \left[P_k F'(z_k,x)\right]^\top P_k^{-\top} \,,$$

the square matrices $I - P_k F'(z_k,x)$ and $I - P_k^\top F'(z_k,x)^\top$ have the same spectrum. Hence the latter has, by Assumption (GC), a spectral norm less than or equal to $\rho < 1$. And if we assume that $P_k \to P_* \in \mathbb{R}^{l\times l}$ then $I - P_*^\top F'(z_*,x)^\top$ has also an ellipsoidal norm arbitrary close to ρ_* as defined in (15.24). Then in a two-phase fashion we may solve the iteration

$$\bar{w}_{k+1} \;=\; \bar{w}_k - P_k^\top \underbrace{\left[F'(z_*,x)^\top \bar{w}_k - f_z(z_*,x)^\top \bar{y}\right]}_{\equiv\, \bar{F}(z_*,x,\bar{w}_k,\bar{y})} \,. \qquad (15.31)$$

Starting from any $\bar{w}_0$ yields also R-linear convergence with

$$R\{\bar{w}_k - \bar{w}_*\} \;\le\; \rho_* \,, \qquad (15.32)$$

The recurrence (15.31) was apparently first analyzed by Christianson [Chr94], albeit with a fixed final preconditioner P_k. For the reasons that we discussed in the subsection "Multistep Contractivity" on page 384, it may sometimes be more advantageous to use several of the late preconditioners P_k, which of course requires additional storage.

Now the question arises of whether the adjoint sensitivity calculations can also be performed in a piggyback fashion, i.e., without setting up a second phase iteration. Using arguments similar to these in the proof of Proposition 15.1 one can show that (15.32) remains true when (15.31) is applied with z_* replaced by z_k. This means that we can propagate the adjoint vectors $\bar{w}_k$ forward without the need to record the intermediates z_k and the corresponding preconditioners. Only each residual evaluation $F(z_k,x)$ must be reverted to yield the adjoint residual $\bar{z}_k \equiv \bar{F}(z_k,x,\bar{w}_k,\bar{y})^\top \in \mathbb{R}^l$ at a comparable computational effort. We may also include the size of $\bar{F}$ in the overall stopping criterion in order to obtain the iteration displayed in Table 15.2.

The adjoint evaluation yielding $\bar{z}_k$ and thus $\bar{w}_{k+1}$ both immediately follow a corresponding forward calculation, so that recording is only required temporarily. What we call here "adjoint fixed point iteration" has been referred to as the "iterative incremental form" of the adjoint sensitivity equation in the aerodynamics literature [NH+92]. The only down side of this method appears to be that we need to somehow separate the preconditioner P_k and subsequently apply its transpose. As for the direct derivatives $\dot{z}_k$ it was shown in [GK05] that the adjoints lag a little behind so that $\|\bar{w}_k - \bar{w}_*\| \sim k\,\|z_k - z_*\|$ despite the common R-factor ρ. The simplified forward recurrence (15.29) with z_k fixed is quite closely related to the recurrence (15.28), which is effectively applied if one applies the forward mode mechanically to the iteration (15.19).

Table 15.2: Adjoint Fixed Point Iteration

fix $x \in \mathbb{R}^n$, $\bar{y} \in \mathbb{R}^m$
initialize $z_0, \bar{w}_0 \in \mathbb{R}^l$
for $k = 0, 1, 2, \ldots$
$\quad [w_k, y_k] = \big[F(z_k, x), f(z_k, x)\big]$
$\quad \bar{z}_k = F'(z_k, x)^\top \bar{w}_k - f_z(z_k, x)^\top \bar{y}$
$\quad$ stop if $\|w_k\|$ and $\|\bar{z}_k\|$ are small
$\quad z_{k+1} = z_k - P_k w_k$
$\quad \bar{w}_{k+1} = \bar{w}_k - P_k^\top \bar{z}_k$
$y = f(z_k, x)$
$\bar{x}^\top = \bar{w}_k^\top F_x(z_k, x) - \bar{y}^\top f_x(z_k, x)$

Mechanical Adjoint Recurrence

Such similarity does not arise if we apply the reverse mode to the whole iteration in a mechanical fashion. Suppose we iterate over $\bar{k}$ steps

$$z_{k+1} \;=\; z_k \;-\; P_k F(z_k, x) \quad \text{for} \quad k = 0 \ldots \bar{k} - 1$$

and then evaluate

$$y \;=\; f(z_{\bar{k}}, x)\,.$$

The adjoint iteration is easily written down, namely,

$$\bar{x}^\top \;=\; \bar{y}^\top f_x(z_{\bar{k}}, x)\,, \quad \bar{z}_{\bar{k}}^\top \;=\; \bar{y}^\top f_z(z_{\bar{k}}, x)$$

and

$$\left.\begin{aligned} \bar{x}^\top &\mathrel{-}= \bar{z}_{k+1}^\top P_k F_x(z_k, x) \\ \bar{z}_k^\top &= \bar{z}_{k+1}^\top \big[I - P_k F'(z_k, x)\big] \\ \bar{P}_k &= -\bar{z}_{k+1} F(z_k, x)^\top \end{aligned}\right\} \quad \text{for} \quad k = \bar{k} - 1 \ldots 0\,. \tag{15.33}$$

Assuming that $\bar{z}_k$ stays bounded, we see that $\bar{P}_k$ converges to zero and can thus be neglected. In other words, we assume for simplicity that the P_k have been deactivated, i.e., are considered as real numbers that have fallen from the sky and are thus unrelated to x and the z_k. Because of our contractivity Assumption (GC) it is clear that $\|\bar{z}_k\| \sim \rho^{\bar{k}-k}$. This means the adjoints $\bar{z}_k$ of early iterates, especially that of z_0, should be small when $\bar{k}$ is sufficiently large. That effect makes sense, since the initial guess z_0 should not have any significant influence on the outcome of the overall computation. Hence the productive part is the incrementation of the parameter adjoint $\bar{x}$. Assuming for a moment that $z_k = z_*$ and $P_k = P$ for all $k \le \bar{k}$ including negative indices, we find using the Neumann expansion that

$$\bar{z}_k^\top = \bar{z}_{\bar{k}}^\top \left[I - P\,F'(z_k, x)\right]^{\bar{k}-k} = \bar{y}^\top f_z(z_*, x)\left[I - P\,F'(z_k, x)\right]^{\bar{k}-k},$$

and thus at the end

$$\bar{y}^\top f_x(z_*, x) - \bar{x}^\top = \sum_{j=0}^{\infty} \bar{z}_{\bar{k}-j}^\top P\,F_x(z_k, x) = \bar{z}_{\bar{k}}^\top \sum_{j=0}^{\infty}\left[I - P\,F'(z_k, x)\right]^j P\,F_x(z_k, x)$$

$$= \bar{y}^\top f_z(z_*, x)\left[P\,F'(z_*, x)\right]^{-1} P\,F_x(z_*, x) = \bar{y}^\top f_z(z_*, x) F'(z_*, x)^{-1} F_x(z_*, x)\,.$$

After bringing $\bar{y}^\top f_x$ onto the right-hand side, we obtain for $\bar{x}^\top$ exactly the same expression as in (15.12).

Without the unrealistic assumption $z_k = z_*$ and $P_k = P$ constant, we can only hope that the resulting $\bar{x}$ is reasonably close to that exact value. In contrast to the solutions $\dot{z}$ and $\bar{w}$, we cannot check the quality of the value of $\bar{x}$ by evaluating some derivative residual that must vanish if it is correct. Of course, since $\|\bar{z}_k\| \leq \rho \|\bar{z}_{k+1}\|$ declines monotonically, one may expect that $\bar{x}$ has reached its proper value when $\|\bar{z}_k\|$ become sufficiently small. Nevertheless, we conclude that the iteration (15.33) really has no mathematical advantage compared to the formula (15.31).

Rule 26

FIXED POINT ITERATIONS CAN AND SHOULD BE ADJOINED BY RECORDING ONLY SINGLE STEPS.

The Case of Nonlinear Preconditioning

The only reason not to apply the iteration (15.31) is again inability or unwillingness to separate out the Jacobian $F'(z_k, x)$ and the preconditioner P_k from the original iteration. Note that in (15.33) the matrix $F'(z_k, x)^\top P_k^\top$ is applied to the vector $\bar{z}_{k+1}$, whereas in (15.31) $P_k^\top F'(z_k, x)^\top$ is applied to $\bar{w}_k \in \mathbb{R}^l$. To overcome the need to separate out P_k, let us assume more generally that

$$z_{k+1} = G_k(z_k, x) \equiv H_k(z_k, x, F(z_k, x))\,,$$

where the $H_k : \mathbb{R}^{l \times n \times l} \to \mathbb{R}^l$ have the fixed point property

$$H_k(z, x, w) = z \iff w = 0 \in \mathbb{R}^l$$

for all (z, x) in some open domain of interest. In other words, H_k maps z into itself exactly when the residual $w = F(z, x)$ vanishes. Provided the H_k are Lipschitz continuously differentiable, we obtain the Taylor expansion

$$\begin{aligned} z_{k+1} &= H_k(z_k, x, F(z_k, x)) \\ &= z_k + H_k'(z_k, x, F(z_k, x))\,F(z_k, x) + o(\|F(z_k, x)\|)\,, \end{aligned}$$

where

$$H'_k(z_k, x, w) \equiv \frac{\partial}{\partial w} H_k(z_k, x, w) \in \mathbb{R}^{l\times l} .$$

By comparison with (15.19) we see that P_k may be identified with the partial Jacobian $-H'_k(z_k, x, F(z_k, x))$. Now deactivating P_k means considering the arguments z_k and x of H as constant and computing $-P_k \dot{w}_k$ as $H'_k \dot{w}_k$ in Table 15.1 and $-P_k^\top \bar{z}_k$ as $(H'_k)^\top \bar{z}_k$ in Table 15.2. These calculations require only a forward or a reverse sweep on H_k. Thus we no longer have to identify the preconditioning matrix P_k but merely need to determine where the components of the residual $F(z_k, x)$ enter into the calculation of the new iterate z_{k+1}. Conceivably one could introduce a new type of variable named "residual" and introduce special rules of differentiation such as $(v * r)' \equiv v * r'$ if r is the new variable type.

15.5 Second-Order Adjoints

In order to obtain second-order implicit derivatives we can differentiate the adjoint sensitivity equation (15.11) once more in the direction $\dot{x}$ to obtain the second-order adjoint equation

$$\begin{aligned} 0 &= \dot{\bar{F}}(z_*, x, \dot{z}_*, \dot{x}, \bar{w}_*, \bar{y}, \dot{\bar{w}}_*) \\ &\equiv F_z(z_*, x)^\top \dot{\bar{w}}_* + \dot{F}_z(z_*, x, \dot{z}_*, \dot{x})^\top \bar{w}_* - \dot{f}_z(z_*, x, \dot{z}_*, \dot{x})^\top \bar{y} \end{aligned} \tag{15.34}$$

where $\dot{F}_z$ and $\dot{f}_z$ are defined by

$$\dot{F}_z \equiv \dot{F}_z(z_*, x, \dot{z}_*, \dot{x}) \equiv F_{zz}(z_*, x)\dot{z}_* + F_{zx}(z_*, x)\,\dot{x} \in \mathbb{R}^{l\times l} \tag{15.35}$$

$$\dot{f}_z \equiv \dot{f}_x(z_*, x, \dot{z}_*, \dot{x}) \equiv f_{zz}(z_*, x)\dot{z}_* + f_{zx}(z_*, x)\,\dot{x} \in \mathbb{R}^{m\times l} \tag{15.36}$$

Since the second-order sensitivity equation is linear in the unknown second-order adjoint vector $\dot{\bar{w}}$, it can be computed by solving a linear system in the Jacobian F_z with the right-hand side given by $\dot{\bar{F}}(z_*, x, \dot{z}_*, \dot{x}, \bar{w}_*, -\bar{y}, 0)$. This second-order residual can be evaluated by a combination of a forward and a reverse sweep as described in section 5.3.

In what one might call a two-phase approach many researchers solve the sensitivity equations separately, after the state z_* has been approximated with satisfactory accuracy. Here we will utilize a piggyback approach, where these linear equations are solved simultaneously with the original state equation. Whatever methods one uses to generate approximate solutions $\dot{z}$, $\bar{w}$, and $\dot{\bar{w}}$ to the sensitivity equations, their quality can be gauged by evaluating the *derivative residuals* $\dot{F}(z_*, \dot{z})$, $\bar{F}(z_*, \bar{w})$, and $\dot{\bar{F}}(z_*, \dot{z}, \bar{w}, \dot{\bar{w}})$ defined in (15.9), (15.10), and (15.34). Here and sometimes in the remainder of this section we omit the argument vectors $x, \dot{x}$, and $\bar{y}$ because they are always selected as constants. The derivative residual vectors can be obtained just as cheaply as the right-hand sides mentioned above and bound the derivative errors as follows. In analogy and as an extension of Lemma 15.1 we obtain the following lemma

Lemma 15.2 (Second-Order Adjoint Consistency Check)
Under Assumption (JR) *and with* $\dot{x} \in \mathbb{R}^n$ *or* $\bar{y} \in \mathbb{R}^m$ *fixed there exist constants* $\delta > 0$ *and* $\gamma < \infty$ *such that with* $\dot{F}$ *defined in* (15.9) *and with* $\dot{\bar{F}}$ *defined in* (15.34)

$$
\begin{aligned}
\|\dot{\bar{w}} - \dot{\bar{w}}_*\| \leq \gamma(\|F(z,x)\| \quad & + \|\dot{F}(z,x,\dot{z},\dot{x})\| \\
+\|\bar{F}(z,x,\bar{w},\bar{y})\| & + \|\dot{\bar{F}}(z,x,\dot{z},\dot{x},\bar{w},\bar{y},\dot{\bar{w}})\|)
\end{aligned}
$$

for all z *with* $\|z - z_*\| < \delta$ *and* $\dot{z}, \bar{w}, \dot{\bar{w}} \in \mathbb{R}^l$ *arbitrary.*

In order to approximate $\dot{\bar{w}}_*$ iteratively we can simply differentiate the adjoint iterate displayed in Table 15.2 directionally and obtain the extended procedure listed in Table 15.3.

Table 15.3: Second-Order Adjoint Fixed Point Iteration

fix $x, \dot{x} \in \mathbb{R}^n, \bar{y} \in \mathbb{R}^m$

initialize $z_0, \dot{z}_0, \bar{w}_0, \dot{\bar{w}}_0 \in \mathbb{R}^l$

for $k = 0, 1, 2, \ldots$

$$
\begin{aligned}
&[w_k, y_k] = [F(z_k, x), f(z_k, x)] \\
&\dot{w}_k = \dot{F}(z_k, x, \dot{z}_k, \dot{x}) \\
&\bar{z}_k = \bar{F}(z_k, x, \bar{w}_k, \bar{y}) \\
&\dot{\bar{z}}_k = \dot{\bar{F}}(z_k, x, \dot{z}_k, \dot{x}, \bar{w}_k, \bar{y}, \dot{\bar{w}}_k)
\end{aligned}
$$

stop if $\|w_k\|, \|\dot{w}_k\|, \|\bar{z}_k\|$, and $\|\dot{\bar{z}}_k\|$ are small

$$
\begin{aligned}
z_{k+1} &= z_k - P_k\, w_k & \dot{z}_{k+1} &= \dot{z}_k - P_k\, \dot{w}_k \\
\bar{w}_{k+1} &= \bar{w}_k - P_k^\top \bar{z}_k & \dot{\bar{w}}_{k+1} &= \dot{\bar{w}}_k - P_k^\top \dot{\bar{z}}_k
\end{aligned}
$$

$$
\begin{aligned}
y_k &= f(z_k, x) & \sigma_k &= \bar{y}^\top y_k - \bar{w}_k^\top w_k \\
\dot{y}_k &= \dot{f}_z(z_k, x, \dot{z}_k, \dot{x}) & \dot{\sigma}_k &= \bar{y}\, \dot{y}_k - \bar{w}_k \dot{w}_k - \dot{\bar{w}}_k\, w_k \\
\bar{x}_k^\top &= \bar{y}^\top f_x(z_k, x) - \bar{w}_k^\top F_x(z_k, x) \\
\dot{\bar{x}}_k^\top &= \bar{y}^\top \dot{f}_x(z_k, x, \dot{z}_k, \dot{x}) - \dot{\bar{w}}_k^\top F_x(z_k, x) - \bar{w}_k^\top \dot{F}_x(z_k, x, \dot{z}_k, \dot{x})
\end{aligned}
$$

As in (15.32) we obtain for $\dot{\bar{w}}_k$ also R-linear convergence with the same factor, namely,

$$
R\{\dot{\bar{w}}_k - \dot{\bar{w}}_*\} \leq \rho_*
$$

Except for the fact that we have only R-linear rather than Q-linear convergence it looks as though in the piggyback fashion the derivative vectors $\dot{z}_k$, $\bar{w}_k$, and $\dot{\bar{w}}$ converge at essentially the same speed as the primal iterates z_k. A closer analysis taking into account a Jordan block of defective eigenvalues shows that $\|\dot{\bar{w}}_k - \dot{\bar{w}}_*\| \sim k^2 \|z_k - z_*\|$ as proven in [GK05]. These relations were verified on a discretized PDE in two spatial dimensions.

As one can see, Table 15.3 contains essentially the union of Table 15.2 with some statements for computing the second-order information and the convergence test involving $\dot{\bar{w}}$. Since the transpose $I - F_z(z_k)^\top P_k^\top$ is also the Jacobian of the fixed point iterations for $\dot{\bar{w}}_k$ its has again the same contraction factor and we obtain the following R-linear convergence result directly from Proposition 15.1 and (15.32).

$$
\begin{aligned}
z_k - z_* &= \mathcal{O}(\|F(z_k)\|) && \sim \rho_*^k \\
y_k - y_* &= \mathcal{O}(\|F(z_k)\|) && \sim \rho_*^k \\
\sigma_k - \bar{y}^\top y_* &= \mathcal{O}(\|F(z_k, \bar{w}_k)\| \, \|\bar{F}(z_k)\| + \|F(z_k)\|^2) && \sim k\, \rho_*^k \rho_0^k \\
\bar{w}_k - \bar{w}_* &= \mathcal{O}(\|F(z_k)\| + \|\bar{F}(z_k, \bar{w}_k)\|) && \sim k\, \rho_0^k \\
\bar{x}_k - \bar{x}_* &= \mathcal{O}(\|F(z_k)\| + \|\bar{F}(z_k, \bar{w}_k)\|) && \sim k\, \rho_0^k \\
\dot{z}_k - \dot{z}_* &= \mathcal{O}(\|F(z_k)\| + \|\dot{F}(z_k, \dot{z}_k)\|) && \sim k\, \rho_0^k \\
\dot{y}_k - \dot{y}_* &= \mathcal{O}(\|F(z_k)\| + \|\dot{F}(z_k, \dot{z}_k)\|) && \sim k\, \rho_0^k \\
\dot{\sigma}_k - \bar{y}^\top \dot{y}_* &= \mathcal{O}\big(\|F(z_k)\| + \|\dot{F}(z_k, \dot{z}_k)\| + \|\bar{F}(z_k, \bar{w}_k)\| \\
&\quad + \|\dot{\bar{F}}(z_k, \dot{z}_k, \bar{w}_k, \dot{\bar{w}}_k)\|\big)^2 && \sim k^2\, \rho_0^{2k} \\
\dot{\bar{x}}_k - \dot{\bar{x}}_* &= \mathcal{O}(\|F(z_k)\| + \|\dot{F}(z_k, \dot{z}_k)\| + \|\bar{F}(z_k, \bar{w}_k)\| \\
&\quad + \|\dot{\bar{F}}(z_k, \dot{z}_k, \bar{w}_k, \dot{\bar{w}}_k)\|) && \sim k^2\, \rho_0^k
\end{aligned}
$$

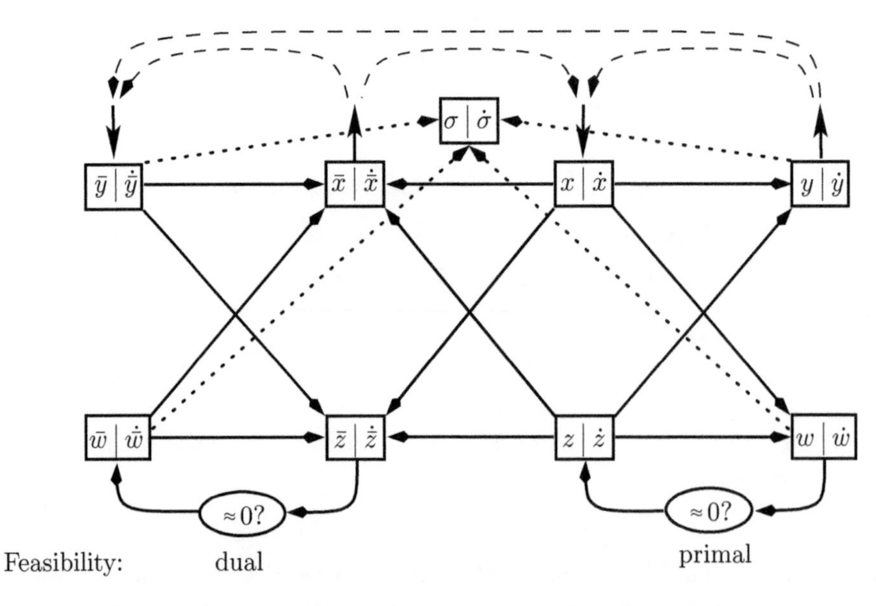

Figure 15.7: Direct and Adjoint Fixed Point Iterations with Directional Derivatives

Extended Computational Graph

In Figure 15.7 we have displayed the dependence relations between the various vector quantities in our iteration loops. The oval conditionals $\approx 0?$ indicate that the input vectors coming from the right are checked for size in a suitable norm. If they are sufficiently small the iteration is terminated; otherwise a suitable multiple of the residual vector is incremented to the node values on the left. The right half displays the original state space iteration with directional derivatives being carried along as well. Here, as indicated by the vertical arrows, the vectors x and $\dot{x}$ are given inputs, whereas y and $\dot{y}$ are the resulting outputs. They may be (re)calculated throughout or only after the state space iteration has terminated. The left half of the graph represents the adjoint quantities as mirror images of the direct ones. The dotted lines represent the dependence of the corrected response σ and its derivative $\dot{\sigma}$ on both direct and adjoint quantities.

15.6 Summary and Outlook

This final chapter concerned the situation where lack of smoothness originates from iterative procedures. In that situation, varying numbers and lengths of steps may cause severe discontinuities and nondifferentiabilities, at least during the early phases of an iteration. However, this roughness is gradually reduced when contractive fixed point methods are applied. Then both the iterates and their derivatives with respect to problem parameters gradually converge to the underlying implicit function and its derivatives. It was demonstrated in [GB$^+$93] that this assertion applies not only to first-order but also to higher-order derivatives.

We refer to the carrying along of derivatives with the original iteration as the piggyback approach. In the alternative two-phase approach, one separately solves the sensitivity equation given by the implicit function theorem after convergence has been achieved. This can be done either by direct linear system solving or by a second iterative phase. In the first case one might as well apply Newton's method to the underlying nonlinear system. In the second case one may try to reuse preconditioners and other information gained about the property of the Jacobian during the nonlinear iterative phase. In this way one may in effect generate a range of options between the pure piggyback and the two-phase approach.

All these observations are similarly valid with regard to the solution of the corresponding adjoint sensitivity equation. Even piggyback is possible and avoids any need to revert the iteration, which does occur if the reverse mode is applied mechanically to an iterative procedure. However, it requires more code modifications than in the case of the forward mode, where black box differentiation is usually close to the simplified derivative recurrence that we recommend.

Differentiating the adjoint code once more forward we obtain second-order adjoints, which may be used to calculate Hessian-vector products in optimization, for example. While the derivative quantities have generally the same

R-factors as the underlying original iteratives they lag behind by a power of the iteration counter. The exponent is exactly the order of the derivative, hence 1 for tangents and gradients and 2 for second-order adjoints.

A key advantage of the adjoint procedure advocated here is that there is no need to store any intermediate states if we are in fact heading to a fixed point. On the other hand, if we have a genuine evolution where the choice of initial conditions and the properties of early iterates are of more than a passing interest, then we must read Chapter 12 concerning trade-offs between temporal and spatial complexity in the reverse mode.

Unfortunately, there are many iterative solvers that cannot be naturally interpreted as reasonably smooth fixed point contractions. This observation applies, for example, to conjugate direction methods, where convergence of derivatives has nevertheless been observed by the author in some numerical experiments. Things are much worse for methods like bisection, which generate a piecewise constant approximation whose derivative is of course zero almost everywhere. Various aspects of this difficulty were examined in [Gru97].

There are many important questions concerning the mathematical meaning of the derivatives obtainable by AD that we have not even mentioned. One that has been examined, for example, in [BB+99, EB99, Ver99] is the automatic differentiation of numerical integrators for ODEs and DAEs. Here the mechanisms for error control and stepsize selection play a similar role to the preconditioners P_k in the iterative solvers considered earlier. They manifestly depend on the problem parameters but could and probably should be deactivated for the sake of an efficient and consistent numerical integration. Here consistency means that in effect the same numerical integrator is applied to the sensitivity equation that can be derived by differentiating the original ODE or DAE with respect to the independent parameters. More generally, one might investigate the relation between differentiation and discretization on PDEs and other infinite-dimensional problems. Many results and observations are scattered throughout the engineering and mathematics literature [Cac81a, Cac81b, BIP88], but there does not seem to exist a comprehensive treatment, especially with regards to nonelliptic PDEs and their adjoints.

15.7 Examples and Exercises

Exercise 15.1 (*Convergence of Derivatives*)
Consider the scalar equation $F(z, x) = 0 \in \mathbb{R}$ and assume that $F : \mathbb{R}^2$ to $\mathbb{R}$ is real analytic in some neighborhood of a solution pair $F(z_*, x_*)$. Furthermore suppose that the partial $F_z(z, x) \equiv= \partial F(z, x)/\partial z$ is nonzero at (z_*, x_*) so that Newton's method converges to a solution $z = z(x)$ for all $x \approx x_*$. Finally, assume that all higher partials of F with respect to z are also nonzero at (z_*, x_*).

a. Write down the simplified Newton, or chord, method where the slope $s = F_z(z_0, x)$ is kept constant throughout the iteration yielding a sequence of iterates z_k for $k = 1, 2, \ldots$

b. Differentiate the iteration derived in part **a** with respect to $x \approx x_*$ treating s as a constant and show that the resulting $z_k' = dz_k/dx$ converges to the implicit derivative $dz/dx = -F_x(z(x),x)/F_z(z(x),x)$. Unless incidentally $s = F_z(z(x),x)$ the rate is linear with the same Q factor as the z_k.
c. Show that the ratio between $\Delta z_k' \equiv z_k' - dz/dx$ and $kF(z_k,x)$ has a fixed limit.
d. Repeat **b** and **c** for the second derivative of $z(x)$ with respect to x with k being replaced by k^2 in **c**.
e. Extend **b** and **c** to the dth derivative with k being replaced by k^d.
f. Verify the theoretical results for the test equation $F(z,x) = z - \sin(z/x)$ with $0.25 < x < 0.5$.

Exercise 15.2 (*R-linear but Not Q-linear*)
Show that $z_k \equiv 0.5^k(1 + (-1)^k/2)$ is R-linearly but not Q-linearly convergent in any norm.

Exercise 15.3 (*Harvesting a Single Natural Population*)
We consider a basic model for harvesting a single species. Let $N(t) \geq 0$ be the population of the species at time t and the rate of change be given by the *conservation equation*

$$\frac{dN}{dt} = f(N,E) \equiv g(N) - y(N,E) \quad \text{with} \quad g(N) = rN(1 - N/K).$$

This autonomous differential equation is based on the logistical population model $g(N)$ suggested by Verhulst in 1836, see [Mur93]. The constant $K > 0$ determines the carrying capacity of the environment. The per capita birthrate is given by $r(1 - N/K)$ for $r > 0$. The mortality rate of the model is increased as a result of harvesting by a yield term $y = y(N,E)$, which depends on the human effort $E \geq 0$ and on the actual population N.

Equilibrium states $N_* = N_*(E)$ of the conservation equation are determined by $f(N_*,E) = 0$ and can easily be computed by Newton's method. In order to maximize the *sustainable yield* $y_* \equiv y(N_*,E)$ (in the equilibrium state) with respect to the human effort E we are interested in the sensitivity $\dot{y}_*$, using the dot to denote differentiation with respect to E from now on.

Consider the problem for $K = 4.2$, $r = 1.7$, $E \in \{0.56, 1.04\}$ using the yield term $y(N,E) = EN\ln(1+N)$.
a. Write a program on your favorite computing platform that computes the equilibrium point N_* applying a simplified Newton's method of the form (15.19) with $P = P_k = -2/3$ starting from $N_0 = K(1 - E/r)$. In the equilibrium point N_* solve the direct sensitivity equation $f_N\dot{N}_* + f_E\dot{E} = 0$ exactly using the input tangent $\dot{E} = 1$ to obtain $\dot{N}_*$. Finally, compute the desired sensitivity $\dot{y}_* = y_N\dot{N}_* + y_E\dot{E}$.
b. Apply the "direct fixed point iteration" (Table 15.1) to the problem using N_0 as given in part **a**, $\dot{N}_0 = 0$, $P = P_k = -2/3$, and again the input tangent $\dot{E} = 1$. Visualize the residuals $|N_j - N_*|$ and $|\dot{N}_j - \dot{N}_*|$ for $j = 0 \ldots 25$ in a logarithmic plot.

c. Write a program similar to that in part **a**. But now solve the adjoint sensitivity equation $\bar{w}_* f_N + \bar{y} f_E = 0$ at the equilibrium point N_* using the adjoint $\bar{y} = 1$ to obtain $\bar{\omega}_*$. Here w is considered to be the return value of $f(N, E)$. Compute $\bar{E}_* = \bar{y} y_E + \bar{w}_* f_E$ and compare the result with the value obtained for $\dot{y}_*$.

d. Apply the "adjoint fixed point iteration" (Table 15.2) to the problem using N_0 given in part **a**, $\bar{w}_0 = 0$, $P = P_k = -2/3$, and the initial adjoint $\bar{y} = 1$. Now visualize the residuals $|N_j - N_*|$ and $|\bar{w}_j - \bar{w}_*|$ for $j = 0 \ldots 25$ in a logarithmic plot.

Epilogue

Like the solution of systems of linear equations, the evaluation of derivatives for composite functions is a classical computational problem whose basic solution has been known for hundreds of years. Yet with every new class of mathematical models and perpetual changes in the characteristics of modern computing platforms, the old problem reappears in a new light, calling at least for adjustments if not innovations in the established solution methodology. In this book we have tried to provide a repository of concepts and algorithmic ideas that can be used to discuss and resolve the differentiation aspects of any nonlinear computational task. Many basic questions remain unresolved, many promising ideas untested, and many almost certain improvements unimplemented.

From a mathematical point of view the most important challenge seems to firmly connect AD with the theory of semi-algebraic and subanalytic functions. That would then allow in principle the rigorous foundation of generalized differentiation concepts and the provision of corresponding derivative objects. The practicality of that approach is not entirely clear since, for example, the composition of the sine function with the reciprocal must be excluded. Another very interesting question is whether direct and adjoint derivatives of iterates converge when these are not generated by an ultimately smooth and contractive fixed point iteration. This question concerns, for example, conjugate gradients and other Krylov subspace methods. It appears currently that in these important cases the piggyback approach discussed in sections 15.3–15.5 may not work at all. Then a separate iterative solver for the direct or adjoint sensitivity equation must be set up by the user or possibly a rather sophisticated AD tool.

So far, the application of AD in the context of PDE-constrained optimization problems is restricted to the discretize-then-optimize, or better discretize-then-differentiate, approach. Currently, the relation to the alternative differentiate-then-discretize, also known as discretize-then-optimize, forms an active research area. Of critical importance here is that gradients and Jacobians of continuous (differential) equations are defined in terms of appropriate inner- or other duality products. In contrast the algebraic differentiation of discretized equations always yields gradient representations with respect to the Euclidean inner product on the "design space" of independent variables. It may therefore need to be transformed to a more appropriate ellipsoidal norm by what is often called a smoothing step. These connections warrant further investigation and may lead to more flexible and convenient AD software in the future.

Another area of current interest is the quantification of sensitivities and moments in the presence of stochastic uncertainty, which typically involves the evaluation of second and higher order derivatives. These are also of great importance in numerical bifurcation and experimental design, where third or even fourth derivatives become indispensable. Moreover, as recently shown by Kubota [BB+08], the evaluation of (sparse) permanents and some combinatorial graph problems can be recast as the task of evaluating mixed derivatives of high order. This observation provides another confirmation that some aspects of (algorithmic) differentiation are NP hard.

From a computer science point of view, the most pressing problem appears to be the identification, storage, and retrieval of those intermediate values or elemental partials that truly need to be passed from the recording sweep of a subroutine call to the corresponding returning sweep, which may take place much later. As the gap between processor speed and memory bandwidth continues to grow, all implementations of the basic reverse mode tend to become memory bound, whether they be based on source transformation or operator overloading. Therefore, to avoid disk storage of temporary data altogether, one may attempt to employ checkpointing and partial recomputation strategies of the kind developed in Chapter 12. Due to the associativity of the chain rule derivative computations typically allow for more concurrency than the underlying simulations. The promise and urgency to exploit this added parallelism grow as multicore processing become more and more prevalent. Therefore, appropriate strategies for the algorithmic differentiation of such codes turns into a significant aspect for the future acceptance of AD.

AD tool development has so far been carried out by small research groups with varying composition and fluent objectives. While a lot has been achieved in this way, it seems doubtful that sufficient progress toward a comprehensive AD system for all of Fortran 95 or C++ can be made in this "academic" fashion. Hence, wider collaboration to develop and maintain a compiler with "-ad" option as in the NAG Fortran compiler is a natural approach. Synergies of the required execution reversal technology can be expected with respect to parallelization and debugging. Operator overloading will continue to provide an extremely flexible and moderately efficient implementation of AD, and its efficiency may benefit from improved compiler optimization on user-defined types, possibly aided by the exploitation of expression templates on the statement level.

Concerning the general problem of evaluating derivatives, quite a few scientists and engineers have recently been convinced that it is simply an impossible task on their increasingly complex multilayered computer models. Consequently, some of them abandoned calculus-based simulation and optimization methods altogether in favor of "evolutionary computing approaches" that are based exclusively on function values and that often draw on rather colorful analogies with "real-life" processes. In contrast, we firmly believe that the techniques sketched in this book greatly enlarge the range of problems to which dreary old calculus methods can still be applied efficiently and even conveniently. In fact, they will be indispensable to the desired transition from system simulation to system optimization in scientific computing.

List of Figures

List of Tables

Assumptions and Definitions

Propositions, Corollaries, and Lemmas

Bibliography

[AB74] R.S. Anderssen and P. Bloomfield, *Numerical differentiation proceedings for non-exact data*, Numer. Math. **22** (1974), 157–182.

[AB04] N. Arora and L.T. Biegler, *A trust region SQP algorithm for equality constrained parameter estimation with simple parameter bounds*, Comput. Optim. Appl. **28**, (2004), 51–86.

[Alk98] B. Alkire, *Parallel computation of Hessian matrices under Microsoft Windows NT*, SIAM News **31**, December 1998, 8–9.

[AC92] J.S. Arora and J.B. Cardosot, *Variational principle for shape design sensitivity analysis*, AIAA **30** (1992), 538–547.

[AC+92] B.M. Averick, R.G. Carter, J.J. Moré, and G.-L. Xue, *The MINPACK-2 Test Problem Collection*, Preprint MCS–P153–0692, ANL/MCS–TM–150, Rev. 1, Mathematics and Computer Science Division, Argonne National Laboratory, Argonne, IL, 1992.

[ACP01] P. Aubert, N. Di Césaré, and O. Pironneau, *Automatic differentiation in C++ using expression templates and application to a flow control system*, Comput. Vis. Sci. **3** (2001) 197–208.

[AH83] G. Alefeld and J. Herzberger, *Introduction to Interval Computations*, Academic Press, New York, 1983.

[AHU74] A.V. Aho, J. Hopcroft, and J.D. Ullman, *The Design and Analysis of Computer Algorithms*, Addison-Wesley, Reading, MA, 1974.

[Bau74] F.L. Bauer, *Computational graphs and rounding error*, SIAM J. Numer. Anal. **11** (1974), 87–96.

[BBC94] M.C. Bartholemew-Biggs, L. Bartholemew-Biggs, and B. Christianson, *Optimization and automatic differentiation in Ada: Some practical experience*, Optim. Methods Softw. **4** (1994), 47–73.

[BB+95] A.W. Bojanczyk, R.P. Brent, and F.R. de Hoog, *Stability analysis of a general Toeplitz system solver*, Numer. Algorithms **10** (1995), 225–244.

[BB+96] M. Berz, C.H. Bischof, G. Corliss, and A. Griewank (eds.), *Computational Differentiation—Techniques, Applications, and Tools*, SIAM, Philadelphia, 1996.

[BB+97] C.H. Bischof, A. Bouaricha, P.M. Khademi, and J.J. Moré, *Computing gradients in large-scale optimization using automatic differentiation*, INFORMS J. Comput. **9** (1997), 185–194.

[BB+99] I. Bauer, H.G. Bock, S. Körkel, and J.P. Schlöder, *Numerical methods for initial value problems and derivative generation for DAE models with application to optimum experimental design of chemical processes*, in Proc. of Scientific Computing in Chemical Engineering II, Hamburg, Springer, Berlin, 1999, pp. 338–345.

[BB+08] C.H. Bischof, H.M. Bücker, P. Hovland, U. Naumann, J. Utke,(eds.), *Advances in Automatic Differentiation*, Lect. Notes Comput. Sci. Eng. **64**, Springer, Berlin, 2008.

[BCG93] C.H. Bischof, G. Corliss, and A. Griewank, *Structured second- and higher-order derivatives through univariate Taylor series*, Optim. Methods Softw. **2** (1993), 211–232.

[BCL01] A. Ben-Haj-Yedder, E. Cances, and C. Le Bris, *Optimal laser control of chemical reactions using automatic differentiation*, in [CF+01], pp. 205–211.

[BC+92] C.H. Bischof, G.F. Corliss, L. Green, A. Griewank, K. Haigler, and P. Newman, *Automatic differentiation of advanced CFD codes for multidisciplinary design*, J. Comput. Syst. in Engrg. **3** (1992), 625–638.

[BC+96] C.H. Bischof, A. Carle, P.M. Khademi, and A. Mauer, *The ADIFOR 2.0 system for the automatic differentiation of Fortran 77 programs*, IEEE Comput. Sci. Engrg. **3** (1996).

[BC+06] M. Bücker, G. Corliss, P. Hovland, U. Naumann, and B. Norris (eds.), *Automatic Differentiation: Applications, Theory, and Implementations*, Lect. Notes Comput. Sci. Eng. **50**, Springer, New York, 2006.

[BCP96] K.E. Brenan, S.L. Campbell, and L.R. Petzold, *Numerical Solution of Initial-Value Problems in Differential-Algebraic Equations*, Classics Appl. Math. **14**, SIAM, Philadelphia, 1996.

[BD95] C.H. Bischof and F. Dilley, *A compilation of automatic differentiation tools*, SIGNUM Newsletter **30** (1995), no. 3, 2–20.

[BD+02] L.S. Blackford, J. Demmel, J. Dongarra, I. Duff, S. Hammarling, G. Henry, M. Heroux, L. Kaufman, A. Lumsdaine, A. Petitet, R. Pozo, K. Remington, and R.C. Whaley, *An updated set of basic linear algebra subprograms (BLAS)*, ACM Trans. Math. Softw. **28** (2002), no. 2, 135–151.

[BDL08] J. Bolte, A. Daniilidis, and A.S. Lewis, *Tame mappings are semismooth*, to appear in Math. Program. (2008).

[Bel07] B.M. Bell, *A Package for C++ Algorithmic Differentiation*, `http://www.coin-or.org/CppAD/`, 2007.

[Ben73] Ch. Bennett, *Logical reversibility of computation*, IBM J. Research and Development **17** (1973), 525–532.

[Ber90a] M. Berz, *Arbitrary order description of arbitrary particle optical systems*, Nuclear Instruments and Methods **A298** (1990), 26–40.

[Ber90b] M. Berz, *The DA Precompiler DAFOR*, Tech. Report, Lawrence Berkeley National Laboratory, Berkeley, CA, 1990.

[Ber91b] M. Berz, *Forward algorithms for high orders and many variables with application to beam physics*, in [GC91], pp. 147–156.

[Ber95] M. Berz, *COSY INFINITY Version 7 Reference Manual*, Tech. Report MSUCL–977, National Superconducting Cyclotron Laboratory, Michigan State University, East Lansing, MI, 1995.

[Ber96] M. Berz, *Calculus and numerics on Levi-Civita fields*, in [BB+96], pp. 19–35.

[Ber98] M. Berggren, *Numerical solution of a flow control problem: Vorticity reduction by dynamic boundary action*, SIAM J. Sci. Comput. **19** (1998), 829–860.

[Bes98] A. Best, *Vergleich von Minimierungsverfahren in der Umformsimulation unter Verwendung des Automatischen Differenzierens*, diploma thesis, Technische Universität Dresden, Germany, 1998.

[Bey84] W.J. Beyn, *Defining equations for singular solutions and numerical applications*, in Numerical Methods for Bifurcation Problems, T. Küpper, H.D. Mittelman, and H. Weber, eds., Internat. Ser. Numer. Math., Birkhäuser, Boston, 1984, pp. 42–56.

[BGL96] M. Berggren, R. Glowinski, and J.L. Lions, *A computational approach to controllability issues for flow-related models.* I: *Pointwise control of the viscous Burgers equation*, Int. J. Comput. Fluid Dyn. **7** (1996), 237–252.

[BGP06] R.A. Bartlett, D.M. Gay, and E.T. Phipps: *Automatic differentiation of C++ codes for large-scale scientific computing*, in Proceedings of ICCS 2006, V. Alexandrov et al., eds., Lecture Notes in Comput. Sci. **3994**, Springer, Berlin, 2006, pp. 525–532.

[BH96] C.H. Bischof and M.R. Haghighat, *Hierarchical approaches to automatic differentiation*, in [BB+96], pp. 83–94.

[BIP88] J. Burns, K. Ito, and G. Prost, *On nonconvergence of adjoint semigroups for control systems with delays*, SIAM J. Control Optim. **26** (1988), 1442–1454.

[BJ+96] C.H. Bischof, W.T. Jones, J. Samareh-Abolhassani, and A. Mauer, *Experiences with the application of the ADIC automatic differentiation tool to the CSCMDO 3-D volume grid generation code*, AIAA Paper 96-0716, Jan. 1996.

[BK+59] L.M. Beda, L.N. Korolev, N.V. Sukkikh, and T.S. Frolova, *Programs for Automatic Differentiation for the Machine BESM*, Tech. Report, Institute for Precise Mechanics and Computation Techniques, Academy of Science, Moscow, 1959.

[BK78] R.P. Brent and H.T. Kung, *Fast algorithms for manipulating formal power series*, Assoc. Comput. Mach. **25** (1978), 581–595.

[BK+97] C.H. Bischof, P.M. Khademi, A. Bouaricha, and A. Carle, *Efficient computation of gradients and Jacobians by dynamic exploitation of sparsity in automatic differentiation*, Optim. Methods Softw. **7** (1997), 1–39.

[BL+01] H.M. Bücker and B. Lang and D. an Mey and C.H. Bischof, *Bringing together automatic differentiation and OpenMP*, in ICS '01: Proceedings of the 15th international conference on Supercomputing, ACM, New York, 2001, 246–251.

[BP97] Å. Björck and V. Pereyra, *Solution of Vandermonde systems of equations*, Math. Comp. **24** (1997), 893–903.

[BPS06] S. Basu, R. Pollack, and M.-F. Roy, *Algorithms in Real Algebraic Geometry*, Springer, New York, 2006.

[BR01] R. Becker and R. Rannacher, *An optimal control approach to a-posteriori error estimation in finite element methods*, Acta Numerica **10** (2001), pp. 1–102.

[BRM97] C.H. Bischof, L. Roh, and A. Mauer, *ADIC: An Extensible Automatic Differentiation Tool for ANSI-C*, Tech. Report ANL/MCS-P626-1196, Mathematics and Computer Science Division, Argonne National Laboratory, Argonne, IL, 1997.

[Bro98] S.A. Brown, *Models for Automatic Differentiation: A Conceptual Framework for Exploiting Program Transformation*, Ph.D. thesis, University of Hertfordshire, Hertfordshire, UK, 1998.

[BRW04] H.M. Bücker, A. Rasch, and A. Wolf, *A class of OpenMP applications involving nested parallelism*, in SAC '04: Proceedings of the 2004 ACM Symposium on Applied Computing, 2004, pp. 220–224.

[BS83] W. Baur and V. Strassen, *The complexity of partial derivatives*, Theoret. Comput. Sci. **22** (1983), 317–330.

[BS92] D. Bestle and J. Seybold, *Sensitivity analysis of constraint multibody systems*, Arch. Appl. Mech. **62** (1992), 181–190.

[BS96] C. Bendsten and O. Stauning, *FADBAD, a Flexible C++ Package for Automatic Differentiation Using the Forward and Backward Methods*, Tech. Report IMM-REP-1996-17, Department of Mathematical Modelling, Technical University of Denmark, Lyngby, Denmark, 1996.

[Cac81a] D.G. Cacuci, *Sensitivity theory for nonlinear systems. I. Nonlinear functional analysis approach*, Math. Phys. **22** (1981), 2794–2802.

[Cac81b] D.G. Cacuci, *Sensitivity theory for nonlinear systems. II. Extensions of additional classes of responses*, Math. Phys. **22** (1981), 2803–2812.

[Cam89] S.L. Campbell, *A computational method for general higher index nonlinear singular systems of differential equations*, IMACS Trans. Sci. Comp. **1.2** (1989), 555–560.

[CC86] T.F. Coleman and J.-Y. Cai, *The cyclic coloring problem and estimation of sparse Hessian matrices*, SIAM J. Algebraic Discrete Methods **7** (1986), 221–235.

[CC94] Y.F. Chang and G.F. Corliss, *ATOMFT: Solving ODEs and DAEs using Taylor series*, Comput. Math. Appl. **28** (1994), 209–233.

[CDB96] B. Christianson, L.C.W. Dixon, and S. Brown, *Sharing storage using dirty vectors*, in [BB+96], pp. 107–115.

[CD+06] W. Castings, D. Dartus, M. Honnorat, F.-X. Le Dimet, Y. Loukili, and J. Monnier, *Automatic differentiation: A tool for variational data assimilation and adjoint sensitivity analysis for flood modeling*, in [BC+06], pp. 250–262.

[Cés99] N. Di Césaré, *Outils pour l'optimisation de forme et le contrôle optimal, application à la méchanique des fluides*, Ph.D. thesis, de l'Université Paris 6, France, 2000.

[CF96] A. Carle and M. Fagan, *Improving derivative performance for CFD by using simplified recurrences*, in [BB+96], pp. 343–351.

[CF+01] G. Corliss, C. Faure, A. Griewank, L. Hascoët, and U. Naumann (eds.), *Automatic Differentiation: From Simulation to Optimization*, Computer and Information Science, Springer, New York, 2001.

[CG97] W.G. Choe and J. Guckenheimer, *Computing periodic orbits with high accuracy*, Comput. Methods Appl. Mech. Engrg., **170** (1999), 331–341.

[CGT92] A.R. Conn, N.I.M. Gould, and P.L. Toint, *LANCELOT, a Fortran package for large-scale nonlinear optimization (release A)*, Comput. Math. **17**, Springer, Berlin, 1992.

[Cha90] R.W. Chaney, *Piecewise C^k-functions in nonsmooth analysis*, Nonlinear Anal. **15** (1990), 649–660.

[Che06] B. Cheng, *A duality between forward and adjoint MPI communication routines*, in Computational Methods in Science and Technology, Polish Academy of Sciences, 2006, pp. 23–24.

[Chr92] B. Christianson, *Reverse accumulation and accurate rounding error estimates for Taylor series coefficients*, Optim. Methods Softw. **1** (1992), 81–94.

[Chr94] B. Christianson, *Reverse accumulation and attractive fixed points*, Optim. Methods Softw. **3** (1994), 311–326.

[Chr01] B. Christianson, *A self-stabilizing Pantoja-like indirect algorithm for optimal control*, Optim. Methods Softw. **16** (2001), 131–149.

[Cla83] F.H. Clarke, *Optimization and Nonsmooth Analysis*, Classics Appl. Math. **5**, SIAM, Philadelphia, 1990.

[CM83] T.F. Coleman and J.J. Moré, *Estimation of sparse Jacobian matrices and graph coloring problems*, SIAM J. Numer. Anal. **20** (1983), 187–209.

[CMM97] J. Czyzyk, M.P. Mesner, and J.J. Moré, *The Network-Enabled Optimization Server*, Preprint ANL/MCS–P615–1096, Mathematics and Computer Science Division, Argonne National Laboratory, Argonne, IL, 1997.

[Coe01] G.C. Cohen, *Higher-Order Numerical Methods for Transient Wave Equations*, Springer, New York, 2001.

[Con78] J.H. Conway, *Elementary Numerical Analysis*, North–Holland, Amsterdam, 1978.

[Cos00] M. Coste, *An Introduction to O-minimal Geometry*, Dip. Mat. Univ. Pisa, Dottorato di Ricerca in Mathematica, Instituti Editoriale e Poligrafici Internazionali, Pisa, 2000.

[Cou81] P. Cousot, *S*emantic foundations of program analysis, in Program Flow Analysis: Theory and Applications, S.S. Muchnick and N.D. Jones, eds., Prentice–Hall, Englewood Cliffs, NJ, 1981, pp. 303–342.

[CPR74] A.R. Curtis, M.J.D. Powell, and J.K. Reid, *On the estimation of sparse Jacobian matrices*, J. Inst. Math. Appl. **13** (1974), 117–119.

[CS+01] D. Casanova, R.S. Sharp, M. Final, B. Christianson, and P. Symonds, *Application of automatic differentiation to race car performance optimisation*, in [CF+01], pp. 113–120.

[CV96] T.F. Coleman and A. Verma, *Structure and efficient Jacobian calculation*, in [BB+96], pp. 149–159.

[CW+80] D.G. Cacuci, C.F. Weber, E.M. Oblow, and J.H. Marable, *Sensitivity theory for general systems of nonlinear equations*, Nuclear Sci. Engrg. **88** (1980), 88–110.

[dB56] F. de Bruno, *Note sur une nouvelle formule de calcule differentiel*, Quart. J. Math. **1** (1856), 359–360.

[DB89] J.C. Dunn and D.P. Bertsekas, *Efficient dynamic programming implementations of Newton's method for unconstrained optimal control problems*, J. Optim. Theory Appl. **63** (1989), 23–38.

[DD+90] J.J. Dongarra, J.J. Du Croz, I.S. Duff, and S.J. Hammarling, *A set of level* 3 *basic linear algebra subprograms*, ACM Trans. Math. Software **16** (1990), 1–17.

[DER89] I.S. Duff, A.M. Erisman, and J.K. Reid, *Direct Methods for Sparse Matrices*, Monogr. Numer. Anal., Oxford University Press, New York, 1989.

[Deu94] A. Deutsch, *Interprocedural may-alias analysis for pointers: Beyond k-limiting*, ACM SIGPLAN Notices, **29** (1994), no. 6, 230–241.

[Dix91] L.C.W. Dixon, *Use of automatic differentiation for calculating Hessians and Newton steps*, in [GC91], pp. 114–125.

[DLS95] M. Dobmann, M. Liepelt, and K. Schittkowski, *Algorithm* 746 *POCOMP: A Fortran code for automatic differentiation*, ACM Trans. Math. Software **21** (1995), 233–266.

[DM48] P.S. Dwyer and M.S. Macphail, *Symbolic Matrix Derivatives*, Ann. Math. Statist. **19** (1948), 517–534.

[DPS89] P.H. Davis, J.D. Pryce, and B.R. Stephens, *Recent Developments in Automatic Differentiation*, Appl. Comput. Math. Group Report ACM-89-1, The Royal Military College of Science, Cranfield, UK, January 1989.

[DR69] S.W. Director and R.A. Rohrer, *Automated network design—the frequency-domain case*, IEEE Trans. Circuit Theory **CT-16** (1969), 330–337, reprinted by permission.

[DS96] J.E. Dennis, Jr., and R.B. Schnabel, *Numerical Methods for Unconstrained Optimization and Nonlinear Equations*, Classics Appl. Math. **16**, SIAM, Philadelphia, 1996.

[Dwy67] P.S. Dwyer, *Some applications of matrix derivatives in multivariate analysis*, J. Amer. Statist. Assoc. **62** (1967), 607–625.

[EB99] P. Eberhard and C.H. Bischof, *Automatic differentiation of numerical integration algorithms*, J. Math. Comp. **68** (1999), 717–731.

[Fat74] R.J. Fateman, *Polynomial multiplication, powers and asymptotic analysis: Some comments*, SIAM J. Comput. **3** (1974), 196–213.

[Fau92] C. Faure, *Quelques aspects de la simplification en calcul formel*, Ph.D. thesis, Université de Nice, Sophia Antipolis, France, 1992.

[FB96] W.F. Feehery and P.I. Barton, *A differentiation-based approach to dynamic simulation and optimization with high-index differential-algebraic equations*, in [BB+96], pp. 239–252.

[FDF00] C. Faure, P. Dutto and S. Fidanova, *Odyssée and parallelism: Extension and validation*, in Proceedings of the 3rd European Conference on Numerical Mathematics and Advanced Applications, Jyväskylä, Finland, July 26–30, 1999, World Scientific, pp. 478–485.

[FF99] H. Fischer and H. Flanders, *A minimal code list*, Theoret. Comput. Sci. **215** (1999), 345–348.

[Fis91] H. Fischer, *Special problems in automatic differentiation*, in [GC91], pp. 43–50.

[FN01] C. Faure and U. Naumann, *Minimizing the Tape Size*, in [CF+01], pp. 279–284.

[Fos95] I. Foster, *Designing and Building Parallel Programs: Concepts and Tools for Parallel Software Engineering*, Addison-Wesley Longman Publishing Co., Inc., Boston, MA, 1995.

[Fra78] L.E. Fraenkel, *Formulae for high derivatives of composite functions*, J. Math. Proc. Camb. Philos. Soc. **83** (1978), 159–165.

[Gar91] O. García, *A system for the differentiation of Fortran code and an application to parameter estimation in forest growth models*, in [GC91], pp. 273–286.

[Gay96] D.M. Gay, *More AD of nonlinear AMPL models: Computing Hessian information and exploiting partial separability*, in [BB+96], pp. 173–184.

[GB+93] A. Griewank, C. Bischof, G. Corliss, A. Carle, and K. Williamson, *Derivative convergence of iterative equation solvers*, Optim. Methods Softw. **2** (1993), 321–355.

[GC91] A. Griewank and G.F. Corliss (eds.), *Automatic Differentiation of Algorithms: Theory, Implementation, and Application*, SIAM, Philadelphia, 1991.

[GC+97] A. Griewank, G.F. Corliss, P. Henneberger, G. Kirlinger, F.A. Potra, and H.J. Stetter, *High-order stiff ODE solvers via automatic differentiation and rational prediction*, in Numerical Analysis and Its Applications, Lecture Notes in Comput. Sci. **1196**, Springer, Berlin, 1997, pp. 114–125.

[Gil07] M.B. Giles, *Monte Carlo Evaluation of Sensitivities in Computational Finance.* Report NA-07/12, Oxford University Computing Laboratory, 2007.

[Gei95] U. Geitner, *Automatische Berechnung von dünnbesetzten Jacobimatrizen nach dem Ansatz von Newsam-Ramsdell*, diploma thesis, Technische Universität Dresden, Germany, 1995.

[Ges95] Gesellschaft für Anlagen- und Reaktorsicherheit mbH, Garching, *ATHLET Programmers Manual, ATHLET Users Manual*, 1995.

[GF02] A. Griewank and C. Faure, *Reduced functions, gradients and Hessians from fixed point iterations for state equations*, Numer. Algorithms **30** (2002), 113–139.

[GGJ90] J. Guddat, F. Guerra, and H.Th. Jongen, *Parametric Optimization: Singularities, Pathfollowing and Jumps*, Teubner, Stuttgart, John Wiley, Chichester, 1990.

[Gil92] J.Ch. Gilbert, *Automatic differentiation and iterative processes*, Optim. Methods Softw. **1** (1992), 13–21.

[GJ79] M.R. Garey and D.S. Johnson, *Computers and Intractability. A Guide to the Theory of NP-completeness*, W.H. Freeman and Company, 1979.

[GJU96] A. Griewank, D. Juedes, and J. Utke, *ADOL—C, a Package for the Automatic Differentiation of Algorithms Written in C/C++*, ACM Trans. Math. Software **22** (1996), 131–167; `http://www.math.tu-dresden.de/~adol-c/`.

[GK98] R. Giering and T. Kaminski, *Recipes for adjoint code construction*, ACM Trans. Math. Software **24** (1998), 437–474.

[GK+06] R. Giering and T. Kaminski, R. Todling, R. Errico, R. Gelaro, and N. Winslow, *Tangent linear and adjoint versions of NASA/GMAO's Fortran* 90 *global weather forecast model*, in [BC+06], pp. 275–284.

[GK05] A. Griewank and D. Kressner, *Time-lag in Derivative Convergence for Fixed Point Iterations*, Revue *ARIMA*, Numéro Special CARI'04, pp. 87–102, 2005.

[GLC85] S. Gomez, A.V. Levy, and A. Calderon, *A global zero residual least squares method*, in Numerical Analysis, Proceedings, Guanajuato, Mexico, 1984, Lecture Notes in Math. 1230, J.P. Hennart, ed., Springer, New York, 1985, pp. 1–10.

[GM96] J. Guckenheimer and M. Myers, *Computing Hopf bifurcations* II: Three examples from neurophysiology, SIAM J. Sci. Comput. **17** (1996), 1275–1301.

[GM97] P. Guillaume and M. Masmoudi, *Solution to the time-harmonic Maxwell's equations in a waveguide: Use of higher-order derivatives for solving the discrete problem*, SIAM J. Numer. Anal. **34** (1997), 1306–1330.

[GMP05] A.H. Gebremedhin, F. Manne, and A. Pothen, *What color is your Jacobian? Graph coloring for computing derivatives*, SIAM Rev. **47** (2005), no. 4, 629–705.

[GMS97] J. Guckenheimer, M. Myers, and B. Sturmfels, *Computing Hopf bifurcations* I, SIAM J. Numer. Anal. **34** (1997), 1–21.

[GN93] J.Ch. Gilbert and J. Nocedal, *Automatic differentiation and the step computation in the limited memory BFGS method*, Appl. Math. Lett. **6** (1993), 47–50.

[GOT03] N. Gould, D. Orban and Ph.L. Toint, *CUTEr, a constrained and unconstrained testing environment*, revisited, ACM Trans. Math. Software, **29** (2003), 373–394.

[GP01] M.B. Giles and N.A. Pierce, *An introduction to the adjoint approach to design*, Flow, Turbulence and Combustion **65** (2001), 393–415.

[GP+96] J. Grimm, L. Potter, and N. Rostaing-Schmidt, *Optimal time and minimum space-time product for reversing a certain class of programs*, in [BB+96], pp. 95–106.

[GP+06] A.H. Gebremedhin, A. Pothen, A. Tarafdar, and A. Walther, *Efficient Computation of Sparse Hessians Using Coloring and Automatic Differentiation*, to appear in INFORMS J. Comput. (2006).

[GR87] A. Griewank and P. Rabier, *Critical points of mixed fluids and their numerical treatment*, in Bifurcation: Analysis, Algorithms, Applications, T. Küpper, R. Reydel, and H. Troger, eds., Birkhäuser, Boston, 1987, pp. 90–97.

[GR89] A. Griewank and G.W. Reddien, *Computation of cusp singularities for operator equations and their discretizations*, J. Comput. Appl. Math. **26** (1989), 133–153.

[GR91] A. Griewank and S. Reese, *On the calculation of Jacobian matrices by the Markowitz rule*, in [GC91], pp. 126–135.

[Gri80] A. Griewank, *Starlike domains of convergence for Newton's method at singularities*, Numer. Math. **35** (1980), 95–111.

[Gri89] A. Griewank, *On automatic differentiation*, in Mathematical Programming: Recent Developments and Applications, M. Iri and K. Tanabe, eds., Kluwer, Dordrecht, The Netherlands, 1989, pp. 83–108.

[Gri90] A. Griewank, *Direct calculation of Newton steps without accumulating Jacobians*, in Large-Scale Numerical Optimization, T.F. Coleman and Y. Li, eds., SIAM, Philadelphia, 1990, pp. 115–137.

[Gri91] A. Griewank, *Achieving logarithmic growth of temporal and spatial complexity in reverse automatic differentiation*, Optim. Methods Softw. **1** (1992), 35–54.

[Gri93] A. Griewank, *Some bounds on the complexity of gradients, Jacobians, and Hessians*, in Complexity in Nonlinear Optimization, P.M. Pardalos, ed., World Scientific, River Edge, NJ, 1993, pp. 128–161.

[Gri94] A. Griewank, *Tutorial on Computational Differentiation and Optimization*, University of Michigan, Ann Arbor, MI, 1994.

[Gri95] A. Griewank, *ODE solving via automatic differentiation and rational prediction*, in Numerical Analysis 1995, Pitman Res. Notes Math. Ser. **344**, D.F. Griffiths and G.A. Watson, eds., Addison-Wesley Longman, Reading, MA, 1995.

[Gri03] A. Griewank, *A mathematical view of automatic differentiation*, Acta Numerica **12** (2003), 321–398.

[Gru97] D. Gruntz, *Automatic differentiation and bisection*, MapleTech **4** (1997), 22–27.

[GS02] M.B. Giles and E. Süli, *Adjoint methods for PDEs: A-posteriori error analysis and postprocessing by duality*, Acta Numerica **11** (2002), 145–236,

[GT+07] A.H. Gebremedhin, A. Tarafdar, F. Manne, and A. Pothen, *New acyclic and star coloring algorithms with application to computing Hessians*, SIAM J. Sci. Comput. **29** (2007), 1042–1072.

[GT82] A. Griewank and Ph.L. Toint, *On the unconstrained optimization of partially separable objective functions*, in Nonlinear Optimization 1981, M.J.D. Powell, ed., Academic Press, London, 1982, pp. 301–312.

[GUW00] A. Griewank, J. Utke, and A. Walther, *Evaluating higher derivative tensors by forward propagation of univariate Taylor series*, Math. Comp. **69** (2000), 1117–1130.

[GW00] A. Griewank and A. Walther, *Revolve: An implementation of checkpointing for the reverse or adjoint mode of computational differentiation*, ACM Trans. Math. Softw. **26** (2000), 19–45.

[GW04] A. Griewank and A. Walther, *On the efficient generation of Taylor expansions for DAE solutions by automatic differentiation*, Proceedings of PARA'04, in J. Dongarra et al., eds., Lecture Notes in Comput. Sci. **3732**, Springer, New York, 2006, 1089–1098.

[GV96] G.H. Golub and C.F. Van Loan, *Matrix Computations*, third ed., Johns Hopkins University Press, Baltimore, 1996.

[Han79] E.R. Hansen, *Global optimization using interval analysis—the one-dimensional case*, J. Optim. Theory Appl. **29** (1979), 331–334.

[HB98] P.D. Hovland and C.H. Bischof, *Automatic differentiation of message-passing parallel programs*, in Proceedings of the First Merged International Parallel Processing Symposium and Symposium on Parallel and Distributed Processing, IEEE Computer Society Press, 1998, pp. 98–104.

[HBG71] G.D. Hachtel, R.K. Bryton, and F.G. Gustavson, *The sparse tableau approach to network analysis and design*, IEEE Trans. Circuit Theory **CT-18** (1971), 111–113.

[Her93] K. Herley, *Presentation at: Theory Institute on Combinatorial Challenges in Computational Differentiation*, Mathematics and Computer Science Division, Argonne National Laboratory, Argonne, IL, 1993.

[HKP84] H.J. Hoover, M.M. Klawe, and N.J. Pippenger, *Bounding fan-out in logical networks*, Assoc. Comput. Mach. **31** (1984), 13–18.

[HNP05] L. Hascoët, U. Naumann, and V. Pascual, *"To be recorded" analysis in reverse-mode automatic differentiation*, Future Generation Comp. Sys. **21** (2005), 1401–1417.

[HNW96] E. Hairer, S.P. Nørsett, and G. Wanner, *Solving Ordinary Differential Equations* I. *Nonstiff Problems*, second revised ed., Computational Mechanics **14**, Springer, Berlin, 1996.

[Hor92] J.E. Horwedel, *Reverse Automatic Differentiation of Modular Fortran Programs*, Tech. Report ORNL/TM-12050, Oak Ridge National Laboratory, Oak Ridge, TN, 1992.

[Hos97] A.K.M. Hossain, *On the Computation of Sparse Jacobian Matrices and Newton Steps*, Ph.D. thesis, Department of Informatics, University of Bergen, Norway, 1997.

[Hov97] P. Hovland, *Automatic Differentiation of Parallel Programs*, Ph.D. thesis, Department of Computer Science, University of Illinois, Urbana, 1997.

[HP04] L. Hascoët, and V. Pascual, *TAPENADE* 2.1 *User's Guide*, Technical report, INRIA 300, INRIA, 2004.

[HR99] R. Henrion and W. Römisch, *Metric regularity and quantitative stability in stochastic programs with probabilistic constraints*, Math. Program. **84** (1999), 55–88.

[HS02] S. Hossain and T. Steihaug, *Sparsity issues in the computation of Jacobian matrices*, in Proceedings of the International Symposium on Symbolic and Algebraic Computing, T. Mora, ed., ACM, New York, 2002, pp. 123–130.

[HW96] E. Hairer and G. Wanner, *Solving Ordinary Differential Equations* II. *Stiff and Differential-Algebraic Problems*, second revised ed., Computational Mechanics **14**, Springer, Berlin, 1996.

[HW06] V. Heuveline and A. Walther, *Online checkpointing for parallel adjoint computation in PDEs: Application to goal oriented adaptivity and flow control*, in Proceedings of Euro-Par 2006, W. Nagel et al., eds., Lecture Notes in Comput. Sci. **4128**, Springer, Berlin, 2006, pp. 689–699.

[Iri91] M. Iri, *History of automatic differentiation and rounding error estimation*, in [GC91], pp. 1–16.

[ITH88] M. Iri, T. Tsuchiya, and M. Hoshi, *Automatic computation of partial derivatives and rounding error estimates with applications to large-scale systems of nonlinear equations*, Comput. Appl. Math. **24** (1988), 365–392.

[JM88] R.H.F. Jackson and G.P. McCormic, *Second order sensitivity analysis in factorable programming: Theory and applications*, Math. Programming **41** (1988), 1–28.

[JMF06] K-W. Joe, D.L. McShan, and B.A. Fraass, *Implementation of automatic differentiation tools for multicriteria IMRT optimization*, in [BC+06], pp. 225–234.

[KB03] M. Knauer and C. Büskens, *Real-time trajectory planning of the industrial robot IRB* 6400, PAMM **3** (2003), 515–516.

[Ked80] G. Kedem, *Automatic differentiation of computer programs*, ACM Trans. Math. Software **6** (1980), 150–165.

[Keh96] K. Kehler, *Partielle Separabilität und ihre Anwendung bei Berechnung dünnbesetzter Jacobimatrizen*, diploma thesis, Technische Universität Dresden, Germany, 1996.

[KHL06] J.G. Kim, E.C. Hunke, and W.H. Lipscomb, *A sensitivity-enhanced simulation approach for community climate system model*, in Proceedings of ICCS 2006, V. Alexandrov et al., eds., Lecture Notes in Comput. Sci. **3994**, Springer, Berlin, 2006, pp. 533–540.

[Kiw86] K.C. Kiwiel, *A method for solving certain quadratic programming problems arising in nonsmooth optimization*, IMA J. Numer. Anal. **6** (1986), 137–152.

[KK+86] H. Kagiwada, R. Kalaba, N. Rasakhoo, and K. Spingarn, *Numerical Derivatives and Nonlinear Analysis*, Math. Concepts Methods Sci. Engrg.**31**, Plenum Press, New York, London, 1986.

[KL91] D. Kalman and R. Lindell, *Automatic differentiation in astrodynamical modeling*, in [GC91], pp. 228–243.

[KM81] U.W. Kulisch and W.L. Miranker, *Computer Arithmetic in Theory and Practice*, Academic Press, New York, 1981.

[KN+84] K.V. Kim, Yu.E. Nesterov, V.A. Skokov, and B.V. Cherkasski, *Effektivnyi algoritm vychisleniya proizvodnykh i èkstremal'nye zadachi*, Èkonomika i Matematicheskie Metody **20** (1984), 309–318.

[Knu73] D.E. Knuth, *The Art of Computer Programming* **1**, Fundamental Algorithms, third ed., Addison-Wesley, Reading, MA, 1997.

[Knu98] D.E. Knuth, *The Art of Computer Programming* **3**, Sorting and Searching, second ed., Addison-Wesley, Reading, MA, 1998.

[Koz98] K. Kozlowski, *Modeling and Identification in Robotics*, in Advances in Industrial Control, Springer, London, 1998.

[KRS94] J. Knoop, O. Rüthing, and B. Steffen, *Optimal code motion: Theory and practice*, ACM Trans. Program. Languages Syst. **16** (1994), 1117–1155.

[KS90] E. Kaltofen and M.F. Singer, *Size Efficient Parallel Algebraic Circuits for Partial Derivatives*, Tech. Report 90-32, Rensselaer Polytechnic Institute, Troy, NY, 1990.

[KT51] H.W. Kuhn and A.W. Tucker, *Nonlinear programming*, in Proceedings of the Second Berkeley Symposium on Mathematical Statistics and Probability, J. Newman, ed., University of California Press, Berkeley, 1951, pp. 481–492.

[Kub98] K. Kubota, *A Fortran* 77 *preprocessor for reverse mode automatic differentiation with recursive checkpointing*, Optim. Methods Softw. **10** (1998), 319–335.

[Kub96] K. Kubota, *PADRE2—Fortran precompiler for automatic differentiation and estimates of rounding error*, in [BB+96], pp. 367–374.

[KW06] A. Kowarz and A. Walther, *Optimal checkpointing for time-stepping procedures*, in Proceedings of ICCS 2006, V. Alexandrov et al., eds., Lecture Notes in Comput. Sci. **3994**, Springer, Berlin, 2006, pp. 541–549.

[KW00] W. Klein and A. Walther, *Application of techniques of computational differentiation to a cooling system*, Optim. Methods Softw. **13** (2000), 65–78.

[Lin76] S. Linnainmaa, *Taylor expansion of the accumulated rounding error*, BIT **16** (1976), 146–160.

[LU08] A. Lyons and J. Utke, *On the practical exploitation of scarsity*, in C. Bischof et al., eds., Advances in Automatic Differentiation, to appear in Lecture Notes in Comput. Sci. Eng., Springer (2008).

[Lun84] V.Yu. Lunin, *Ispol'zovanie algoritma bystrogo differentsirovaniya v zadache utochneniya znachenij faz strukturnykh faktorov*, Tech. Report UDK 548.73, Naychnyj Tsenter Biolongicheskikh Nauk, AN SSSR, Pushchino, 1984.

[Mar57] H.M. Markowitz, *The elimination form of the inverse and its application*, Management Sci. **3** (1957), 257–269.

[Mar96] K. Marti, *Differentiation formulas for probability functions: The transformation method*, Math. Programming **75** (1996), 201–220.

[Mat86] Yu.V. Matiyasevich, *Real numbers and computers*, Cybernet. Comput. Mach. **2** (1986), 104–133.

[MG+99] J. Marotzke, R. Giering, K.Q. Zhang, D. Stammer, C. Hill, and T. Lee, *Construction of the adjoint MIT ocean general circulation model and application to Atlantic heat transport sensitivity*, J. Geophys. Res. **104** (1999), no. C12, p. 29.

[Mic82] M. Michelson, *The isothermal flash problem, part* I. *Stability*, Fluid Phase Equilibria **9** (1982), 1–19.

[Mic91] L. Michelotti, *MXYZPTLK: A C++ hacker's implementation of automatic differentiation*, in [GC91], pp. 218–227.

[Min06] A. Miné, *Field-sensitive value analysis of embedded C programs with union types and pointer arithmetics*, in LCTES '06: Proceedings of the 2006 ACM SIGPLAN/SIGBED Conference on Language, Compilers, and Tool Support for Embedded Systems, ACM, New York, 2006, pp. 54–63.

[MIT] MITgcm, *The MIT General Circulation Model*, `http://mitgcm.org`.

[MN93] Mi.B. Monagan and W.M. Neuenschwander, *GRADIENT: Algorithmic differentiation in Maple*, in Proceedings of ISSAC '93, ACM Press, New York, 1993, pp. 68–76.

[Mog00] T.A. Mogensen, *Glossary for Partial Evaluation and Related Topics*, Higher-Order Symbolic Comput. **13** (2000), no. 4.

[Moo66] R.E. Moore, *Interval Analysis*, Prentice–Hall, Englewood Cliffs, NJ, 1966.

[Moo79] R.E. Moore, *Methods and Applications of Interval Analysis*, SIAM, Philadelphia, 1979.

[Moo88] R.E. Moore, *Reliability in Computing: The Role of Interval Methods in Scientific Computations*, Academic Press, New York, 1988.

[Mor85] J. Morgenstern, *How to compute fast a function and all its derivatives, a variation on the theorem of Baur-Strassen*, SIGACT News **16** (1985), 60–62.

[Mor06] B.S. Mordukhovich, *Variational Analysis and Generalized Differentiation.* I: *Basic Theory*, Grundlehren Math. Wiss. **330**, Springer, Berlin, 2006.

[MPI] MPI, *The Message Passing Interface Standard*, `http://www.mcs.anl.gov/mpi`.

[MR96] M. Monagan and R.R. Rodoni, *Automatic differentiation: An implementation of the forward and reverse mode in Maple*, in [BB+96], pp. 353–362.

[MRK93] G.L. Miller, V. Ramachandran, and E. Kaltofen, *Efficient parallel evaluation of straight-line code and arithmetic circuits*, Computing **17** (1993), 687–695.

[MS93] S.E. Mattsson and G. Söderlind, *Index reduction in differential-algebraic equations using dummy derivatives*, SIAM J. Sci. Comput. **14** (1993), 677–692.

[Mun90] F.S. Munger, *Applications of Definor Algebra to Ordinary Differential Equations*, Aftermath, Golden, CO, 1990.

[Mur93] J.D. Murray, *Mathematical Biology*, second ed., Biomathematics, Springer, Berlin, 1993.

[Nau99] U. Naumann, *Efficient Calculation of Jacobian Matrices by Optimized Application of the Chain Rule to Computational Graphs*, Ph.D. thesis, Technische Universität Dresden, Germany, 1999.

[Nau04] U. Naumann, *Optimal accumulation of Jacobian matrices by elimination methods on the dual computational graph*, Math. Program. **99** (2004), 399–421.

[Nau06] U. Naumann, *Optimal Jacobian accumulation is NP-complete*, Math. Program. **112** (2008), 427–441.

[NAW98] J.C. Newman, K.W. Anderson, and D.L. Whitfield, *Multidisciplinary Sensibility Derivatives Using Complex Variables*, Tech. Report MSSU-COE-ERC-98-08, Mississippi State University, Mississippi State, MS, 1998.

[Ned99] N.S. Nedialkov, *Computing Rigorous Bounds on the Solution of an Initial Value Problem for an Ordinary Differential Equation*, Ph.D. thesis, University of Toronto, Toronto, ON, 1999.

[Nei92] R.D. Neidinger, *An efficient method for the numerical evaluation of partial derivatives of arbitrary order*, ACM Trans. Math. Softw. **18** (1992), 159–173.

[NH+92] P.A. Newman, G.J.W. Hou, H.E. Jones, A.C. Taylor, and V.M. Korivi, *Observations on Computational Methodologies for Use in Large-Scale, Gradient-Based, Multidisciplinary Design Incorporating Advanced CFD Codes*, Techn. Mem. 104206, NASA Langley Research Center, 1992, AVSCOM Technical Report 92-B-007.

[NP05] N.S. Nedialkov and J.D. Pryce, *Solving differential-algebraic equations by Taylor series. I: Computing Taylor coefficients*, BIT **45** (2005), 561–591.

[NR83] G.N. Newsam and J.D. Ramsdell, *Estimation of sparse Jacobian matrices*, SIAM J. Algebraic Discrete Methods **4** (1983), 404–418.

[NS96] S.G. Nash and A. Sofer, *Linear and nonlinear programming*, McGraw–Hill Series in Industrial Engineering and Management Science, McGraw–Hill, New York, 1996.

[NU+04] U. Naumann, J. Utke, A. Lyons, and M. Fagan, *Control Flow Reversal for Adjoint Code Generation*, in Proceedings of SCAM 2004 IEEE Computer Society, 2004, pp. 55–64.

[Oba93] N. Obayashi, *A numerical method for calculating the higher order derivatives used multi-based number*, Bull. Coll. Lib. Arts **29** (1993), 119–133.

[Obl83] E.M. Oblow, *An Automated Procedure for Sensitivity Analysis Using Computer Calculus*, Tech. Report ORNL/TM-8776, Oak Ridge National Laboratory, Oak Ridge, TN, 1983.

[OMP] OpenMP, *The OpenMP Specification for Parallel Programming*, `http://www.openmp.org`.

[OR70] J.M. Ortega and W.C. Rheinboldt, *Iterative Solution of Nonlinear Equations in Several Variables*, Academic Press, New York, 1970.

[OVB71] G.M. Ostrovskii, Yu.M. Volin, and W.W. Borisov, *Über die Berechnung von Ableitungen*, Wiss. Z. Tech. Hochschule für Chemie **13** (1971), 382–384.

[OW97] M. Overton and H. Wolkowicz (eds.), *Semidefinite Programming*, North-Holland, Amsterdam, 1997.

[Pfe80] F.W. Pfeiffer, *Some Advances Related to Nonlinear Programming*, Tech. Report **28**, SIGMAP Bulletin, ACM, New York, 1980.

[Pon82] J.W. Ponton, *The numerical evaluation of analytical derivatives*, Comput. Chem. Eng. **6** (1982), 331–333.

[PR97] J.D. Pryce and J.K. Reid, *AD01: A Fortran 90 code for automatic differentiation*, Tech. Report RAL-TR-97, Rutherford-Appleton Laboratories, Chilton, UK, 1997.

[Pry01] J. Pryce, *A simple structural analysis method for DAEs*, BIT **41** (2001), 364–394.

[PT79] M.J.D. Powell and Ph.L. Toint, *On the estimation of sparse Hessian matrices*, SIAM J. Numer. Anal. **16** (1979), 1060–1074.

[Ral81] L.B. Rall, *Automatic Differentiation: Techniques and Applications*, Lecture Notes in Comput. Sci. **120**, Springer, Berlin, 1981.

[Ral84] L.B. Rall, *Differentiation in Pascal-SC: Type GRADIENT*, ACM Trans. Math. Softw. **10** (1984), 161–184.

[RBB07] A. Rasch, H. M. Bücker, and C. H. Bischof, *Automatic computation of sensitivities for a parallel aerodynamic simulation*, in Proceedings of the International Conference on Parallel Computing (ParCo2007), Jülich, Germany, 2007.

[RDG93] N. Rostaing, S. Dalmas, and A. Galligo, *Automatic differentiation in Odyssée*, Tellus **45A** (1993), 558–568.

[RH92] L.C. Rich and D.R. Hill, *Automatic differentiation in MATLAB*, Appl. Numer. Math. **9** (1992), 33–43.

[Rho97] A. Rhodin, *IMAS integrated modeling and analysis system for the solution of optimal control problems*, Comput. Phys. Comm. **107** (1997), 21–38.

[RO91] L. Reichel and G. Opfer, *Chebyshev-Vandermonde systems*, Math. Comp. **57** (1991), 703–721.

[RR99] K.J. Reinschke and K. Röbenack, *Analyse von Deskriptorsystemen mit Hilfe von Berechnungsgraphen*, Z. Angew. Math. Mech. **79** (1999), 13–16.

[R-S93] N. Rostaing-Schmidt, *Différentiation automatique: Application à un problème d'optimisation en météorologie*, Ph.D. thesis, Université de Nice, Sophia Antipolis, France, 1993.

[PS07] *PolySpace*, `http://www.mathworks.com/products/polyspace/`.

[RT78] D.J. Rose and R.E. Tarjan, *Algorithmic aspects of vertex elimination on directed graphs*, SIAM J. Appl. Math. **34** (1978), 176–197.

[Saa03] Y. Saad, *Iterative Methods for Sparse Linear Systems*, second ed., SIAM, Philadelphia, 2003.

[Sch65] H. Schorr, *Analytic differentiation using a syntax-directed compiler*, Comput. J. **7** (1965), 290–298.

[Sch94] S. Scholtes, *Introduction to Piecewise Differentiable Equations*, Preprint 53/1994, Institut für Statistik und Mathematische Wirtschaftstheorie, Universität Karlsruhe, 1994.

[SH98] T. Steihaug and S. Hossain, *Computing a sparse Jacobian matrix by rows and columns*, Optim. Methods Softw. **10** (1998), 33–48.

[SH03] R. Serban and A.C. Hindmarsh, *CVODES: An ODE Solver with Sensitivity Analysis Capabilities*, Tech. Report UCRL-JP-20039, Lawrence Livermore National Laboratory, Livermore, CA, 2003.

[SH05] M. Hinze and J. Sternberg, *A-Revolve: An adaptive memory and run-time-reduced procedure for calculating adjoints; with an application to the instationary Navier-Stokes system*, Optim. Methods Softw. **20** (2005), 645–663.

[Shi93] D. Shiriaev, *Fast Automatic Differentiation for Vector Processors and Reduction of the Spatial Complexity in a Source Translation Environment*, Ph.D. thesis, Institut für Angewandte Mathematik, Universität Karlsruhe, Germany, 1993.

[SKH06] M.M. Strout, B. Kreaseck, and P.D. Hovland, *Data-flow analysis for MPI programs*, in Proceedings of ICPP '06, IEEE Computer Society, 2006, 175–184.

[SMB97] T. Scott, M.B. Monagan, and J. Borwein (eds.), *MapleTech: Functionality, Applications, Education*, Vol. 4, Birkhäuser, Boston, 1997.

[SO98] M. Snir and S. Otto, *MPI-The Complete Reference: The MPI Core*, MIT Press, Cambridge, MA, 1998.

[Spe80] B. Speelpenning, *Compiling Fast Partial Derivatives of Functions Given by Algorithms*, Ph.D. thesis, University of Illinois at Urbana, Champaign, 1980.

[SS77] R.W.H. Sargent and G.R. Sullivan, *The development of an efficient optimal control package*, in Proceedings of the 8th IFIP Conference on Optimization Technology 2, 1977.

[SS+91] R. Seydel, F.W. Schneider, T. Kupper, and H. Troger (eds.), *Bifurcation and Chaos: Analysis, Algorithms, Applications*, Proceedings of the Conference at Würzburg, Birkhäuser, Basel, 1991.

[Sta85] *IEEE Standard for Binary Floating-Point Arithmetic*, ANS, New York, 1985.

[Sta97] O. Stauning, *Automatic Validation of Numerical Solutions*, Ph.D. thesis, Department of Mathematical Modelling, Technical University of Denmark, Lyngby, Denmark, October 1997, Technical Report IMM-PHD-1997-36.

[Ste96] B. Steensgaard, *Points-to analysis in almost linear time*, in Symposium on Principles of Programming Languages, ACM Press, New York, 1996, pp. 32–41.

[Str86] B. Stroustrup, *The C++ Programming Language*, Addison—Wesley, Reading, MA, 1986.

[Stu80] F. Stummel, *Rounding error analysis of elementary numerical algorithm*, in Fundamentals of Numerical Computation, Comput. Suppl. **2**, Springer, Vienna, 1980, pp. 169–195.

[SW85] D.F. Stubbs and N.W. Webre, *Data Structures with Abstract Data Types and Pascal*, Texts Monogr. Comput. Sci. Suppl. **2**, Brooks/Cole, Pacific Grove, CA, 1985.

[Sym07] W.W. Symes, *Reverse time migration with optimal checkpointing*, Geophys. **72** (2007), SM213–SM221.

[SZ92] H. Schramm and J. Zowe, *A version of the bundle idea for minimizing a nonsmooth function: Conceptual idea, convergence analysis, numerical results*, SIAM J. Optim. **2** (1992), 121–152.

[Tad99] M. Tadjouddine, *La différentiation automatique*, Ph.D. thesis, Université de Nice, Sophia Antipolis, France, 1999.

[Tal08] O. Talagrand, *Data Assimilation in Meteorology And Oceanography*, Academic Press Publ., 2008.

[Tar83] R.E. Tarjan, *Data Structures and Network Algorithms*, CBMS-NSF Regional Conf. Ser. in Appl. Math. **44**, SIAM, Philadelphia, 1983.

[Tha91] W.C. Thacker, *Automatic differentiation from an oceanographer's perspective*, in [GC91], pp. 191–201.

[Tip95] F. Tip, *A survey of program slicing techniques*, J. Progr. Lang. **3** (1995), 121–189.

[TKS92] S. Ta'asan, G. Kuruvila, and M.D. Salas, *Aerodynamic design and optimization in* One Shot, in Proceedings of the 30th AIAA Aerospace Sciences Meeting & Exhibit, AIAA 92-0025, 1992.

[TR+02] E. Tijskens, D. Roose, H. Ramon, and J. De Baerdemaeker, *Automatic differentiation for solving nonlinear partial differential equations: An efficient operator overloading approach*, Numer. Algorithms **30** (2002), 259–301.

[UH+08] J. Utke, L. Hascoët, C. Hill, P. Hovland, and U. Naumann, *Toward Adjoinable MPI*, Preprint ANL/MCS-P1472-1207, 2007, Argonne National Laboratory, Argonne, IL, 2008.

[Utk96a] J. Utke, *Efficient Newton steps without Jacobians*, in [BB+96], pp. 253–264.

[Utk96b] J. Utke, *Exploiting Macro- and Micro-structures for the Efficient Calculation of Newton Steps*, Ph.D. thesis, Technische Universität Dresden, Germany, 1996.

[vdS93] J.L.A. van de Snepscheut, *What Computing Is All About*, Texts Monogr. Comput. Sci. Suppl. **2**, Springer, Berlin, 1993.

[VD00] D.A. Venditti and D.L. Darmofal, *Adjoint error estimation and grid adaptation for functional outputs: Application to quasi-one-dimensional flow*, J. Comput. Phys. **164** (2000), 204–227.

[Vel95] T.L. Veldhuizen, *Expression templates*, C++ Report **7** (1995), no. 5, pp. 26–31.

[Ver99] A. Verma, *Structured Automatic Differentiation*, Ph.D. thesis, Cornell University, Ithaca, NY, 1999.

[VO85] Yu.M. Volin and G.M. Ostrovskii, *Automatic computation of derivatives with the use of the multilevel differentiating technique*—I: *Algorithmic basis*, Comput. Math. Appl. **11** (1985), 1099–1114.

[Wal99] A. Walther, *Program Reversal Schedules for Single- and Multi-Processor Machines*, Ph.D. thesis, Technische Universität Dresden, Germany, 1999.

[Wan69] G. Wanner, *Integration gewöhnlicher Differentialgleichnugen, Lie Reihen, Runge-Kutta-Methoden* **XI**, B.I-Hochschulskripten, no. 831/831a, Bibliogr. Inst. , Mannheim-Zürich, Germany, 1969.

[War75] D.D. Warner, *A Partial Derivative Generator*, Computing Science Technical Report, Bell Laboratories, 1975.

[Wen64] R.E. Wengert, *A simple automatic derivative evaluation program*, Comm. ACM **7** (1964), 463–464.

[Wer82] P.J. Werbos, *Application of advances in nonlinear sensitivity analysis*, in System Modeling and Optimization: Proceedings of the 19th IFIP Conference New York, R.F. Drenick and F. Kozin, eds., Lecture Notes in Control Inform. Sci. **38**, Springer, New York, 1982, pp. 762–770.

[Wer88] P.J. Werbos, *Generalization of backpropagation with application to a recurrent gas market model*, Neural Networks **1** (1988), 339–356.

[WG04] A. Walther and A. Griewank, *Advantages of binomial checkpointing for memory-reduced adjoint calculations*, in Numerical Mathematics and Advanced Applications: Proceedings of ENUMATH 2003, M. Feistauer et al., eds., Springer, Berlin, 2004, pp. 834–843.

[WG99] A. Walther and A. Griewank, *Applying the checkpointing routine* **treeverse** *to discretizations of Burgers' equation*, in High Performance Scientific and Engineering Computing, H.-J. Bungartz, F. Durst, and C. Zenger, eds., Lect. Notes Comput. Sci. Eng. **8**, Springer, Berlin, 1999, pp. 13–24.

[Wil65] G.J.H. Wilkinson, *The Algebraic Eigenvalue Problem*, Clarendon Press, Oxford, UK, 1965.

[WN+95] Z. Wang, I.M. Navon, X. Zou, and F.X. Le Dimet, *A truncated Newton optimization algorithm in meteorology applications with analytic Hessian/vector products*, Comput. Optim. Appl. **4** (1995), 241–262.

[Wol82] P. Wolfe, *Checking the calculation of gradients*, ACM Trans. Math. Softw. **8** (1982), 337–343.

[WO+87] B.A. Worley, E.M. Oblow, R.E. Pin, J.E. Maerker, J.E. Horwedel, R.Q. Wright, and J.L. Lucius, *Deterministic methods for sensitivity and uncertainty analysis in large-scale computer models*, in Proceedings of the Conference on Geostatistical, Sensitivity, and Uncertainty Methods for Ground-Water Flow and Radionuclide Transport Modelling, B.E. Buxton, ed., Battelle Press, 1987, pp. 135–154.

[Wri08] P. Wriggers, *Nonlinear Finite Element Methods*, Springer, Berlin, Heidelberg, 2008.

[Yos87] T. Yoshida, *Derivatives of probability functions and some applications*, Ann. Oper. Res. **11** (1987), 1112–1120.

Index